T0296729

Data Fusion Techniques and Applications for Smart Healthcare

Intelligent Data-Centric Systems

Data Fusion Techniques and Applications for Smart Healthcare

Edited by

Amit Kumar Singh
National Institute of Technology Patna
Department of Computer Science & Engineering
Bihar, India

Stefano Berretti
University of Florence
Media Integration and Communication Center
Florence, Italy

Series Editor Fatos Xhafa
Universitat Politècnica Catalunya, Barcelona, Spain

Academic Press is an imprint of Elsevier
125 London Wall, London EC2Y 5AS, United Kingdom
525 B Street, Suite 1650, San Diego, CA 92101, United States
50 Hampshire Street, 5th Floor, Cambridge, MA 02139, United States

ISBN: 978-0-443-13233-9

For information on all Academic Press publications
visit our website at https://www.elsevier.com/books-and-journals

Publisher: Mara Conner
Editorial Project Manager: Emily Thomson
Production Project Manager: Omer Mukthar
Cover Designer: Mark Rogers

Typeset by VTeX

Contents

List of contributors

Amrit Kumar Agrawal
Department of Computer Science and Engineering, School of Engineering and Technology, Sharda University, Greater Noida, India

Ramsha Ahmed
Department of Biomedical Engineering, Khalifa University, Abu Dhabi, United Arab Emirates
Healthcare Engineering Innovation Center (HEIC), Khalifa University, Abu Dhabi, United Arab Emirates

Arefeh Amiri
Shahid Madani Hospital, Lorestan University of Medical Sciences, Khorramabad, Iran

Ashima Anand
Department of Computer Science and Engineering, Thapar Institute of Engineering and Technology, Patiala, Punjab, India

Javier Andreu-Perez
Centre for Computational Intelligence, University of Essex, Colchester, United Kingdom

Marios Antonakakis
School of Electrical and Computer Engineering, Technical University of Crete, Chania, Greece

Divyanshu Awasthi
Electronics and Communication Engineering Department, Motilal Nehru National Institute of Technology, Prayagraj, Uttar Pradesh, India

Gaurav Bhatnagar
Indian Institute of Technology Jodhpur, Jodhpur, Rajasthan, India

Mou Dasgupta
Department of Computer Application, National Institute of Technology Raipur, Raipur, India

Jorge Dias
Khalifa University Center for Autonomous Robotic Systems (KUCARS), Khalifa University, Abu Dhabi, United Arab Emirates
Department of Electrical Engineering and Computer Science, Khalifa University, Abu Dhabi, United Arab Emirates

Mohammad Bagher Dowlatshahi
Department of Computer Engineering, Faculty of Engineering, Lorestan University, Khorramabad, Iran

Mavis Gezimati
Centre for Smart Information and Communication Systems, Department of Electrical and Electronic Engineering Science, University of Johannesburg, Johannesburg, South Africa

Shubhangi Goyal
Department of Electronics and Communication Engineering, Indian Institute of Technology, Roorkee, India

Ankit Gupta
Interactive Technologies Institute, ARDITI, LarSYS, Funchal, Portugal
Universidade da Madeira, Engineering and Exact sciences, Funchal, Portugal

Bilal Hassan
Khalifa University Center for Autonomous Robotic Systems (KUCARS), Khalifa University, Abu Dhabi, United Arab Emirates
Department of Electrical Engineering and Computer Science, Khalifa University, Abu Dhabi, United Arab Emirates

Taimur Hassan
Department of Electrical, Computer, and Biomedical Engineering, Abu Dhabi University, Abu Dhabi, United Arab Emirates

Sakshi Indolia
Department of Computer Science, Banasthali Vidyapith, Rajasthan, India
Centre for Artificial Intelligence, Banasthali Vidyapith, Rajasthan, India

Nazanin Zahra Joodaki
Department of Computer Engineering, Faculty of Engineering, Lorestan University, Khorramabad, Iran

Aditya Kahol
Indian Institute of Technology Jodhpur, Jodhpur, Rajasthan, India

Prasad Kanhegaonkar
Department of Computer Science & Engineering, Indian Institute of Technology Indore, Indore, India

Ali Khan
College of Mathematics and Computer Science, Zhejiang Normal University, Jinhua, China

Abhinav Kumar
Indian Institute of Technology (BHU), Department of Computer Science & Engineering, Varanasi, Uttar Pradesh, India

Zhihan Lyu
Uppsala University, Department of Game Design, Uppsala, Sweden

L. Meenachi
Department of Information Technology, Dr. Mahalingam College of Engineering and Technology, Pollachi, Tamilnadu, India

Fernando Morgado-Dias
Interactive Technologies Institute, ARDITI, LarSYS, Funchal, Portugal
Universidade da Madeira, Engineering and Exact sciences, Funchal, Portugal

Swati Nigam
Department of Computer Science, Banasthali Vidyapith, Rajasthan, India
Centre for Artificial Intelligence, Banasthali Vidyapith, Rajasthan, India

Jaya Pathak
Department of Computer Science & Information Systems, Birla Institute of Technology & Science, Pilani, Rajasthan, India

Surya Prakash
Department of Computer Science & Engineering, Indian Institute of Technology Indore, Indore, India

Amitesh Singh Rajput
Department of Computer Science & Information Systems, Birla Institute of Technology & Science, Pilani, Rajasthan, India

S. Ramakrishnan
Department of Information Technology, Dr. Mahalingam College of Engineering and Technology, Pollachi, Tamilnadu, India

Antonio G. Ravelo-García
Interactive Technologies Institute, ARDITI, LarSYS, Funchal, Portugal
Institute of Technological Development and Innovation in Communications, Universidad de Las Palmas de Gran Canaria, Las Palmas, Spain

Sangram Ray
Department of Computer Science and Engineering, National Institute of Technology Sikkim Ravangla, Sikkim, India

Srikireddy Dhanunjay Reddy
Department of Electronics and Communication Engineering, Indian Institute of Technology, Roorkee, India

Tharun Kumar Reddy
Department of Electronics and Communication Engineering, Indian Institute of Technology, Roorkee, India

Dipanwita Sadhukhan
Department of Computer Science and Engineering, National Institute of Technology Sikkim Ravangla, Sikkim, India

Mohamed L. Seghier
Department of Biomedical Engineering, Khalifa University, Abu Dhabi, United Arab Emirates
Healthcare Engineering Innovation Center (HEIC), Khalifa University, Abu Dhabi, United Arab Emirates

Anshul Sharma
National Institute of Technology, Patna, Department of Computer Science & Engineering, Patna, Bihar, India

Amit Kumar Singh
Department of Computer Science and Engineering, National Institute of Technology Patna, Bihar, India

Ghanshyam Singh
Centre for Smart Information and Communication Systems, Department of Electrical and Electronic Engineering Science, University of Johannesburg, Johannesburg, South Africa

Kedar Nath Singh
Department of Computer Science and Engineering and Information Technology, Jaypee Institute of Information Technology Noida, UP, India

Om Prakash Singh
Department of Computer Science and Engineering, Indian Institute of Information Technology Bhagalpur, Bihar, India

Rajiv Singh
Department of Computer Science, Banasthali Vidyapith, Rajasthan, India
Centre for Artificial Intelligence, Banasthali Vidyapith, Rajasthan, India

Sanjay Kumar Singh
Indian Institute of Technology (BHU), Department of Computer Science & Engineering, Varanasi, Uttar Pradesh, India

Vinay Kumar Srivastava
Electronics and Communication Engineering Department, Motilal Nehru National Institute of Technology, Prayagraj, Uttar Pradesh, India

Anurag Tiwari
Electronics and Communication Engineering Department, Motilal Nehru National Institute of Technology, Prayagraj, Uttar Pradesh, India

Sanjeev Kumar Varun
Department of Electronics and Communication Engineering, Indian Institute of Technology, Roorkee, India

Ramana Vinjamuri
Department of Computer Science and Electrical Engineering, University of Maryland, College Park, MD, United States

Naoufel Werghi
Khalifa University Center for Autonomous Robotic Systems (KUCARS), Khalifa University, Abu Dhabi, United Arab Emirates
Department of Electrical Engineering and Computer Science, Khalifa University, Abu Dhabi, United Arab Emirates
Center for Cyber-Physical Systems (C2PS), Khalifa University, Abu Dhabi, United Arab Emirates

Michelis Zervakis
School of Electrical and Computer Engineering, Technical University of Crete, Chania, Greece

Huiyu Zhou
School of Computing and Mathematical Sciences, University of Leicester, Leicester, United Kingdom

Preface

1 Introduction

Medical data exist in several formats, ranging from structured data and medical reports to 1D signals, 2D images, and 3D volumes or even higher-dimensional data such as temporal 3D sequences. Healthcare experts can perform an auscultation and produce a report in text format; an electrocardiogram can be made and printed in time series format; X-ray imaging can be performed and the results can be saved as an image; a volume can be provided through angiography; temporal information can be given by echocardiograms; and 4D information can be extracted through flow MRI. Another typical source of variability is the existence of data from different time points, such as pre- and posttreatment. This high and diverse amount of information needs to be organized and mined in an appropriate way so that meaningful information can be extracted. In recent times, multimodal medical data fusion can combine salient information into a single source to ensure better diagnostic accuracy and assessment.

Considering the above context, this book covers cutting-edge research from both academia and industry with a particular emphasis on recent advances in algorithms and applications that involve combining multiple sources of medical information.

This book can be used as a reference for practicing engineers, scientists, and researchers. It can also be useful for senior undergraduate and graduate students and practitioners from government and industry as well as healthcare technology professionals working on state-of-the-art information fusion solutions for healthcare applications. In particular, the book is mainly directed to:

- researchers and scientists in academia and industry working on data processing and security solutions for smart healthcare and medical data and systems;
- experts and developers who want to understand and realize the aspects (opportunities and challenges) of using emerging techniques and algorithms for designing and developing more secure systems and methods for e-health applications;
- PhD students documenting their research and looking for appropriate security solutions to specific issues regarding healthcare applications.

Overall, we selected 17 chapters on various themes related to data fusion techniques and applications for smart healthcare. In the remainder of this chapter, we first summarize the main contribution of each chapter in the book (see Section 2), and then we draw provide concluding comments and remarks.

2 Summary of book chapters

In the following, we summarize the content and contributions of the selected book chapters, so as to provide the reader with an overview of the book material and its organization.

Chapter 1: Retinopathy screening from OCT imagery via deep learning

Optical coherence tomography (OCT) is a noninvasive ophthalmic technique used to diagnose different retinal diseases based on image texture and geometric features. In this chapter, Ahmed et al. propose a deep learning framework to discern four forms of retinal degeneration using OCT. The authors started from the observation that manual processing of OCT scans is time consuming and operator-dependent and might limit early prognosis and medication for eye conditions. These limitations of manual processing naturally demand for automated methods. In particular, the model proposed in this chapter is a lightweight network that combines the atrous spatial pyramid pooling (ASPP) mechanism with deep residual learning. Based on the shortcut connections and dilated atrous convolutions in ASPP, the proposed model exploits the multiscale retinal features in OCT scans for disease prediction. A total of 108,309 OCT scans were used to train the model and 1000 OCT scans were employed to test its performance. The reported simulation results revealed that the method outperforms other cutting-edge approaches in a multiclass classification task, achieving an accuracy of 98.90% with a true positive rate of 97.80% and a true negative rate of 99.27%.

Chapter 2: Multisensor data fusion in Digital Twins for smart healthcare

In this chapter, Zhihan Lv discusses the application value of multisensor data fusion when combined with digital numbers in intelligent medical care. The innovation of this work mainly lies in four aspects. First, the monitoring and alarm system and the support vector machine (SVM) algorithm are established and implemented based on Digital Twins technology. Second, the dynamic chaotic firefly (DCF) algorithm is used to optimize its penalty factor and Gaussian kernel function (KF). The optimized SVM is combined with Dempster–Shafer evidence theory (DSET) to establish a multisymbol parameter data fusion model. Third, the software and hardware platforms are built for human health monitoring, which combine the collected human physiological parameters with the multisign parameter data fusion model. Finally, the collected data of different human postures and behaviors are analyzed, providing an experimental basis for human health monitoring and efficient evaluation in later intelligent medical care (IMC). Results demonstrate that the recognition accuracy of the multisign parameter data fusion model reported here is over 90% under different conditions. The recognition accuracy, monitoring accuracy, and decision making of different sign parameters by the model are all superior to those by SVM and DSET. Under the three postures of standing, sitting, and lying, the detection accuracy of the included model is higher than 96% when the table and chair shielding is 0%, 30%, and 50%. The results demonstrate the feasibility of the multisymbol parameter data fusion model based on a data fusion algorithm and Digital Twins technology for human health monitoring.

Chapter 3: Deep learning for multisource medical information processing

Deep learning algorithms play a significant role in healthcare, showing the potential to revolutionize it with the development of smart clinical decision support systems. In fact, the development of computer-aided diagnostics (CAD) systems is crucial to aid clinicians with a second diagnostic opinion due to the subjectivity of manual diagnosis and handcrafted features. While significant progress has been made in multimodal image data fusion tasks, the application of deep learning for processing of healthcare images and acoustic information has not yet been fully explored. It is through multisource information fusion and computational modeling that outcomes of interest such as treatment targets and drug development ultimately facilitate patient-level decision making in care facilities and homes. Such a phenomenon has attracted interest in healthcare multisource data fusion studies. The application of machine learning and deep learning algorithms provides insight into various aspects of healthcare such as drug discovery, clinical trials, phenotyping, and surgical techniques. This is crucial in providing support to practitioners and health centers to enable precise, efficient, and evidence-based medicine. Moreover, unimodal deep models are less robust, face misclassification, and suffer from system complexity. In this chapter, Gezimati and Singh focus on multisource medical information processing streamlined to disease classification and prediction tasks. In particular, they propose a framework and algorithms for multimodal deep learning models in the classification task of acoustic and image type multimodal datasets for lung cancer. In the chapter, first, the data fusion techniques are reported, and then a deep learning-based multimodal data fusion framework is proposed for multisource image and acoustic data processing. As an additional contribution, existing challenges in multisource information processing systems based on deep learning are identified and perspectives are given with the aim of paving a roadmap for future research.

Chapter 4: Robust watermarking algorithm based on multimodal medical image fusion

The volume of big data has drastically increased for medical applications over recent years. Such data are shared by cloud providers for storage and further processing. Medical images contain sensitive information, and these images are shared with healthcare workers, patients, and in some scenarios researchers for diagnostic and study purposes. Presently, multimodal image fusion is the technique of merging information from two or more image modalities into a single composite image that is better suited for diagnosis and assessment. However, an increasingly serious concern is the illegal copying, modification, and forgery of fused medical records. In this chapter, Singh et al. propose a robust and secure watermarking algorithm based on multimodal medical image fusion. First, nonsubsampled shearlet transform (NSST)-based fusion is used to fuse MRI and CT scans and thus obtain a fused mark image. This fused mark image contains much information and is better suited for diagnosis and assessment than an individual image. Furthermore, the combination of integer wavelength transform (IWT), QR, and singular value decomposition (SVD) is utilized to perform an imperceptible marking of the fused image within the cover media.

Additionally, an efficient encryption algorithm is run by utilizing a 3D chaotic map on a marked image to ensure better security. Experimental outcomes on Kaggle and Open-i datasets show better resistance against a wide range of attacks. Lastly, the reported results indicate that the proposed algorithm outperforms other state-of-the-art techniques.

Chapter 5: Fusion-based robust and secure watermarking method for e-healthcare applications

The fusion of medical images provides the benefits of multiple images in a distinct image and a better clinical experience. Transmission of these fused images is encouraged for better diagnosis and treatment by healthcare professionals. Also, accommodating large volumes of patient records in cloud-based healthcare applications has become common practice. However, it leads to exposure, illegal distribution, and privacy and security concerns of these records. Based on these considerations, in this chapter Anand and Singh present a fusion-based robust and secure watermarking solution using a combination of robust imperceptible marking and encryption. In this method, a fused image is generated by a transform-based fusion method. This fused image is then concealed in the carrier image using watermarking to resolve ownership issues, if any. Further, histogram of oriented gradients (HOG) is used to calculate the values of gain factors to maintain a balanced relation between robustness and visual quality of the marked carrier image. In addition, encryption of the marked image is performed to provide better security. The encrypted image is then stored in a cloud environment for better accessibility. Finally, an experimental and comparative evaluation of the proposed framework is presented to show its versatility, robustness, and imperceptibility.

Chapter 6: Recent advancements in deep learning-based remote photoplethysmography methods

Health monitoring of an individual is guided by physiological parameters that can be estimated by a photoplethysmography (PPG) signal using contact-based or contactless approaches. In particular, contactless approaches are more advantageous than contact-based approaches. In addition, conventional contactless approaches are based on assumptions, which are not required for deep learning methods. Based on this, in this chapter, Gupta et al. review deep learning-based remote PPG (rPPG) signal extraction methods. In doing this, four main contributions are given: First, various compressed and uncompressed datasets used in this domain are presented; second, the region of interest selection methods are summarized and analyzed, followed by rPPG signal extraction methods based on deep learning architecture baselines; finally, the limitations of the existing methods are highlighted with recommendations for future studies.

Chapter 7: Federated learning in healthcare applications

In this chapter, Kanhegaonkar and Prakash highlight the major challenges and design considerations in federated learning related to the healthcare field. Federated learning, also known as collaborative learning, uses a number of dispersed edge devices or servers to run the training algorithms, without exchanging local data samples. It differs from previous approaches in that it does not make the assumption that local data samples are evenly distributed, as is the case with more conventional decentralized systems. It also differs from traditional centralized machine learning techniques, which demand that all local datasets be uploaded to a single server. Federated learning helps in handling crucial data-related challenges like heterogeneous data, privacy, access rights, security, etc. Because of the privacy and secrecy concerns of medical data, sharing or exchange of diagnostic data across different entities is undesirable. Moreover, there are multiple formats for collecting and storing medical data. This results in insufficient and imbalanced data, making model building and training a challenging task. Further, the collected medical diagnostic data are generally heterogeneous in terms of their statistical properties. This reduces the model's capability to generalize well in the medical domain. Federated learning provides fusion-based secure, robust, cost-effective, and privacy-preserving solutions to all these challenges, where knowledge obtained from different decentralized sources of data is fused to build a strong classification model. A detailed discussion of these issues and a possible scope for future research are given in this chapter.

Chapter 8: Riemannian deep feature fusion with autoencoder for MEG depression classification in smart healthcare applications

Major depression disorder (MDD) is a common and severe illness that alters the emotional behavior of the patient and affects how one feels, thinks, and acts in a negative manner. The number of patients suffering from different stages of depression has increased globally, posing a severe challenge to the existing smart healthcare systems. One of the diagnostic methods is magnetoencephalography (MEG). The MEG data collected from MDD patients and healthy subjects are subjected to a series of steps, including preprocessing, feature extraction, and classification. This research is mostly based on structured datasets. However, classification becomes complicated when the dataset is unlabeled. In this chapter, Reddy et al. propose a method of depression classification using a combination of Riemannian geometry, transfer learning, and feature-level deep fusion using autoencoders. The proposed method also participated in the BIOMAG 2022 challenge. Competition results indicate that the proposed method of deep transfer feature fusion overcame the baseline deep learning-based approach by 5.5% in terms of accuracy, thereby providing a potentially useful tool to smart healthcare lawmakers and users.

Chapter 9: Source localization of epileptiform MEG activity towards intelligent smart healthcare: a retrospective study

Epilepsy is a chronic noncommunicable neurological disorder. Around 80% of epileptic patients live in low- or middle-income countries. According to the World Health Organization (WHO) factsheet on epilepsy published in February 2022, over 50 million people are affected by epilepsy worldwide. Approximately 25% of epileptic patients have drug-resistant epilepsy (DRE), despite the growing number of antiseizure drugs. Patients with extreme DRE may undergo resective seizure surgery, for which identification of focal or generalized epileptic seizures in the spatiotemporal domain is a tedious task because of various reasons such as artifacts, mimickers, etc. Researchers used a combination of different modalities such as EEG-MRI, MEG-MRI, etc., for better localization of epileptic spikes. EEG has a higher temporal resolution but lags behind in spatial resolution, and MEG is of a similar nature. So during the resective surgery, there is a requirement for high spatial as well as high temporal resolution. To overcome these limitations, in this chapter Varun et al. use novel MEG-MRI modality fusion techniques to get a better spatiotemporal resolution. To this end, the BIOMAG-2022 Epilepsy data of two patients were used, where all data are provided as resting-state MEG and MRI signals. For localization of interictal epileptic spikes, kurtosis beamforming with linear constrained minimum variance (LCMV) was applied on the BIOMAG dataset.

Chapter 10: Early classification of time series data: overview, challenges, and opportunities

In this chapter, Sharma et al. review early classification approaches, considering univariate and multivariate time series and also prospecting future research directions. A time series is an ordered sequence of measurements, called data points, recorded over time. Generally, the term time series refers to univariate time series (UTSs), where only one variable is measured, such as the temperature of a room or the electrical activity of a patient's heart (electrocardiogram). If two or more variables are measured, the time series is called a multivariate time series (MTS). For example, in monitoring a patient's health, multiple variables such as temperature, pulse rate, blood pressure, and oxygen rate may be measured. In human activity monitoring, multiple sensors can be attached to different parts of the human body, and decisions can be made by fuzzing data collected through the sensors. Usually, time series are classified when a complete data sequence becomes available. However, time-sensitive applications greatly benefit from early classification. For instance, if a patient's disease is detected early by a series of medical observations, the cost of therapy and the length of the recovery period can be reduced. Additionally, an early diagnosis could save the patient's life by giving health practitioners more time to treat them. The primary aim of early classification is to classify the time series as early as possible with desirable accuracy. Several approaches have been developed to solve early classification problems in various domains, including patient monitoring, human activity recognition, drought prediction, and industrial monitoring.

Chapter 11: Deep learning-based multimodal medical image fusion

Medical practitioners often have to work with images which come from various modalities, ranging from X-ray-based CT images to radio wave-based MRI scans. Each image modality provides different information. Multimodal image fusion is the process of merging images of different modalities to obtain a single image that carries almost all the complementary as well as the redundant details to form an image containing much more information. This process, where a single image carrying information of different modalities is produced, is rather useful for medical practitioners and researchers for analyzing a patient's body to detect lesions (if any) and to make a correct diagnosis. Feature extraction plays the key role when it comes to image fusion for multimodal image data, and with that in mind convolutional neural networks have been extensively used in the image fusion literature. However, not many of the deep learning-based models have been specifically designed for medical images. Based on the above considerations, in this chapter Kahol and Bhatnagar first present a comprehensive review of some of the works that have been done recently in the field of multimodal image fusion. Then, inspired by several of the methods discussed, an unsupervised deep learning-based medical image fusion architecture incorporating multiscale feature extraction is proposed. Extensive experiments on various multimodal medical images are conducted to analyze the performance and stability of the proposed technique.

Chapter 12: Data fusion in Internet of Medical Things: towards trust management, security, and privacy

The advent of the Internet of Medical Things (IoMT) has led to a massive revolution in disease monitoring management approaches, improving diagnosis and treatment procedures in order to reduce medical expenditure and facilitate immediate treatment. Through the participation of a huge number of wireless sensor medical devices in IoMT, it engenders a range of various real-time health datasets, which are large, multisource, heterogeneous, and scarce. Moreover, the data transferred through IoMT are mostly prone to security and privacy flaws because of the huge number of sensor devices, which transmit sensitive medical data wirelessly over public channels. The absence of security awareness among healthcare users, e.g., medical workers, patients, etc., can result in different fatal security threats, which eventually jeopardize the lives of patients. Consequently, this situation demands sufficient data security and ensuring patients' privacy in IoMT. However, it is practically infeasible to provide privacy and security due to the enormous volumes of data transmitted in IoMT. In this regard, data fusion is one of the most efficient processes to reduce the size and dimension of data to optimize the data traffic and to acquire real-time health information. In this chapter, Sadhukhan et al. thoroughly investigate various state-of-the-art techniques for IoMT data fusion with considerable attention to trust management, privacy, and secrecy. A brief summary of the foremost advantages, challenges, and limitations is provided, along with a comparative performance study of prevailing data fusion methods for IoMT. Additionally, a comprehensive discussion on the issues on trust management, security, and privacy of data fusion in IoMT is presented to highlight future research directions in this domain.

Chapter 13: Feature fusion for medical data

With the advancement of technology, different imaging models and multimedia content in medical science have been increasingly used for diagnosis, treatment, and education. In addition, another common problem in the medical area is the volume of data, which makes the process longer and require heavy computations. To improve the quality of images from one or more models and create images with more valuable features and higher quality, methods to combine images and their many features are needed. Feature fusion methods are categorized into feature-level and decision-level methods. In this chapter, Joodaki et al. review the feature fusion methods in medical images, data, and background theories. In other words, each image has some features that alone are not visible. These hidden features are visualized by combining several images from different modalities, such as CT and MRI. On the other hand, this process can be expensive because sometimes a lot of time and experience are required. The fusion strategy combines several features and creates a new set of features that contains more practical knowledge that can be applied for a more exact diagnosis. This technique is used for integrating judgments received from several feature sets to create global findings. Medical images with more relevant and valuable features support the diagnosis process.

Chapter 14: Review on hybrid feature selection and classification of microarray gene expression data

Microarray gene expression data are widely used in identifying the classes in cancer data for diagnosis. Classification of microarray gene expression data based on selected features is one of the predominant healthcare applications in biomedical research. Relevant features are selected from the dataset by searching a subset of features and evaluating the subset to select the optimal one. In this chapter, Meenachi and Ramakrishnan review different techniques involved in feature selection, hybridization of feature selection techniques, and data classification based on reduced features, and their performance is analyzed using different metrics. Feature selection can be applied in datasets that are labeled or not labeled. It is used in identifying the feature subset that is optimal from the given dataset. Such reduced feature dataset does not have any negative impact on the classification accuracy. A metaheuristic search algorithm is used in feature selection. These can be categorized as population-based and neighborhood search techniques. The searched feature subsets are evaluated with a classification algorithm to select the best subset. To select the population-based, global optimal features, evolutionary search algorithms such as genetic algorithm (GA) and differential evolution (DE) and swarm intelligence algorithms such as ant colony optimization (ACO) and particle swarm optimization (PSO) are employed. The neighborhood-based tabu search algorithm is used to find the neighborhood's best features. Classifiers like nearest neighbor, support vector machine, fuzzy rough nearest neighbor, etc., are used to evaluate the subsets of features and select the optimal subset. Results of several feature selection algorithms are studied.

Chapter 15: MFFWmark: multifocus fusion-based image watermarking for telemedicine applications with BRISK feature authentication

The growing use of the internet poses significant difficulties for the copyright protection of images. By storing, transmitting, and processing data, watermarking systems can prevent interference and safeguard the copyright of digital multimedia contents. Imperceptibility, robustness, and reliability are important characteristics of image watermarking. For medical applications, multimedia data volumes have vividly expanded. When two medical images are fused, the combined data of both images are transformed at the same time, reducing the data volume. In this chapter, Tiwari et al. propose an image watermarking scheme based on the integer wavelet transform (IWT) and dual decomposition. In the proposed scheme, Schur decomposition (SD) and singular value decomposition (SVD) are used to decompose the IWT-processed cover image. Two watermark images, (1) a brain MRI and (2) a brain CT scan, are fused using a multifocus image fusion technique in the DCT domain and embedded into the DICOM ultrasound image of a liver using IWT and multiple decomposition. The watermark image is fused using two fusion techniques, with and without consistency verification, and the performance of the proposed scheme is compared for both fused watermarks. The scheme proposed in this chapter is tested under various attacks, such as filtering, image compression, and checkmark attacks. The performance is evaluated using different performance parameters like peak signal-to-noise ratio (PSNR), normalized correlation coefficient (NCC), and structural similarity index measurement (SSIM). The authentication of watermarked images is performed using binary robust invariant scalable keypoints (BRISK) features.

Chapter 16: Distributed information fusion for secure healthcare

Recent years have seen a significant increase in the demand for cutting-edge healthcare systems. With the rising potential of artificial intelligence and big data technology, all sectors, especially the healthcare sector, have greatly benefited. Huge amounts of privacy-sensitive clinical data are being generated from several sources. When processing these enormous amounts of diverse healthcare data, the problem of data heterogeneity emerges. The data vary with respect to the patient population, environment, data source, size, complexity, medical procedures, and treatment protocols at individual medical centers. This creates the need for a central knowledge base in the healthcare setting. Federated learning-based fusion techniques can be beneficial to acquire knowledge from these distributed data. This will bring the distributed data together into a single view that can help hospitals and health workers to obtain new insights and helps secure patients' personal information and safeguards them from information leakage. If the distribution of data among the classes is skewed or biased, the distribution is said to be imbalanced. In this chapter, Pathak and Rajput discuss problems associated with imbalanced and heterogeneous healthcare data and their effects on machine learning models and propose methods to improve data fairness in a distributed healthcare system using federated learning.

Chapter 17: Deep learning for emotion recognition using physiological signals

Emotions are a crucial aspect of social interaction among humans. Generally, human emotions are expressed visibly, for example through facial expressions and hand gestures; however, emotions can be easily hidden. To capture real emotions of a person, physiological signals such as heart rate variability, breaths per minute, electroencephalography (EEG), the galvanic skin response, and electromyography (EMG) can be analyzed. In recent years, the analysis of electrical activity in the human brain captured through EEG has gained popularity because of its wide range of potential applications in healthcare (depression, sleep disorders, epilepsy, Alzheimer), human–computer interaction, surveillance systems, entertainment, and police interrogations. Several works on feature fusion-based analysis of EEG signals for emotion recognition have been published; however, deep models have not been well explored in the context of fusion-based emotion recognition. Motivated by aforementioned applications of physiological signals and deep learning, in this chapter, Indolia et al. provide a background study of the EEG-based emotion recognition techniques and propose a fusion approach for EEG-based emotion recognition using bidirectional long short-term memory (Bi-LSTM) and fast Fourier transform (FFT). To illustrate the effectiveness of the proposed integrated method, experimental and comparative analyses of deep learning models using the DEAP and SEED benchmark datasets are provided.

3 Conclusions

In conclusion, this book provides a unique overview of data fusion techniques and applications for smart healthcare:

- A broad scope covering trends in healthcare in terms of medical data fusion is presented, with a focus on identifying challenges, solutions, and new directions, written by experts in the field.
- State-of-the-art data fusion techniques for smart healthcare applications are discussed.
- This book is useful for senior undergraduate and graduate students, scientists, researchers, practitioners from government, industry/healthcare professionals, and others demanding state-of-the-art solutions for medical data fusion.

Overall, this area of research is rapidly expanding due to the increasing number of sensors and diagnostic instruments, creating an urgent need for innovative fusion solutions between different data; the final goal remains that of deriving a better understanding of the data, supporting diagnosis in healthcare applications.

Stefano Berretti[a] and Amit Kumar Singh[b]
[a]University of Florence, Media Integration and Communication Center, Florence, Italy
[b]National Institute of Technology Patna, Department of Computer Science & Engineering, Bihar, India

CHAPTER 1

Retinopathy screening from OCT imagery via deep learning

Ramsha Ahmed[a,b], Bilal Hassan[c,d], Ali Khan[e], Taimur Hassan[f], Jorge Dias[c,d], Mohamed L. Seghier[a,b], and Naoufel Werghi[c,d,g]

[a]*Department of Biomedical Engineering, Khalifa University, Abu Dhabi, United Arab Emirates*
[b]*Healthcare Engineering Innovation Center (HEIC), Khalifa University, Abu Dhabi, United Arab Emirates*
[c]*Khalifa University Center for Autonomous Robotic Systems (KUCARS), Khalifa University, Abu Dhabi, United Arab Emirates*
[d]*Department of Electrical Engineering and Computer Science, Khalifa University, Abu Dhabi, United Arab Emirates*
[e]*College of Mathematics and Computer Science, Zhejiang Normal University, Jinhua, China*
[f]*Department of Electrical, Computer, and Biomedical Engineering, Abu Dhabi University, Abu Dhabi, United Arab Emirates*
[g]*Center for Cyber-Physical Systems (C2PS), Khalifa University, Abu Dhabi, United Arab Emirates*

1.1 Introduction

1.1.1 Background and motivation

Vision impairment is the third leading cause of disability globally, affecting over a billion people [1]. Maculopathy (a collection of illnesses that damage the macula region in the human retina) significantly contributes to visual impairment and blindness [2,3]. The macula is the part of the eye that creates a sharp, central vision. However, maculopathy causes this vision to be distorted because of a buildup of vascular fluid under the macula as a result of damage to blood vessels [4,5].

Eye diseases that commonly affect the macula include age-related macular degeneration (AMD), choroidal neovascularization (CNV), and diabetic macular edema (DME) [3]. Wet AMD and dry AMD are the two manifestations of AMD. Most people with wet AMD have CNV and related retina symptoms, while people with dry AMD have drusen. CNV is characterized by the development of aberrant blood vessels in the retina's choroid layer. DME is a buildup of fluid in the macula area caused by leaking blood vessels [3]. In diabetic retinopathy, DME commonly develops in around a quarter of the patients [6]. Nevertheless, these eye conditions can be treated if diagnosed and medicated early enough, as maculopathy can cause permanent visual loss [7–9].

Data Fusion Techniques and Applications for Smart Healthcare. https://doi.org/10.1016/B978-0-44-313233-9.00007-2

Optical coherence tomography (OCT) is a noninvasive imaging modality that provides high-resolution images of the retina, allowing for early detection and monitoring of retinal diseases [10–14]. In a cross-sectional view, the eye's retina reveals a complex, multilayered structure. This type of cross-sectional image of the retina is acquired by OCT using light waves [15]. The characteristics and severity of retinal diseases can be deduced from the thickness map and contiguity values of the retina's various layers [16]. However, the interpretation of OCT images can also be challenging and requires specialized training. Fig. 1.1 shows the OCT-based visualization of different retinal conditions.

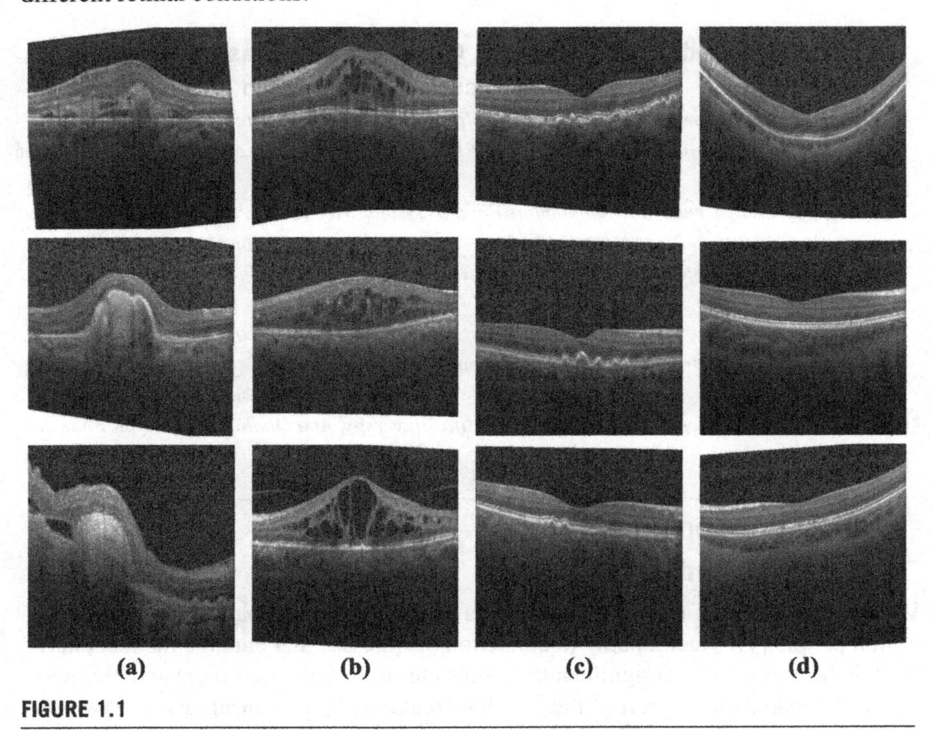

FIGURE 1.1

OCT scans of different retinal conditions. (a) CNV. (b) DME. (c) AMD. (d) Normal.

The traditional screening method for retinal diseases involves manual inspection of eye scans by ophthalmologists or trained graders, which can be time consuming, subjective, and prone to errors. To address the challenges in early detection and diagnosis of retinal diseases that can lead to vision loss or blindness, we propose a deep learning-based approach to automatically screen for retinal diseases from OCT images, which can potentially improve the accuracy and efficiency of retinopathy screening. The motivation behind this work is to develop a reliable and automated tool for detecting and monitoring retinopathy, which can ultimately help prevent vision loss and improve patient outcomes by aiding medical practitioners in their prognosis [17].

1.1.2 Related works

Recent years have seen an increased interest in retinal image processing, focusing on OCT-based automated methods [18]. Retinal OCT imaging has traditionally relied on machine learning methods for various functions, including lesion [19] or retinal layer [20] segmentation, denoising [21], and classification or detection of retinal diseases [22]. However, new and more sophisticated artificial intelligence (AI) techniques such as deep learning have been developed as new and powerful tools for various applications [23–34], including retinal OCT imaging [35,36], demonstrating exceptional performance. It learns the distinctive features automatically and gets classification results that are equivalent to or better than those obtained using standard machine learning approaches [37–39]. Also, when trained extensively and optimally, deep learning frameworks can remarkably match human specialists' performance in identifying retinal diseases [40,41].

Many deep learning techniques have been presented in the literature to classify retinal disorders using automated image analysis of OCT scans. In [42], the authors proposed a deep learning framework to automatically classify different retinal conditions based on domain adaptation using Inception V3. In another work [43], the authors proposed a deep learning framework using a generative adversarial network (GAN)-based classifier to predict AMD and DME. Similarly, in [44] a deep learning architecture comprised of a densely connected neural network (DenseNet) and a recurrent neural network (RNN) to predict neovascular AMD using retinal OCT imaging is introduced. Further, in [45] the authors presented a convolutional neural network (CNN) with constituent residual learning units, called Optic-Net, to identify different retinal diseases, including DME, CNV, and drusen, and normal OCT scans. In [46], the authors proposed a multiinstance learning-based CNN method for retinal OCT classification.

Furthermore, in [47] a self-supervised learning approach called angular contrastive distillation for retinopathy screening and grading is presented. It uses a deep neural network that learns to represent retinal images in a low-dimensional space without the need for labeled data and achieves high accuracy. In another study [48], the authors suggested a hybrid CNN and transformers approach for categorizing retinal diseases. The technique was developed using labeled and unlabeled data, increasing accuracy and efficiency. In areas with limited resources, both of these approaches may help with the diagnosis and management of retinal disorders. Another method for diagnosing retinal disorders [49] combines deep learning CNN (DL-CNN) and image processing techniques. Further, in [50] a technique is proposed for detecting retinal diseases using a coherent CNN (CCNN) architecture. The CCNN achieves high accuracy in the identification and classification of retinal diseases after being trained on a sizable dataset of labeled OCT images. Similar to this, a strategy is proposed in [51] for classifying retinal disorders utilizing pretrained CNNs that have been optimized on a labeled OCT dataset, achieving high accuracy in disease classification. Additionally, a number of techniques described in the literature [52,53] used the transfer learning strategy to classify retinal diseases.

To improve the performance and clinical applications of deep learning models in retinal OCT imaging, it is crucial to address several research gaps and challenges. These challenges include the absence of extensive and diverse datasets, the need for interpretability and explainability of deep learning models, and concerns related to overfitting and generalization, as well as the issues of robustness and reliability. Another challenge is the lack of clinical validation and adoption. Additionally, the existing deep learning models are computationally expensive, which makes them impractical for clinical use. Overcoming these challenges will require further research and development efforts.

1.1.3 Key contributions

In this chapter, we introduce a novel deep learning framework designed for the automated detection of four forms of retinal degeneration: AMD, CNV, DME, and normal, using OCT scans. Our framework utilizes a lightweight network that combines the atrous spatial pyramid pooling (ASPP) mechanism with deep residual learning to capture multiscale retinal features and predict diseases accurately. Our approach stands out by leveraging the distinctive retinal features related to each disease. For example, drusen are present in AMD patients, while subretinal fluid (SRF) and CNV membrane occur in CNV disease. In DME, intraretinal fluid (IRF) leaks, exudates, or cystic spaces are detected, while normal OCT scans show closely intact fovea and retinal layers. By capturing these subtle variations in retinal features with a minimalistic and easy-to-train deep neural network architecture, our framework enables more accurate diagnosis and prognosis of eye conditions, thus leading to improved patient outcomes. Moreover, our proposed method overcomes the challenges of model complexity and significant computational resource requirements, which can be challenging to implement in low-resource settings or in real-time applications. Overall, our framework presents a significant advancement in the field of ophthalmic diagnosis and has the potential to enhance the early detection and treatment of retinal diseases. The notable contributions of the proposed work are threefold:

1. Our deep learning-based retinal OCT classification framework incorporates residual learning with the ASPP mechanism. It captures more profound and high-resolution feature maps distributed in the OCT scans to predict various retinal conditions, including AMD, CNV, DME, and normal eyes.
2. The proposed model is a lightweight CNN model using fewer parameters (around 0.28 million) to train, unlike popular pretrained architectures, substantially reducing the model's training and inference time.
3. Extensive experimentation confirmed the superiority of our model with respect to other competitive architectures. Our model achieves a mean accuracy of 98.90%, with a true positive rate (TPR) of 97.80% and a true negative rate (TNR) of 99.27% for the classification of retinal diseases.

The remainder of this chapter is organized as follows. Section 1.2 elaborates on the dataset details and the proposed deep learning framework for multiclass retinal disease classification, Section 1.3 presents the details on evaluation metrics and experimental results, Section 1.4 discusses the findings of our proposed work, and Section 1.5 summarizes the conclusion and future scope of our research.

1.2 Retinal OCT classification framework

In this section, we first present the details of the dataset used in this work. Next, we describe the proposed deep learning-based network architecture and constituent segments, the loss function used, and the training performance.

1.2.1 Dataset description

A publicly available OCT imaging dataset (referred to as the Zhang dataset[1]) [53] was used for the experimental analysis in this investigation. The dataset is comprised of 109,309 OCT scans with separate training (108,309 scans) and testing (1000 scans) subsets. Further, it is based on four discrete retina states, including AMD, CNV, DME, and normal, collected using the Spectralis imaging machine. In addition, we split the training subset in a ratio of 80:20 to accordingly train and validate the proposed deep learning model. The model was tested using the entire testing subset (1000 scans). Table 1.1 presents the dataset details and the split strategy adopted for training, validating, and testing the deep learning model in the proposed research.

Table 1.1 Dataset details.

Subset	CNV	DME	AMD	Normal	Total
Training	29,764	9078	6893	40,912	86,647
Validation	7441	2270	1723	10,228	21,662
Testing	250	250	250	250	1000

1.2.2 Deep learning architecture

We propose a 52-layer deep learning architecture based on a directed acyclic graph (DAG) network, as shown in Fig. 1.2. It is constructed as a lightweight network that uses only roughly 0.28 million parameters and has a network depth of 14 layers (convolutional layers). Furthermore, the proposed predictive model incorporates an $ASPP$ module with residual connections to improve feature learning competency in identifying the retinal conditions with high accuracy compared to existing deep learning models that merely feed forward and employ standard convolutional networks.

The model receives the candidate grayscale OCT scan as input with 299 × 299 dimensions. In the proposed deep learning framework, the first five blocks (N^1, N^2,

[1] Zhang OCT dataset [53]: https://data.mendeley.com/datasets/rscbjbr9sj/3.

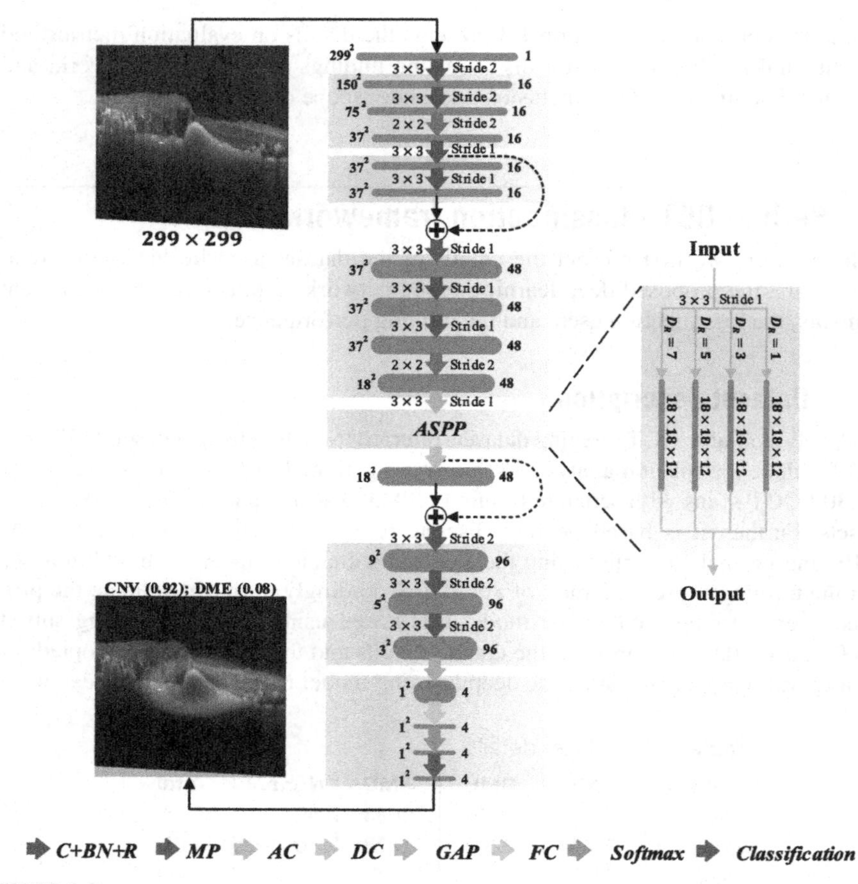

FIGURE 1.2

Block diagram of the proposed CNN model for classification of retinal diseases using OCT scans. The different color arrows represent the different operations performed in the model to produce the following feature maps shown in gray color (please note that the colors may only appear in the web/online version of the article whereas in the print version, the content may appear as gray). The proposed network employs atrous convolutions in the ASPP block with varying dilation rates to capture multiscale feature extraction. The resultant multiscale features are fused using the depthwise concatenation operation.

N^3, $ASPP$, and N^4) constitute the contracting path of the network, which learns the unique features among the four different classes (AMD, CNV, DME, and normal) investigated in this study. Furthermore, the last block (N^5) classifies the input OCT scan using a softmax confidence score. As observed in Fig. 1.2, the model begins as an elongated stem structure to generate feature maps via convolution (C) and max pooling (MP) layers on a kernel of size 3×3 and 2×2, respectively. Furthermore, batch normalization (BN) and rectified linear activation (R) are applied to each convolutional layer in the network to reduce nonlinearity and truncate negative pixels, allowing the model to identify variations in underlying features.

The proposed model is further enhanced by residual skip connections to prevent gradient degradation, enabling explicit learning of multiscale features and fast network convergence [28]. The $ASPP$ block is another constituent segment of our network that leverages high-resolution feature maps from candidate OCT scans and preserves more spatial information [29] to classify the retinal conditions with high accuracy. Moreover, there are four 3×3 atrous convolution (AC) layers in the $ASPP$ block, where the dilation rates (D_r) are increased to 1, 3, 5, and 7, respectively. The expression for AC is given as

$$F_{out}(\alpha, \beta) = \sum_{x=1}^{\alpha} \sum_{y=1}^{\beta} F_{in}(\alpha + D_r \times x, \beta + D_r \times y) k(x, y), \tag{1.1}$$

where $F_{in}(\alpha, \beta)$ are the input and $F_{out}(\alpha, \beta)$ are the output feature maps. Further, α is the length and β is the width of the feature maps. The dilation rate is represented by D_r, and $k(x, y)$ shows the convolutional kernel.

The kernel size is normally fixed in a standard convolution. Furthermore, selecting a single kernel size large enough to capture sufficient information on the underlying features for distinguishing between various retinal conditions is challenging. On the other hand, using a multikernel convolutional method may be an option. However, because of large convolutional kernels, the number of network parameters and complexity can considerably increase in a multikernel approach, resulting in high inference and convergence times [54]. Because of this, we employed AC operations in the proposed deep learning framework. AC allows flexible aggregation of multiscale contextual information compared to the standard convolution. Also, it is possible to adaptably increase the kernel size ($k \times k$) to a new receptive field $k + (k-1)(r-1)$ by altering AC's dilation rate D_r. In this way, the network will keep the spatial resolution the same as before, which means that the computational complexity and number of parameters are not increased while executing a wider and more context-aware filtering [54]. Thus, the improved classification performance of the proposed framework compared to prior deep learning-based methods is due to the inclusion of $ASPP$ and AC.

Furthermore, as seen in Fig. 1.2, BN and R activations follow the AC in the $ASPP$ block. Next, all the convolutional branches in $ASPP$ are combined using depthwise concatenation (DC) into a single output giving the same-sized features as the $ASPP$ input. From here, the output is sent to the N^5 block, comprising global average pooling (GAP), fully connected (FC), softmax, and classification operations for predicting true retinal conditions in the candidate OCT scans.

1.2.3 Loss function

The proposed deep learning framework is trained with a loss function based on the binary cross-entropy loss (L_e). This loss function shows how closely the predicted and actual classes resemble each other. L_e is calculated as follows:

$$L_e = -\sum_{i=1}^{U}\sum_{j=1}^{V} \gamma_{ij} \log(\rho_{ij}), \tag{1.2}$$

where V represents the class and U denotes the number of samples in V. The prediction probability of the ith sample in class j is given by ρ_{ij} and γ_{ij} indicates whether a given ith sample is a member of class j.

1.2.4 Training performance

To accurately classify underlying retinal diseases, the hyperparameters are empirically fine-tuned for the training process. Moreover, we applied different transformations on the raw OCT scans to augment the training data in order to prevent the model from overfitting. These transformations include reflection, translation, and scaling. Further, we used the stochastic gradient descent with momentum (SGDM) solver to train the proposed deep learning framework over 20,280 iterations across 30 epochs. Fig. 1.3 depicts the proposed model's training graphs for accuracy and cross-entropy loss. The data augmentations used and the network hyperparameters specified to train the network are shown in Table 1.2.

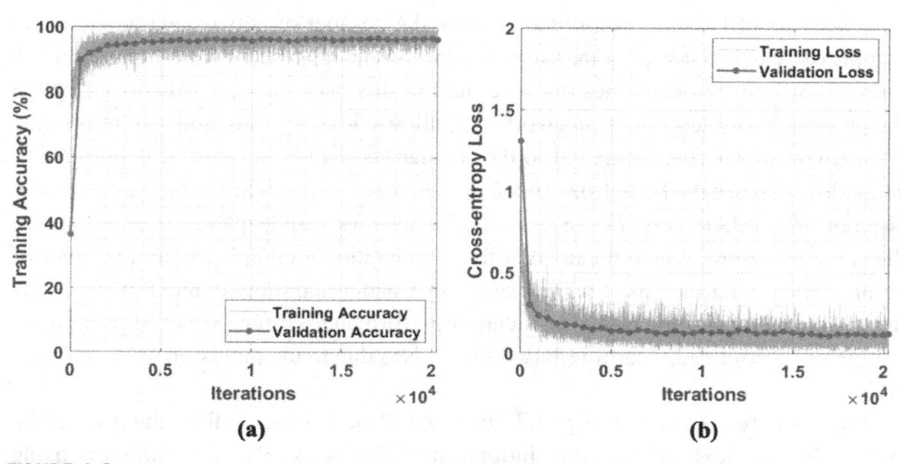

FIGURE 1.3

Training performance curves of the proposed deep learning model. (a) Training and validation accuracy plots. (b) Training and validation loss plots.

Table 1.2 Details of data augmentation and hyperparameter selection for training the proposed deep learning model.

Process	Parameter	Value
Augmentation	Reflection (x- and y-axes)	50% probability
	Translation (x- and y-axes)	[−30 to 30] pixels
	Scaling	[0.7 to 1.3] times
Training	Training OCT scans	86,647
	Validation OCT scans	21,662
	Number of epochs	30
	Minibatch size	128
	Total iterations	20,280
	Validation frequency	500
	Solver	SGDM
	Momentum	0.9
	Loss function	Cross-entropy
	Initial learning rate	0.01

1.3 Simulation results

In this section, we discuss the performance metrics used for evaluating the proposed deep learning framework, the system specifications, and the experimental results in detail.

1.3.1 Performance metrics

We evaluated the proposed deep learning framework using various performance indicators, including classification accuracy (M_{ca}), recall or TPR (M_{tpr}), specificity or TNR (M_{tnr}), precision or positive predictive value (PPV) (M_{ppv}), and negative predictive value (NPV) (M_{npv}). These evaluation metrics are expressed as follows:

$$M_{ca} = \frac{T_p + T_n}{T_p + T_n + F_p + F_n}, \tag{1.3}$$

$$M_{tpr} = \frac{T_p}{T_p + F_n}, \tag{1.4}$$

$$M_{tnr} = \frac{T_n}{T_n + F_p}, \tag{1.5}$$

$$M_{ppv} = \frac{T_p}{T_p + F_p}, \tag{1.6}$$

$$M_{npv} = \frac{T_n}{T_n + F_n}, \tag{1.7}$$

where T_p indicates true positive instances (predicted correctly) and T_n represents true negatives (accurately ruled out); similarly, F_p indicates false positives and F_n represents false negative instances. For instance, when the retinal disease identified by the predictive model is not the actual one, it is referred to as F_p. Likewise, F_n shows the wrong prediction when the retinal disease is truly present, but the model fails to detect or identify it.

1.3.2 System specification

The proposed deep learning framework was implemented with MATLAB® R2022a on a Windows system with a 64-bit OS, an Intel Core i7-11700 @2.5GHz processor, 32 GB of memory, and an Nvidia GeForce RTX 3090 graphics processor. In addition, we ran a series of experiments to test its ability to predict the various retinal conditions on OCT imaging and found it to be accurate. The experimental results are elaborated on in the following subsection in detail.

1.3.3 Experimental results

We first looked at the trained deep learning framework's disease classification performance on the entire testing subset. It includes 250 OCT scans each for AMD, CNV, DME, and normal, as listed in Table 1.1. To determine the ability of the proposed model to classify candidate OCT scans, we used the five performance metrics mentioned earlier. The classification results are displayed in Table 1.3. As observed, the classification performance of the proposed model is high, with an overall mean classification accuracy (M_{ca}) of 98.90%, an M_{tpr} of 97.80%, and an M_{tnr} of 99.27%.

Table 1.3 Performance evaluation of the proposed deep learning model for the classification of retinal diseases using different metrics.

Metric	CNV	DME	AMD	Normal	Combined
T_p (#)	250	241	238	249	978
T_n (#)	736	750	749	743	2978
F_p (#)	14	0	1	7	22
F_n (#)	0	9	12	1	22
M_{ca} (%)	98.60	99.10	98.70	99.20	98.90
M_{tpr} (%)	100	96.40	95.20	99.60	97.80
M_{tnr} (%)	98.13	100	99.87	99.07	99.27
M_{ppv} (%)	94.70	100	99.58	97.27	97.80
M_{npv} (%)	100	98.81	98.42	99.87	99.27

Next, Fig. 1.4 shows the model's disease classification results on OCT scans selected randomly from each class. The results are demonstrated using the gradient-weighted class activation mapping (Grad-CAM) approach [55]. Furthermore, according to the results displayed, the red hue denotes the area in the candidate OCT scan

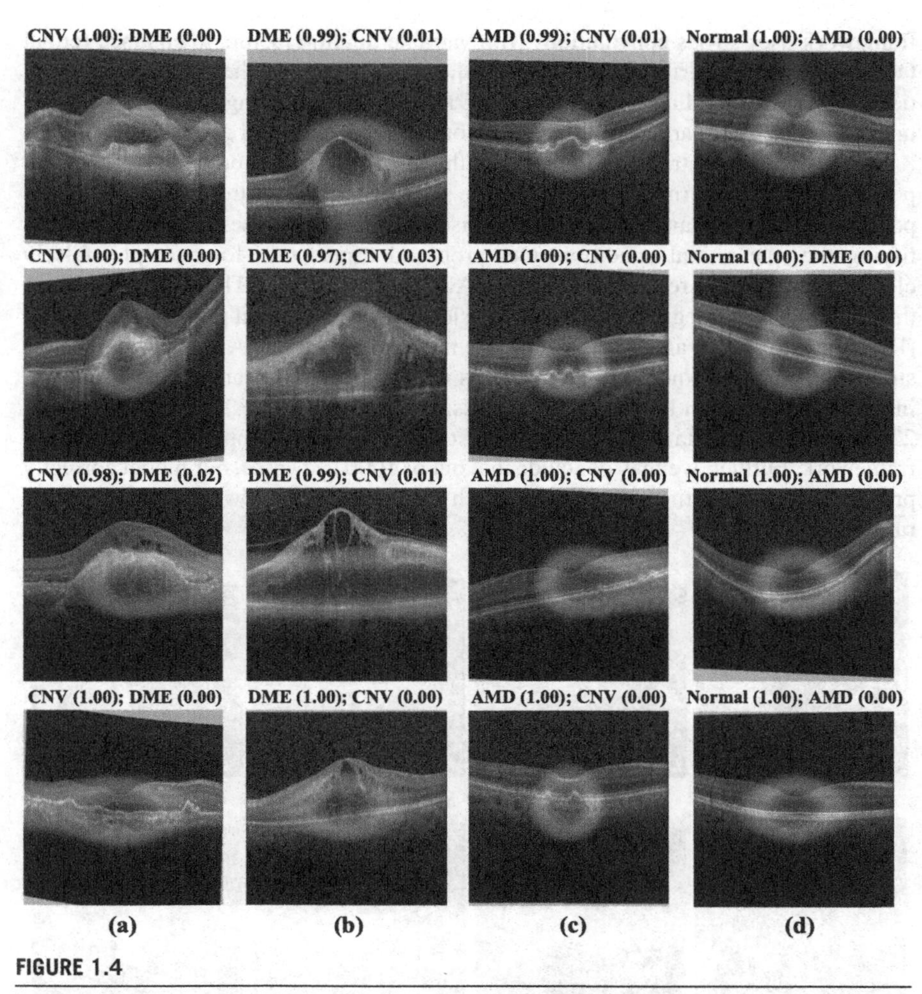

FIGURE 1.4

Depiction of correct classification results using class activation maps and confidence scores. The OCT scans were randomly selected from the test set and are correctly classified by the proposed deep learning model. (a) CNV OCT scans. (b) DME OCT scans. (c) AMD OCT scans. (d) Normal OCT scans. The dark red (dark gray in print version) regions denote the parts of the scan which largely influenced the model's decision.

which significantly impacts the network's decision-making process towards the class label predicted. Also, it indicates that our model relies heavily on the associated retina symptoms or lesions prominent in the neurosensory retina for disease prediction. For example, AMD is characterized by drusen and CNV is characterized by SRF and the presence of CNV membrane. Further, IRF leaks are found in DME, while all retinal layers seem to be fully intact in normal OCT scans. Our proposed deep learning framework uses atrous convolutions with variable dilation factors to enhance retinal OCT classification performance. It allows the network to learn the distinctive properties of the corresponding lesions in the OCT scans for predicting retinal diseases by understanding their variations in delimitation, contour, size, and geometry.

Additionally, illustrated in Fig. 1.5 are the misclassification outcomes of the proposed deep learning framework. The figure shows how closely the model's incorrect prediction resembles another ophthalmic disorder, which confuses the model in identifying the true retinal condition. Our proposed framework, for example, falsely classifies the normal retinal condition as AMD in Fig. 1.5(a). This failure is due to the slightly bulging region observed on the left side of the retinal pigment epithelium (RPE) layer on the scan, which caused the model to label it as AMD. Such misclassification results are known as F_n instances with respect to the target class and as F_p instances with respect to the predicted class. As shown in Table 1.3, there were only 22 F_n and 22 F_p instances out of 1000 test cases using the proposed deep learning framework. Further, we may conclude that out of 1000 test cases, our model correctly predicted (T_p) 978 retinal conditions, which validates its very low false classification rate.

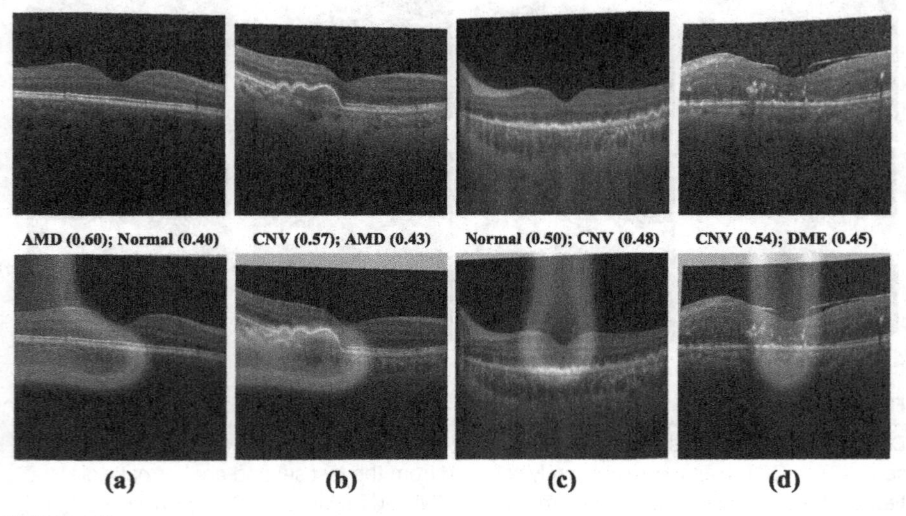

FIGURE 1.5

Depiction of false classification results by the proposed deep learning model. (a) The normal eye scan is incorrectly classified as AMD. The AMD scans are wrongly classified as CNV in (b) and as normal in (c). (d) The DME scan is falsely classified as CNV. The confidence scores in the case of false classifications are observed to be low compared to correct classification results. Please note that the colors in this figure may only appear in the web/online version of the article whereas they will appear as gray in the print version.

Next, the confusion matrix of the predictive model is presented in Fig. 1.6, summarizing the correct and incorrect predictions broken down by each class. The results

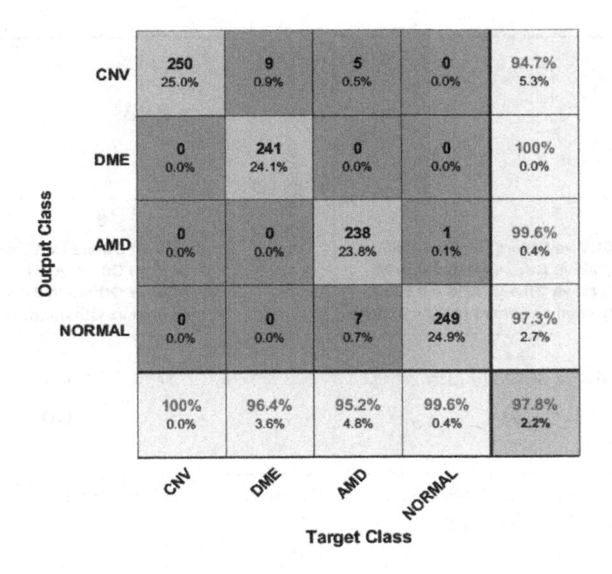

FIGURE 1.6

Confusion matrix to demonstrate the retinal disease classification results by the proposed deep learning model. The trained model faced the most confusion for the AMD class, with 12 false negatives (5 as CNV, 7 as normal).

reveal how the proposed classification model makes incorrect predictions for each retinal condition. Further, the T_p instances, i.e., the model's correct predictions, are represented by the diagonal entries of the matrix. As observed, our proposed framework gives the best predictive results for CNV without any misclassification. In contrast, with 12 F_n and 1 F_p cases, it causes the most uncertainty in AMD. Besides, according to the final diagonal value from the confusion matrix, the model has an overall hit rate of 97.80% and a miss rate of 2.20%.

Further, the receiver operating characteristic (ROC) and precision recall (PR) graphs are generated by altering the classification threshold to demonstrate the proposed model's ability to predict various retinal diseases. The classification threshold is set between zero and one with a 0.01 increase in step size to construct these curves, as exhibited in Fig. 1.7. Fig. 1.7(a) and Fig. 1.7(b) show that the variation in threshold has no negative impact on the network's performance. It achieves high values for mean average precision (mAP) and area under the curve (AUC) metrics for each retinal condition.

Finally, we compared our proposed framework to several known deep learning algorithms for classifying the different retinal conditions using the Zhang dataset [53]. Table 1.4 provides the mean quantitative results in terms of various performance indicators, considering all four classes investigated in this study. With an M_{ca} of 98.90%, an M_{tpr} of 97.80%, and an M_{tnr} of 99.27% (Table 1.4), our model surpasses the existing methods in the multiclass disease classification task between normal individuals

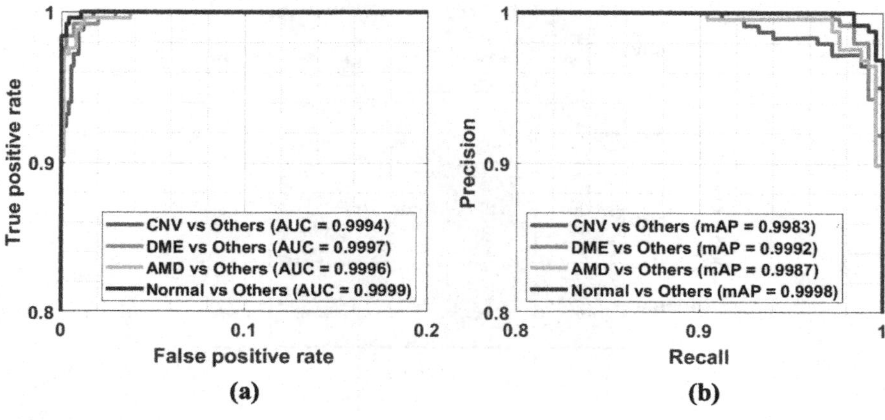

FIGURE 1.7

The performance of the proposed deep learning model for retinal disease classification using (a) ROC and (b) PR curves.

Table 1.4 Performance comparison of the proposed deep learning model with other state-of-the-art methods for the classification of retinal diseases. The best and second-best results are indicated with bold and underlined, respectively.

Method	M_{ca}	M_{tpr}	M_{tnr}	M_{ppv}	M_{npv}
Transfer learning [53]	96.60	97.80	97.40	-	-
MacularNet [8]	92.60	88.52	-	90.27	-
TS-SSL(ve) [56]	96.20	-	-	-	-
LSGAN-ResNet50 [4]	-	81.46	91.17	80.36	-
LSGAN-DenseNet121 [4]	-	86.22	93.10	85.91	-
LightOCT [57]	96.00	94.50	-	-	-
[58]	97.40	96.53	-	97.40	-
IFCNN [9]	87.30	82.20	83.65	-	-
AS model [2]	97.25	94.50	-	94.50	-
RRI-Net [3]	98.80	97.60	99.20	97.61	99.20
[47]	95.90	97.20	99.13	97.39	99.07
LLCT [48]	98.85	97.65	-	97.83	-
[50]	95.63	94.40	96.03	88.81	98.09
[49]	97.14	94.47	98.16	95.61	-
ResNet34 [51]	98.20	98.25	-	98.25	-
ResNet101 [51]	98.40	98.75	-	98.25	-
Proposed	**98.90**	**97.80**	**99.27**	**97.80**	**99.27**

and those with AMD, DME, and CNV. The novel integration of residual learning and multiscale pyramid pooling in the proposed deep learning framework allows it to capture high-resolution feature maps from OCT scans and preserve more spatial information. The architecture of our proposed model is also minimalist; consequently, it is simple to train. In addition, based on its high feature learning capability with residual connections and ASPP, it can deal with the wide range of retina symptoms or lesions related to the disease classification problem being investigated in this work.

1.4 Discussion

This section reports the outcomes and findings of the proposed research for classifying retinal disorders using OCT images. The proposed deep learning model was trained to classify four retinal conditions: AMD, CNV, DME, and normal. The performance of the model was assessed using various performance indicators.

The research showed that the proposed deep learning framework performed extremely well in classifying each of the four classes, with an overall mean classification accuracy of 98.90%. Moreover, we visualized the results of the model using the Grad-CAM technique, which revealed that the model depended substantially on linked retina symptoms or lesions conspicuous in the neurosensory retina for disease prediction. The study also demonstrated how the proposed framework misclassified some OCT scans, providing examples of erroneous predictions and their causes.

Furthermore, the performance of the proposed framework is compared to many known deep learning methods for the classification of investigated retinal diseases. In the multiclass disease classification task, the proposed model outperformed existing techniques. The better performance of the proposed model is attributed to the integration of residual learning and multiscale pyramid pooling, allowing it to collect high-resolution feature maps from OCT scans while preserving more spatial information. Besides, the minimalistic architecture of the proposed model also makes it easy to train and capable of handling a wide range of retina symptoms or lesions associated with the disease classification problem investigated in this research. In addition, the research findings show that the superior performance of the proposed model may aid in the early detection and diagnosis of retinal illnesses, potentially leading to better treatment outcomes for those affected.

1.5 Conclusion and future scope

An automated framework based on deep residual learning and ASPP was developed to predict retinal diseases using OCT imaging. Our proposed model is structured as a lightweight deep CNN network. It leverages dilated AC with broad receptive fields to capture high-resolution feature maps from OCT scans to distinguish between different classes of retinal conditions (AMD, CNV, DME, and normal). It was trained and bench-marked on a publicly available dataset of more than 100,000 OCT scans. The classification performance of the proposed framework was demonstrated through a series of experiments and it was shown to be superior to other state-of-the-art methods. However, some potential limitations of the work include data bias, generalization, and clinical applicability. Further validation and testing in a clinical setting would be necessary to assess the practicality and effectiveness of the proposed model. In the future, we intend to extend our method to include boundary segmentation of retinal layers and their associated retina symptoms or lesions in the neurosensory retina. This extension will enable detecting more subtle traits and anomalies automatically and with greater certainty, hence empowering ophthalmologists with an early retinal disorder detection tool.

References

[1] B. Hassan, R. Ahmed, B. Li, A. Noor, Z.u. Hassan, A comprehensive study capturing vision loss burden in Pakistan (1990-2025): findings from the Global Burden of Disease (GBD) 2017 study, PLoS ONE 14 (5) (2019) e0216492.

[2] L. Ashok, V. Latha, K. Sreeni, Detection of macular diseases from optical coherence tomography images: ensemble learning approach using VGG-16 and Inception-V3, in: Proceedings of International Conference on Communication and Computational Technologies, Springer, 2021, pp. 101–116.

[3] B. Hassan, S. Qin, R. Ahmed, RRI-Net: classification of multi-class retinal diseases with deep recurrent residual inception network using OCT scans, in: 2020 IEEE International Symposium on Signal Processing and Information Technology (ISSPIT), IEEE, 2020, pp. 1–6.

[4] X. He, L. Fang, H. Rabbani, X. Chen, Z. Liu, Retinal optical coherence tomography image classification with label smoothing generative adversarial network, Neurocomputing 405 (2020) 37–47.

[5] B. Hassan, T. Hassan, Fully automated detection, grading and 3D modeling of maculopathy from OCT volumes, in: 2019 2nd International Conference on Communication, Computing and Digital Systems (C-CODE), IEEE, 2019, pp. 252–257.

[6] J. Kim, L. Tran, Retinal disease classification from OCT images using deep learning algorithms, in: 2021 IEEE Conference on Computational Intelligence in Bioinformatics and Computational Biology (CIBCB), IEEE, 2021, pp. 1–6.

[7] B. Hassan, R. Ahmed, B. Li, O. Hassan, T. Hassan, Automated retinal edema detection from fundus and optical coherence tomography scans, in: 2019 5th International Conference on Control, Automation and Robotics (ICCAR), IEEE, 2019, pp. 325–330.

[8] S.S. Mishra, B. Mandal, N.B. Puhan, MacularNet: towards fully automated attention-based deep CNN for macular disease classification, SN Computer Science 3 (2) (2022) 1–16.

[9] L. Fang, Y. Jin, L. Huang, S. Guo, G. Zhao, X. Chen, Iterative fusion convolutional neural networks for classification of optical coherence tomography images, Journal of Visual Communication and Image Representation 59 (2019) 327–333.

[10] B. Hassan, R. Ahmed, B. Li, Automated foveal detection in OCT scans, in: 2018 IEEE International Symposium on Signal Processing and Information Technology (ISSPIT), IEEE, 2018, pp. 419–422.

[11] X. He, L. Fang, M. Tan, X. Chen, Intra-and inter-slice contrastive learning for point supervised OCT fluid segmentation, IEEE Transactions on Image Processing 31 (2022) 1870–1881.

[12] I. Lains, J.C. Wang, Y. Cui, R. Katz, F. Vingopoulos, G. Staurenghi, D.G. Vavvas, J.W. Miller, J.B. Miller, Retinal applications of swept source optical coherence tomography (OCT) and optical coherence tomography angiography (OCTA), Progress in Retinal and Eye Research 84 (2021) 100951.

[13] T.K. Yoo, J.Y. Choi, H.K. Kim, Feasibility study to improve deep learning in OCT diagnosis of rare retinal diseases with few-shot classification, Medical & Biological Engineering & Computing 59 (2) (2021) 401–415.

[14] A. Vivekanand, N. Werghi, H. Al-Ahmad, Multiscale roughness approach for assessing posterior capsule opacification, IEEE Journal of Biomedical and Health Informatics 18 (6) (2014) 1923–1931.

[15] B. Hassan, R. Ahmed, B. Li, Computer aided diagnosis of idiopathic central serous chorioretinopathy, in: 2018 2nd IEEE Advanced Information Management, Communicates, Electronic and Automation Control Conference (IMCEC), IEEE, 2018, pp. 824–828.

[16] G. Altan, DeepOCT: an explainable deep learning architecture to analyze macular edema on OCT images, International Journal of Engineering Science and Technology 34 (2022) 101091.

[17] R. Bhadra, S. Kar, Retinal disease classification from optical coherence tomographical scans using multilayered convolution neural network, in: 2020 IEEE Applied Signal Processing Conference (ASPCON), IEEE, 2020, pp. 212–216.

[18] B. Hassan, T. Hassan, R. Ahmed, S. Qin, N. Werghi, Automated segmentation and extraction of posterior eye segment using OCT scans, in: 2021 International Conference on Robotics and Automation in Industry (ICRAI), IEEE, 2021, pp. 1–5.

[19] T. Kurmann, S. Yu, P. Márquez-Neila, A. Ebneter, M. Zinkernagel, M.R. Munk, S. Wolf, R. Sznitman, Expert-level automated biomarker identification in optical coherence tomography scans, Scientific Reports 9 (1) (2019) 1–9.

[20] M. Christopher, A. Belghith, R.N. Weinreb, C. Bowd, M.H. Goldbaum, L.J. Saunders, F.A. Medeiros, L.M. Zangwill, Retinal nerve fiber layer features identified by unsupervised machine learning on optical coherence tomography scans predict glaucoma progression, Investigative Ophthalmology & Visual Science 59 (7) (2018) 2748–2756.

[21] J. Cheng, D. Tao, Y. Quan, D.W.K. Wong, G.C.M. Cheung, M. Akiba, J. Liu, Speckle reduction in 3D optical coherence tomography of retina by A-scan reconstruction, IEEE Transactions on Medical Imaging 35 (10) (2016) 2270–2279.

[22] B. Hassan, T. Hassan, B. Li, R. Ahmed, O. Hassan, Deep ensemble learning based objective grading of macular edema by extracting clinically significant findings from fused retinal imaging modalities, Sensors 19 (13) (2019) 2970.

[23] B. Hassan, R. Ahmed, B. Li, O. Hassan, An imperceptible medical image watermarking framework for automated diagnosis of retinal pathologies in an eHealth arrangement, IEEE Access 7 (2019) 69758–69775.

[24] R. Ahmed, Y. Chen, B. Hassan, L. Du, CR-IoTNet: machine learning based joint spectrum sensing and allocation for cognitive radio enabled IoT cellular networks, Ad Hoc Networks 112 (2021) 102390.

[25] B. Hassan, R. Ahmed, B. Li, O. Hassan, T. Hassan, Autonomous framework for person identification by analyzing vocal sounds and speech patterns, in: 2019 5th International Conference on Control, Automation and Robotics (ICCAR), IEEE, 2019, pp. 649–653.

[26] G. Sidra, N. Ammara, H. Taimur, H. Bilal, A. Ramsha, Fully automated identification of heart sounds for the analysis of cardiovascular pathology, in: Applications of Intelligent Technologies in Healthcare, Springer, 2019, pp. 117–129.

[27] R. Ahmed, Y. Chen, B. Hassan, Optimal spectrum sensing in MIMO-based cognitive radio wireless sensor network (CR-WSN) using GLRT with noise uncertainty at low SNR, AEÜ. International Journal of Electronics and Communications 136 (2021) 153741.

[28] R. Ahmed, Y. Chen, B. Hassan, Deep learning-driven opportunistic spectrum access (OSA) framework for cognitive 5G and beyond 5G (B5G) networks, Ad Hoc Networks 123 (2021) 102632.

[29] R. Ahmed, Y. Chen, B. Hassan, Deep residual learning-based cognitive model for detection and classification of transmitted signal patterns in 5G smart city networks, Digital Signal Processing 120 (2022) 103290.

[30] R. Alkadi, F. Taher, A. El-Baz, N. Werghi, A deep learning-based approach for the detection and localization of prostate cancer in T2 magnetic resonance images, Journal of Digital Imaging 32 (5) (2019) 793–807.

[31] I. Reda, A. Shalaby, F. Khalifa, M. Elmogy, A. Aboulfotouh, M. Abou El-Ghar, E. Hosseini-Asl, N. Werghi, R. Keynton, A. El-Baz, Computer-aided diagnostic tool for early detection of prostate cancer, in: 2016 IEEE International Conference on Image Processing (ICIP), IEEE, 2016, pp. 2668–2672.

[32] E. Al Hadhrami, M. Al Mufti, B. Taha, N. Werghi, Transfer learning with convolutional neural networks for moving target classification with micro-Doppler radar spectrograms, in: 2018 International Conference on Artificial Intelligence and Big Data (ICAIBD), IEEE, 2018, pp. 148–154.

[33] T. Hassan, M. Shafay, S. Akçay, S. Khan, M. Bennamoun, E. Damiani, N. Werghi, Meta-transfer learning driven tensor-shot detector for the autonomous localization and recognition of concealed baggage threats, Sensors 20 (22) (2020) 6450.

[34] B. Hassan, T. Hassan, R. Ahmed, N. Werghi, J. Dias, SIPFormer: segmentation of multiocular biometric traits with transformers, IEEE Transactions on Instrumentation and Measurement (2022).

[35] B. Hassan, S. Qin, T. Hassan, R. Ahmed, N. Werghi, Joint segmentation and quantification of chorioretinal biomarkers in optical coherence tomography scans: a deep learning approach, IEEE Transactions on Instrumentation and Measurement 70 (2021) 1–17.

[36] B. Hassan, S. Qin, R. Ahmed, SEADNet: deep learning driven segmentation and extraction of macular fluids in 3D retinal OCT scans, in: 2020 IEEE International Symposium on Signal Processing and Information Technology (ISSPIT), IEEE, 2020, pp. 1–6.

[37] A. Khan, B. Hassan, S. Khan, R. Ahmed, A. Abuassba, DeepFire: a novel dataset and deep transfer learning benchmark for forest fire detection, Mobile Information Systems 2022 (2022).

[38] S. Mazhar, G. Sun, A. Bilal, B. Hassan, Y. Li, J. Zhang, Y. Lin, A. Khan, R. Ahmed, T. Hassan, AUnet: a deep learning framework for surface water channel mapping using large-coverage remote sensing images and sparse scribble annotations from OSM data, Remote Sensing 14 (14) (2022) 3283.

[39] R. Ahmed, Y. Chen, B. Hassan, L. Du, T. Hassan, J. Dias, Hybrid machine learning-based spectrum sensing and allocation with adaptive congestion-aware modeling in CR-assisted IoV networks, IEEE Internet of Things Journal (2022).

[40] B. Hassan, S. Qin, T. Hassan, M.U. Akram, R. Ahmed, N. Werghi, CDC-Net: cascaded decoupled convolutional network for lesion-assisted detection and grading of retinopathy using optical coherence tomography (OCT) scans, Biomedical Signal Processing and Control 70 (2021) 103030.

[41] B. Hassan, S. Qin, R. Ahmed, T. Hassan, A.H. Taguri, S. Hashmi, N. Werghi, Deep learning based joint segmentation and characterization of multi-class retinal fluid lesions

on OCT scans for clinical use in anti-VEGF therapy, Computers in Biology and Medicine 136 (2021) 104727.

[42] Z. Li, K. Cheng, P. Qin, Y. Dong, C. Yang, X. Jiang, Retinal OCT image classification based on domain adaptation convolutional neural networks, in: 2021 14th International Congress on Image and Signal Processing, BioMedical Engineering and Informatics (CISP-BMEI), IEEE, 2021, pp. 1–5.

[43] V. Das, S. Dandapat, P.K. Bora, A data-efficient approach for automated classification of OCT images using generative adversarial network, IEEE Sensors Letters 4 (1) (2020) 1–4.

[44] D. Romo-Bucheli, U.S. Erfurth, H. Bogunović, End-to-end deep learning model for predicting treatment requirements in neovascular AMD from longitudinal retinal OCT imaging, IEEE Journal of Biomedical and Health Informatics 24 (12) (2020) 3456–3465.

[45] S.A. Kamran, S. Saha, A.S. Sabbir, A. Tavakkoli, Optic-net: a novel convolutional neural network for diagnosis of retinal diseases from optical tomography images, in: 2019 18th IEEE International Conference on Machine Learning and Applications (ICMLA), IEEE, 2019, pp. 964–971.

[46] R. Tennakoon, G. Bortsova, S. Ørting, A.K. Gostar, M.M. Wille, Z. Saghir, R. Hoseinnezhad, M. de Bruijne, A. Bab-Hadiashar, Classification of volumetric images using multi-instance learning and extreme value theorem, IEEE Transactions on Medical Imaging 39 (4) (2019) 854–865.

[47] T. Hassan, Z. Li, M.U. Akram, I. Hussain, K. Khalaf, N. Werghi, Angular contrastive distillation driven self-supervised scanner independent screening and grading of retinopathy, Information Fusion 92 (2023) 404–419.

[48] H. Wen, J. Zhao, S. Xiang, L. Lin, C. Liu, T. Wang, L. An, L. Liang, B. Huang, Towards more efficient ophthalmic disease classification and lesion location via convolution transformer, Computer Methods and Programs in Biomedicine 220 (2022) 106832.

[49] A. Tayal, J. Gupta, A. Solanki, K. Bisht, A. Nayyar, M. Masud, DL-CNN-based approach with image processing techniques for diagnosis of retinal diseases, Multimedia Systems (2021) 1–22.

[50] P.K. Upadhyay, S. Rastogi, K.V. Kumar, Coherent convolution neural network based retinal disease detection using optical coherence tomographic images, Journal of King Saud University: Computer and Information Sciences 34 (10) (2022) 9688–9695.

[51] K. Karthik, M. Mahadevappa, Convolution neural networks for optical coherence tomography (OCT) image classification, Biomedical Signal Processing and Control 79 (2023) 104176.

[52] S.P.K. Karri, D. Chakraborty, J. Chatterjee, Transfer learning based classification of optical coherence tomography images with diabetic macular edema and dry age-related macular degeneration, Biomedical Optics Express 8 (2) (2017) 579–592.

[53] D.S. Kermany, M. Goldbaum, W. Cai, C.C. Valentim, H. Liang, S.L. Baxter, A. McKeown, G. Yang, X. Wu, F. Yan, et al., Identifying medical diagnoses and treatable diseases by image-based deep learning, Cell 172 (5) (2018) 1122–1131.

[54] J. Zhang, G. Zhu, Z. Wang, Multi-column Atrous convolutional neural network for counting metro passengers, Symmetry 12 (4) (2020) 682.

[55] R.R. Selvaraju, M. Cogswell, A. Das, R. Vedantam, D. Parikh, D. Batra, Grad-cam: visual explanations from deep networks via gradient-based localization, in: Proceedings of the IEEE International Conference on Computer Vision, 2017, pp. 618–626.

[56] Y. Zhang, M. Li, Z. Ji, W. Fan, S. Yuan, Q. Liu, Q. Chen, Twin self-supervision based semi-supervised learning (TS-SSL): retinal anomaly classification in SD-OCT images, Neurocomputing 462 (2021) 491–505.

[57] A. Butola, D.K. Prasad, A. Ahmad, V. Dubey, D. Qaiser, A. Srivastava, P. Senthilkumaran, B.S. Ahluwalia, D.S. Mehta, Deep learning architecture "LightOCT" for diagnostic decision support using optical coherence tomography images of biological samples, Biomedical Optics Express 11 (9) (2020) 5017–5031.

[58] P.D. Barua, W.Y. Chan, S. Dogan, M. Baygin, T. Tuncer, E.J. Ciaccio, N. Islam, K.H. Cheong, Z.S. Shahid, U.R. Acharya, Multilevel deep feature generation framework for automated detection of retinal abnormalities using OCT images, Entropy 23 (12) (2021) 1651.

CHAPTER

2 Multisensor data fusion in Digital Twins for smart healthcare

Zhihan Lyu
Uppsala University, Department of Game Design, Uppsala, Sweden

2.1 Introduction

In contemporary society, marked by rapid socio-economic growth, people have experienced notable improvements in their living standards. However, certain lifestyle choices and dietary behaviors continue to pose significant challenges to human health. For instance, indulging in late-night roadside stall barbecues and adhering to irregular meal schedules due to work obligations, coupled with prolonged sedentary behavior without engaging in regular physical exercise, contribute to the prevalence of suboptimal health conditions [1,2]. Therefore, the utilization of multisensor data fusion (MSDF) to obtain more comprehensive and accurate health status information has become a crucial challenge in the field of intelligent medical care (IMC). The rapid development of sensor technology has facilitated real-time monitoring and recording of patients' physiological parameters, including heart rate, body temperature, and blood pressure. However, relying on a single sensor alone may result in incomplete and less accurate health information due to issues such as noise and errors, thereby hindering a comprehensive evaluation of the body's health status [3]. In the era of metadata, researchers in the field of human health have placed significant emphasis on leveraging wireless sensor networks comprising multisource sensors, metadata analysis, and Internet of Things (IoT) technologies to enable intelligent and personalized health assessments of individuals' physical well-being [4–6]. The use of MSDF to obtain more comprehensive and accurate health status information has become an important challenge in the field of IMC.

In the IMC field, multisource sensors play a crucial role in analyzing the impact of health monitoring through data fusion and feature extraction. MSDF methods can be categorized into three main aspects: data-level fusion, feature-level fusion, and decision-level fusion. Data-level fusion involves integrating data measured by multiple sensors and extracting features from the fused data to accomplish human health detection or recognition tasks. Feature-level fusion involves extracting features from the original data collected by multiple sensors using independent information processing algorithms, enabling human health or behavior recognition based on the fused features. Decision-level fusion involves independent signal processing algo-

Data Fusion Techniques and Applications for Smart Healthcare. https://doi.org/10.1016/B978-0-44-313233-9.00008-4

rithms employed by multiple sensors to obtain judgment results, followed by fusion decisions utilizing techniques such as Dempster–Shafer evidence theory (DSET) or weighted average, ultimately yielding the final decision result [7,8]. The integration of data from multiple sensors in intelligent healthcare presents numerous challenges, encompassing variations in data characteristics and sampling frequencies among different sensors, as well as concerns surrounding noise, missing data, and inaccuracies within multisensor data. However, the application of Digital Twins (DTs) technology within the medical domain offers a potential solution. By facilitating the mapping of multisensor human physiological data acquired in physical space to a virtual environment, DTs technology effectively surmounts obstacles associated with extracting information from multiple sensors and fusing it for decision-making purposes. This technological approach enables the monitoring and analysis of human physiological parameters within the virtual realm, thereby facilitating the evaluation of human health [9]. Moreover, integrating deep learning into the MSDF method can enhance the fusion and analysis of data from multiple sources, enabling a more precise assessment of human health status [10,11].

The integration of data from multiple sensors in intelligent healthcare presents numerous challenges, encompassing variations in data characteristics and sampling frequencies among different sensors, as well as concerns surrounding noise, missing data, and inaccuracies within multisensor data. However, the application of DTs technology within the medical domain offers a potential solution. By facilitating the mapping of multisensor human physiological data acquired in physical space to a virtual environment, DTs technology effectively surmounts obstacles associated with extracting information from multiple sensors and fusing it for decision-making purposes. This technological approach enables the monitoring and analysis of human physiological parameters within the virtual realm, thereby facilitating the evaluation of human health.

This study makes several noteworthy contributions, which are outlined as follows:

- An MSDF method is proposed based on DTs technology, enabling comprehensive monitoring and prediction of patients' health status.
- A dynamic chaotic firefly (DCF) algorithm is developed to optimize the parameter selection of support vector machines (SVMs), thereby enhancing the accuracy of classification and prediction.
- The improved SVM algorithm is combined with DSET to achieve decision-level fusion of data from multiple sources. This integration enhances the reliability and robustness of decision making by effectively fusing information from different sources, ensuring more informed and accurate decisions.
- The effectiveness of the proposed model is verified under different conditions, demonstrating a recognition accuracy of over 90% and a high detection accuracy in various occlusion scenarios.
- Innovative research directions and experimental references are provided for health monitoring and decision making in the field of IMC and valuable insights and practical references are offered that contribute to improving the efficiency and reliability of healthcare systems, ultimately enhancing patient care.

The present chapter is organized into several sections, including the introduction, literature review, SVM optimization, DSET analysis, construction of a decision-level fusion DTs model for multisensor medical data, experimental verification, results, discussion, and conclusion sections.

Introduction: Section 2.1 provides an overview of the current state of human health and the background of intelligent monitoring in IMC. It highlights the motivation, innovation points, and contributions of the present study.

Literature review: Section 2.2 summarizes recent studies related to health monitoring in IMC and medical MSDF, aiming to understand the current development status in this field.

SVM optimization: Section 2.3 focuses on the optimization of SVM and introduces the DCF algorithm for enhancing the performance of the SVM algorithm.

DSET analysis: In Section 2.4, the uncertainty of medical information based on multisensor physiological data is analyzed using DSET, which reflects the application of DSET in IMC.

Construction of decision-level fusion DTs model: Section 2.5 describes the construction of a decision-level fusion DTs model for multisensor medical data. This model combines the improved SVM algorithm and DSET to facilitate and improve data fusion and analysis.

Experimental verification: Section 2.6 explains the construction of a hardware and software platform for human health monitoring based on the established model. It provides detailed information on the experimental environment and the data acquisition process, and the performance of the constructed model is validated and evaluated.

Results: Section 2.7 compares the algorithm proposed in this study with the SVM algorithm and DSET. The comparison is based on recognition accuracy, monitoring accuracy, and detection precision for different human postures and behaviors. The results are discussed and analyzed.

Discussion: Section 2.8 discusses the experimental results and compares them with the findings of other researchers in related fields. It highlights the strengths and novelties of the current study.

Conclusion: Section 2.9 provides a summary of the methodology and results. The limitations of the study are explained and future prospects for further research are discussed.

2.2 Literature review

As living conditions improve, people's awareness of their health has increased. This has led to the growth of IMC as individuals place greater emphasis on monitoring and improving their health. Numerous scientific researchers have dedicated significant efforts to the intelligent advancement of the medical field. Subhan et al. (2023) conducted a comprehensive investigation into the application of wearable medical Internet of Medical Things (IoMT) devices driven by artificial intelligence (AI) in healthcare systems. Their study provided a detailed overview of the methods and findings related to the implementation of this technology in the healthcare domain [12]. In a similar vein, Yıldırım et al. (2023) proposed a fog-cloud architecture-based IoMT framework designed for system implementation and development in healthcare monitoring. The results of their research showcased the effectiveness of the framework in enabling efficient medical monitoring and data transmission, thus holding promise for enhancing the efficiency and reliability of healthcare systems [13]. Ktari et al. (2022) proposed an IoMT-based platform for monitoring patients' health. In addition, they integrated blockchain technology as a security system to address the requirements of medical confidentiality and information privacy. Through evaluation and analysis, the platform demonstrated effectiveness as a cost-effective solution for maintaining secure electronic health records [14]. Elhence et al. (2022) introduced an advanced intelligent medical condition consultation framework that ensures privacy using AI and blockchain. Within this framework, patients can anonymously submit medical queries on the blockchain network, and naive Bayes and logistic regression algorithms are employed to classify the queries into relevant domains. The evaluation revealed that the framework model can be assessed based on reputation, expertise, details, and usage of supporting documents, thereby offering affordable and accessible smart healthcare solutions [15].

Medical sensors or IoMT devices play a crucial role in the development of intelligent healthcare. However, due to the intricate nature of diseases, the diagnosis often demands the integration of multimodal medical signals. Consequently, numerous researchers have focused their efforts on exploring the field of medical MSDF. Kapoor et al. (2023) proposed a remote patient safety and health monitoring system based on energy-efficient IoT sensors, which ensured patient privacy and data security [16]. Jaber and Idrees (2021) presented a method for energy-efficient multisensor data sampling and decision fusion in wireless body sensor networks to monitor patient health risks. The proposed approach involves assessing the patient's risk level and making decisions accordingly. The authors observed significant energy savings by evaluating the method's performance using a real health dataset while maintaining data accuracy and integrity. Moreover, the network model demonstrated the capability to appropriately detect emergencies and make informed decisions [17]. Szunerits and Zhang (2023) summarized the successes and challenges in the development of medical diagnostic sensors [18]. Lu et al. (2021) proposed a processing method that utilizes MSDF and machine learning to identify the jumping phase of the human body. The authors synthesized Euler angles into phase angles and employed sample entropy of the electromyography signal and the standard deviation of the acceleration signal to detect the onset and offset of the effective segment. The average recognition rate of surface electromyography (sEMG) was found to be 91.76% when comparing the performance with that of state-of-the-art machine learning classifiers. Additionally, when combining sEMG and foot switch signals, the average accuracy increased to 98.70% [19]. Vishwakarma and Bhuyan (2022) proposed an optimization algorithm for fusing multisensor source images in the medical field. The method optimizes the weights of the Karhunen–Loeve transform based on relevant information from the source medical images, enhancing the interpretation of the fused images. Qualitative and quantitative evaluation of the fused medical images demonstrated an average increase of 1% in peak signal-to-noise ratio and 2.2% in structural similarity values compared to the image denoising method. Moreover, the image fusion metric showed an average improvement of 9.04% compared to existing state-of-the-art methods [20].

In conclusion, the application of emerging technologies, such as fog-cloud architecture in IoMT and blockchain, has shown promise in health monitoring within the field of smart healthcare. However, certain challenges still persist in IoMT systems, including the limitations of a single data acquisition method and low diagnostic efficiency. Moreover, while data fusion and analysis techniques for multiple sensors have been utilized to some extent in the medical field, research in the emerging interdisciplinary domain of medical data fusion remains relatively limited. To address these challenges, this study aims to incorporate machine learning algorithms to overcome the limitations of a single data acquisition method and enhance efficiency. Additionally, the integration of DTs technology supports the collection, storage, and analysis of large-scale sensor data in the medical domain, enabling real-time medical decision support. The primary objective of this research is to contribute to the innovative study and analysis of MSDF in the field of smart healthcare, ultimately advancing the development of smart healthcare.

2.3 SVM optimization

In the field of IMC, MSDF plays a crucial role in classifying and analyzing individuals' health status. Various classification algorithms are commonly used, including neural networks, decision tree, naive Bayes, and SVM. However, neural networks require numerous parameters, such as topology, weights, and thresholds, making it difficult to interpret the results and prolonging the training process. The decision tree algorithm is prone to overfitting and overlooks attribute correlations. The naive Bayes algorithm relies on prior probabilities and classification decision error rates [21–24]. Given these considerations, we selected SVM as the classifier for its comprehensive advantages.

The conventional mathematical SVM model can be transformed into a convex quadratic optimization problem to obtain a linear classification machine through the process of maximizing hard margin learning [25]. In cases where the training data are nonlinearly inseparable, a kernel function (KF) is employed to map the input space to a feature space, enabling the study of a linear SVM in a high-dimensional feature space through soft margin maximization. Fig. 2.1 illustrates the optimal classification lineoid for linearly separable data.

A linearly dissociable dataset $T = \{(x_1, y_1), \cdots, (x_n, y_n)\}$ is provided in Fig. 2.1, where (x_i, y_i) represent the sample point, $x_i \in R^n$, $y_i \in \{+1, -1\}$, and $i = 1, 2, \cdots, n$. Fig. 2.1 depicts the optimal lineoid classification line for binary classification of human health status. The sample data in "normal" and "abnormal" states are represented by green (gray in print version) dots and black five-pointed stars, respectively. H de-

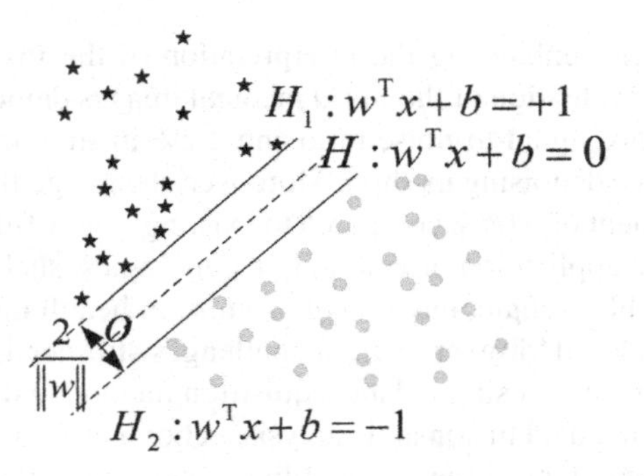

FIGURE 2.1

Linearly separable optimal classification lineoid of SVM.

notes the optimal boundary, while H_1 and H_2 denote the data points closest to H and parallel to it after the binary classification. The interval between H_1 and H_2, as well as H, is referred to as the classification interval. Therefore, when monitoring human health status, it is essential to identify a method that ensures minimal empirical risk and accurately separates the samples into "normal" and "abnormal" states. Simultaneously, the optimal classification lineoid aids in reducing practical risk. The classification lineoid function of the linearly separable SVM can be expressed as follows:

$$w \cdot x + b = 0. \tag{2.1}$$

In Eq. (2.1), the normal vector of the lineoid is represented by w and b corresponds to the offset. The following condition must be satisfied to achieve accurate sample classification and maximize the gap:

$$y_i (w \cdot x + b) \geq 1. \tag{2.2}$$

By applying the above conditions, the classification interval can be determined. Consequently, the optimal lineoid for the training dataset T can be written as follows:

$$\begin{aligned} &\min_{a,b} \Phi(w) = \frac{\|w\|^2}{2} \\ &s.t. \quad y_i (w \cdot x + b) \geq 1, i = 1, 2, \cdots, n. \end{aligned} \tag{2.3}$$

Eq. (2.3) represents a quadratic optimization issue that can be transformed into a Lagrange dual problem. To address this, the following Lagrange function is introduced:

$$\begin{aligned} &L(w, b, \alpha) = \frac{\|w\|^2}{2} - \sum_{i=1}^{n} \alpha_i [y_i (w \cdot x + b) - 1] \\ &s.t. \quad \alpha_i \geq 1, i = 1, 2, \cdots, n. \end{aligned} \tag{2.4}$$

In Eq. (2.4), α_i refers to the Lagrange multipliers. The partial derivation of the Lagrange function is solved according to the extreme value condition in Eq. (2.5):

$$\begin{aligned} \frac{\partial L}{\partial b} &= 0 \rightarrow \sum_{i=1}^{n} y_i \alpha_i = 0 \\ \frac{\partial L}{\partial w} &= 0 \rightarrow w = \sum_{i=1}^{n} y_i \alpha_i x_i . \end{aligned} \tag{2.5}$$

Combining Eq. (2.5) and Eq. (2.4) we obtain the dual problem regarding the original problem:

$$\begin{aligned} &\min_{\alpha} \frac{1}{2} \sum_{i=1}^{n} \sum_{j=1}^{n} y_i y_j \alpha_i \alpha_j \left(x_i \cdot x_j\right) - \sum_{i=1}^{n} \alpha_i \\ &s.t. \quad \sum_{i=1}^{n} y_i \alpha_i = 0, \alpha_i y_i \geq 0, i = 1, 2, \cdots, n. \end{aligned} \tag{2.6}$$

The optimal solution $\alpha^* = \left(\alpha_1^*, \cdots, \alpha_n^*\right)^{\mathrm{T}}$ is obtained by solving Eq. (2.6). Then, w^* and b^* are calculated to reach the optimal classification function:

$$f(x) = \operatorname{sgn}\left[\sum_{i=1}^{n} \alpha_i^* y_i \left(x_i \cdot x\right) + b^*\right]. \tag{2.7}$$

In Eq. (2.7), sgn refers to the sign function and b^* signifies the classification threshold. According to Eq. (2.7), the optimal categorization surface of the SVM just relies on the corresponding sample point (x_i, y_i) in the training data when $\alpha_i^* > 0$. Besides, these samples are distributed on the boundary of the interval. Hence, when $\alpha_i^* > 0$, the corresponding sample points $x_i \in R^n$ are usually called support vectors.

In the context of binary classification problems in actual health states, SVMs often encounter a common issue: the presence of indistinct boundaries, which can lead to a decrease in classification accuracy. To address this, we introduced a slack variable (ξ_i) and an error penalty factor (C) to mitigate this effect. Consequently, Eq. (2.3) can be transformed into Eqs. (2.8) and (2.9) as follows:

$$\Phi(w, \xi) = \frac{1}{2} w^T w + C \sum_{i=1}^{N} \xi_i , \tag{2.8}$$

$$y_i \left(w^T \cdot x_i + b\right) \geq 1 - \xi_i , i = 1, 2, \cdots, n. \tag{2.9}$$

The range of the Lagrange multiplier α_i is further limited as follows:

$$0 \leq \alpha_i \leq C, i = 1, 2, \cdots, n. \tag{2.10}$$

The penalty factor (C) is a coefficient used to control the influence of sample points that cross the boundary and ensure the accuracy of SVM classification. However, selecting an appropriate value for C is crucial. If C is set to a large value, the model may suffer from overfitting even with a few error points. If C is too small, the number of error points increases, leading to a model that deviates significantly from the actual data. Therefore, choosing a suitable value for C is essential to maintain the accuracy of SVM classification.

Moreover, the KF in the SVM algorithm is examined. There are several commonly used KFs, including the linear KF, polynomial KF, Gaussian radial KF, and sigmoid KF. The KF works by applying a nonlinear function to map the input data into a high-dimensional space. Among these options, the Gaussian KF offers benefits such as low computational complexity, wide applicability, and efficient calculations. The Gaussian KF is represented as follows:

$$K\left(x_i, x_j\right) = \exp\left(-\left\|x_i, x_j\right\|^2 / \sigma^2\right). \tag{2.11}$$

Simultaneously, the KF employed in this study has the advantageous property of having only one parameter, namely the coefficient σ, which contributes to its versatility. Hence, the Gaussian radial KF is chosen as the KF for the SVM. By utilizing this KF in the high-dimensional feature space, it becomes possible to achieve nonlinear classification on the optimal classification surface, provided a KF satisfies the following condition:

$$K\left(x_i, x_j\right) = \Phi\left(x_i\right) \cdot \Phi\left(x_j\right). \tag{2.12}$$

In Eq. (2.12), $\left\{\Phi_j(x)\right\}_{j=1}^{m}$ refers to a nonlinear set. Then, the optimal lineoid in the high-dimensional feature space can be defined as

$$\sum_{j=1}^{m} w_j \varphi_j(x) + b = 0. \tag{2.13}$$

Let

$$\begin{cases} w = \left[b, w_1, w_2, \cdots, w_m\right]^T \\ \Phi(x) = \left[1, \varphi_1(x), \varphi_2(x), \cdots, \varphi_m(x)\right]^T. \end{cases} \tag{2.14}$$

Subsequently, we have

$$w^T \Phi(x) = 0. \tag{2.15}$$

The dual problem can be expressed as

$$Q(\alpha) = \sum_{i=1}^{n} \alpha_i - \frac{1}{2} \sum_{i=1}^{n} \sum_{j=1}^{n} y_i y_j \alpha_i \alpha_j K\left(x_i, x_j\right). \tag{2.16}$$

If α_i^* is the optimal solution, we have

$$w^* = \sum_{i=1}^{n} \alpha_i^* y_i \Phi(x_i). \tag{2.17}$$

Eq. (2.7) can be converted into

$$f(x) = \mathrm{sgn}\left[\sum_{i=1}^{n} \alpha_i^* y_i K(x_i, x) + b^*\right]. \tag{2.18}$$

In this study, the DCF algorithm is chosen as the optimization search algorithm for determining the optimal penalty factor C and Gaussian KF coefficients σ in SVM [26,27]. The standard firefly algorithm, while commonly used, exhibits certain limitations in the later stages, including a low discovery rate, suboptimal solution convergence, and slow computational speed. Hence, an improved variant of the firefly algorithm, referred to as the DCF algorithm, is adopted in this study. The DCF algorithm specifically addresses and enhances the following four aspects, which are thoroughly analyzed in this study.

The first improvement lies in optimizing the movement of the brightest fireflies. Initially, within the initialized population, the brightest firefly is not influenced by other fireflies. Therefore, its position movement is determined by Eq. (2.19):

$$k_i^{t+1} = k_i^t + \beta(rand - 1/2). \tag{2.19}$$

In Eq. (2.19), k_i represents the position information of any firefly, β refers to the step factor, and *rand* stands for a random factor, uniformly distributed on (0, 1). If $rand > 0.5$, the value of X_m is $u_{\max} - k_i^t$; if $rand < 0.5$, the value is $u_{\min} - k_i^t$. Variables $u_{\max}$ and $u_{\min}$ signify the boundaries of the upper and lower regions in the search area, respectively.

Secondly, the population is initialized using a chaotic sequence, which imparts the initial population with properties of randomness, regularity, and boundedness. In this study, the improved logistic map is employed as the chaotic map, and the mapping is described by Eq. (2.20):

$$X_{n+1} = 1 - 2 * X_n^2, n = 1, 2, 3, \cdots. \tag{2.20}$$

In Eq. (2.20), $X_n \in (-1, 1)$. The initial population is generated using chaotic mapping as follows.

(a) An M-dimensional individual is randomly generated for an individual of N in the M-dimensional space, that is, $L = (l_1, l_2, \cdots, l_n)$, where $l_i \in [-1, 1]$, $1 \le i \le d$.

(b) The improved method mentioned above is utilized to generate the remaining $M - 1$ vectors.

(c) The above M variables are mapped to the search space according to Eq. (2.21):

$$x_{id} = U_{\min d} + (1 + y_{id})(U_{\max d} + U_{\min d}) i. \tag{2.21}$$

In Eq. (2.21), $U_{\max d}$ and $U_{\min d}$ represent the upper and lower bounds of the dth dimension of the search space, y_{id} denotes the dth dimension of the ith individual generated by Eq. (2.21), and x_{id} refers to the coordinate of the ith firefly in the dth dimension of the search space.

Thirdly, a dynamic population improvement is implemented. During the course of the algorithm, fireflies with low fitness may cease to move. Therefore, 30% of individuals with poor ranking are removed based on their fitness ranking to prevent falling into a local optimum. Simultaneously, an equal number of individuals are randomly regenerated to preserve population diversity and explore the optimal solution space. This dynamic adjustment ensures the algorithm's ability to search for the best solution.

Fourthly, a dynamic search space improvement is implemented to expedite convergence as the algorithm progresses. The current population's search space is gradually contracted to focus the search on the most promising regions. This regional shrinkage is achieved using Eqs. (2.22) and (2.23):

$$k_{\min,i} = \max\left\{k_{\max,i}, k_{\max light,i} - r\left(k_{\max,i} - k_{\min,i}\right)\right\}, 0 < r < 1, \tag{2.22}$$

$$k_{\max,i} = \min\left\{k_{\max,i}, k_{\max light,i} - r\left(k_{\max,i} - k_{\min,i}\right)\right\}, 0 < r < 1. \tag{2.23}$$

In Eqs. (2.22) and (2.23), $k_{\max light,i}$ refers to the value of the ith dimension variable of the brightest firefly in the group.

In the SVM algorithm, parameter selection is a crucial optimization process. Thus, the present study employs the DCF algorithm to enhance the optimization of SVM parameters. Fig. 2.2 illustrates the specific optimization process of the DCF algorithm.

As depicted in Fig. 2.2, the SVM model implemented in this study utilizes the DCF algorithm, specifically the chaotic firefly algorithm, to identify an optimal set of parameters that guarantee a high classification accuracy for SVM. The primary parameters subjected to optimization in this study are the penalty factor C and the KF coefficient. Fig. 2.3 provides a detailed overview of the SVM parameter optimization and classification process.

As depicted in Fig. 2.3, the multimodal physiological data of the human body acquired through multiple sensors undergo preprocessing and are subsequently divided into test samples and training samples. The chaotic firefly algorithm is employed, and the penalty factor C and KF coefficient σ are selected as parameters for optimization. The training samples are utilized to train the classifier and obtain the optimal model. Finally, the performance of the model is evaluated using the test samples.

2.4 DSET analysis

The health state reliability of each physiological parameter is monitored by collecting physiological parameters through the sensor terminal. The data fusion process using DSET solves the reliability function, enabling a comprehensive understanding of the

1 **start**
2 Input: parameters such as light intensity absorption coefficient δ, maximum attraction factor ω_0, and step factor β
3 Output: optimal KF coefficient σ and penalty factor C
4 Parameter initialization to generate the initial population size of firefly n, d_{ij} refers to the distance between firefly i and firefly j
5 Calculate the brightness S, attraction $\omega(r)$
6 $S \leftarrow S_0 e^{-\delta d_{ij}}$
7 $\omega(r) \leftarrow \omega_0 e^{-\delta d_{ij}^2}$
8 The brightest firefly moves according to Eq. (2.19)
9 According to the set number of iterations, when the threshold is reached, calculate the population fitness, remove the 30% fireflies with poor ranking, and randomly generate the remaining fireflies
10 Shrink the current search area according to Eq. (2.22) and Eq. (2.23)
11 if the largest number of repetitions or the inquiry precision do
12 Otherwise return to step 5
13 Until the largest number of repetitions or the inquiry precision is satisfied
14 end if
15 **end**

FIGURE 2.2

Optimization process of the DCF algorithm.

final health state of the human body. Decision rules are then applied to determine the overall health state based on the fused information [28,29]. Fig. 2.4 visually presents the uncertainty representation of DSET for obtaining physiological information based on medical multisensor data.

In Fig. 2.4, *pl* represents the likelihood function and *Bel* signifies the reliability function. The composition rules of DSET are analyzed in detail. In DSET, synthesis rules are applied to determine the integrated evidence results. By combining different synthesis rules, it becomes possible to address indeterminate questions, assess the level of associations among various pieces of evidence, and obtain accurate results [30].

When there are two pieces of evidence, the combination rules are as follows.

There are an identification framework Θ and two sets of evidence. The primary belief distribution function is m_1 and m_2, and the key factor is A_i and B_j, respectively. The Dempster–Shafer synthesis rule is divided into two steps: conjunction and

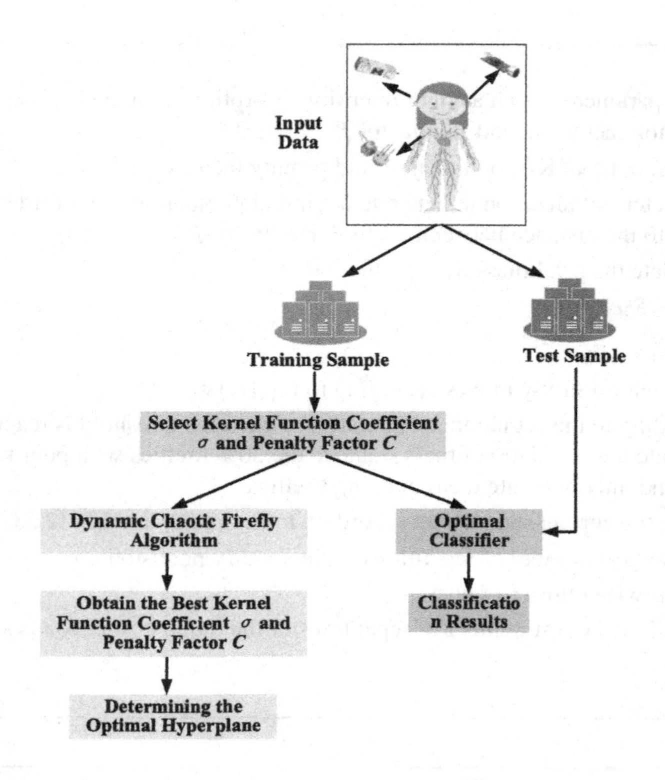

FIGURE 2.3

SVM parameter optimization and classification process.

normalization. A conjunction can be described as

$$m(A)^* = \sum_{\substack{A_i \cap B_j = A \\ A \subseteq \Omega, A \neq \phi}} m_1(A_i)\, m_2(B_j), \tag{2.24}$$

$$k = \sum_{A_i \cap B_j = A} m_1(A_i)\, m_2(B_j), \tag{2.25}$$

where k refers to the degree of conflict used to measure evidence and $k \in [0, 1]$.

Eq. (2.26) indicates normalization:

$$m(A) = m(A)^* + \frac{m(A)^*}{1-k} k, k \neq 1. \tag{2.26}$$

We have

$$m(A) = \begin{cases} \frac{\sum_{A_i \cap B_j = A} m_1(A_i) m_2(B_j)}{1-k}, A \neq \phi \\ 0, A = \phi \end{cases}. \tag{2.27}$$

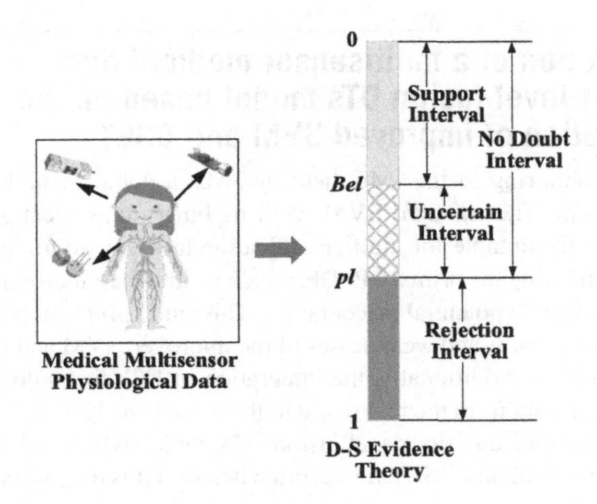

FIGURE 2.4

Schematic representation of uncertainty of medical information based on multisensor physiological data under DSET.

The above synthesis rule can be written as $m_1 \oplus m_2$.

When there are multiple shreds of evidence, the combination rules are as follows.

The primary probability distribution function (PPDF) is $m_1, m_2, \cdots, m_n$. At this time, the synthesis rule is expressed as

$$m(A) = \frac{\sum_{\cap A_i = A} \prod_{1 \le i \le N} m_i(A_i)}{1-k}, A \ne \phi. \tag{2.28}$$

The conflict coefficient k of Eq. (2.25) becomes

$$k = \sum_{\cap A_i = A} \prod_{1 \le i \le N} m_i(A_i), k \ne 1. \tag{2.29}$$

DSET is widely recognized as an effective approach for decision-level fusion; however, it has certain limitations, such as the presence of high evidence conflict [31,32]. To address this issue, this study integrates the improved SVM with DSET to perform decision-level fusion, enabling accurate data fusion decisions. SVM is employed for multiclassification, obtaining the basic probability distribution of different evidence. Subsequently, multievidence synthesis is conducted using combination rules, facilitating human health monitoring in the context of IMC.

2.5 Construction of a multisensor medical data decision-level fusion DTs model based on the combination of improved SVM and DSET

Human health monitoring in the IMC field involves a complex multiclass pattern recognition problem. The standard SVM, with its binary classification output of 1 or −1, is not directly suitable for multiclassification tasks. In addition, DSET has a drawback in constructing the primary PPDF, which is often based on experience or incomplete data, leading to potential inaccuracies. This study proposes a fusion method that combines the strengths and weaknesses of the improved SVM and DSET to overcome these challenges. Additionally, the integration of DTs technology enables the mapping of medical data from the real world to the digital world. Fig. 2.5 presents the multisensor medical data decision-level fusion DTs model, which utilizes the combination of improved SVM and DSET for accurate health status diagnosis of the human body.

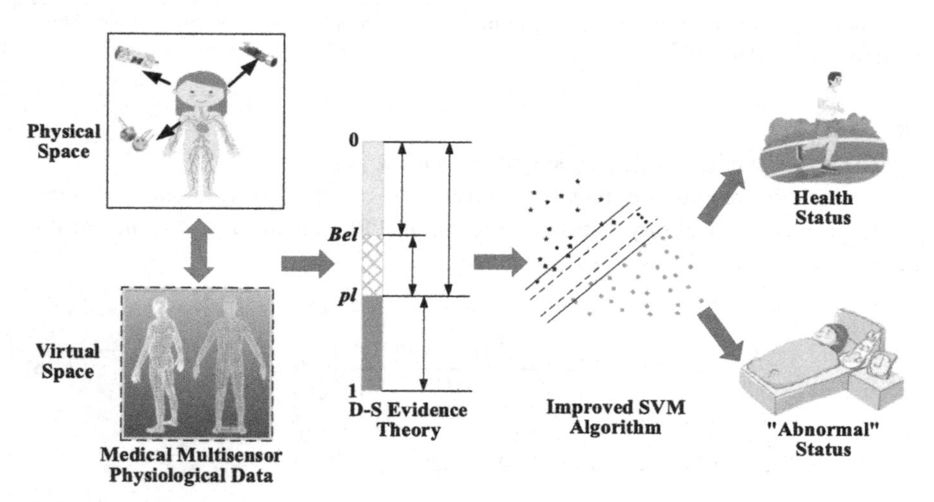

FIGURE 2.5

Multisensor medical data decision-level fusion DTs model based on the improved SVM and DSET.

As depicted in Fig. 2.5, the construction of the basic probability assignment (BPA) function using the fusion method of the improved SVM and DSET involves the following steps.

First, the essential data samples are collected through sensors and input into the improved SVM for training, resulting in the creation of multiple binary classifiers.

Second, two multiclassification methods, namely, the "one-to-many" classification algorithm and the "one-to-one" classification algorithm, are employed for analysis.

By following these steps, the fusion method combines the outputs of the binary classifiers to construct the BPA function, which facilitates accurate multiclass classification in the health monitoring system.

Assuming the utilization of the "one-to-many" classification algorithm for analysis, let us consider a sample data source. A portion of the dataset is selected as the training set, denoted by $Q = \{x_i, y_i\}$, where n represents the number of data instances in the training set, which are distributed across m ($m > 2$) classes. The primary approach involves generating multiple binary classifications to achieve multiclassification. This is achieved by constructing M binary classifiers from the m classes. The procedure begins by isolating one of the classes, while the remaining $m - 1$ classes are grouped together as a single class, resulting in a binary classification problem. Mathematically, this can be represented as follows:

$$\min \frac{1}{2}\left\|w^i\right\|^2 + \frac{1}{2}C\sum_{j=1}^{m}\xi_j^i$$

$$subject\begin{cases}\left(w^i\right)^T\varphi\left(x_j\right)+b^i \geq 1-\xi_j^i,\ y_j=i\\ \left(w^i\right)^T\varphi\left(x_j\right)+b^i \geq -1+\xi_j^i,\ y_j\neq i\\ \xi_j^i\geq 0,\ j=1,2,\cdots,m.\end{cases} \tag{2.30}$$

Then, m decision functions can be obtained by solving Eq. (2.30):

$$\begin{cases} w_1^T\varphi(x)+b_1\\ w_2^T\varphi(x)+b_2\\ \cdots\\ w_m^T\varphi(x)+b_m.\end{cases} \tag{2.31}$$

The optimal lineoid of the ith class and the remaining classes can be obtained as follows:

$$f(x) = w_i^T\varphi(x) + b_i = 0. \tag{2.32}$$

The "one-to-one" classification algorithm follows a similar approach to the first multiclassification method, where multiple binary classifiers are created to enable multiclassification. Suppose the training set consists of m classes, which are paired together to form $m(m-1)/2$ classifiers. The sample x to be classified is then input into these $m(m-1)/2$ classifiers. Subsequently, the results obtained from each classifier are combined using a voting method. For instance, in determining whether the sample vector x belongs to class i or j, it is assigned to class i if the number of votes is $C(i)+1$ and to class j if the number of votes is $C(j)+1$. By simultaneously utilizing $m(m-1)/2$ groups of classifiers for classification, the final result is determined based on the classification outcome with the highest number of votes among the four categories of samples.

This study employs the "one-to-one" classification algorithm for classification and voting of the classification results, as discussed earlier. Within the recognition framework, the voting function $C(x)$ is determined, which accumulates a total number of votes equal to $t(t+1)/2$.

Third, the aforementioned individual votes and the total number of votes are utilized to construct a basic PPDF:

$$m(A_s) = 2C(A_s) / (t(t+1)). \tag{2.33}$$

In Eq. (2.33), A_s refers to a nonempty identification frame and $s = 1, 2, \cdots, t$.

Fourth, the synthesis rules of DSET are applied to combine the individual evidence and obtain the final results of the human health status.

2.6 Experimental verification

A software and hardware platform is established for human health monitoring to validate the performance of the proposed multisensor medical data decision-level fusion DTs model based on the improved SVM and DSET. The platform is comprised of three main components: information collection, information transfer, and monitoring. The information collection part is composed of end nodes using the STM32 Microprocessor Unit, including a LoRa radio communication component and various induction components. The information transfer module consists of an STM32-based Microprocessor Unit equipped with a LoRa radio communication component and a WIFI component. The monitoring center is primarily established on a service center, and the Web service center is configured to facilitate matching surveillance actions.

The platform consists of software and hardware units, which are described below. The hardware module is comprised of end nodes and rendezvous nodes. These nodes utilize LoRa technology to exchange information, enabling wireless data transmission. Once the rendezvous node acquires the data, they are transmitted to the cloud service center using the MATT communication protocol. The data are then monitored through a web page. The software module includes the end node information collection module, the rendezvous node, and the cloud server.

The performance of the proposed model is evaluated by comparing its recognition accuracy and monitoring performance with those of the SVM algorithm and DSET. The evaluation is conducted based on various metrics, including F1 value, recall, precision, and accuracy. The three models are applied to analyze different human posture and behavior data in three postures: standing, sitting, and lying. Additionally, the data are analyzed under different occlusion levels, specifically when the table and chair are 0%, 30%, and 50% occluded. Lastly, ablation experiments are conducted to analyze the mean absolute error (MAE) and root mean square error (RMSE) of various algorithms. These performance metrics are used to assess the accuracy and precision of the algorithms under investigation.

2.7 Results

Fig. 2.6 and Fig. 2.7 illustrate the F1 value, recall, precision, and accuracy of the proposed model, the SVM algorithm, and DSET, showcasing their recognition and monitoring performance. The analysis is performed by repeating the experiments multiple times to ensure reliable results, thus minimizing the impact of accidental errors.

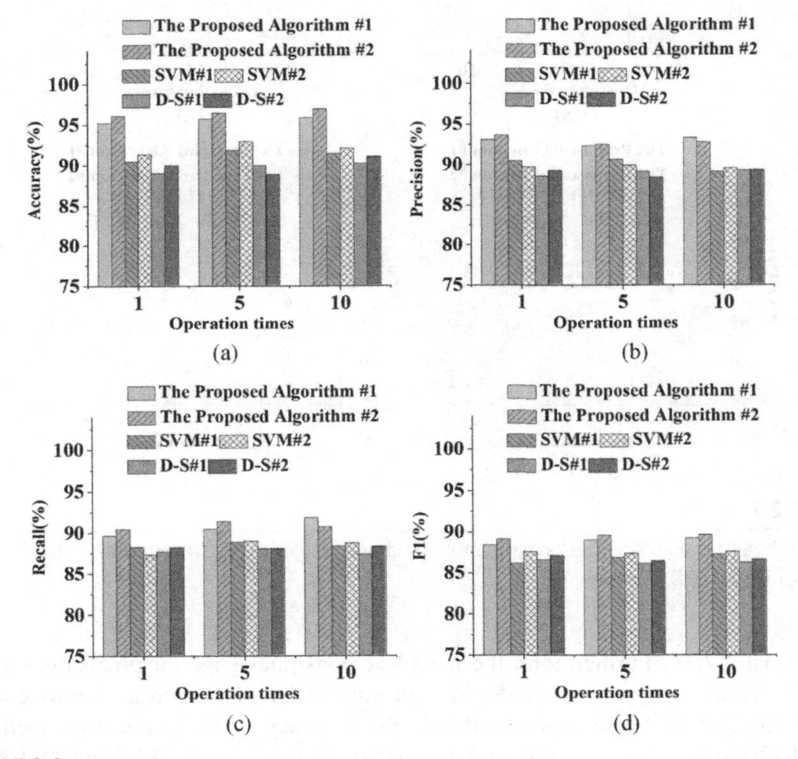

FIGURE 2.6

Human health status recognition performance using different algorithms. (a) Accuracy. (b) Precision. (c) Recall. (d) F1 value.

In Fig. 2.6, the proposed model achieves a recognition accuracy of 96.95%, which is at least 4.81% higher than that of other algorithms. Moreover, the model also demonstrates superior precision, recall, and F1 values compared to the SVM algorithm and the DSET algorithm, surpassing them by at least 3.95%. These results indicate that the multisensor medical data decision-level fusion DTs model based on the improved SVM and DSET exhibits outstanding recognition performance for human health status.

In Fig. 2.7, it is shown that the proposed model exhibits a monitoring accuracy of 96.56% for physiological health parameters, surpassing the other two algorithms

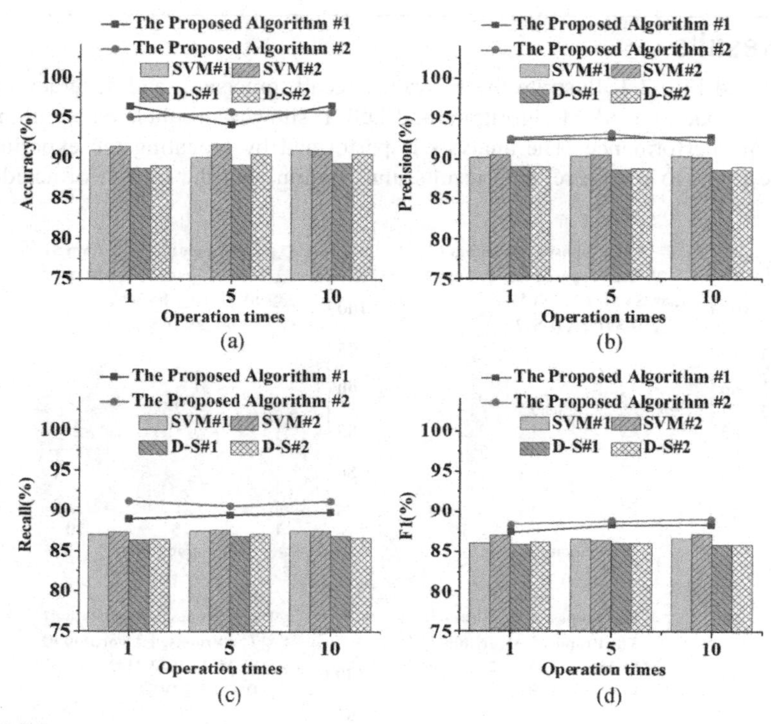

FIGURE 2.7

Human health monitoring performance using different algorithms. (a) Accuracy. (b) Precision. (c) Recall. (d) F1 value.

by at least 4.70%. Furthermore, the model demonstrates superior precision, recall, and F1 values compared to the SVM algorithm and DSET, with an improvement of at least 2.66%. These results highlight the accuracy of the multisensor medical data decision-level fusion DTs model based on the improved SVM and DSET in monitoring physiological parameters related to human health and its ability to make reliable decisions.

Furthermore, the monitoring performance of the fusion DTs model based on the improved SVM and DSET algorithm, SVM algorithm, and DSET is evaluated using diverse human posture and behavior data. Fig. 2.8 illustrates the detection accuracy for standing, sitting, and lying postures under 0% occlusion. Moreover, Fig. 2.9 showcases the detection accuracy for standing posture under 0%, 30%, and 50% occlusion conditions for tables and chairs. These analyses provide insights into the model's performance under different scenarios and occlusion levels.

As depicted in Fig. 2.8, when there is no occlusion of the desk and chair, the proposed model demonstrates high detection accuracy across various postures. Specifically, the model achieves an accuracy of 97.55% for the standing posture, 97.63%

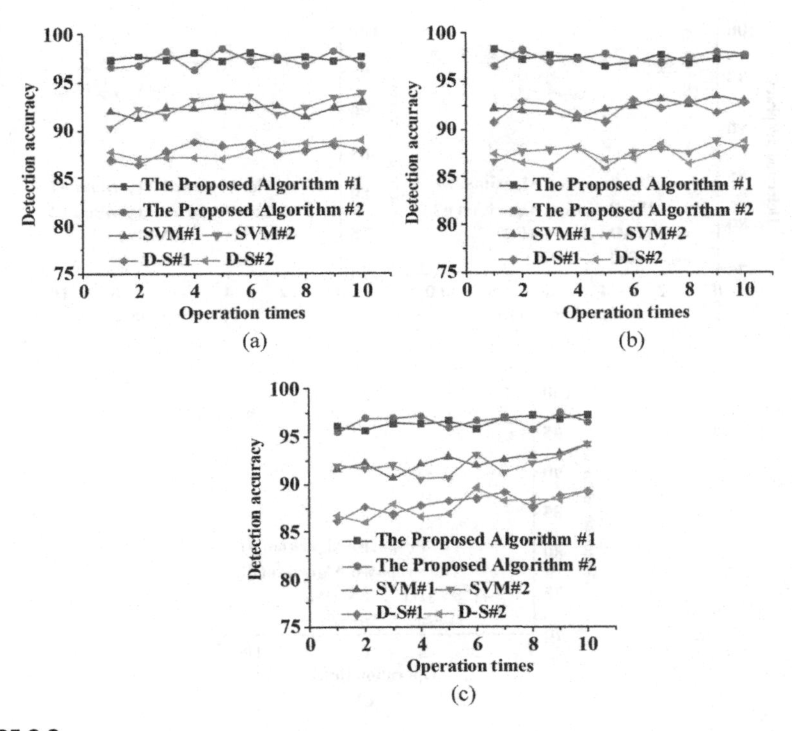

FIGURE 2.8

Results of the detection accuracy of each algorithm with different human postures. (a) Standing. (b) Sitting. (c) Lying.

for the sitting posture, and 97.32% for the lying posture, based on 10 repeated operations. These accuracies surpass the 97% threshold, indicating that different poses have minimal impact on the detection performance. Notably, the model's detection accuracy significantly outperforms that of the SVM and DSET algorithms. Consequently, the multisensor medical data decision-level fusion DTs model, employing the optimized SVM and DSET, accurately detects physiological parameters related to human health across different postures.

In Fig. 2.9, the model's detection accuracy is evaluated under different occlusion conditions. Specifically, when the number of operations is 10, the reported model achieves detection accuracies of 97.93%, 97.09%, and 96.40% for 0%, 30%, and 50% occlusion of tables and chairs, respectively. Notably, despite a decrease in detection accuracy with increasing occlusion, the model maintains accuracies above 96% across various occlusion conditions. Moreover, the model consistently outperforms the SVM algorithm and DSET in terms of accuracy. Consequently, the multisensor medical data decision-level fusion DTs model, leveraging the optimized SVM and

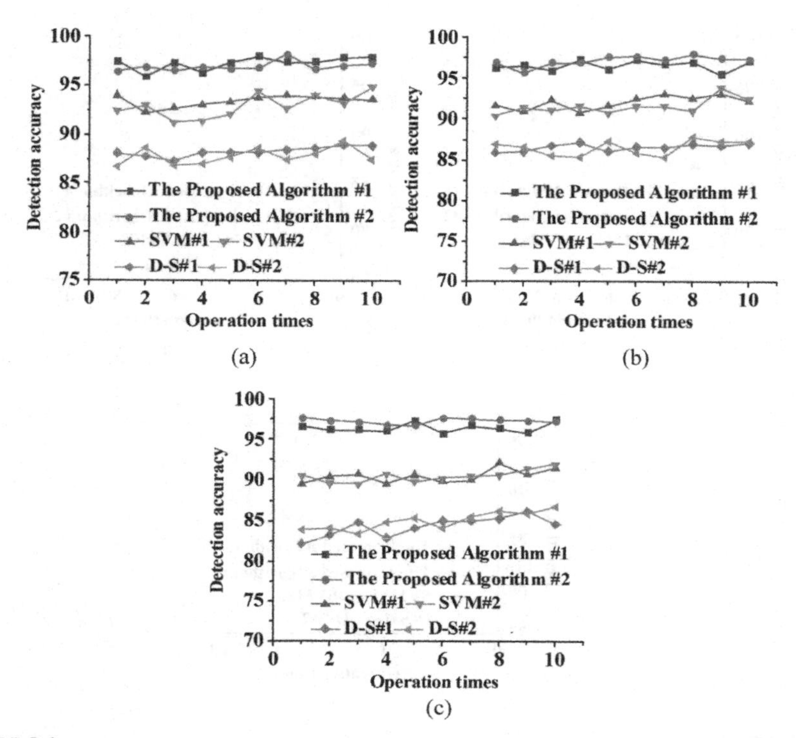

FIGURE 2.9

Detection accuracy results of each algorithm under different occlusion conditions. (a) 0% occlusion of tables and chairs. (b) 30% occlusion of tables and chairs. (c) 50% occlusion of tables and chairs.

DSET, accurately detects physiological parameters associated with human health under diverse occlusion conditions.

Ablation experiments are conducted to analyze the MAE and RMSE of various algorithms, assessing the performance of the model. The results of these experiments are presented in Fig. 2.10, providing valuable insights into the accuracy and precision of the different algorithms.

Fig. 2.10 depicts the results of the ablation experiment, which analyzed the MAE and RMSE values for each algorithm over 10 iterations. The analysis reveals a gradual stabilization of the MAE and RMSE values for all algorithms. Among the algorithms studied, the proposed model algorithm exhibits the lowest MAE and RMSE values, followed by SVM and DSET. Notably, after 10 iterations, the proposed model algorithm reaches a stable MAE of 1.129 and a stable RMSE of 1.275. These findings indicate that the proposed model algorithm outperforms the other algorithms in terms of prediction accuracy for detecting human health status.

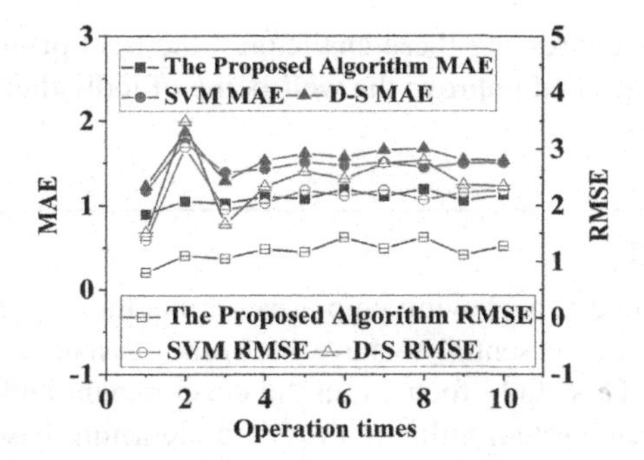

FIGURE 2.10

Ablation experiment results with different algorithms.

2.8 Discussion

This study employs the DCF algorithm to optimize the penalty factor and Gaussian KF. Additionally, the optimized SVM is integrated with the DSET to create a multisymbolic parameter data fusion model. Subsequently, a software and hardware platform for human health monitoring is established to analyze the model's performance. Through a comparative analysis with the SVM algorithm and DSET, the algorithm utilized in this study achieves a recognition accuracy of 96.95% and a monitoring accuracy of 96.56%. Furthermore, the results obtained from the ablation experiment highlight the superior performance of the proposed model algorithm in detecting human health status, exhibiting lower errors compared to alternative algorithms. This outcome can be attributed to the incorporation of optimized SVM and DSET, which enables the model to leverage their respective advantages. Similar successful outcomes in the accurate multiclassification of multisensory data for smart healthcare were observed in a study conducted by Vishwakarma and Bhuyan (2022). Moreover, the model achieves a detection accuracy exceeding 96% across various human postures and occlusion scenarios. These results indicate that the model reported here is robust against variations in human posture and occlusion during the acquisition of medical data, offering advantages over both the SVM and DSET algorithms. Additionally, the validation process aligns with the methodology proposed by Divya and Peter (2022) [33], further substantiating the feasibility of the reported data fusion model for human health monitoring.

In conclusion, the integration of MSDF and DTs technology in intelligent healthcare holds great promise for enhancing the accuracy and predictive capacity of human health status. Through algorithm optimization and model design, this research showcases the innovative application of these technologies and offers valuable insights for the advancement of intelligent healthcare. It paves the way for the realization of personalized and precision medicine. However, further research is necessary to overcome challenges related to data fusion and to address concerns surrounding privacy and data security. By addressing these challenges, the field of intelligent healthcare can continue to advance and improve the well-being of individuals.

2.9 Conclusion

In the modern era, there is a growing emphasis on maintaining and monitoring personal health status. It is essential to have real-time, dynamic information of our physical well-being. This study focuses on the development and implementation of a monitoring and alarm system utilizing the SVM algorithm based on DTs technology. The penalty factor and Gaussian KF are optimized using the DCF algorithm. Furthermore, the optimized SVM is integrated with DSET to establish a decision-level fusion DTs model for multisensor medical data. Experimental analysis reveals that the established model achieves a recognition accuracy exceeding 90% across different states. Additionally, the detection accuracy of the included model surpasses 96% for three postures (standing, sitting, and lying) and three occlusion scenarios (0%, 30%, and 50% occlusion while using a desk and chair). These findings provide valuable insights for human health monitoring in the context of IMC, serving as an experimental reference for future studies. However, this study has several limitations. Firstly, the use of data fusion algorithms for human health monitoring is still in its early stages. Secondly, the experiments are conducted in a controlled laboratory environment, lacking validation in real clinical settings, which may result in a gap between the findings and actual application scenarios. Therefore, in future research, several improvements can be made. Firstly, it is possible to include a wider range of monitoring parameters to better differentiate between human health states. Secondly, the model can be ported to more advanced hardware platforms and miniaturized for deployment in wearable devices, which is a direction for future development. Additionally, leveraging big data platforms for data management and evaluation could be helpful. By conducting further research and development, we expect to provide more accurate and reliable experimental references for human health monitoring and support future clinical practice.

References

[1] J. Zhou, Q. Wu, Z. Wang, Effect of self-employment on the sub-health status and chronic disease of rural migrants in China, BMC Public Health 21 (1) (2021) 1–12.

[2] Y. Xue, G. Liu, Y. Feng, M. Xu, L. Jiang, Y. Lin, J. Xu, Mediating effect of health consciousness in the relationship of lifestyle and suboptimal health status: a cross-sectional study involving Chinese urban residents, BMJ Open 10 (10) (2020) e039701–e039708.

[3] T.H. Fan, E.S. Rosenthal, Physiological monitoring in patients with acute brain injury: a multimodal approach, Critical Care Clinics 39 (1) (2023) 221–233.

[4] V. Hayyolalam, M. Aloqaily, Ö. Özkasap, M. Guizani, Edge intelligence for empowering IoT-based healthcare systems, IEEE Wireless Communications 28 (3) (2021) 6–14.

[5] W. Tu, B. Jiang, L. Kong, Comments on "measuring housing vitality from multi-source big data and machine learning", Journal of the American Statistical Association 117 (539) (2022) 1060–1062.

[6] V.A. Memos, K. Psannis, Z. Lv, A secure network model against bot attacks in edge-enabled industrial Internet of things, IEEE Transactions on Industrial Informatics 18 (11) (2022) 7998–8006.

[7] X. Peng, R. Krishankumar, K.S. Ravichandran, A novel interval-valued fuzzy soft decision-making method based on CoCoSo and CRITIC for intelligent healthcare management evaluation, Soft Computing 25 (6) (2021) 4213–4241.

[8] B. Wu, W. Qiu, W. Huang, G. Meng, J. Huang, S. Xu, A multi-source information fusion approach in tunnel collapse risk analysis based on improved Dempster–Shafer evidence theory, Scientific Reports 12 (1) (2022) 1–17.

[9] R. Aluvalu, S. Mudrakola, A.C. Kaladevi, M.V.S. Sandhya, C.R. Bhat, The novel emergency hospital services for patients using digital twins, Microprocessors and Microsystems 98 (2023) 104794.

[10] Y. Lu, H. Wang, F. Hu, B. Zhou, H. Xi, Effective recognition of human lower limb jump locomotion phases based on multi-sensor information fusion and machine learning, Medical & Biological Engineering & Computing 59 (4) (2021) 883–899.

[11] S.R. Stahlschmidt, B. Ulfenborg, J. Synnergren, Multimodal deep learning for biomedical data fusion: a review, Briefings in Bioinformatics 23 (2) (2022) 1–15.

[12] F. Subhan, A. Mirza, M.B.M. Su'ud, M.M. Alam, S. Nisar, U. Habib, M.Z. Iqbal, AI-enabled wearable medical Internet of things in healthcare system: a survey, Applied Sciences 13 (3) (2023) 1394–1406.

[13] E. Yıldırım, M. Cicioğlu, A. Çalhan, Fog-cloud architecture-driven Internet of Medical Things framework for healthcare monitoring, Medical & Biological Engineering & Computing 61 (2023) 1133–1147.

[14] J. Ktari, T. Frikha, N. Ben Amor, L. Louraidh, H. Elmannai, M. Hamdi, IoMT-based platform for E-health monitoring based on the blockchain, Electronics 11 (15) (2022) 2314–2333.

[15] A. Elhence, V. Kohli, V. Chamola, B. Sikdar, Enabling cost-effective and secure minor medical teleconsultation using artificial intelligence and blockchain, IEEE Internet of Things Magazine 5 (1) (2022) 80–84.

[16] B. Kapoor, B. Nagpal, M. Alharbi, Secured healthcare monitoring for remote patient using energy-efficient IoT sensors, Computers & Electrical Engineering 106 (2023) 108585–108591.

[17] A.S. Jaber, A.K. Idrees, Energy-saving multisensor data sampling and fusion with decision-making for monitoring health risk using WBSNs, Software, Practice & Experience 51 (2) (2021) 271–293.

[18] S. Szunerits, X. Zhang, The successes and challenges in the development of sensors for medical diagnostics, Sensors & Diagnostics 2 (1) (2023) 10–11.

[19] Y. Lu, H. Wang, F. Hu, B. Zhou, H. Xi, Effective recognition of human lower limb jump locomotion phases based on multi-sensor information fusion and machine learning, Medical & Biological Engineering & Computing 59 (4) (2021) 883–899.

[20] A. Vishwakarma, M.K. Bhuyan, A curvelet-based multi-sensor image denoising for KLT-based image fusion, Multimedia Tools and Applications 81 (4) (2022) 4991–5016.

[21] X. Zhan, H. Long, F. Gou, X. Duan, G. Kong, J. Wu, A convolutional neural network-based intelligent medical system with sensors for assistive diagnosis and decision-making in non-small cell lung cancer, Sensors 21 (23) (2021) 7996–8019.

[22] X. Zhou, W. Liang, I. Kevin, K. Wang, H. Wang, L.T. Yang, Q. Jin, Deep-learning-enhanced human activity recognition for Internet of healthcare things, IEEE Internet of Things Journal 7 (7) (2020) 6429–6438.

[23] A.Y. Saleh, C.K. Chin, V. Penshie, H.R.H. Al-Absi, Lung cancer medical images classification using hybrid CNN-SVM, International Journal of Advances in Intelligent Informatics 7 (2) (2021) 151–162.

[24] T. Huynh-The, C.H. Hua, N.A. Tu, D.S. Kim, Physical activity recognition with statistical-deep fusion model using multiple sensory data for smart health, IEEE Internet of Things Journal 8 (3) (2020) 1533–1543.

[25] T. Ni, J. Zhu, J. Qu, J. Xue, Labeling privacy protection SVM using privileged information for COVID-19 diagnosis, ACM Transactions on Internet Technology 22 (3) (2021) 1–21.

[26] M. Webber, R.F. Rojas, Human activity recognition with accelerometer and gyroscope: a data fusion approach, IEEE Sensors Journal 21 (15) (2021) 16979–16989.

[27] D. Jovanovic, M. Antonijevic, M. Stankovic, M. Zivkovic, M. Tanaskovic, N. Bacanin, Tuning machine learning models using a group search firefly algorithm for credit card fraud detection, Mathematics 10 (13) (2022) 2272–2302.

[28] R.A. Hamid, A.S. Albahri, O.S. Albahri, A.A. Zaidan, Dempster–Shafer theory for classification and hybridised models of multi-criteria decision analysis for prioritisation: a telemedicine framework for patients with heart diseases, Journal of Ambient Intelligence and Humanized Computing 13 (9) (2022) 4333–4367.

[29] B. Liu, X. Bi, L. Gu, J. Wei, B. Liu, Application of a Bayesian network based on multi-source information fusion in the fault diagnosis of a radar receiver, Sensors 22 (17) (2022) 6396–6413.

[30] L. Liu, A.M. Olteanu-Raimond, L. Jolivet, A.L. Bris, L. See, A data fusion-based framework to integrate multi-source VGI in an authoritative land use database, International Journal of Digital Earth 14 (4) (2021) 480–509.

[31] Y. Chen, Z. Hua, Y. Tang, B. Li, Multi-source information fusion based on negation of reconstructed basic probability assignment with padded Gaussian distribution and belief entropy, Entropy 24 (8) (2022) 1164–1189.

[32] M. Kumar, S. Chand, A secure and efficient cloud-centric internet-of-medical-things-enabled smart healthcare system with public verifiability, IEEE Internet of Things Journal 7 (10) (2020) 10650–10659.

[33] R. Divya, J.D. Peter, Smart healthcare system-a brain-like computing approach for analyzing the performance of detectron2 and PoseNet models for anomalous action detection in aged people with movement impairments, Complex & Intelligent Systems 8 (4) (2022) 3021–3040.

CHAPTER

Deep learning for multisource medical information processing

3

Mavis Gezimati and Ghanshyam Singh
Centre for Smart Information and Communication Systems, Department of Electrical and Electronic Engineering Science, University of Johannesburg, Johannesburg, South Africa

3.1 Introduction and motivation

Next-generation healthcare is all about connected and smart healthcare. It is patient-centric and focuses on providing precision medicine. It will be enabled by the implementation of advanced technologies from the Fourth Industrial Revolution (4IR) enabling technologies for healthcare application. The 4IR technologies are characterized by fusion of technologies ranging from the Internet of Things (IoT), wearable technology, artificial intelligence (AI), big data, blockchain, and augmented reality (AR) to robotics, nanotechnology, and 3D printing and are paving a way to a new era of innovative developments for next-generation healthcare. The application of 4IR technologies to the healthcare industry is called Healthcare 4.0 and it integrates the healthcare cyber physical systems (HCPSs), Internet of Medical Things (IoMT), fog and cloud computing, blockchain, AI, big data analytics, 3D printing, virtual reality technologies, etc. This will revolutionize the healthcare system with increased interoperability, smartness, and interconnectivity. HCPS Healthcare 4.0 is comprised of biosensors, computers, storage, communication, interfaces, and bioactuators. It can enable real-world observations in real-time and patient monitoring before, during, and after treatment procedures, such as cancer surgery. The sources of healthcare information include medical imaging modalities, laboratory values, omics, clinical notes, and wearable and mobile devices [1–8].

AI, particularly machine learning and deep learning, plays a crucial role in providing algorithmic decision support in the healthcare sphere. Such decision support systems make use of various data, such as medical images, clinical exam scores, lab test values, and biosensing data, to provide diagnosis support. The human decision making capacity is subjective and often limited by failure to consider additive scoring systems that require admixture of positive and negative hallmarks prior to confirmatory labeling [4]. In other words, human-based decision making may fail to consider relative weighting of disparate inputs of data as well as potentially nonlinear relationships. Instead, algorithmic decision-making support is able to offload such tasks and ideally yield more precise results, which is the promise of precision medicine. Even-

Data Fusion Techniques and Applications for Smart Healthcare. https://doi.org/10.1016/B978-0-44-313233-9.00009-6

tually precision medicine is aimed at creating a medical model that can customize healthcare – decisions, practices, treatments, etc. – that is personalized and tailored to the patient conditions. Machine learning and deep learning modeling is able to provide this through tracking patients' health trajectories whereby the diagnosis and treatments incorporate heterogeneous and unique information, which contrasts with the conventional one-drug-fits-all treatment model [9], [10].

The application of machine learning, particularly deep learning, in medical information systems is becoming more and more extensive and is proving to provide more solutions in disease prediction, prognosis evaluation, auxiliary diagnosis, drug research and development, medical image recognition, health management, and other aspects. The main goal is to establish the relationship between preventive/therapeutic options and health. Deep learning techniques have been widely used to enhance the performance of medical information technology through providing support for medical data processing, reducing the number of variables (dimensionality reduction) and increasing prediction accuracy and efficiency with reduced human intervention. The application of deep learning in medical information technology has been commonly realized in reconstruction and denoising tasks which are intended for image and spectrum preprocessing to remove unwanted or irrelevant information and reduce the number of variables from spectral and image parameters, thus increasing the efficiency of data analysis [11]. Deep learning methods are also useful for multivariate quantitative and qualitative analysis of data for highly precise classification, detection, recognition, identification, and related tasks like characterization, prediction, and analysis of samples for computer-assisted diagnosis [12]. Deep learning is proving to provide accurate and fast diagnosis, mostly based on the capability to automatically extract features and automatic learning, achieving excellent results in lesion segmentation, disease classification, and disease prediction [13]. Deep learning is also useful for multimodal data fusion in health modeling for more accurate representation of populations, demographics, etc. Also, it facilitates drug development and improves clinical decision support, which in turn brings improvements in disease prediction, clinical research, and patient outcomes that offer positive impacts in moral, economic, and ethical ramifications by reducing suffering and saving human lives.

Recently, deep learning has been investigated in multisource medical information processing rather than the conventional training on single source datasets. Conventional deep learning model-based diagnosis focuses mostly on unimodal data sources such as electronic health records (EHRs) or electronic medical records (EMRs) and medical imaging data including X-ray images, MRIs, CT scans, and ultrasound images. However, diagnosis of not all diseases can rely on a single data source and the use of single source datasets is rather task-specific than robust. Further, with the advances in technology, the healthcare community is now made up of multiple heterogeneous information sources and types including EMRs, radiology data, patient health monitoring devices, etc. The application of deep learning for processing and fusion of multisource medical information will enable the development of connected and smart systems for next-generation healthcare [14], [15], [24–30], [16–23].

It is through information fusion and computational modeling that outcomes of interest such as treatment targets and drug development ultimately facilitate improved patient-level decision making in care facilities and homes. Such a phenomenon has attracted interest in healthcare data fusion studies. The application of machine learning and deep learning algorithms provides insight into various components of healthcare such as drug discovery, clinical trials, phenotyping, and surgical techniques. This is crucial in providing support to practitioners and health centers to provide the most precise and efficient evidence-based medicine possible. Unimodal deep learning models are often less robust and suffer misclassification challenges. Here, we propose a framework and algorithms for multimodal deep learning models for classifying acoustic and image type multimodal datasets.

More work is needed to address data management, accessibility, findability, reuse of digital datasets, and interoperability [31].

In this chapter we describe the design of a multimodal data fusion framework based on deep learning for processing of information from medical image features and acoustic data to improve medical diagnosis and patient care. We describe data fusion techniques to integrate acoustic and pixel data. We conduct a systematic review of deep learning for multimodal data fusion and present current knowledge, identify existing challenges, and provide a road map for future research in multimodal medical data fusion systems. The outline of the chapter is shown in Fig. 3.1.

Contributions of this chapter can be summarized as follows:

- We introduce the aspects of medical information and 4IR enabling technologies including machine learning and deep learning.
- The architectures and techniques for multimodal data fusion based on deep learning as well as multimodal medical information sources are reported.
- The fundamentals of deep learning are summarized.
- The design of a deep learning-based multimodal data fusion framework for critical lung cancer patient care is described.
- Existing challenges in deep learning-based multimodal data fusion systems are identified and prospective opinions are given.
- We cover the fundamentals of deep learning in the healthcare application context.

3.2 Background, definitions, and notations

The application of AI, particularly deep learning algorithms, is rapidly expanding and shows great potential to revolutionize healthcare. They are playing a significant role in the development of smart clinical decision support systems for disease diagnosis, prediction, prognosis, health management, and drug research and development. The development of computer-aided diagnostics (CAD) systems is crucial to support clinicians with a second diagnostic opinion for more accurate, fast, precise, and personalized care compared to the subjectivity and time-intensive procedures of manual diagnosis and handcrafted features. Significant research progress has been made in developing deep learning-based systems; however, more focus has been placed on the

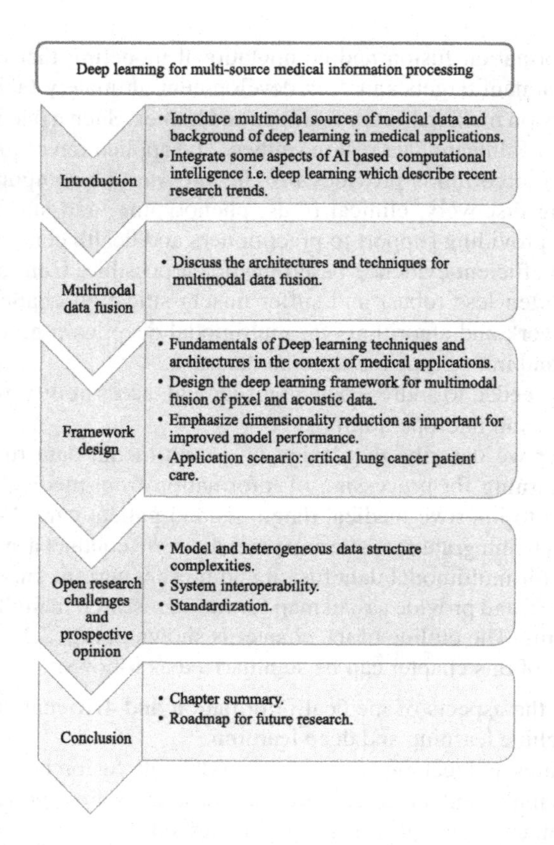

FIGURE 3.1

Chapter outline.

development of CAD systems that are based on image data, including X-ray images, MRIs, CT scans, and ultrasound images, for image classification, segmentation, disease detection, and prediction tasks. The existing deep learning models consider pixel value information without data informing about the clinical context. Yet in practice, the healthcare community is now made up of multiple heterogeneous data sources and types, including contextual data from EHRs, structured laboratory data, imaging pixel data, unstructured narrative data, and sometimes temporal data. The sources of image pixel data are also increasing from emerging potentially medical imaging modalities like terahertz (THz), infrared, and microwave-based imaging techniques. Next-generation healthcare relies on synthesis of information and should improve clinical decision making, diagnostic accuracy, and patient outcomes through seamless fusion of imaging and nonimaging data for personalized care. Thus, the application of deep learning for processing and fusion of multisource medical information will enable the development of connected and smart systems for next-generation healthcare systems.

It is also significant to point out that the terms multisource and multimodal are used interchangeably to denote heterogeneous data/information acquired from heterogeneous sources or modalities. Also, the terms classification and prediction have been used to denote the deep learning-based processing tasks for classification and prediction.

3.2.1 Data fusion

Data fusion is fostered by information theory and it is a methodology whereby disparate data sources or types are merged for creating an information state based on the complementarity of the data sources [32]. The application of machine learning-based modeling has led to the concept of multimodal machine learning also known as multiview machine learning whereby an algorithmic framework is used to integrate disparate data sources in order to maximize on the complementary and unique information. Data harmonization entails the use of machine learning for nullification of different data sources for its quality improvement and utilization. Data fusion is the mechanism for performing data fusion for multimodal machine learning and comes in three broad categories: early, intermediate/hybrid, and late fusion. Thus, machine learning is expected to support data fusion efforts to result in improvement of predictive power [33], [34] to provide more reliable and accurate modeling results even in potentially low-validity settings. Multimodal data fusion brings the salient property of robustness through the dependence on a variety of informational factors as compared to the task specificity of single data types.

The three main data fusion types used in machine learning and deep learning are early (input or data level), intermediate (joint or hybrid or layer level), and late (decision level). In early fusion, multiple data sources are converted to the same space of information. Such process often results in numerical conversion or vectorization from an alternative state. In other words, early fusion performs the integration of information from different modalities before giving them to a network such that the feature extraction and classification of multimodal datasets are performed from the integrated training set. Medical images entail features that can undergo numerical conversion such as structural, volume, and area calculations [35], [36]. The extracted features can then be concatenated with some other additional measurements from sources of structured data and fed into an individual classifier. The transformation of all data to the same feature space can be achieved using conversion methodologies such as deep learning, canonical correlation analysis, principal component analysis, and nonnegative matrix factorization [37].

In intermediate data fusion, a stepwise set of models are utilized, and it offers the greatest latitude in model architecture. One or more modalities are independently given to the network after that fusion of the intermediate representations is performed in the fully connected layer of the network in the case of deep learning-based data fusion. A three-stage deep neural network has been proposed in [38], where the first stage consists of feature selection using a softmax classifier for individual and independent modalities. The second and third stages are made up of a combination of the selected features to establish a set of further refined features which are then fed to a Cox-net for performing latent feature representation of Alzheimer's disease. For both layer- and input-level fusion, fusion is conducted at the feature level, where the various modality features are merged into a single feature vector and each sample for all the modalities has to be available in the training set. This results in superior performance and better decisions; however, it is difficult to achieve in practice with medical data. Nevertheless, in contrast to early fusion, for intermediate fusion the features that distinguish each data type are combined to produce a new data representation that is more expressive than separate representations of the data sources.

Each modality is a single input for the training of a single neural network in decision-level fusion (late fusion), where the resulting network outputs are integrated after classification. Late data fusion typically trains multiple models where each model corresponds to an incoming data source. In this decision-level fusion, there is no need for all the modalities for each sample to be present; thus, the exploitation of the corresponding modality's unique information is improved since the search space is smaller than in input- and layer-level fusion methods [35]. This is related to ensemble learning, which uses multiple learning algorithms to provide better performance than individual models. In ensemble learning, however, multiple algorithms are typically applied to the same dataset, but in late data fusion, multimodal machine learning methods use multiple sets of a particular data type (for example, CT scans, X-ray images, MRIs) or across data types (e.g., images, text, audio). These take symbolic representations as data sources and they are combined to get a decision that is more accurate. Typically, Bayesian methods are employed at this level for offering support to the voting process between model sets into a global decision. There has been wide progress in multitask deep learning based on late data fusion [39–45].

An illustrative schematic of the three subtypes of data fusion based on deep learning for multimodal image data is represented in Fig. 3.2. In deep learning, information flows into a single model in early fusion and in a stepwise fashion in intermediate fusion, where outputs of one model become input of the next model, and in late fusion, unique data types or sources undergo a separate modeling after which ensemble and/or voting is performed.

Table 3.1 presents the attributes of data fusion techniques, it could be seen that late fusion has the most outstanding attributes of the three. Moreover, in late fusion there is no need for all the modalities for each sample to be present, i.e., it can handle missing data; thus, the exploitation of the corresponding modality's unique information is improved since the search space is smaller than in input- and layer-level fusion methods [35].

3.2.2 Sources of healthcare information and the role of multimodal fusion

Fig. 3.3 shows examples of the generation of various multimodal healthcare data from healthcare centers, homes, etc., to provide disparate data and the cyclical information

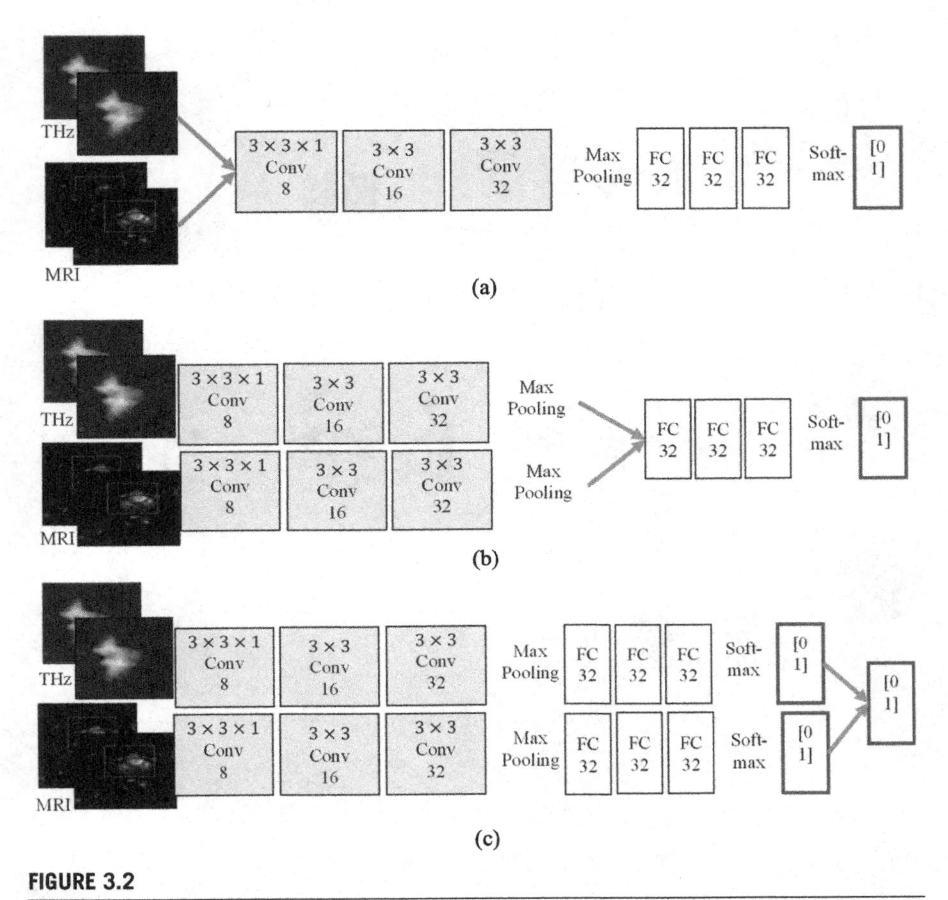

FIGURE 3.2

Three multimodal fusion strategies. (a) Input-level fusion. (b) Layer-level fusion. (c) Decision-level fusion.

Table 3.1 Attributes of data fusion strategies [32].

Attribute	Early	Intermediate/joint	Late/decision
Scalable	No	Yes	Yes
Improved accuracy	Yes	Yes	Yes
Need for multiple models	No	Yes	Yes
Interaction effects across sources	Yes	Yes	No
Multiple models voting	No	Yes	Yes
Implemented in health	Yes	Yes	Yes

flow to the information base, where it can be algorithmically modeled and transformed. The multimodal information includes medical image data, clinical notes, lab values, omics data, etc. It is through information fusion and computational modeling

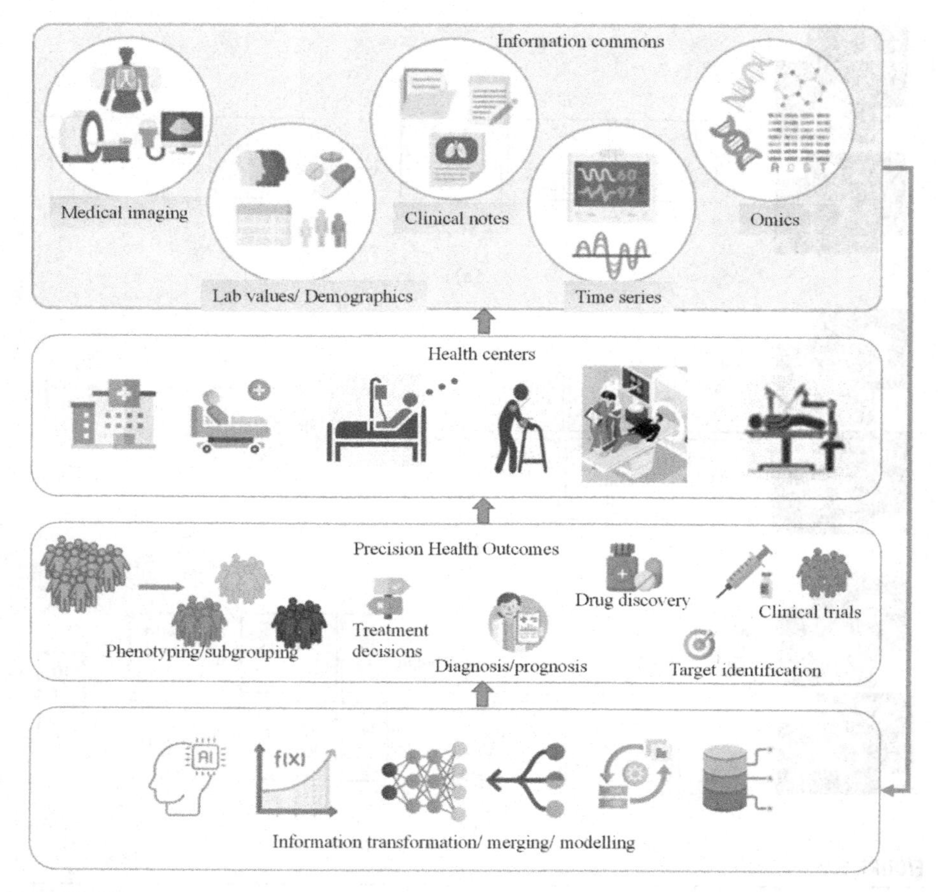

FIGURE 3.3

Sources of multimodal healthcare data and the role of computational modeling and information fusion.

that outcomes of interest such as treatment targets and drug development ultimately facilitate improved patient level decision making in care facilities and homes. Such a phenomenon has attracted interest in healthcare data fusion studies. The application of machine learning and deep learning algorithms provides insight into various outcomes of healthcare, such as drug discovery, clinical trials, phenotyping, and surgical techniques. This is crucial in providing support to practitioners and health centers to achieve the most precise and efficient evidence-based medicine possible. Unimodal deep learning models are often less robust and suffer misclassification challenges. Here, we propose the framework and algorithms for multimodal deep learning models for classifying acoustic and image type multimodal datasets.

3.3 Literature review and state-of-the-art

The application of deep learning for multimodal medical data fusion has been widely investigated. However, the majority of studies have been mainly limited to exploring the fusion of multimodal images as well as the fusion of image and contextual data from EHRs. For example, a systematic review has been reported of the deep learning-based fusion of imaging and EHR healthcare data [46]. A recent review reported the fusion of imaging and omics data [47] and a more inclusive review highlighted the application of machine learning for fusion of all current information types or sources [32]. Deep learning algorithms have rapidly shown promising results for knowledge extraction from multiple data modalities [48–50]. Deep learning-based multimodal image fusion has been widely investigated, for example CT, MRI, and PET data [51].

Based on the data reported in a recent study [32], the fusion of different data types has been previously investigated for different health conditions such as cancer, Alzheimer's disease, heart failure, infectious diseases (like COVID-19), skin lesions and endocrine conditions such as diabetes etc. Multimodality within the same data type, for instance images from multiple imaging modalities, has been widely explored, but here we are considering multimodal fusion of data across different modalities and different data types. As shown in Fig. 3.4, most of the studies have investigated the fusion of imaging and EHR data (n = 52), while fusion of EHR, omics, and textual data as well as omics and textual data have been least investigated. Most investigations have been conducted in the neurology category (n = 50); particularly, Alzheimer's disease data fusion studies account for the most published papers (n = 22), while hematology, dermatology, and drug/medication have been least investigated. Such information is a useful resource to fellow researchers for identification of lagging areas.

Further, of the three data fusion types, early fusion was the most used in published papers, and of the early data fusion studies, the modalities were imaging and EHR in 28 studies, EHR and textual data in 15 studies, imaging and genomic data in 10 studies, imaging, EHR, and genomic data in 9 papers, imaging and time series data in 2 studies, genomic and EHR data in 2 studies, imaging and textual data in 1 study, and EHR, transcriptomics, genomics, and insurance claims data in 1 study. Among intermediate data fusion studies, 14 used imaging and EHR data, 11 used textual and EHR data, 6 of which used CNN, long short-term memory (LSTM) networks, or knowledge-guided CNN. A few studies used intermediate fusion of genomic and imaging data and few studies used genomic and textual data, imaging, EHR and textual data, time series and imaging data, EHR, imaging, and genomics data, time series, imaging, and textual data, EHR, imaging, and time series data, genomics and EHR data, and EHR, textual, and time series data.

Few studies (n = 20) have investigated late fusion, 7 of which studied late fusion of EHR and imaging modalities and 3 studied genomic, imaging, and EHR data; EHR and textual data were used in 1 study, 2 studies used time series and imaging data, and few studies used imaging, textual and EHR data, textual, genomic, and EHR data, imaging, time series, and EHR data, genomics and imaging data, genomics

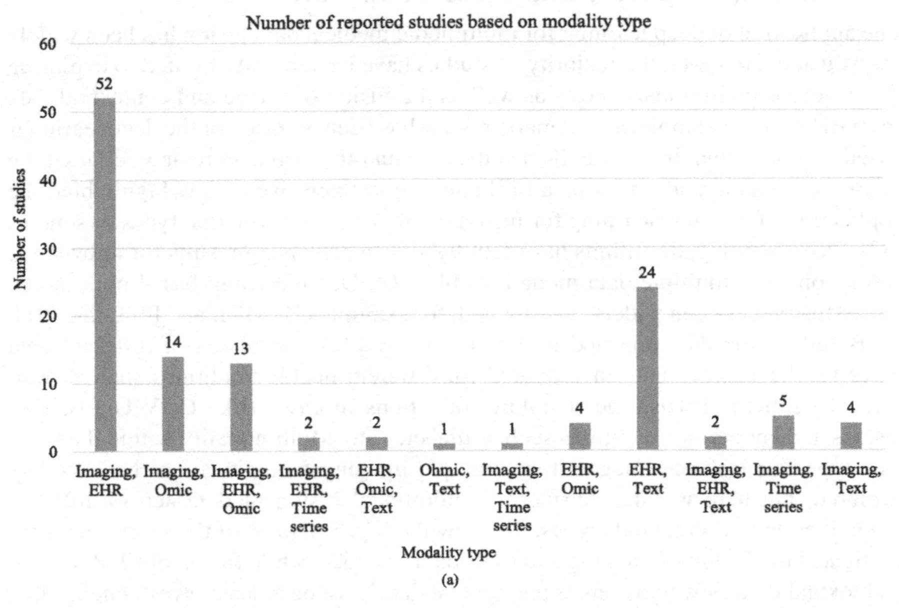

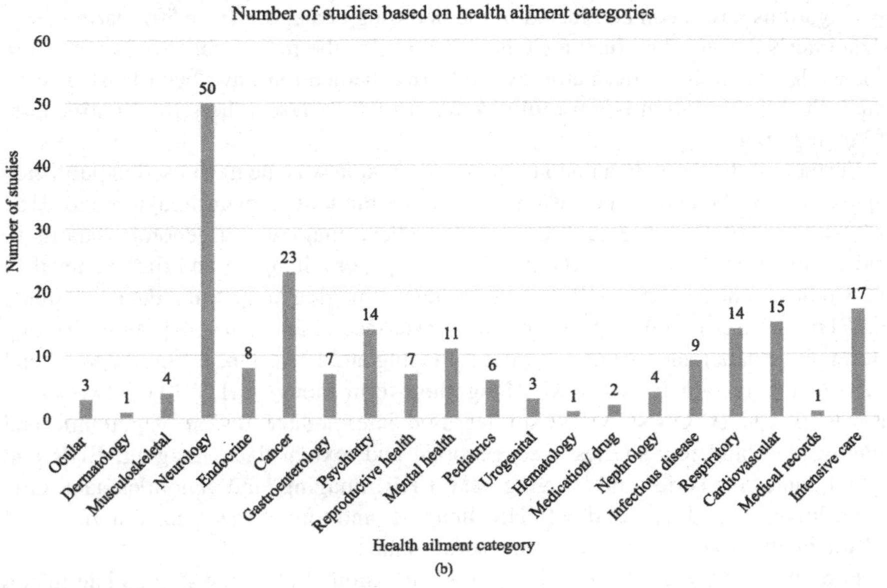

FIGURE 3.4

Multimodal fusion studies based on (a) modality type and (b) health ailment category.

and EHR data, and imaging and textual data. In these studies, different classifier algorithms were independently applied on datasets including CNN, random forest, logistic regression, multilayer perceptron, and stochastic gradient descent classifier. The final outputs were combined using different schemes, like multiplication, summation, and ranking [32]. Lastly, mixed fusion was studied in 2 papers, for example the use of seven different fusion architectures including early, intermediate, and late fusion was studied in [52], [53]. More recently, the application of AI-driven fusion of healthcare information to facilitate improved and effective decision making was studied [54]. The study focused on bibliometric analysis of previously reported studies using structural topic modeling and provided technological and scientific research perspectives with implications and insights for the development of this technology. CNN has been used for multimodal image fusion of MRI and CT images and showed high performance based on a low error and high correlation values [55]. A comprehensive overview of the integration of AI and multimodal healthcare information [56], multilevel and multitype, self-generated knowledge for segmentation tasks [57], biomedical imaging data [58], and multilesion recognition in medical images [59] has also been reported.

3.4 Problem definition

Conventional deep learning model-based diagnosis focuses mostly on unimodal data sources such as EHRs or EMRs and medical images including X-ray images, MRIs, CT scans, and ultrasound images. However, the diagnosis of not all diseases can rely on a single data source and the use of single source datasets is rather task-specific than robust. Further, with the advances in technology, the healthcare community is now made up of multiple heterogeneous information sources and types, including EMRs, radiology data, and patient monitoring devices. The application of deep learning for processing and fusion of multisource medical information will enable the development of connected and smart systems for next-generation healthcare [14], [15], [24–30], [16–23]. It is through information fusion and computational modeling that outcomes of interest such as treatment targets and drug development ultimately improve patient level decision making in care facilities and homes. Such a phenomenon has attracted interest in healthcare data fusion studies. The application of machine learning and deep learning algorithms provides insight into various outcomes of healthcare such as drug discovery, clinical trials, phenotyping, and surgical techniques. This is crucial in providing support to practitioners and health centers to achieve the most precise and efficient evidence-based medicine possible. Unimodal deep learning models are often less robust and suffer misclassification challenges. Here, we propose a framework and algorithms for multimodal deep learning models for classifying acoustic and image type multimodal datasets. More work is needed to address data management, accessibility, findability, reuse of digital datasets, and interoperability [31]. Further, it can be deduced from the previously reported studies that the fusion of image and acoustic information has not yet been explored in many studies; moreover, the implementation of deep learning-based late fusion has been least explored.

3.5 Proposed solution

Cough, voice breaks, lung infection, and shortness of breath are some of the symptoms of diseases like lung cancer and COVID-19. It is crucial to monitor these symptoms in patients diagnosed with lung cancer. Further remote monitoring can be enabled using IoT to offer immediate assessment and facilitate accessibility. The monitoring of patients' response to treatment can be achieved by monitoring the symptoms. Machine learning and deep learning techniques are particularly useful in analyzing retrieved data for both diagnosis and monitoring.

Vocal attributes of patients with respiratory ailments have been shown to have distinguishing features when compared to healthy subjects. Using suitable techniques for signal processing, the features associated with symptomatic vocal traits such as speech, cough, and breathing can be extracted. The extracted features can then be used to train a deep learning model for performing data analysis and classification. Moreover, conventional modalities such as MRI, CT, and X-ray imaging are capable of providing salient spatial information about the cancer in the form of scans which can be used for diagnosis and monitoring. Conventional CAD methods usually rely on a single type of data such as images from a single modality, which may lead to a compromised or subjective diagnostic perspective. More robust and precise diagnostics require fusion of multiple input modalities, and hence fusion strategies have been developed to integrate the complementary information from different modalities and different symptom types. This enables increased accuracy and reliability of the developed systems. Thus, in this work, we propose a framework of a multimodal information fusion system with the use of six inputs, i.e., MRIs, CT scans, X-ray images, and speech, cough, and breathing sounds. The multimodal system with an ensemble of trained deep learning-based models leads to increased system complexity and robustness with reduced false diagnosis. As previously reported, there are three fusion techniques: early fusion, hybrid fusion, and late fusion. Early fusion usually suffers from difficulties in time synchronization and data scarcity, and hybrid fusion suffers from the same factors. Here, late fusion uses random forest for decision making. The late fusion strategy is the most suitable for smaller-sized datasets. A dynamic retraining method is used for updating the parameters of the trained random forest to deal with noisy environments.

As shown in Fig. 3.5, the proposed framework is comprised of three major modules: an audio data module, an image data module, and a multimodal fusion module. The acquired patient data are forwarded in real-time to their respective classifiers in order to obtain predictions of those particular data. The CNN model for audio data is trained using speech sounds, cough sounds, and breathing sounds. The second CNN-based model for imaging data is trained on MRIs, CT scans, and X-ray chest image

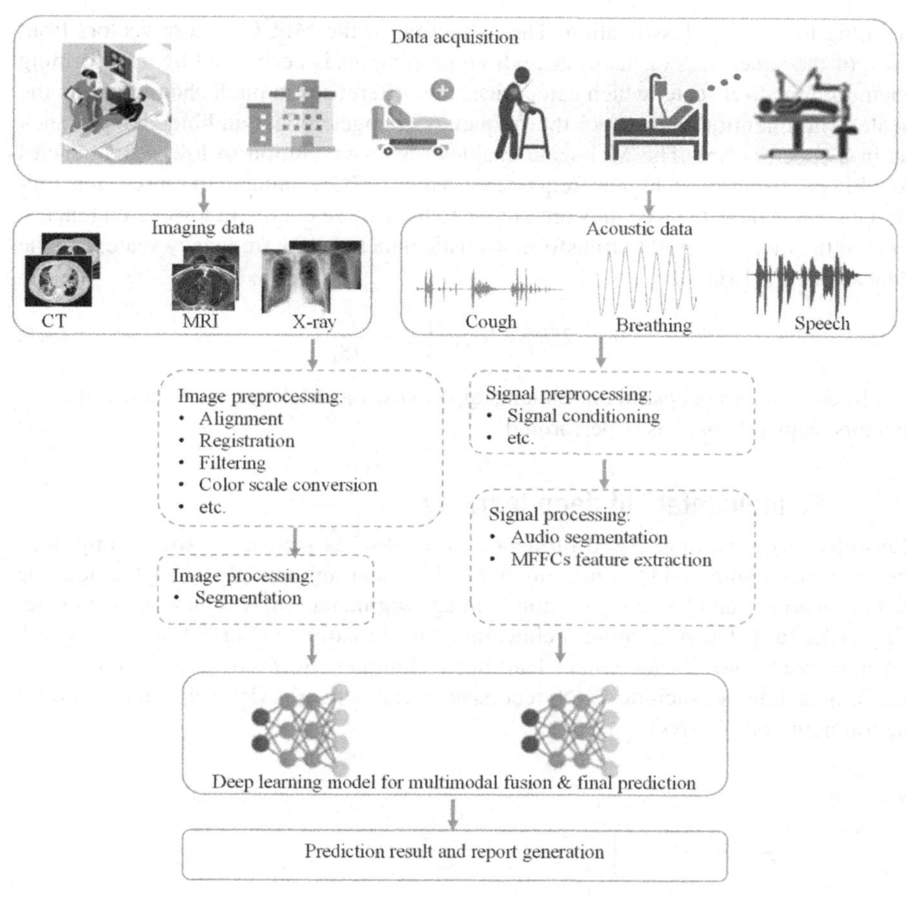

FIGURE 3.5

Proposed multimodal data fusion framework.

data. The dynamic random forest algorithm is then used for the late fusion of the predicted outcomes of six separately trained models to obtain the final decision.

3.5.1 Classification of acoustic data

The acquired raw audio waveforms are firstly preprocessed including filtering, etc., segmented for dimensionality reduction, and then used as input to the deep learning model. The deep learning model is capable of automatic feature extraction; thus, there is no need for manually extracting the features. However, initial feature extraction of audio data can be necessary to reduce the dimensionality of the features. The adoption of the MFCC feature is proposed that can be directly input to the deep learning model for classification. The extraction of the MFCC feature vectors from each of the subclasses of the three audio type samples is performed by transforming them into the Mel scale, which categorizes the differences in pitch changes along the scale. The intention is to make the frequency changes reflect audible changes such as in a spectrogram. The Mel scale enables higher resolution in lower frequencies and lower resolution in higher frequencies. The MFCC is suitable to represent symptomatic acoustic data since they are known to have more energy in lower frequencies. One of the methods used to transform acoustic data from the frequency scale f to the Mel scale m is [31]

$$m = 2595 * log_{10}\left(1 + \frac{f}{500}\right). \tag{3.1}$$

To compute the cepstral coefficients also known as Mel frequency cepstral coefficients, cepstral analysis is performed.

3.5.2 Fundamentals of deep learning

Recently, deep learning based on neural networks has shown to provide improved performance compared to conventional machine learning models through automatic feature learning and backpropagation in image segmentation, recognition, and detection tasks [60]. Deep learning architectures are broadly categorized as supervised, unsupervised, and reinforcement learning techniques. Some of the common deep learning techniques include CNN, recursive neural networks (RvNNs), and recurrent neural networks (RNNs).

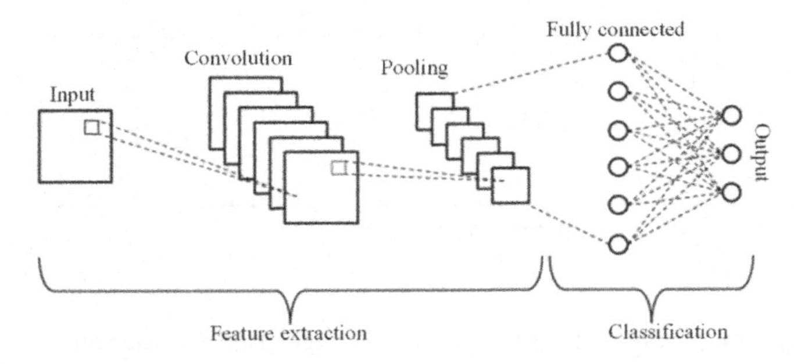

FIGURE 3.6

The architecture of CNN.

The most commonly employed and famous deep learning network is the CNN, which has been extensively used in a wide range of applications [61–63]. CNN is a deep neural network that has been widely used conventionally for image analysis and it has been more recently realized to effectively analyze sequential data, for example sound and natural language processing. The typical architecture of the CNN is shown in Fig. 3.6. It mainly involves input, convolution, and pooling layers. The convolution operation, i.e., a piecewise multiplication, is performed in the convolution layer consisting of various kernels or filters for extracting feature maps from the input data. The pooling operation is used for reducing the dimensionality of the obtained feature maps from the convolution. Activation functions such as Rectified Linear Unit (ReLU), leaky ReLU, etc., are used to introduce nonlinearity, i.e., transfer the gradient during training using backpropagation. This may however cause loss of learned information from initial layers in sequential CNN. A more suitable neural network for sequential data, for example speech processing, is RNN; however, it is prone to suffering from gradient exploding and vanishing problems.

The main salient property of CNN is its ability to automatically learn relevant image features without human supervision. The architecture of CNNs is shown in Fig. 3.3. It mimics the activity of human and animal neurons. CNN layers can be summarized as input layer, convolution layer, pooling layer, fully connected layer, and output layer. A 3D convolution of say $3 \times 3 \times 3$ is applied to an input image so as to compute the output image with characteristic representations of the input image. When different convolution filters are applied, many output images are obtained, termed feature maps or channels, where each feature map represents its input modeling result. Pooling is then applied to compute the average or the maximum of pixel values, thereby reducing dimension and increasing the modeling invariance by small signal change [64]. At the end of the CNN network, the fully connected layers are connected to determine the final output. Backpropagation is used to determine the parameters such as filter coefficients.

The most common networks based on CNN include AlexNet, GoogleNet, VGG, Inception-ResNet-v2, and ResNet. CNNs have been investigated for various medical imaging tasks. An extensive review of all deep learning techniques and their architectures has been published [61].

3.5.3 Deep learning-based image and acoustic data fusion

As shown in Fig. 3.7, the proposed CNN model for classifying acoustic data is firstly trained on the three acoustic data types separately. It is comprised of the input layer and three convolution layers. Three parallel convolution layers are involved in the first stage, which are comprised of different filters of sizes 8, 32, and 64 for feature learning. The second stage is a concatenation layer that is used to merge the extracted features by the parallel convolution layers, and another convolution layer is introduced. Three parallel convolution layers are contained in the third stage similar to the first stage. A concatenation is then performed, followed by flattening of features; two fully connected layers with the ReLU activation function are used in the first and softmax is used in the second dense layer. Max pooling is performed after each convolution layer for dimensionality reduction. Following the max pooling, batch normalization is performed to normalize the current layer activations before forwarding the result to the next layer for speeding up the training. The following training parameters are proposed: 70 epochs, batch size 32, Adam optimizer, and sparse cross-entropy as the loss function. The model output classifies the audio samples as diseased or healthy, i.e., positive or negative. Parallel convolutions with

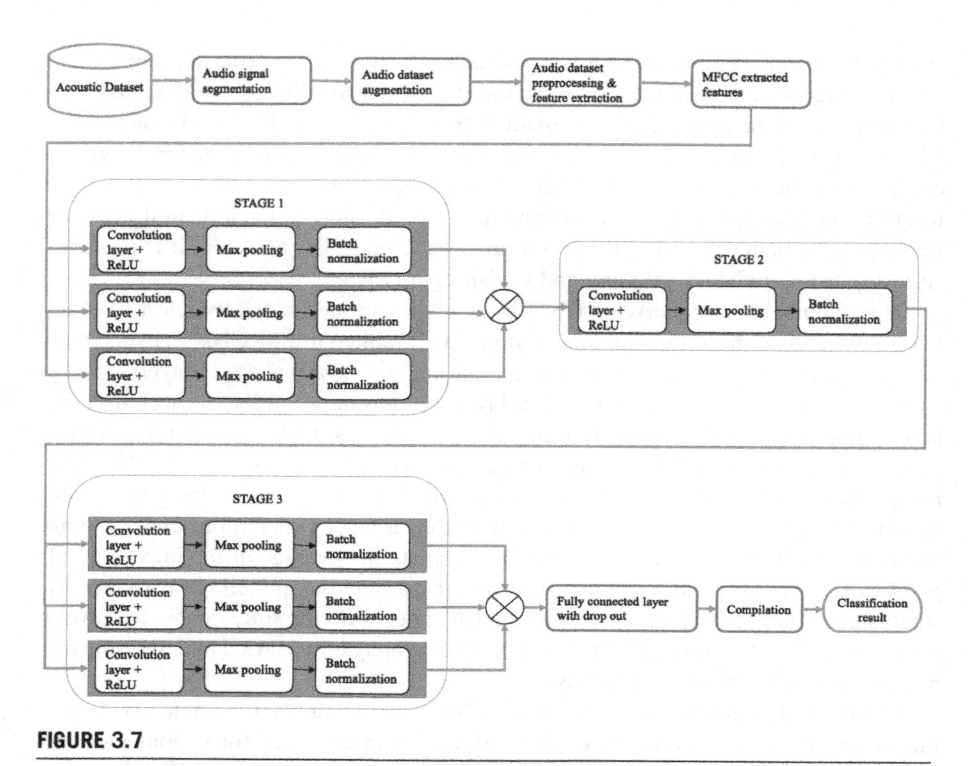

FIGURE 3.7

Architecture of the proposed CNN model for sequential data.

various filter sizes are crucial for learning important features from initial stages that may be missed easily in the sequential network; also it enables fast training due to the reduced system complexity.

Imaging techniques have been known to provide significant information pertaining to various diseases such as cancer and COVID-19. MRI, X-ray and CT imaging modalities have been widely advanced and used in this regard. The integration of these imaging modalities and deep learning is recently showing huge potential for wide-scale implementation of clinical decision support systems. The development of multimodal deep learning systems will further bring robustness of clinical decision support systems. Deep learning techniques such as CNN, ResNet, DenseNet, MobileNet, You Only Look Once (YOLO), AlexNet, Region-based CNN (RCNN), and U-Net have been used widely for image analysis [9]. The details of these network models have been widely reported [61], [62], [65].

Fig. 3.8 shows the proposed CNN model for image data classification. The model is based on the YOLOv3 network and is made up of 12 convolution layers with 7 max pooling layers for dimensionality reduction. The ReLU activation function and batch normalization are performed for normalizing the activations. The softmax layer

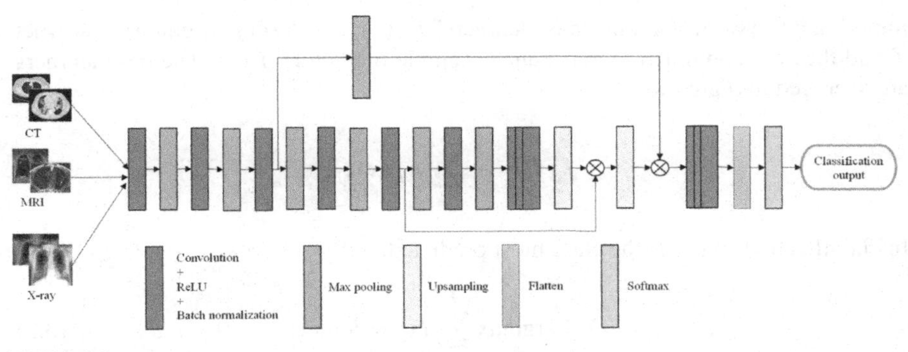

FIGURE 3.8

CNN architecture for image data classification.

is used for flattening the extracted feature maps into a feature vector and for final prediction.

3.5.4 Random forest for multimodal fusion

Random forest is a machine learning-based algorithm and is a tree-based ensemble where there is a connection of each tree to a collection of random variables [66] and the average trees for decision are trained through separate parts of a training set for variance reduction. The random forest technique is effective in handling missing data. Random subsets of the training data are exploited; thus, they are based on the bagging technique.

The mathematical concept behind random forest can be expressed as follows.

Assume we have a certain unknown joint distribution $P_{XY}(X, Y)$, where $X = (X_1, \ldots, X_P)^T$ is a p-dimensional vector of random variables and Y is a real-valued response. Random forest aims at finding the prediction function $f(X)$ so as to predict Y. The function $f(X)$ minimizes the loss function expected value $L(Y, f(X))$ such that $E_{XY}(L(Y, f(X)))$, with the subscripts denoting the expectation of X and Y's joint distribution. $L(Y, f(X))$ measures the closeness of $f(X)$ to Y, penalizing $f(X)$ values far from Y. An example of L for regression is $L(Y, f(X)) = (Y - f(X))^2$, denoting squared error loss, and an example for classification is zero one loss, where

$$L(Y, f(X)) = I(Y \neq f(X)) = \begin{cases} O & if\ Y = f(X), \\ 1 & otherwise. \end{cases} \tag{3.2}$$

The regression function is obtained from minimizing $E_{XY}(L(Y, f(X)))$ for squared error loss giving a conditional expectation $f(x) = E(Y \mid X = x)$, which is the regression function. For classification, the set of γ possible values denoting Y and minimizing $E_{XY}(L(Y, f(X)))$ for zero one loss results in

$$f(x) = \underset{y \in \gamma}{\operatorname{argmax}}\, P(Y = y \mid X = x), \tag{3.3}$$

implying a Bayes rule. Using "base learners" $h_1(x), \ldots, h_J(x)$ ensembles construct f and they are combined to make an "ensemble predictor" $f(x)$. The base learners are averaged in regression:

$$f(x) = \frac{1}{J}\sum_{j=1}^{J} h_j(x). \tag{3.4}$$

In classification, $f(x)$ is the class most predicted:

$$f(x) = \underset{y \in \mathcal{Y}}{\operatorname{argmax}} \sum_{j=1}^{J} I(y = h_j(x)). \tag{3.5}$$

Knowledge of tree types is required for understanding of the random forest algorithm. The jth base learner is for instance a tree $h_j(X, \theta_j)$, where θ_j is such that $j = 1, \ldots, J$.

To make the system dynamic with no increased complexity, the fusion model goes through retraining with misclassified unimodal prediction scores. Therefore, the system is made dynamic without having to increase the model complexity. A fragment of the multimodal data fusion algorithm is shown in Algorithm 3.1.

3.5.5 Performance evaluation metrics

The performance of deep learning or machine learning models is evaluated in terms of accuracy, sensitivity, specificity, etc. Some of the metrics used for evaluation of the performance of deep learning models include a confusion matrix, receiver operation characteristic curves (ROCs) and the area under the curve (AUC), the coefficient of determination, and k-fold and leave-one-out cross-validation [10].

3.6 Challenges and prospective opinion

Previously reported studies have identified common challenges associated with data fusion implementations, including model and heterogeneous data structure complexities, single site cohorts, retrospective data, small sample size, missing data, imbalanced samples, model interpretation, feature engineering, and confounding factors. Most of the samples are mostly collected from a single academic medical center or hospital. Small datasets often cause poor model fitting and poor generalizability. Missing data and sample imbalances lead to biased models and performance metrics that are misleading; however, the late data fusion technique has been shown to handle missing data well. Most studies used shallow network machine learning models which require manual feature extraction and are often task-specific. To alleviate the feature engineering challenge, deep learning neural networks are recommended that are capable of performing automatic feature extraction that is rather robust. Challenges of combinatory information include increased complexity of specifying the

Algorithm 3.1: Multimodal data fusion

Input: validation dataset X,

Parameters: validation set $X = \{X_1, X_2, X_3, \ldots\ldots\ldots, X_n\}$, test set $X_i = \{X_a, X_b, X_c, X_d, X_e, X_f\}$, cough dataset X_a, speech dataset X_b, breathing dataset X_c, CT dataset X_d, X-ray dataset X_e, MRI dataset X_f, looping variables i, j, prediction vector P, prediction vector for training, testing, and misclassification = P_x, P_y, and F respectively, decision tree classification model DTC, number of trees, features, and test subjects are denoted by n.

Output: Classification Y.

1: Initialize: k=6. Estimator S_n =12 (Decision tree classes). n=300 (assumed), min samples split =2
2: for i=1 to n do
3: for j= 1 to k do
4: $P(i, j) = P(i, j) + predict(x_i, m_j)$
5: end for
6: end for
7: Split dataset P into P_x(70%) and P_y(30%)
8: Build the DTC with maximum features = sqrt (n)
9: Train DTC using training set P_x
10: Y = Test P_y with trained model DTC
11: for i=1 to 30 do
12: if Y_i is incorrect prediction then
13: Append P_i to F
14: end if
15: end for
16: Retrain DTC with F to update learned parameters
17: return Y

model and reduced interpretability of results [67], [68]. Multimodal data sources and file formats are rarely uniform, especially with clinical data. As an example, the datasets can have different units of measure and naming conventions or represent different local biases of the population. Extra care must be taken to search and correct for dataset differences and assess their degree of interoperability. For example, the establishment of interoperability has been suggested for IoMT and EHR technologies in [69–71], and intrasite normalization has been performed by [72] to integrate CT and PCR lab values to make the values comparable across sites. Harmonization – a balance to allow similar information to work together and retain information correspondence (data purity) – is required in data fusion [73]. Harmonization techniques are used in successful fusion to ensure quality control in integration or fusion processes, and clinical harmonization requires the amalgamation of multidisciplinary research spanning biology, medicine, computer science, etc. [31]. More specifically, the challenges associated with implementation of deep learning-based multimodal information processing in healthcare are discussed below.

3.6.1 Imbalanced data

Data imbalance is a challenge associated with applying deep learning models in multimodal medical information fusion. Biological data tend to be commonly imbalanced as generally there are more numerous negative samples than positive ones [74]. When a deep learning model is trained over imbalanced datasets, undesirable results may be obtained. To solve the imbalanced dataset limitation in multimodal fusion applications, the following techniques can be used: employing the correct criteria for result prediction and evaluating the loss such as the AUC as the criteria and resultant loss. Weighted cross-entropy loss should also be employed which ensures good performance of the model with small classes. Large classes can be downsampled and small classes upsampled. To make the model handle imbalanced datasets, some methods such as constructing models for every hierarchical level can be employed since biological systems have a hierarchical label space [61], [75].

3.6.2 Overfitting

Due to the vast number of correlated and complex parameters as well as a lack of training data, deep learning models often have high chances of resulting in overfitting of data during the training stage, which reduces the performance of the model on testing data [76]. Overfitting impedes generalization of the classifier to new samples. High complexity and flexibility of a deep learning model are also considered to bring high risks of overfitting [75]. When proposing deep learning for multimodal data fusion applications, this problem should be accurately handled and considered by developing techniques that handle the problem. Deep learning models can overcome overfitting through implied bias of the training process [76], [77]. Some techniques have been reported to ease overfitting, including batch normalization, weight decay, and dropout [61]. Model input-based techniques such as data augmentation and corruption and model output-based techniques that regularize the model through penalizing the overconfident outputs have been proposed [78]. To avoid overfitting by increasing the amount of training data through augmentation, data augmentation techniques which are data-space solutions incorporate a couple of methods for improving size and attributes of training datasets [79]. The data augmentation techniques include but are not limited to flipping, rotation, color space augmentation, translation, noise injection, and cropping. When such techniques are used, deep learning networks can perform better.

3.6.3 Interpretability of data

The deep learning techniques are in fact interpretable though occasionally analyzed as a black box. A method is however required to interpret deep learning results to obtain valuable patterns and motifs that are recognizable to the network. In disease prediction or diagnosis tasks, this will be helpful to enhance accuracy of prediction outcomes which are the basis of the model decision. Scores of importance for each portion of a particular example can be given for example through perturbation- or backpropagation-based techniques to enhance accuracy [80].

3.6.4 Uncertainty scaling

When employing deep learning techniques, the final prediction label is required together with the label of the score of confidence for each inquiry from the model to achieve prediction. The measure of how confident the model is in its prediction is the so-called score of confidence and it is a significant attribute in preventing belief of misleading and unreliable predictions which reduces resources and time consumed in providing misleading prediction outcomes in various application scenarios [81], [82]. In multimodal information fusion and related applications, uncertainty scaling is crucial for evaluating automated clinical decisions and improve reliability of deep learning-based disease diagnosis. Due to overconfident prediction output of different deep learning models, the score of probability, e.g., from the softmax output of a deep learning network, is more often incorrectly scaled and thus requires postscaling for a reliable probability score. Several techniques can be used to output correct probability scores such as histogram binning, Bayesian binning into quantiles, legendary Platt scaling, and isotonic regression. Temperature scaling has been reported to achieve superior performance for deep learning techniques [61].

3.6.5 Model compression

Deep learning models require intensive computational and memory requirements for obtaining well-trained models, because of the large number of parameters and huge complexity of the models. Healthcare is one of the most data-intensive fields, which reduces the implementation of deep learning in machines with limited computational power. Additional computation power is required to comply with the vast sizes of heterogeneous data in healthcare. Modern hardware-based parallel processing technologies have been proposed such as field programmable gate array (FPGA) and graphics processing units (GPUs) to alleviate the computational limitations associated with deep learning [83], [84]. Techniques for compressing deep learning models to reduce the computational issues have also been designed such as parameter pruning, knowledge distillation, the use of compact convolution filters, and estimation of information parameters for preservation using low-rank factorization [61].

3.6.6 Other challenges

Other issues requiring proper attention that are associated with the implementation of deep learning algorithms and multimodal information processing include catastrophic forgetting, the vanishing gradient problem, the exploding gradient problem, and underspecification [61]. The deep learning methods enable visualization of results in a manner that can be conveyed to medical practitioners.

The machine learning and deep learning models in the imaging application domain are also vulnerable to attacks that can either lead to slight result discrepancies or the consequences can be lethal in applications where safety is critical. Such attacks can be subjected to the deep learning models at the edge, fog, or cloud layers of the system and they include adversarial attacks, neural-level Trojans, hardware attacks, and intellectual property (IP) stealing. Adversarial attacks are crafted adversarial machine learning attacks that compromise model performance in various machine learning applications, and the THz imaging application is no exception. Approaches to develop machine learning models that are adversarial-robust have been reported that can restrain adversarial examples and perturbations to ensure model security and integrity [79]. Such adversarial robust approaches include modifying the training and testing data, modifying the features/parameters learned by the training model, and the use of additional auxiliary models to enhance the main model's robustness.

For the effective modeling of medical tasks, the challenges associated with processing heterogeneous observational data from real-world clinical databases must be considered. These potential challenges include nonstandardized data structures, small and incomplete datasets, preserving patient data privacy, cost-effective annotation processes, multimodal data, and irregular health trajectories [15]. Despite the massive clinical potential of multimodal information processing models, the technology is still at an early stage and still associated with a lot of limitations. The alleviation of current limitations can pave the way for future research.

3.6.7 Prospective opinion

A more illustrative representation of existing challenges and future prospects is summarized in Fig. 3.9. The multimodal fusion implementation limitations are stratified by their location in the workflow, i.e., issues associated with underlying data, modeling issues arising from the data, and issues associated with how the data are ported back to the healthcare system for translational decision support. The several gaps outlined and prospects for future research will facilitate and expedite the multimodal data fusion field.

To address the data unavailability limitations in multimodal information processing, we suggest four approaches that could facilitate academic research. First, we propose the employment of transfer learning based on deep learning whereby one pretrained deep learning network model is used as a starting point for developing a training model. This can be achieved by fine-tuning a pretrained model such as AlexNet, GoogleNet, and ResNet to learn a new task such that the acquired knowledge from one domain (source) gets transferred to the target domain even when there is a disjoint feature space and data distribution of source and target. The network retraining using transfer learning is easier and faster than developing and training the model from scratch. Moreover, it enables less and imbalanced training data usage and reduces computing resources and training time [9]. Secondly, data augmentation tasks can be performed to increase the training dataset size through image rotation,

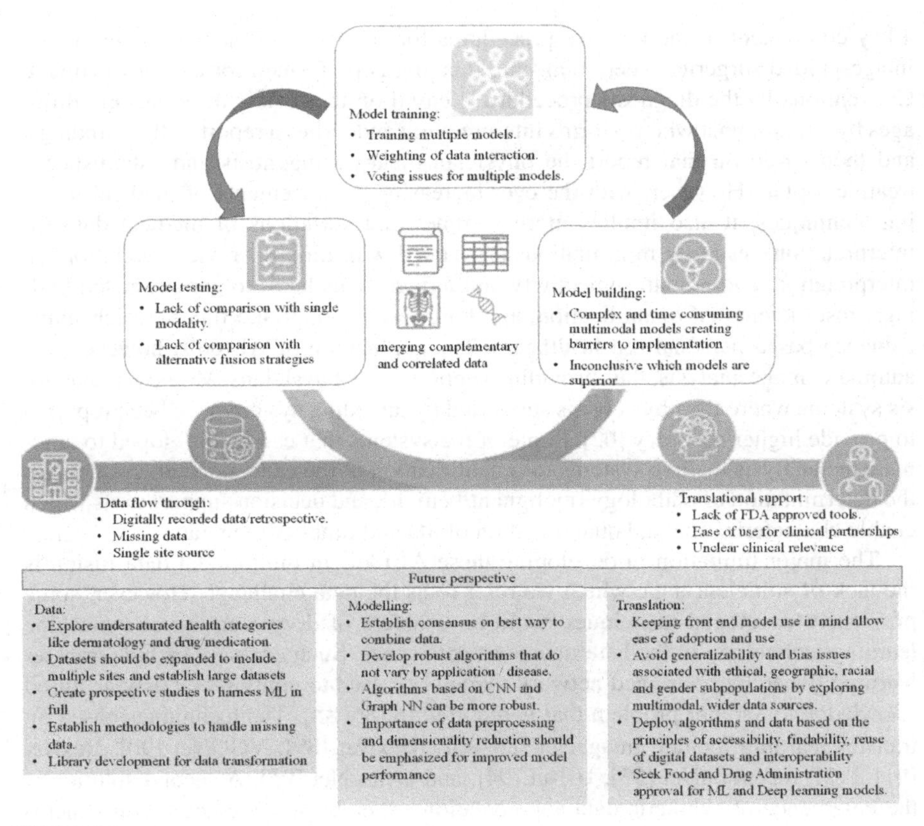

FIGURE 3.9

Limitations to multimodal data fusion in healthcare and proposed future directions.

scaling, translation, mirroring, etc. This improves the model performance and accuracy while the original data label is not changed. Thirdly, generation of synthetic datasets through simulations can increase training dataset volumes required for deep learning-based simulation [85]. Lastly, the implementation of multimodal data fusion can also be considered to alleviate the data shortage challenges in data-driven medical studies by taking advantage of the automatic feature learning capability of deep learning models. This does not only address the dataset shortages but also enhances model complexity and makes the deep learning model robust rather than task-specific.

The developments in deep learning-based multimodal information processing have made significant progress in various applications and trigger a plethora of promising research directions, particularly in the deep learning-enabled multimodal data fusion. The advances in multimodal data fusion technology play a significant role in medical imaging for cancer diagnosis, treatment, and follow-up to verify the success of a treatment. Additionally, medical image and time series data are now a key component of the invasive procedures for surgical and therapy planning and image-guided surgeries where real-time imaging is performed for cancer treatment. Conventionally, the diagnosis procedure is based on the review of the acquired images by a radiologist who performs interpretation and writes a report of their findings, and then based on that report the physician makes a diagnosis and established a treatment plan. However, with the ever-increasing advancements of medical imaging techniques, it also implies more volumes and variations of medical data for interpretation, resulting in limitations associated with time for review, variations in interpretation, and human subjectivity and fatigue. This leads to compromised findings, insufficiency of quantification, and long result turnaround times, which limits evidence-based personalized healthcare. The application of AI tools, however, can automate image analysis, thus providing support to the physicians. Moreover, diagnosis systems where the physician is supported by an aiding system have been reported to provide higher accuracy [62]. Some of the systems that can be developed to automate the analysis include systems for quantification of the cancer extent, systems for the determination of pathology (malignant/benign), and decision support tools which enable characterization and quantification of 3D and time-varying data.

The major limitation in developing these AI tools in multimodal data fusion is the lack of sufficient data, which we refer to as the data challenge. However, some previously developed techniques could be enablers of deep learning and machine learning technology in the heterogeneous data space. Such enablers include transfer learning, whereby pretrained network models are used to apply previously acquired knowledge to another problem that is reported in [86–88]. The techniques based on transfer learning include ImageNet [86–89], AlexNet [89], VGGNet [90], ResNet [91], [92], InceptionNet [93], U-Net [94], and DenseNet [95]. A second solution is the emergence of synthetic data augmentation, whereby schemes based on generative modeling, for instance generative adversarial networks and variational encoders, are used to synthesize data to increase the training dataset [96] and thus improve the performance of the model. The development of integrated learning for domain adaptation models capable of discriminating heterogeneous feature spaces of different and multiple domains for cross-modality cancer image analysis can also be applied [97], [98]. Another solution would be the adoption of the novel federated learning to combat limitations associated with data privacy, data access rights, data sharing, and data security so as to facilitate academic research. Federated learning uses distributed computing and strategies of data aggregation so that a robust and common algorithmic model can be constructed without transfer of the data, i.e., the algorithm is trained across decentralized devices without exchanging data, in contrast to uploading datasets to a centralized server [62], [99–106].

The cutting-edge quantum-inspired deep learning approaches promise to resolve the limitations of deep learning based on parallel computers and GPU hard drive expenses in the future, where classical data are transformed to a quantum state by the quantum routine, and after the quantum operations the classical data are retrieved [107]. The deep learning models address the limitation of scalability in conventional machine learning methods, which are task-specific or scenario-dependent, enabling adaptation to continuous updates in multimodal data. The problem of long training times that are obstructing current deep learning systems from effectively operating in real-time may be solved by developing real-time deep learning techniques with shorter training times and improving the training process. The successful operation and deployment of next-generation multimodal systems for cancer applications calls for better localization and sensing performance. The reinforcement learning-based techniques will be useful in interactive learning frameworks, for example in indoor patient monitoring situations and multipurpose platforms.

The novel neural network structures such as the broad learning system (BLS) that consist of enhancement nodes and feature nodes and are based on pseudoinverse theory and compressed sensing could also offer new opportunities in future multimodal data analysis. Compared to the popular deep neural networks, BLS networks are capable of incremental learning and have the ability to remodel the system without tedious retraining processes. Thus, higher modeling speed, higher regression accuracy, and better generalization can be achieved for solving various tasks [108], [109]. Unified multimodal data fusion systems that can support various applications ranging from in-home health to virtual reality services and residential security with high reliability, high data rates, and low latencies are expected to pave the way for next-generation research frontiers [64].

3.7 Conclusion

Information fusion and computational modeling play crucial roles in processes such as treatment and drug development to ultimately facilitate patient-level decision making in care facilities and homes. Healthcare data fusion studies have attracted much interest. The application of machine learning and deep learning algorithms provides insight into various components of healthcare, such as drug discovery, clinical trials, phenotyping, and surgical techniques. This is crucial in providing support for practitioners and health centers to provide the most precise and efficient evidence-based medicine possible. Here, we have proposed a framework and optimized the algorithms for multimodal deep learning models for classifying acoustic and image multimodal datasets. Such multimodal data fusion is useful in the diagnosis and monitoring of diseases like lung cancer. We note that most studies have focused on the fusion of multimodal image datasets and the fusion of image and EHR data. The fusion of image and acoustic data has only been investigated for COVID-19. The development of deep learning-enabled multimodal information fusion systems will facilitate the development of computer-aided detection and diagnosis systems for clinical decision support. Thus, we identify the existing limitations inhibiting the wide-scale application of deep learning in multimodal healthcare data fusion. Advanced or state-of-the-art AI technologies such as BLS, federated learning, and reinforcement learning are needed for the development of computer-aided techniques that enable automatic diagnosis, detection, prevention, and treatment. Advanced fusion of multisensor data, multiview fusion, and multiatlas fusion based on AI should be explored to facilitate the advancement and development of smart and integrated healthcare systems. Further, solutions could be explored that defend against adversarial machine learning attacks. Possible directions for future research are illustrated in Fig. 3.9.

References

[1] G. Aceto, V. Persico, A. Pescapé, Industry 4.0 and health: internet of things, big data, and cloud computing for healthcare 4.0, Journal of Industrial Information Integration 18 (2020) 100129.

[2] J. Chanchaichujit, A. Tan, F. Meng, S. Eaimkhong, An introduction to healthcare 4.0, in: Healthcare 4.0, 2019, pp. 1–15.

[3] G. Yang, Z. Pang, M. Jamal Deen, M. Dong, Y.T. Zhang, N. Lovell, A.M. Rahmani, Homecare robotic systems for healthcare 4.0: visions and enabling technologies, IEEE Journal of Biomedical and Health Informatics 24 (9) (2020) 2535–2549.

[4] P. Centobelli, R. Cerchione, E. Esposito, E. Riccio, Enabling technological innovation in healthcare: a knowledge creation model perspective, in: Proceedings of the 2021 IEEE Technology and Engineering Management Conference - Europe, TEMSCON-EUR 2021, 2021, pp. 1–6.

[5] J. Al-Jaroodi, N. Mohamed, E. Abukhousa, Health 4.0: on the way to realizing the healthcare of the future, IEEE Access 8 (2020) 211189–211210.

[6] C. Chute, T. French, Introducing care 4.0: an integrated care paradigm built on industry 4.0 capabilities, International Journal of Environmental Research and Public Health 16 (12) (2019) 2247.

[7] A. Kumar, R. Krishnamurthi, A. Nayyar, K. Sharma, V. Grover, E. Hossain, A novel smart healthcare design, simulation, and implementation using healthcare 4.0 processes, IEEE Access 8 (2020) 118433–118471.

[8] M. Wehde, Healthcare 4.0, IEEE Engineering Management Review 47 (3) (2019) 24–28.

[9] M. Gezimati, G. Singh, Transfer learning for breast cancer classification in terahertz and infrared imaging, in: Proceedings of the 5th International Conference on Artificial Intelligence, Big Data, Computing and Data Communication Systems, icABCD 2022, 2022, pp. 1–6.

[10] M. Gezimati, G. Singh, Internet of things enabled framework for terahertz and infrared cancer imaging, Optical and Quantum Electronics 26 (2023) 1–17.

[11] L. Afsah-Hejri, E. Akbari, A. Toudeshki, T. Homayouni, A. Alizadeh, R. Ehsani, Terahertz spectroscopy and imaging: a review on agricultural applications, Computers and Electronics in Agriculture 177 (2020) 105628.

[12] Y. Jiang, G. Li, H. Ge, F. Wang, L. Li, X. Chen, M. Lu, Y. Zhang, Machine learning and application in terahertz technology: a review on achievements and future challenges, IEEE Access 10 (2022) 53761–53776.

[13] D. Yuanchuan, D. Hang, L. Shi, L. Kailin, F. Yijie, Auxiliary diagnosis study of integrated electronic medical record text and ct images, Journal of Intelligent Systems 31 (1) (2022) 753–766.

[14] S. Shamshirband, M. Fathi, A. Dehzangi, A.T. Chronopoulos, H. Alinejad-Rokny, A review on deep learning approaches in healthcare systems: taxonomies, challenges, and open issues, Journal of Biomedical Informatics 113 (2021) 103627.

[15] S. Rabhi, Optimized deep learning-based multimodal method for irregular medical timestamped data, Neural and Evolutionary Computing (2022) 1–48.
[16] D. Lahat, T. Adali, C. Jutten, Multimodal data fusion: an overview of methods, challenges, and prospects, Proceedings of the IEEE 103 (9) (2015) 1449–1477.
[17] C.E.L. M'Sabah, A. Bouziane, Y. Ferdi, A survey on deep learning methods for cancer diagnosis using multimodal data fusion, in: Proceeedings of the 2021 9th E-Health and Bioengineering Conference, EHB 2021, 2021, pp. 1–4.
[18] M.D. Hssayeni, B. Ghoraani, Multi-modal physiological data fusion for affect estimation using deep learning, IEEE Access 9 (2021) 21642–21652.
[19] T. Liu, J. Huang, T. Liao, R. Pu, S. Liu, Y. Peng, A hybrid deep learning model for predicting molecular subtypes of human breast cancer using multimodal data, IRBM 43 (1) (2022) 62–74.
[20] K. Jin, Y. Yan, M. Chen, J. Wang, X. Pan, X. Liu, M. Liu, L. Lou, Y. Wang, J. Ye, Multimodal deep learning with feature level fusion for identification of choroidal neovascularization activity in age-related macular degeneration, Acta Ophthalmologica 100 (2) (2022) e512–e520.
[21] Y. Liu, Y. Shi, F. Mu, J. Cheng, C. Li, X. Chen, Multimodal mri volumetric data fusion with convolutional neural networks, IEEE Transactions on Instrumentation and Measurement 71 (2022) 1–15.
[22] R. Bokade, A. Navato, R. Ouyang, X. Jin, C.-A. Chou, S. Ostadabbas, A.V. Mueller, A cross-disciplinary comparison of multimodal data fusion approaches and applications: accelerating learning through trans-disciplinary information sharing, Expert Systems with Applications 165 (2021) 113885.
[23] R. Walambe, P. Nayak, A. Bhardwaj, K. Kotecha, Employing multimodal machine learning for stress detection, Journal of Healthcare Engineering 2021 (2021) 9356452.
[24] S. El-Sappagh, H. Saleh, R. Sahal, T. Abuhmed, S.M.R. Islam, F. Ali, E. Amer, Alzheimer's disease progression detection model based on an early fusion of cost-effective multimodal data, Future Generations Computer Systems 115 (2021) 680–699.
[25] J. Gao, T. Lyu, F. Xiong, J. Wang, W. Ke, Z. Li, Predicting the survival of cancer patients with multimodal graph neural network, IEEE/ACM Transactions on Computational Biology and Bioinformatics 19 (2) (2021) 699–709.
[26] A. Holzinger, M. Dehmer, F. Emmert-Streib, R. Cucchiara, I. Augenstein, J. Del Ser, W. Samek, I. Jurisica, N. Díaz-Rodríguez, Information fusion as an integrative cross-cutting enabler to achieve robust, explainable, and trustworthy medical artificial intelligence, Information Fusion 79 (2022) 263–278.
[27] H. Yu, L.T. Yang, X. Fan, Q. Zhang, A deep residual computation model for heterogeneous data learning in smart internet of things, Applied Soft Computing 107 (2021) 107361.
[28] J. Yang, J. Ju, L. Guo, B. Ji, S. Shi, Z. Yang, S. Gao, X. Yuan, G. Tian, Y. Liang, Prediction of her2-positive breast cancer recurrence and metastasis risk from histopathological images and clinical information via multimodal deep learning, Computational and Structural Biotechnology Journal 20 (2022) 333–342.
[29] Z. Yang, P. Baraldi, E. Zio, A multi-branch deep neural network model for failure prognostics based on multimodal data, Journal of Manufacturing Systems 59 (2021) 42–50.
[30] R. Gao, S. Zhao, K. Aishanjiang, H. Cai, T. Wei, Y. Zhang, Z. Liu, J. Zhou, B. Han, J. Wang, Deep learning for differential diagnosis of malignant hepatic tumors based on multi-phase contrast-enhanced ct and clinical data, Journal of Hematology & Oncology 14 (1) (2021) 1–7.

[31] A. Manocha, M. Bhatia, A novel deep fusion strategy for Covid-19 prediction using multimodality approach, Computers & Electrical Engineering 103 (2022) 108274.

[32] A. Kline, H. Wang, Y. Li, S. Dennis, M. Hutch, Z. Xu, F. Wang, F. Cheng, Y. Luo, Multimodal machine learning in precision health: a scoping review, npj Digital Medicine 5 (1) (2022) 1–14.

[33] F.S. Ahmad, Y. Luo, R.M. Wehbe, J.D. Thomas, S.J. Shah, Advances in machine learning approaches to heart failure with preserved ejection fraction, Heart Failure Clinics 18 (2) (2022) 287–300.

[34] E. Molino-Minero-Re, A.A. Aguileta, R.F. Brena, E. Garcia-Ceja, Improved accuracy in predicting the best sensor fusion architecture for multiple domains, Sensors 21 (21) (2021) 7007.

[35] F. Behrad, M. Saniee Abadeh, An overview of deep learning methods for multimodal medical data mining, Expert Systems with Applications 200 (2022) 117006.

[36] Y. Gupta, R.K. Lama, G.-R. Kwon, A.D.N. Initiative, Prediction and classification of Alzheimer's disease based on combined features from apolipoprotein-e genotype, cerebrospinal fluid, mr, and fdg-pet imaging biomarkers, Frontiers in Computational Neuroscience 13 (2019) 72.

[37] X. Bi, X. Hu, H. Wu, Y. Wang, Multimodal data analysis of Alzheimer's disease based on clustering evolutionary random forest, IEEE Journal of Biomedical and Health Informatics 24 (10) (2020) 2973–2983.

[38] T. Zhou, K. Thung, X. Zhu, D. Shen, Effective feature learning and fusion of multimodality data using stage-wise deep neural network for dementia diagnosis, Human Brain Mapping 40 (3) (2019) 1001–1016.

[39] Z. Xu, J. Chou, X.S. Zhang, Y. Luo, T. Isakova, P. Adekkanattu, J.S. Ancker, G. Jiang, R.C. Kiefer, J.A. Pacheco, Identifying sub-phenotypes of acute kidney injury using structured and unstructured electronic health record data with memory networks, Journal of Biomedical Informatics 102 (2020) 103361.

[40] D. Zhang, C. Yin, J. Zeng, X. Yuan, P. Zhang, Combining structured and unstructured data for predictive models: a deep learning approach, BMC Medical Informatics and Decision Making 20 (1) (2020) 1–11.

[41] S. El-Sappagh, T. Abuhmed, S.M.R. Islam, K.S. Kwak, Multimodal multitask deep learning model for Alzheimer's disease progression detection based on time series data, Neurocomputing 412 (2020) 197–215.

[42] H. Yang, L. Kuang, F. Xia, Multimodal temporal-clinical note network for mortality prediction, Journal of Biomedical Semantics 12 (1) (2021) 1–14.

[43] Z. Zhang, P. Chen, M. Sapkota, L. Yang, Tandemnet: distilling knowledge from medical images using diagnostic reports as optional semantic references, in: Proceedings of the International Conference on Medical Image Computing and Computer-Assisted Intervention, 2017, pp. 320–328.

[44] X. Wang, Y. Peng, L. Lu, Z. Lu, R.M. Summers, Tienet: text-image embedding network for common thorax disease classification and reporting in chest x-rays, in: Proceedings of the IEEE Conference on Computer Vision and Pattern Recognition, 2018, pp. 9049–9058.

[45] S.A. Qureshi, S. Saha, M. Hasanuzzaman, G. Dias, Multitask representation learning for multimodal estimation of depression level, IEEE Intelligent Systems 34 (5) (2019) 45–52.

[46] S.-C. Huang, A. Pareek, S. Seyyedi, I. Banerjee, M.P. Lungren, Fusion of medical imaging and electronic health records using deep learning: a systematic review and implementation guidelines, npj Digital Medicine 3 (1) (2020) 1–9.

[47] W. Huang, K. Tan, J. Hu, Z. Zhang, S. Dong, A review of fusion methods for omics and imaging data, IEEE/ACM Transactions on Computational Biology and Bioinformatics 14 (8) (2022) 1–19.

[48] N. Algiriyage, R. Prasanna, K. Stock, E.E.H. Doyle, D. Johnston, Multi-source multimodal data and deep learning for disaster response: a systematic review, SN Computer Science 3 (1) (2022) 1–29.

[49] O.P. Jena, B. Bhushan, U. Kose, Machine Learning and Deep Learning in Medical Data Analytics and Healthcare Applications, CRC Press, 2022, pp. 1–292.

[50] S. Amal, L. Safarnejad, J.A. Omiye, I. Ghanzouri, J.H. Cabot, et al., Use of multimodal data and machine learning to improve cardiovascular disease care, Frontiers in Cardiovascular Medicine 9 (840262) (2022) 1–11.

[51] B. Rajalingam, R. Priya, Multimodal medical image fusion based on deep learning neural network for clinical treatment analysis, International Journal of ChemTech Research 11 (06) (2018) 160–176.

[52] S.-C. Huang, A. Pareek, R. Zamanian, I. Banerjee, M.P. Lungren, Multimodal fusion with deep neural networks for leveraging ct imaging and electronic health record: a case-study in pulmonary embolism detection, Scientific Reports 10 (1) (2020) 1–9.

[53] S. El-Sappagh, J.M. Alonso, S.M. Islam, A.M. Sultan, K.S. Kwak, A multilayer multimodal detection and prediction model based on explainable artificial intelligence for Alzheimer's disease, Scientific Reports 11 (1) (2021) 1–26.

[54] X. Chen, H. Xie, Z. Li, G. Cheng, M. Leng, F.L. Wang, Information fusion and artificial intelligence for smart healthcare: a bibliometric study, Information Processing & Management 60 (1) (2023) 103113.

[55] A. Mergin, M.S.G. Premi, Shearlet transform-based novel method for multimodality medical image fusion using deep learning, International Journal on Computational Intelligence and Applications (2023) 2341006.

[56] H. Wang, G. Barone, A. Smith, Current and future role of data fusion and machine learning in infrastructure health monitoring, Structure and Infrastructure Engineering (2023) 1–30.

[57] C. Yu, S. Li, D. Ghista, Z. Gao, H. Zhang, J. Del Ser, L. Xu, Multi-level multi-type self-generated knowledge fusion for cardiac ultrasound segmentation, Information Fusion 92 (2023) 1–12.

[58] S. Nazir, D.M. Dickson, M.U. Akram, Survey of explainable artificial intelligence techniques for biomedical imaging with deep neural networks, Computers in Biology and Medicine (2023) 106668.

[59] H. Jiang, Z. Diao, T. Shi, Y. Zhou, F. Wang, W. Hu, X. Zhu, S. Luo, G. Tong, Y.-D. Yao, A review of deep learning-based multiple-lesion recognition from medical images: classification, detection and segmentation, Computers in Biology and Medicine (2023) 106726.

[60] H. Liu, N. Vohra, K. Bailey, M. El-Shenawee, A.H. Nelson, Deep learning classification of breast cancer tissue from terahertz imaging through wavelet synchro-squeezed transformation and transfer learning, Journal of Infrared, Millimeter, and Terahertz Waves 43 (1–2) (2022) 48–70.

[61] L. Alzubaidi, J. Zhang, A.J. Humaidi, A. Al-Dujaili, Y. Duan, O. Al-Shamma, J. Santamaría, M.A. Fadhel, M. Al-Amidie, L. Farhan, Review of deep learning: concepts, cnn architectures, challenges, applications, future directions, Journal of Big Data 8 (1) (2021) 1–74.

[62] S.K. Zhou, H. Greenspan, C. Davatzikos, J.S. Duncan, B. Van Ginneken, A. Madabhushi, J.L. Prince, D. Rueckert, R.M. Summers, A review of deep learning in medical imaging: imaging traits, technology trends, case studies with progress highlights, and future promises, Proceedings of the IEEE 109 (5) (2021) 820–838.

[63] S.K. Koul, P. Kaurav, Machine learning and biomedical sub-terahertz/terahertz technology, Sub-Terahertz Sensing Technology for Biomedical Applications (2022) 199–239.

[64] S. Helal, H. Sarieddeen, H. Dahrouj, T.Y. Al-Naffouri, M.S. Alouini, Signal processing and machine learning techniques for terahertz sensing: an overview, IEEE Signal Processing Magazine 39 (5) (2022) 42–62.

[65] Y. Zhang, J.M. Gorriz, Z. Dong, Deep learning in medical image analysis, Journal of Imaging 7 (4) (2021) 221–248.

[66] H. Almeida, M.J. Meurs, L. Kosseim, G. Butler, A. Tsang, Machine learning for biomedical literature triage, PLoS ONE 9 (12) (2014) e115892.

[67] Y. Li, X. Wu, P. Yang, G. Jiang, Y. Luo, Machine learning applications in diagnosis, treatment and prognosis of lung cancer, arXiv preprint, arXiv:2203.02794, 2022, pp. 1–43.

[68] E. Blasch, T. Pham, C.-Y. Chong, W. Koch, H. Leung, D. Braines, T. Abdelzaher, Machine learning/artificial intelligence for sensor data fusion–opportunities and challenges, IEEE Aerospace and Electronic Systems Magazine 36 (7) (2021) 80–93.

[69] J.N.S. Rubí, P.R. de L. Gondim, Interoperable internet of medical things platform for e-health applications, International Journal of Distributed Sensor Networks 16 (1) (2020) 1–16.

[70] S. Meliá, S. Nasabeh, S. Luján-Mora, C. Cachero, Mosiot: modeling and simulating iot healthcare-monitoring systems for people with disabilities, International Journal of Environmental Research and Public Health 18 (12) (2021) 6357.

[71] L.C. Kourtis, O.B. Regele, J.M. Wright, G.B. Jones, Digital biomarkers for Alzheimer's disease: the mobile/wearable devices opportunity, npj Digital Medicine 2 (1) (2019) 1–9.

[72] A. Colubri, M.-A. Hartley, M. Siakor, V. Wolfman, A. Felix, T. Sesay, J.G. Shaffer, R.F. Garry, D.S. Grant, A.C. Levine, Machine-learning prognostic models from the 2014–16 Ebola outbreak: data-harmonization challenges, validation strategies, and mhealth applications, EClinicalMedicine 11 (2019) 54–64.

[73] A. Jamshidi, J.-P. Pelletier, J. Martel-Pelletier, Machine-learning-based patient-specific prediction models for knee osteoarthritis, Nature Reviews: Rheumatology 15 (1) (2019) 49–60.

[74] J.M. Johnson, T.M. Khoshgoftaar, Survey on deep learning with class imbalance, Journal of Big Data 6 (1) (2019) 1–54.

[75] Y. Li, S. Wang, R. Umarov, B. Xie, M. Fan, L. Li, X. Gao, DEEPre: sequence-based enzyme ec number prediction by deep learning, Bioinformatics 34 (5) (2018) 760–769.

[76] Q. Xu, M. Zhang, Z. Gu, G. Pan, Overfitting remedy by sparsifying regularization on fully-connected layers of cnns, Neurocomputing 328 (2019) 69–74.

[77] X. Xu, X. Jiang, C. Ma, P. Du, X. Li, S. Lv, L. Yu, Q. Ni, Y. Chen, J. Su, G. Lang, Y. Li, H. Zhao, J. Liu, K. Xu, L. Ruan, J. Sheng, Y. Qiu, W. Wu, A deep learning system to screen novel coronavirus disease 2019 pneumonia, Engineering 6 (10) (2020) 1122–1129.

[78] G. Pereyra, G. Tucker, J. Chorowski, Ł. Kaiser, G. Hinton, Regularizing neural networks by penalizing confident output distributions, in: Proceedings of the 5th International Conference on Learning Representations, ICLR 2017, 2019, pp. 1–12.

[79] Mavis Gezimati, Ghanshyam Singh, Open research challenges and opportunities in terahertz imaging and sensing for cancer detection, in: Proceedings of the International Conference on Microwave, Antenna and Communication, MAC2023, 2023, pp. 1–6.
[80] T. Ching, D.S. Himmelstein, B.K. Beaulieu-Jones, A.A. Kalinin, B.T. Do, G.P. Way, E. Ferrero, P.M. Agapow, M. Zietz, M.M. Hoffman, W. Xie, G.L. Rosen, B.J. Lengerich, J. Israeli, J. Lanchantin, S. Woloszynek, A.E. Carpenter, A. Shrikumar, J. Xu, Opportunities and obstacles for deep learning in biology and medicine, Journal of the Royal Society Interface 15 (141) (2018) 20170387.
[81] T. Nair, D. Precup, D.L. Arnold, T. Arbel, Exploring uncertainty measures in deep networks for multiple sclerosis lesion detection and segmentation, Medical Image Analysis 59 (2020) 101557.
[82] L. Herzog, E. Murina, O. Dürr, S. Wegener, B. Sick, Integrating uncertainty in deep neural networks for mri based stroke analysis, Medical Image Analysis 65 (2020) 101790.
[83] Z. Min, Public welfare organization management system based on fpga and deep learning, Microprocessors and Microsystems 80 (2021) 103333.
[84] O. Al-Shamma, M.A. Fadhel, R.A. Hameed, L. Alzubaidi, J. Zhang, Boosting convolutional neural networks performance based on fpga accelerator, Advances in Intelligent Systems and Computing 940 (2020) 509–517.
[85] Mavis Gezimati, Ghanshyam Singh, M. Gezimati, G. Singh, Curved synthetic aperture radar for near-field terahertz imaging, IEEE Photonics Journal 15 (3) (2023) 1–13.
[86] Y. Bar, I. Diamant, L. Wolf, S. Lieberman, E. Konen, H. Greenspan, Chest pathology detection using deep learning with non-medical training, in: Proceedings of the International Symposium on Biomedical Imaging, vol. 2015-July, 2015, pp. 294–297.
[87] H.C. Shin, H.R. Roth, M. Gao, L. Lu, Z. Xu, I. Nogues, J. Yao, D. Mollura, R.M. Summers, Deep convolutional neural networks for computer-aided detection: cnn architectures, dataset characteristics and transfer learning, IEEE Transactions on Medical Imaging 35 (5) (2016) 1285–1298.
[88] V. Gulshan, L. Peng, M. Coram, M.C. Stumpe, D. Wu, A. Narayanaswamy, S. Venugopalan, K. Widner, T. Madams, J. Cuadros, R. Kim, R. Raman, P.C. Nelson, J.L. Mega, D.R. Webster, Development and validation of a deep learning algorithm for detection of diabetic retinopathy in retinal fundus photographs, JAMA. Journal of the American Medical Association 316 (22) (2016) 2402–2410.
[89] A. Krizhevsky, I. Sutskever, G.E. Hinton, ImageNet classification with deep convolutional neural networks, Communications of the ACM 60 (6) (2017) 84–90.
[90] K. Simonyan, A. Zisserman, Very deep convolutional networks for large-scale image recognition, in: Proceedings of the 3rd International Conference on Learning Representations, ICLR 2015, 2015, pp. 1–14.
[91] Z. Guo, Y. Sun, M. Jian, X. Zhang, Deep residual network with sparse feedback for image restoration, Applied Sciences (Switzerland) 8 (12) (2018) 2417.
[92] B. Rezazadeh, P. Asghari, A.M. Rahmani, Computer-aided methods for combating Covid-19 in prevention, detection, and service provision approaches, Neural Computing & Applications (2023) 1–40.
[93] C. Szegedy, W. Liu, Y. Jia, P. Sermanet, S. Reed, D. Anguelov, D. Erhan, V. Vanhoucke, A. Rabinovich, Going deeper with convolutions, in: Proceedings of the IEEE Computer Society Conference on Computer Vision and Pattern Recognition, vol. 07, 2015, pp. 1–9.
[94] Z. Zhu, C. Liu, D. Yang, A. Yuille, D. Xu, V-nas: neural architecture search for volumetric medical image segmentation, in: Proceedings of the International Conference on 3D Vision (3DV 2019), vol. 7, 2019, pp. 240–248.

[95] G. Huang, Z. Liu, L. Van Der Maaten, K.Q. Weinberger, Densely connected convolutional networks, in: Proceedings of the 30th IEEE Conference on Computer Vision and Pattern Recognition, CVPR 2017, 2017, pp. 2261–2269.
[96] M. Frid-Adar, I. Diamant, E. Klang, M. Amitai, J. Goldberger, H. Greenspan, GAN-based synthetic medical image augmentation for increased cnn performance in liver lesion classification, Neurocomputing 321 (2018) 321–331.
[97] Q. Dou, C. Ouyang, C. Chen, H. Chen, P.A. Heng, Unsupervised cross-modality domain adaptation of convnets for biomedical image segmentations with adversarial loss, in: Proceedings of the International Joint Conference on Artificial Intelligence, 2018, pp. 691–697.
[98] C. Huang, H. Han, Q. Yao, S. Zhu, S.K. Zhou, 3D U^2-Net: A 3D Universal U-Net for Multi-domain Medical Image Segmentation, Lecture Notes in Computer Science (Including Subseries Lecture Notes in Artificial Intelligence and Lecture Notes in Bioinformatics), vol. 11765 LNCS, 2019, pp. 291–299.
[99] Q. Yang, Y. Liu, T. Chen, Y. Tong, Federated machine learning: concept and applications, ACM Transactions on Intelligent Systems and Technology 10 (2) (2019) 1–19.
[100] J. Konečný, H.B. McMahan, F.X. Yu, P. Richtárik, A.T. Suresh, D. Bacon, Federated learning: strategies for improving communication efficiency, arXiv preprint, arXiv:1610.05492, 2016, pp. 1–16.
[101] Y. Zeng, Y. Mu, J. Yuan, S. Teng, J. Zhang, J. Wan, Y. Ren, Y. Zhang, Adaptive federated learning with non-iid data, Computer Journal 71 (7) (2022) 1–13.
[102] E. Bagdasaryan, A. Veit, Y. Hua, D. Estrin, V. Shmatikov, How to backdoor federated learning, in: Proceedings of the International Conference on Artificial Intelligence and Statistics, 2018, pp. 2938–2948.
[103] M.J. Sheller, G.A. Reina, B. Edwards, J. Martin, S. Bakas, Multi-Institutional Deep Learning Modeling Without Sharing Patient Data: a Feasibility Study on Brain Tumor Segmentation, Lecture Notes in Computer Science (Including Subseries Lecture Notes in Artificial Intelligence and Lecture Notes in Bioinformatics), vol. 11383 LNCS, 2019, pp. 92–104.
[104] W. Li, F. Milletarì, D. Xu, N. Rieke, J. Hancox, W. Zhu, M. Baust, Y. Cheng, S. Ourselin, M.J. Cardoso, A. Feng, Privacy-Preserving Federated Brain Tumour Segmentation, Lecture Notes in Computer Science (Including Subseries Lecture Notes in Artificial Intelligence and Lecture Notes in Bioinformatics), vol. 11861 LNCS, 2019, pp. 133–141.
[105] H.T. Nguyen, L.T. Nguyen, Fingerprints classification through image analysis and machine learning method, Algorithms 12 (11) (2019) 241.
[106] X. Li, Y. Gu, N. Dvornek, L.H. Staib, P. Ventola, J.S. Duncan, Multi-site fmri analysis using privacy-preserving federated learning and domain adaptation: abide results, Medical Image Analysis 65 (2020) 101765.
[107] W. Wang, M. Yousaf, D. Liu, A. Sohail, A comparative study of the genetic deep learning image segmentation algorithms, Symmetry 14 (10) (2022) 1977.
[108] C. Zhang, S. Ding, L. Guo, J. Zhang, Broad learning system based ensemble deep model, Soft Computing 26 (15) (2022) 7029–7041.
[109] L. Zhang, J. Li, G. Lu, P. Shen, M. Bennamoun, S.A.A. Shah, Q. Miao, G. Zhu, P. Li, X. Lu, Analysis and variants of broad learning system, IEEE Transactions on Systems, Man, and Cybernetics: Systems 52 (1) (2022) 334–344.

CHAPTER

Robust watermarking algorithm based on multimodal medical image fusion

4

Om Prakash Singh[a], Kedar Nath Singh[b], Amit Kumar Singh[c], Amrit Kumar Agrawal[d], and Huiyu Zhou[e]

[a]*Department of Computer Science and Engineering, Indian Institute of Information Technology Bhagalpur, Bihar, India*

[b]*Department of Computer Science and Engineering and Information Technology, Jaypee Institute of Information Technology Noida, UP, India*

[c]*Department of Computer Science and Engineering, National Institute of Technology Patna, Bihar, India*

[d]*Department of Computer Science and Engineering, School of Engineering and Technology, Sharda University, Greater Noida, India*

[e]*School of Computing and Mathematical Sciences, University of Leicester, Leicester, United Kingdom*

4.1 Introduction

Due to the rapid growth of high-speed internet and communication technology, multimedia content (image, text, audio, and video) has been easily disseminated through various online platforms [1]. However, multimedia content can be easily stored, shared, and tampered with by any unauthorized user [2]. Due to this, intruders can easily violate the security and privacy of multimedia information [3]. The concept of image fusion has been widely used in various applications such as healthcare and remote sensing [4]. Multimodal image fusion is a procedure that integrates two or more medical images into a single composite image that can then be used for medical diagnosis and assessment [5]. Following the global COVID-19 pandemic, most medical diagnoses have been made through online assessment. The transmission of a fused image over a public medium has become a critical task [6]. To address the abovementioned issues, one of the most well-known data-hiding concepts is watermarking, which is utilized to conceal secret information in multimedia objects in order to achieve copyright and identification [7]. The main motive of digital watermarking is to establish a secure system while maintaining essential features such as robustness, capacity, and visual quality [8].

Data Fusion Techniques and Applications for Smart Healthcare. https://doi.org/10.1016/B978-0-44-313233-9.00010-2

The generalized block diagram of a watermarking framework is listed in Fig. 4.1. A watermarking framework has two functions: the embedding and extraction of secret information. In the embedding process, secret data (a watermark image) is concealed in a multimedia object (cover image) with the help of the embedding function. After the embedding process, a marked image is obtained, and then it is transmitted through an open channel. Due to the availability of noise in the transmission medium, the marked image may be distorted. Further, using the extraction function, the extraction process is then carried out to obtain the recovery mark image.

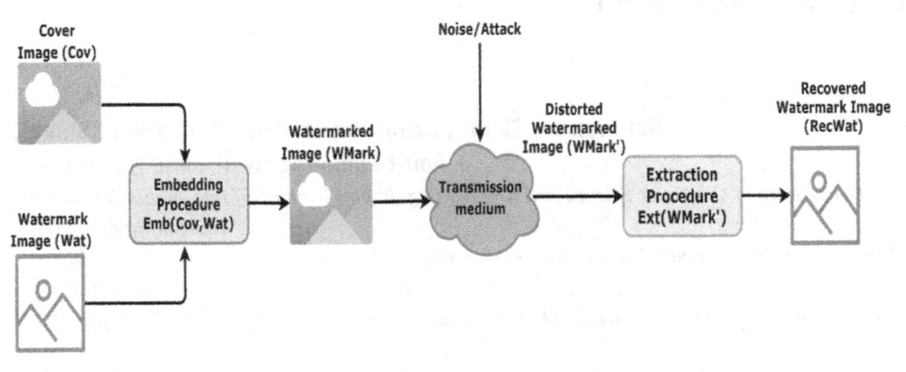

FIGURE 4.1

Generalized block diagram of watermarking framework.

The embedding and extraction procedure of the watermark system is described as follows:

$$Embedding(Cov, Wat) = WMark, \tag{4.1}$$

$$Extraction(WMark') = RecWat, \tag{4.2}$$

where Cov and Wat are the cover and mark images, respectively, $WMark$ and $WMark'$ are the marked and distorted mark images, respectively, and $RecWat$ is the recovery mark image.

4.2 Literature survey

This literature survey highlights some recent state-of-the-art techniques that are based on data-hiding techniques. To achieve the copyright protection of multimedia content, a data-hiding framework based on the integer wavelength transform (IWT), and singular value decomposition (SVD) has been discussed [9]. Firstly, the authors applied the stationary wavelet transform on multifocus images to obtain the fused image, which can act as a secret image. Further, a hybrid optimization framework was utilized to determine the optimal factor, which can maintain a good trade-off between invisibility and robustness. Moreover, the secret image was embedded inside the host image using IWT-SVD decomposition. However, the computational cost of this scheme was high. In [10], a scheme was introduced that can provide the security of medical data in the field of tele-healthcare applications. Firstly, a key-based encryption approach is utilized on the host image to enhance additional security. Then, a nonsubsampled contourlet transform (NSCT) is performed on multimodal images to obtain a fused image, which can act as a mark image. The mark image is concealed inside an encrypted host image using redundant discrete wavelet transform (RDWT) and randomized singular value decomposition (RSVD). Lastly, a denoising operation is performed on the recovery mark image to enhance the additional robustness of this scheme. In [11], Bhardwaz et al. presented a reversible data-hiding framework for medical data that aimed to improve the payload of the system. First, the authors encrypted both the host and the mark image using Paillier cryptography to ensure the security of this scheme. Further, the encrypted mark image was concealed inside an encrypted host media to offer better visual quality. This scheme offers high robustness along with a better payload. In a similar way, Vaidya et al. [12] proposed a scheme that ensures the privacy of medical content. In the embedding process, host media was transformed through multiple transforms (IWT-DWT), which offered high resistance along with better visual quality. Further, Arnold's cat map was applied to the marked image to ensure the robustness of this scheme. However, the security of this scheme was not sufficient. Sayah et al. [13] mentioned a data-hiding concept that ensures the integrity of secret information. Prior to the embedding operation, patient information was compressed using DWT to offer a high payload. Further, the RC4 encryption scheme was utilized on the marked image to provide additional security. This method offers a high payload along with acceptable visual quality, and it is highly resistant to a wide range of attacks. Priyanka et al. [14] introduced a DWT-SVD-based dual watermarking approach for e-healthcare applications. Initially, the host media is transformed into two parts; the region of interest (ROI) part of the host image is used for embedding. Patient information is compressed using Hamming code. The compressed data, along with the QR code, are hidden inside the selected part of the host media. Furthermore, the coding scheme is utilized on marked images to ensure high security. It offers a high payload along with more resistance against selected attacks. To ensure the security of patient data, a data-hiding framework in a transform domain is listed in [15]. Firstly, the fused mark image is obtained using nonsubsampled shearlet transform (NSST). Before the embedding, the fused mark is encrypted using Arnold's cat map to provide additional security. Encrypted data are embedded into a transformed host image using wavelet transform and RSVD. This scheme offers high robustness against listed attacks.

Although the above watermarking approaches have made significant advances in the image fusion environment, most of them have limited performance and cannot efficiently offer a balanced relationship between invisibility and watermark robustness. In addition, the security of most of the watermarking techniques is not very satisfactory. Objectives along with limitations of the existing state-of-the-art techniques are presented in Table 4.1.

Table 4.1 Identified issues of existing works.

Existing work	Objective	Methodology	Remarks
[9]	Develop an optimization-based data-hiding scheme in transform domain	IWT, SVD, PSO, and FA	Computational cost is high for determining the optimal value
[10]	Propose a fusion-based robust data-hiding scheme for healthcare systems	NSCT, RDWT, RSVD, and DcNN	The embedding capacity is low
[11]	Implement the data-hiding framework for e-healthcare	Transform domain and Paillier encryption	The computational time for the embedding process is high
[12]	Design a robust watermarking scheme for fingerprint images	IWT, DWT, LBP, and Arnold transform	The security analysis is limited
[13]	Develop secure watermarking for telemedicine	DWT and RC4 encryption	This scheme tested the results against limited attacks.
[14]	Introduce robust and secure watermarking for IoMT	DWT-SVD and LSB	The payload of this scheme is very low
[15]	Develop a data-hiding approach for securing healthcare systems	NSST, RSVD, PCA, Fr-DTCWT, and Arnold transform	Arnold transform is not sufficient to provide high security.

In this work, a robust and secure data-hiding approach is developed to secure the e-healthcare system. The major contribution of this work is highlighted below.

- **The significance of image fusion**: First, multimodal images, such as MRI and PET images, are fused using NSST [16]. The fused image contains rich information about the MRI and PET images. In healthcare, the fused image is preferred over individual images for diagnostic and assessment purposes.
- **Hiding fused image using IWT-QR-SVD**: Data hiding is performed using IWT-QR-SVD, which ensures better robustness along with high visual quality. IWT offers better reconstruction features and it requires significantly less memory and computational time than DWT [9]. SVD has been widely used in data-hiding approaches due to its various advantages, including high stability and robustness [17].
- **Enhanced security**: The 3D chaotic map-based encryption approach has been utilized on marked images to enhance the security of this scheme.
- **High robustness**: The proposed scheme offers high robustness against a wide range of attacks.

The remainder of this chapter is organized as follows. Firstly, we introduce the background information related to NSST, IWT, QR, and SVD. The next section describes multimodal fusion, the encryption of marked images, and the embedding and recovery procedure of fused mark images. Furthermore, we demonstrate the results of the proposed method. Lastly, we provide conclusions and future remarks.

4.3 Background information

4.3.1 NSST

NSST is a second-generation wavelet transform in the field of image processing [16]. NSST has several properties, such as shift invariance, multiscaling, and multidirectionality, which overcome the limitations of first-generation wavelet transforms such as DWT and LWT [16]. An NSST operation can be performed in two stages using a nonsubsampled pyramid filter bank (NSPFB) and a nonsubsampled direction filter bank (NSDFB) [17]. Initially, NSPFB is utilized to achieve multiscale properties. Further, NSDFB is applied to display the image in multiple directions, which show detailed information about the image. In Fig. 4.2, NSST was applied to transform the cover image into lower and directional subbands, which are crucial for embedding purposes.

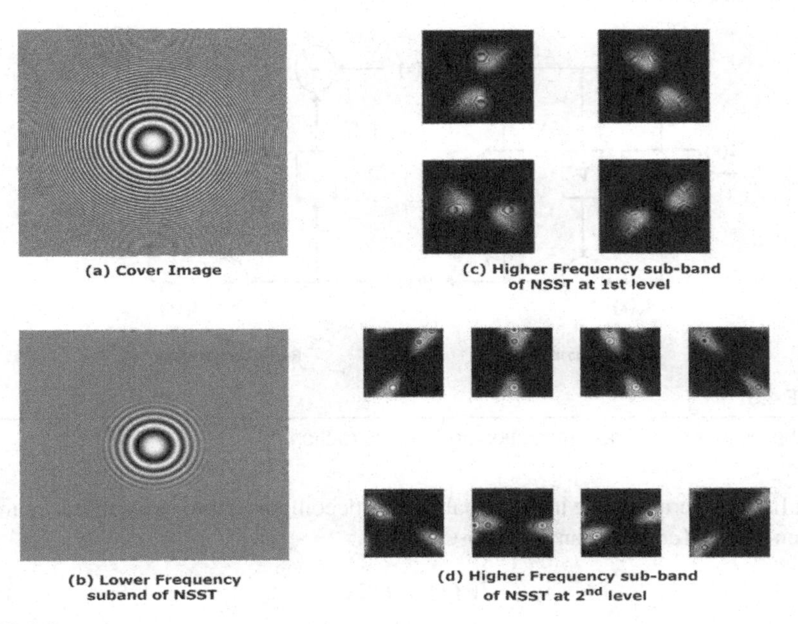

FIGURE 4.2

(a) Cover image. (b) Lower coefficients of NSST. (c, d) High-frequency coefficients of NSST at the first and second levels.

The decomposition and reconstruction filters of NSPFB are denoted as $D(z)$ and $R(z)$, respectively. The perfect reconstruction of NSPFB is achieved through the following equation:

$$D_0(z)\, R_0(z) + D_1(z)\, R_1(z) = 1. \tag{4.3}$$

In NSDFB, decomposition filters are denoted as $U_0(z)$ and $U_1(z)$, respectively. Synthesis filters are described as $W_0(z)$ and $W_1(z)$. They provide a reconstructed image which has a similar size to the original image using the following equation:

$$U_0(z)\,W_0(z)+U_1(z)\,W_1(z)=1. \tag{4.4}$$

4.3.2 IWT

To solve the limitations of DWT, Sweldnes introduced IWT in 1998 [9]. Recently, IWT has become a powerful approach that is used in several applications, such as image compression, watermarking, and pattern recognition [18]. It falls under the class of second-generation wavelet transforms, which have several advantages over conventional first-generation wavelet transforms. Compared to DWT, the split and merge operations of this transform reduce the computational complexity by 50%. The lifting approach is performed through three basic steps, split, predict, and update, as described in Fig. 4.3.

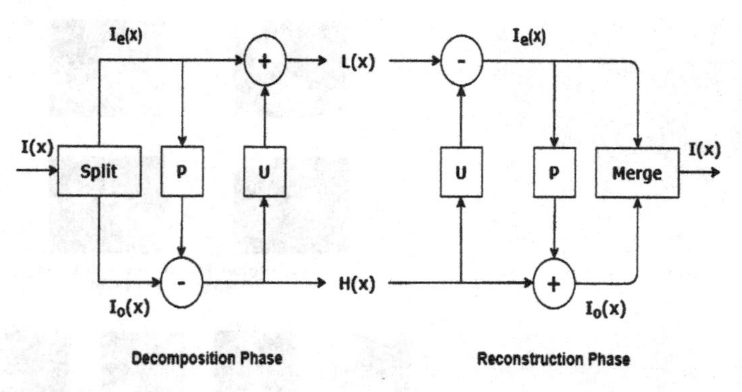

FIGURE 4.3

IWT phases in terms of decomposition and reconstruction.

Split: In this operation, the input signal $I(x)$ is decomposed into two subsets, namely, an even subset $Ie(x)$ and an odd subset $Io(x)$:

$$Ie(x)=I(2x), \tag{4.5}$$

$$Io(x)=I(2x+1). \tag{4.6}$$

Predict: This operation is also termed dual lifting. In this step, the odd subset can be predicted from the neighboring even subset. The error is calculated as the difference between the odd subset and the even subset with the help of the prediction operator P. The high-pass coefficient $H(x)$ is also known as error and is calculated as follows:

$$H(x)=Io(x)-P[Ie(x)]. \tag{4.7}$$

Update: This step is also known as primal lifting. The low-pass coefficient $L(x)$ is obtained by updating the value of the even subset with the help of the upgrade operator U. The update operation is given as follows:

$$L(x) = Ie(x) + U[H(x)]. \tag{4.8}$$

IWT is performed on a brain image to obtain the approximation (LL), horizontal (HL), vertical (LH), and diagonal (HH) subbands, as shown in Fig. 4.4. In this figure, we can observe that the maximum energy of the input image is available in the approximation (LL) subband. In the watermarking process, the LL subband offers good robustness against various types of noise and attacks. However, it does not offer better visual quality. If embedding is done in the diagonal (HH) subband, it offers better visual quality, but it does not resist against attacks.

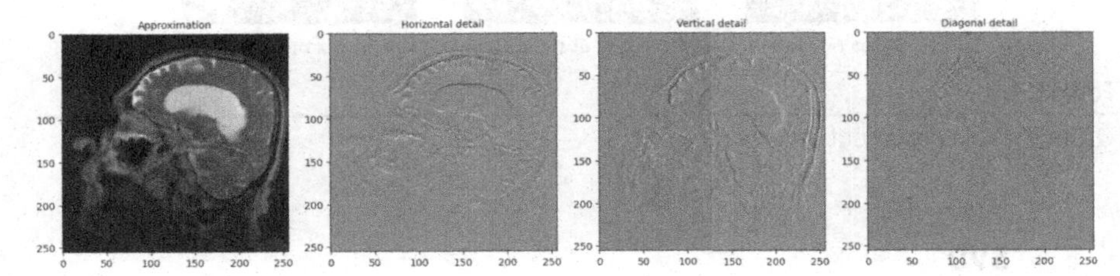

FIGURE 4.4

IWT decomposition of a brain image.

4.3.3 QR decomposition

QR decomposition has been utilized in various fields of image processing, including image compression and image encryption [19]. It is defined as follows:

$$A_{m,n} = Q_{m,m} \times R_{m,n}, \tag{4.9}$$

where the matrix A is decomposed into two matrices Q and R. Matrices Q and R are orthogonal and upper triangular matrices, respectively. Fig. 4.5 indicates the QR factorization of the selected image. For example, the pixel value of matrix A with size 4×4 obtained from the input brain image is given as follows:

$$A = \begin{bmatrix} 169 & 157 & 167 & 160 \\ 130 & 33 & 131 & 160 \\ 80 & 44 & 126 & 133 \\ 83 & 87 & 126 & 115 \end{bmatrix}, \tag{4.10}$$

$$Q = \begin{bmatrix} -0.6972 & 0.4900 & 0.4589 & -0.2514 \\ -0.5363 & -0.7718 & 0.1711 & 0.2955 \\ -0.3301 & -0.1649 & -0.6753 & -0.6386 \\ -0.3424 & 0.3701 & -0.5514 & 0.6646 \end{bmatrix}, \tag{4.11}$$

$$R = \begin{bmatrix} -242.3840 & -171.4800 & -271.4329 & -280.6497 \\ 0 & 76.4043 & 6.5793 & -24.4594 \\ 0 & 0 & -55.5236 & -52.4344 \\ 0 & 0 & 0 & -1.4508 \end{bmatrix}. \tag{4.12}$$

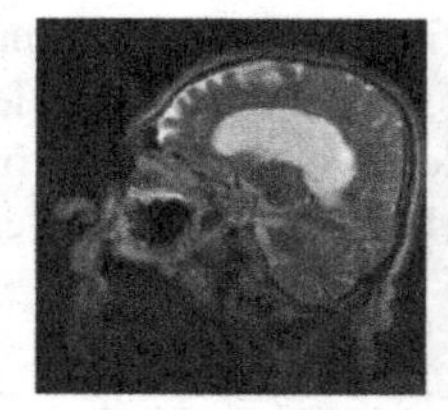
(a) Input Image

(b) Q matrix of Input Image

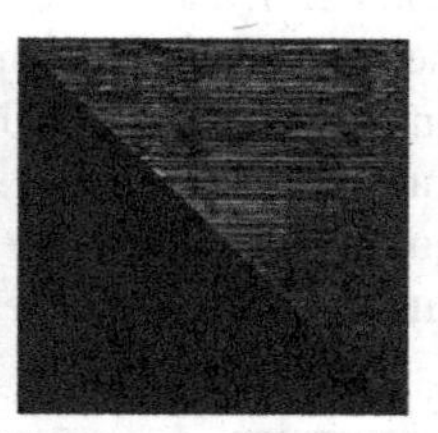
(c) R matrix of Input Image

FIGURE 4.5

QR factorization of input image.

4.3.4 SVD

SVD is a subtype of matrix decomposition in linear algebra [20]. It can be applied in various image processing applications. It is represented as follows:

$$A = U \times S \times V, \tag{4.13}$$

where matrices U and V have orthonormal columns known as the left and right singular vectors of A, respectively, and S is a diagonal matrix with real nonnegative eigenvalues in a nonincreasing manner. SVD is a popular choice for watermarking techniques due to the following reasons [17]:

- It is highly stable since no significant changes in visual quality are observed upon applying minor changes to its singular matrix, hence confirming its high imperceptibility.
- It can be applied on matrices of any size, either square or rectangular.
- SVD has good energy compaction properties, making it suitable for the watermarking process.

For illustration purposes, the pixel value of matrix A with size 4×4 obtained from the input image is presented below:

$$A = \begin{bmatrix} 104.6010 & 42.2519 & 56.5573 & 106.6036 \\ 100.1120 & 43.6523 & 63.5101 & 141.4542 \\ 83.2060 & 81.4629 & 45.2070 & 118.1840 \\ 24.4301 & 56.1044 & 83.4849 & 93.2716 \end{bmatrix}, \tag{4.14}$$

$$U = \begin{bmatrix} -0.5002 & 0.3804 & -0.1682 & -0.7585 \\ -0.5730 & 0.2440 & -0.4903 & 0.6096 \\ -0.5180 & 0.0334 & 0.8370 & 0.1729 \\ -0.3912 & -0.8906 & -0.1752 & -0.1522 \end{bmatrix}, \tag{4.15}$$

$$S = \begin{bmatrix} 327.1756 & 0 & 0 & 0 \\ 0 & 61.3820 & 0 & 0 \\ 0 & 0 & 34.4322 & 0 \\ 0 & 0 & 0 & 16.9048 \end{bmatrix}, \tag{4.16}$$

$$V = \begin{bmatrix} -0.4962 & 0.7404 & -0.0381 & -0.4518 \\ -0.3415 & -0.3120 & 0.8616 & -0.2088 \\ -0.3658 & -0.5921 & -0.4872 & -0.5275 \\ -0.7094 & -0.0624 & -0.1369 & 0.6885 \end{bmatrix}. \tag{4.17}$$

4.4 Proposed methodology

The proposed method has been categorized into four major parts: (a) multimodal image fusion, (b) the embedding and recovery process of a fused mark image, (c) encryption, and (d) the decryption process of a marked image. A complete flowchart diagram of the image fusion, and proposed method is shown in Fig. 4.6 and Fig. 4.7,

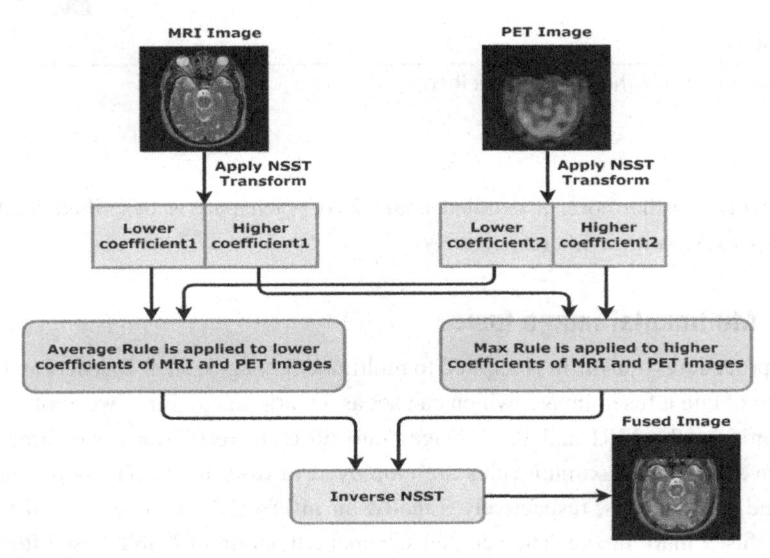

FIGURE 4.6

NSST-based image fusion approach.

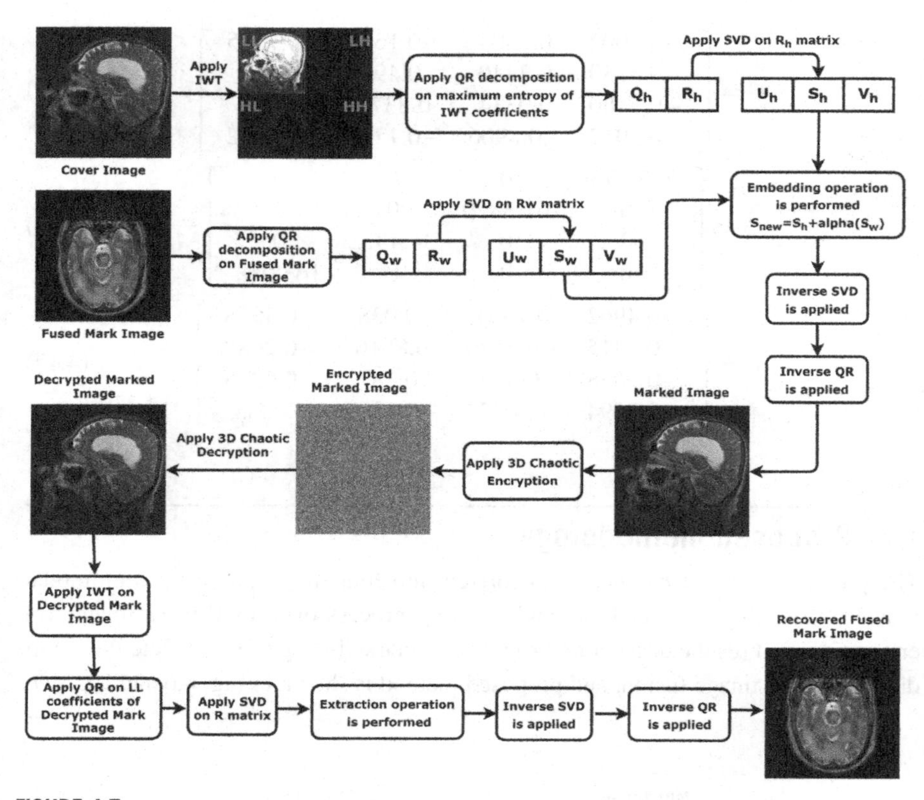

FIGURE 4.7

Complete diagram of the proposed method.

respectively. Furthermore, a detailed analysis of each part is described in Algorithm 4.1 to Algorithm 4.6, respectively.

4.4.1 Multimodal image fusion

In this procedure, transform is applied to multimodal images, such as MRIs and PET scans, to obtain a fused image, which can act as a mark image. First, we apply NSST to decompose the MRI and PET images into different resolutions and directions. Then, average and maximum rules are employed to fuse the NSST coefficients of MRI and PET images, respectively. Finally, an inverse NSST is performed to obtain the fused mark image. The detailed schematic diagram of NSST-based fusion is shown in Fig. 4.6. Detailed information on the multimodal image fusion approach is given in Algorithm 4.1.

Algorithm 4.1: Multimodal image fusion

Input: MRI_{Img}, PET_{Img}
Output: $Fused_{Img}$
Begin
1. $Img_1 = imresize(MRI_{Img}, [256, 256]);$
2. $Img_2 = imresize(PET_{Img}, [256, 256]);$
3. $NSST_{coef1} = NSST\left(Img_1, P_{filter}, D_{filter}, level = 1\right);$
4. $NSST_{coef2} = NSST\left(Img_1, P_{filter}, D_{filter}, level = 1\right);$
5. $Low_{coef1} = NSST_{coef1}\{1\};\ Low_{coef2} = NSST_{coef2}\{1\};$
6. $[M_1, M_2] = Size(Low_{coef1})]$
7. $[M_3, N_3, O_3, P_3, Q_3] = zeros([M_1, M_2]);$
8. *for* a: 1 to M_1 *do*
9. *for* b: 1 to M_2 *do*
10. $M_3\,(a, b) = [M_1\,(a, b) + M_2(a, b)]/2;$
11. *end for*
12. *end for*
13. $High_{coef11} = NSST_{coef1}\{1, 1\};\ High_{coef12} = NSST_{coef1}\{1, 2\};$
14. $High_{coef13} = NSST_{coef1}\{1, 3\};\ High_{coef14} = NSST_{coef1}\{1, 4\};$
15. $High_{coef21} = NSST_{coef1}\{1, 1\};\ High_{coef22} = NSST_{coef1}\{1, 2\};$
16. $High_{coef23} = NSST_{coef1}\{1, 3\};\ High_{coef24} = NSST_{coef1}\{1, 4\};$
17. *for* a: 1 to M_1 *do*
18. *for* b: 1 to M_2 *do*
19. $N_3\,(a, b) = max[High_{coef11}\,(a, b) + High_{coef21}\,(a, b)];$
20. $O_3\,(a, b) = max[High_{coef12}\,(a, b) + High_{coef22}\,(a, b)];$
21. $P_3\,(a, b) = max[High_{coef13}\,(a, b) + High_{coef23}\,(a, b)];$
22. $Q_3\,(a, b) = max[High_{coef14}\,(a, b) + High_{coef24}\,(a, b)];$
23. *end for*
24. *end for*
25. $NSST_{coefnew} = [M_3, N_3, O_3, P_3, Q_3];$
26. $Fused_{Img} = NSST(NSST_{coefnew}, P_{filter}, D_{filter}, level = 1)$

Return $Fused_{Img}$

4.4.2 Embedding and recovery procedure of fused mark image

Firstly, the host image is transformed using IWT, and then the maximum entropy is chosen from the IWT coefficients of the host image. QR and SVD decomposition is then performed on the maximum entropy coefficients of the host image. Similarly, a fused image is also transformed using QR and SVD decomposition. Furthermore, the fused mark image is concealed inside the IWT coefficient of the host image by modifying the singular value of the fused mark image. Lastly, inverse SVD, QR, and IWT are utilized to obtain the marked image. In the recovery process, IWT is utilized to decompose the marked image. In addition, QR and SVD decomposition is also performed on a selected IWT coefficient of the marked image to obtain the recovery mark image.

Algorithm 4.2: Embedding procedure

Input: $Cov_{Img}, Fused_{Img}, Gain_{val}$
Output: $Marked_{Img}$
Begin
1. $Cov_{Img} = imresize(Cov_{Img}, [512, 512]);$
2. $Fused_{Img} = imresize(Fused_{Img}, [256, 256]);$
3. $[LL_c, LH_c, HL_c, HH_c] = IWT\left(Cov_{Img}, 'Int2Int'\right);$
4. $LL_1 = entropy\,(LL_c)\,;\,LL_2 = entropy\,(LH_c);$
5. $LL_3 = entropy\,(HL_c) : LL_4 = entropy\,(HH_c);$
6. $LL_{new} = maximum\,(LL_1, LL_2, LL_3, LL_4);$
7. $[Q_{new}, R_{new}] = QR\,(LL_{new});$
8. $[U_{new}, S_{new}, V_{new}] = SVD\,(R_{new});$
9. $[Q_{wat}, R_{wat}] = QR\left(Fused_{Img}\right);$
10. $[U_{wat}, S_{wat}, V_{wat}] = SVD\,(R_{wat});$
11. $S_{11} = S_{new} + Gain_{val} \times S_{wat};$
12. $S_{12} = U_{new} \times S_{11} \times (V_{new})^T;$
13. $R_{11} = Q_{new} \times S_{12};$
14. $Marked_{Img} = InverseIWT\left(R_{11}, LH_c, HL_c, HH_c, 'Int2Int'\right);$

Return $Marked_{Img}$

Algorithm 4.3: Recovery procedure

Input: $Dec_{Img}, Gain_{val}$
Output: Rec_{wat}
Begin
1. $[LL_m, LH_m, HL_m, HH_m] = IWT\left(Dec_{Img}, 'Int2Int'\right);$
2. $[Q_m, R_m] = QR\,(LL_m);$
3. $[U_m, S_m, V_m] = SVD\,(R_m);$
4. $S_{m1} = (S_m - S_{new})/Gain_{val};$
5. $S_{12} = U_{wat} \times S_{m1} \times (V_{wat})^T;$
6. $Rec_{wat} = Q_{wat} \times S_{12};$

Return Rec_{wat}

4.4.3 Encryption procedure

Encryption of the marked image is performed by utilizing the Lorenz map [21]. The Lorenz map is a popular chaotic system, notably for its picture encryption. It is defined as

$$\begin{cases} X_{n+1} = \rho\left(Y_n - X_n\right), \\ Y_{n+1} = \left(\alpha - Z_n\right) X_n - Y_n, \\ Z_{n+1} = X_n \times Y_n - \beta \times Z_n, \end{cases} \tag{4.18}$$

where the chaos of the Lorenz map is controlled by parameters ρ, α, and β and X_n, Y_n, Z_n and X_{n+1}, Y_{n+1}, Z_{n+1} are the previous and next sequence of the map, respectively. The Lorenz map exhibits chaotic characteristics when $\rho = 10$, $\alpha = 28$, and $\beta = 8/3$. Its positive Lyapunov exponents illustrate the sensitivity of the map to initial values as well as chaotic characteristics. A sophisticated confusion and diffusion process is adopted to encrypt the marked image. The proposed encryption scheme has three processes, i.e., key initialization, confusion, and diffusion.

Key initialization

A key of 256 bits is chosen to be the secret key, and this key is then used for initializing the encryption and decryption keys. To generate initial parameters, 8 parts of 32 bits each are taken, which can be described as follows:

$$K_i = \sum_{j=0}^{8} 2^{8-j} b_{8j+i}, \tag{4.19}$$

where b_i is the bit at index 0 in the key. After these parts are created, they are combined with other parts using Eqs. (4.20) and (4.21) to form three parameters that will serve as initial parameters to the Lorenz map. However, there is a possibility that a transient effect may occur. These parameters are iterated 20 times using the Lorenz map in order to mitigate this effect. The key initialization process is summarized in Algorithm 4.4. We have

$$P = K1 \oplus K2 \oplus K3 \oplus K4 \oplus K5 \oplus K6 \oplus K7 \oplus K8, \tag{4.20}$$

$$P_i = \begin{cases} K1\ AND\ K2 \oplus P & i = 1, \\ K4\ OR\ K5 \oplus P & i = 2, \\ K7\ AND\ K8\ OR\ K6 \oplus P & i = 3. \end{cases} \tag{4.21}$$

The proposed encryption process involves two rounds of confusion and diffusion operations. Confusion is the process of shifting the pixels from their original location. Hence, the pixels are shuffled along the matrix, and the image becomes unreadable by perception. In the proposed algorithm, confusion is done using a matrix generated by the Lorenz map. In this process, image pixels are selected sequentially using sorted indices of the Lorenz map. This process can be represented as

$$X_c(i) = X(R(i)), \tag{4.22}$$

where $R(i)$ is the sorted indices of the Lorenz map and $X_c(i)$ is the confused image sequence. The diffusion process entails changing the pixel intensity near random pixels, thus mitigating any information attackers can gain. To create an avalanche effect, we have taken previous rows and columns to diffuse into the current row and

Algorithm 4.4: Key initialization

Input: Key (256 bits)
Output: Initial parameters P1, P2, P3
Begin
1. $K_1,\ K_2,\ K_3,\ K_4,\ K_5,\ K_6,\ K_7,\ K_8 = Initialize\ from\ Eq.$ (4.1)
2. $P = K1 \oplus K2 \oplus K3 \oplus K4 \oplus K5 \oplus K6 \oplus K7$
3. $P1 = K1\ AND\ K2 \oplus P$
4. $P2 = K4\ OR\ K5 \oplus P$
5. $P3 = K7\ AND\ K8\ OR\ K6 \oplus P$
6. $for\ i = 0\ to\ n = 20\ do$
7. $\quad P1,\ P2,\ P3 = LZ(P1,\ P2,\ P3)$
8. $endfor$

Return P1, P2, P3

column, along with the row and column of the Lorenz matrix R. After this process, pixel values are in the range of [0, 768). Hence, we need to map these values to the range [0, 256). We have used a mod operation to map these values to their respective ranges. This process is described as follows:

$$X(i,:) = mod\left(X(i,:) + X(i-1,:) + R(i,:), 256\right), \tag{4.23}$$

$$X(:,i) = mod\left(X(:,i) + X(:,i-1) + R(:,i), 256\right), \tag{4.24}$$

where $X(:,i)$, $X(i,:)$, $R(:,i)$, and $R(i,:)$ represent the i_{th} row and column of the confused image and the Lorenz matrix. These processes are iterated twice to achieve adequate security against crucial and critical attacks. The overall process of encryption is described in Algorithm 4.5.

4.4.4 Decryption process

The decryption process is the exact opposite of the encryption procedure. First, the cypher image is dediffused row by row and column by column. The complimentary equations to the diffusion method described in the encryption scheme are given as Eqs. (4.25) and (4.26). This process maps pixels to their original values, hence removing the effect of diffusion. After dediffusing, the image is deconfused, yielding the original plain image. We have

$$X(i,:) = mod\left(X(i,:) - X(i-1,:) - R(i,:), 256\right), \tag{4.25}$$

$$X(:,i) = mod\left(X(:,i) - X(:,i-1) - R(:,i), 256\right), \tag{4.26}$$

where $X(:,i)$, $X(i,:)$, $R(:,i)$, and $R(i,:)$ represent the i_{th} row and column of the encrypted image and the Lorenz matrix. These processes are applied two times to reconstruct the image from the encrypted image. The overall process of decryption is illustrated in Algorithm 4.6.

Algorithm 4.5: Encryption

Input: $Marked_{Img}$, $Lorenz\ matrix$
Output: Enc_marked_{Img}

Begin
1. $W, H = Size(Marked_{Img})$
2. $img = Flatten(Marked_{Img})$
3. $K = Flatten(mat)$
4. $[\sim, indices] = sort(K)$
5. $for\ r = 0\ to\ 2\ do$
6. $\quad img = img(indices)$
7. $\quad Img = reshape(img, [w, h])$
8. $\quad for\ k = 0\ to\ 2\ do$
9. $\quad\quad for\ i = 0\ to\ w\ do$
10. $\quad\quad\quad Img(i, :) = (img(i, :) + img(i - 1, :) + mat(i, :))\ mod\ 256$
11. $\quad\quad endfor$
12. $\quad\quad for\ i = 0\ to\ h\ do$
13. $\quad\quad\quad Img(:, i) = (img(:, i) + img(:, i - 1) + mat(:, i - 1))\ mod\ 256$
14. $\quad\quad endfor$
15. $\quad endfor$
16. $endfor$

Return Enc_marked_{Img}

Algorithm 4.6: Decryption process

Input: Enc_marked_{Img}, $Lorenz\ matrix$
Output: Dec_marked_{Img}

Begin
1. $W, H = Size(img)$
2. $img = Flatten(img)$
3. $K = Flatten(mat)$
4. $[\sim, indices] = sort(K)$
5. $[\sim, indices] = sort(indices)$
6. $for\ r = 0\ to\ 2\ do$
7. $\quad for\ k = 0\ to\ 2\ do$
8. $\quad\quad for\ i = h\ to\ 0\ do$
9. $\quad\quad\quad Img(:, i) = (img(:, i) - img(:, i - 1) - mat(:, i - 1))\ mod\ 256$
10. $\quad\quad endfor$
11. $\quad\quad for\ i = w\ to\ 0\ do$
12. $\quad\quad\quad Img(i, :) = (img(i, :) - img(i - 1, :) - mat(i, :))\ mod\ 256$
13. $\quad\quad endfor$
14. $\quad endfor$
15. $\quad img = img(indices)$
16. $\quad Img = reshape(img, [w, h])$
17. $endfor$

Return Dec_marked_{Img}

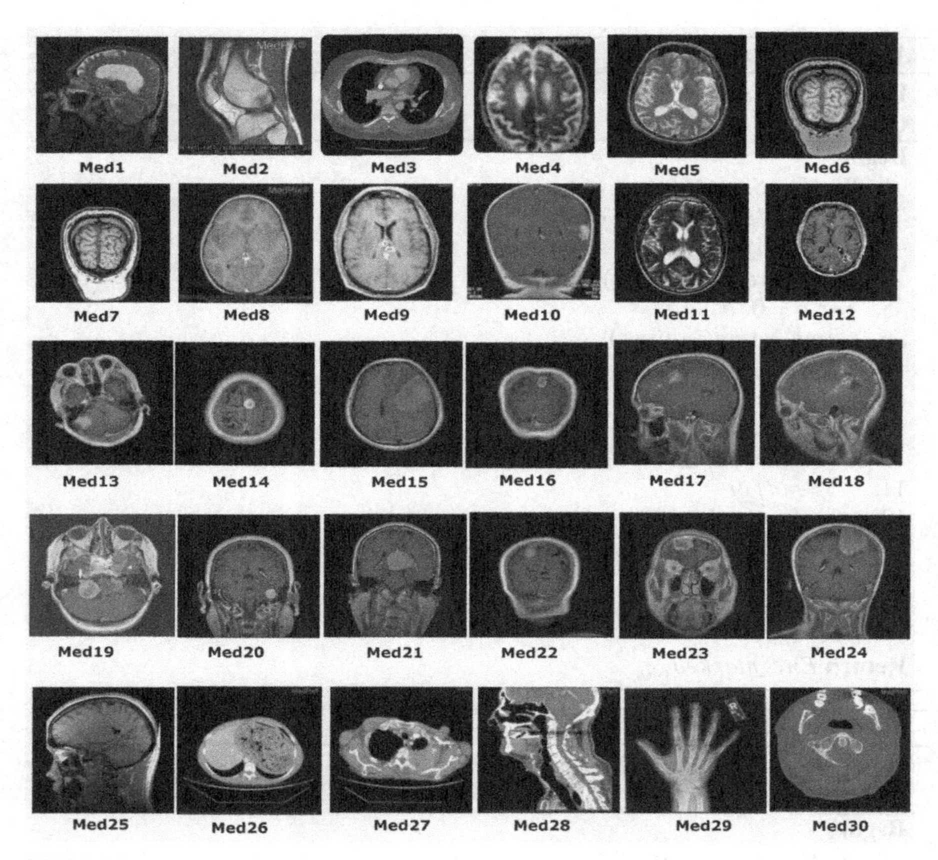

FIGURE 4.8

Sample cover image for experiment purposes.

4.5 Results

In this section, we briefly describe various experimental analyses to measure the effectiveness of the proposed method. The entire experimental analysis was simulated on a personal computer with an i5 processor and 8 GB of RAM using the MATLAB® 2018a environment. To evaluate the performance of the proposed method, we used grayscale images sized 512×512 and 256×256 for the cover and fused mark images, respectively. Fig. 4.8 indicates the cover images used for the experimental analysis. In addition, we have simulated all the results of the proposed method on the standard Kaggle and Open-i datasets, respectively [22,23]. To measure the visual quality along with the robustness and security of the proposed method, we evaluated the results using objective, robustness, statistical, differential, key space, noise, and computational cost analyses.

Table 4.2 Objective analysis of the proposed method at various gain values.

Gain value	PSNR (dB)	SSIM	NC
0.001	63.8638	0.9998	1.0000
0.01	57.8590	0.9984	1.0000
0.03	55.2707	0.9869	1.0000
0.05	47.5315	0.9647	1.0000
0.07	45.3324	0.9483	1.0000
0.09	43.8024	0.9126	1.0000
0.1	42.7057	0.9068	1.0000

4.5.1 Objective analysis (PSNR-SSIM-NC)

Two metrics – peak signal-to-noise ratio (PSNR) and the structural similarity index (SSIM) – were utilized to measure the visual quality features of a watermark system, and the robustness feature was measured using normalized correlation (NC) [24]. From Table 4.2, it can be observed that the PSNR, SSIM, and NC scores of the proposed method are recorded at different gain values. The PSNR score is ≥42.7057 dB and the SSIM score is close to the ideal value 1.000, indicating that this scheme has good invisibility performance. Furthermore, the robustness score is found to be NC = 1.000 (without attacks), indicating that this scheme has high resistance features. The highest PSNR and SSIM scores are recorded as 63.8638 dB and 0.9998, respectively, at gain value = 0.001. Table 4.3 indicates the objective analysis of the proposed method for 10 different medical images at gain value = 0.03. From this table, we obtained the acceptable invisibility and robustness scores for 10 different cover images. We also compared the average score (PSNR-SSIM-NC) for two differ-

Table 4.3 Objective analysis for different cover images (gain value = 0.03).

Tested image	PSNR (dB)	SSIM	NC
Med1	57.2675	0.9970	1.0000
Med2	56.4481	0.9969	1.0000
Med3	57.3470	0.9970	1.0000
Med4	55.7624	0.9967	1.0000
Med5	53.4494	0.9969	1.0000
Med6	57.9032	0.9968	1.0000
Med7	53.2985	0.9968	1.0000
Med8	55.2096	0.9968	1.0000
Med9	56.6453	0.9969	1.0000
Med10	54.4678	0.9969	1.0000
Avg. of Kaggle dataset	53.2460	0.9924	1.0000
Avg. of Open-i dataset	54.8924	0.9936	1.0000

ent datasets, Kaggle and Open-i, at gain value = 0.03. An objective analysis of the proposed method was additionally conducted with 30 different medical images, as shown in Figs. 4.9 and 4.10.

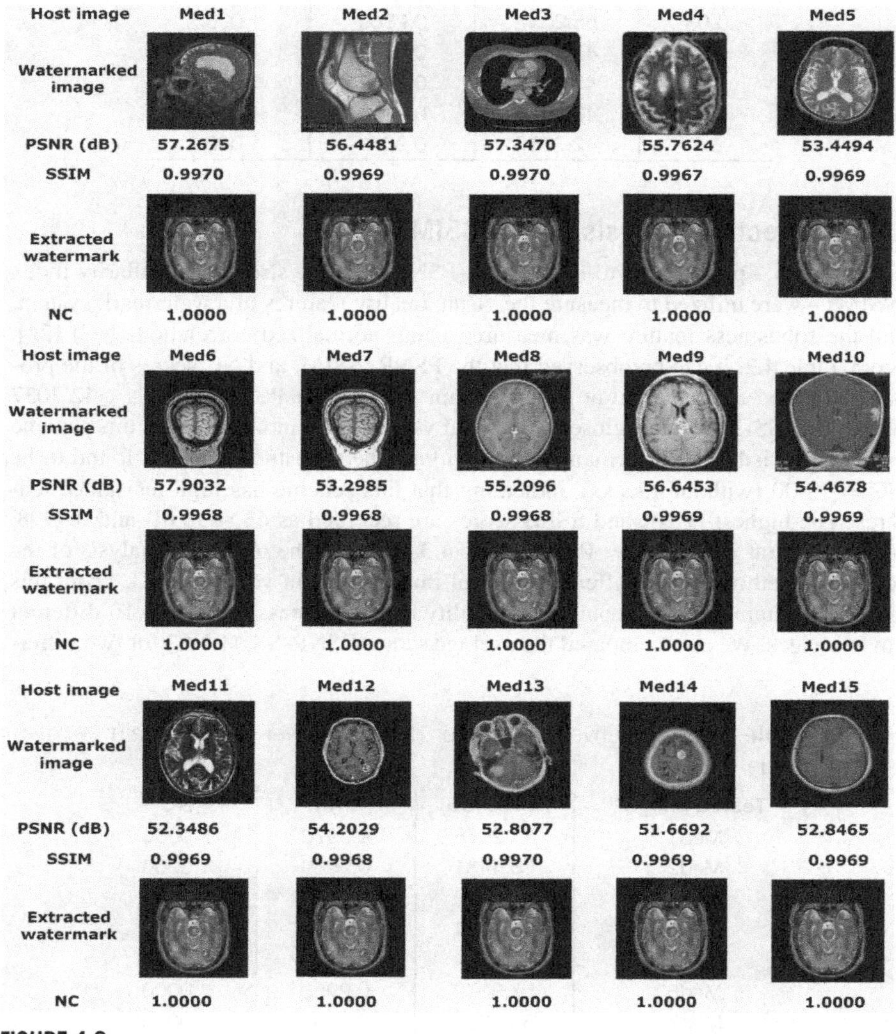

FIGURE 4.9

Objective analysis (PSNR-SSIM-NC) of the proposed method for 15 different medical images (med1–med15).

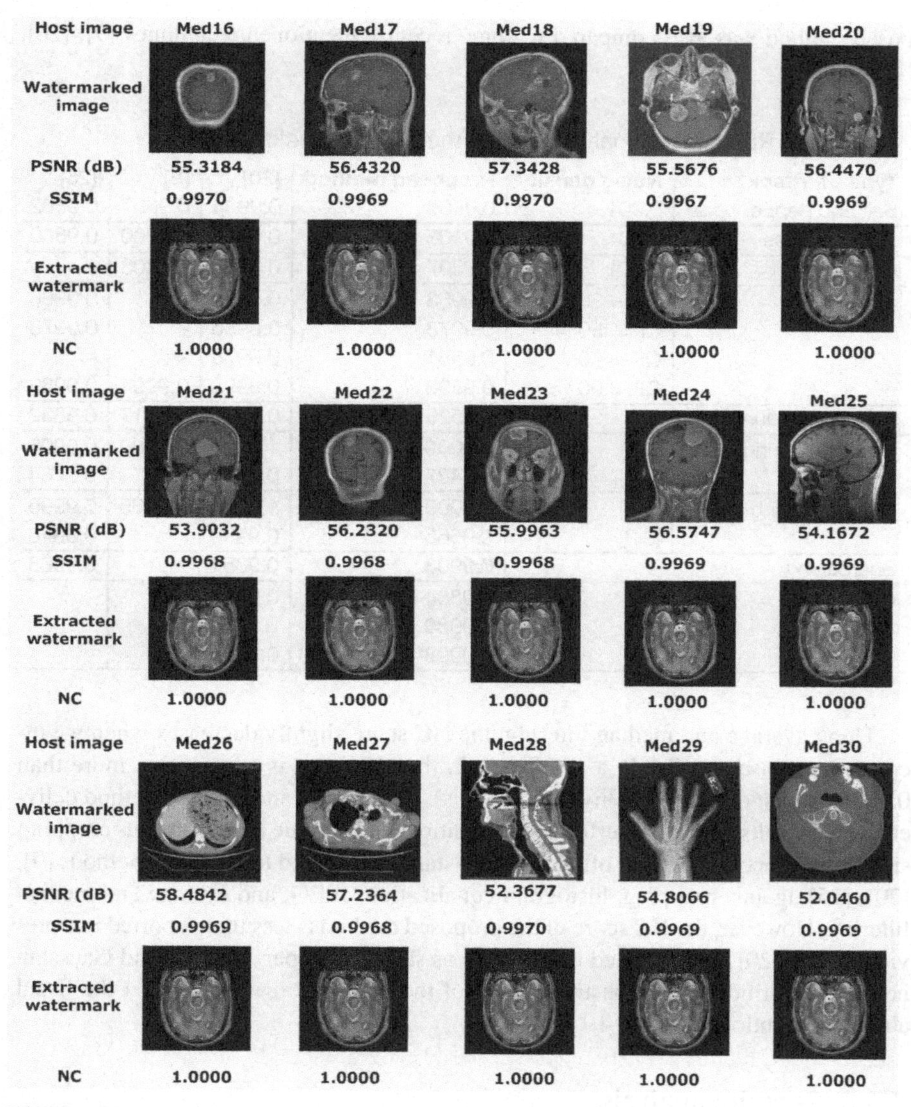

FIGURE 4.10

Objective analysis (PSNR-SSIM-NC) of the proposed method for 15 different medical images (med16–med30).

4.5.2 Robustness analysis

To measure the robustness of the proposed system, we evaluated the NC score against the listed attacks in Table 4.4. According to this table, it can be concluded that the proposed method is highly resistant to the listed attacks. The NC score of the proposed method was also compared to some recently mentioned techniques [9], [20], [25].

Table 4.4 Robustness analysis against mentioned attacks.

Type of attack	Noise density	Proposed method	[20]	[9]	[25]
Salt and pepper noise	0.001	0.9806	0.9934	0.9687	0.9362
Speckle noise	0.001	0.9906	0.9936	0.9800	0.9360
Gaussian noise	0.001	0.9897	0.9921	0.9803	-
JPEG	QF = 30	0.9953	0.9910	-	0.9966
	QF = 50	0.9973	0.9956	-	0.9978
	QF = 70	0.9981	0.9978	-	-
	QF = 90	0.9998	0.9998	0.9994	0.9989
Histogram equalization		0.9526	0.9432	0.9117	0.9832
Average filtering	[1 1]	1.0000	1.0000	0.9986	0.9995
	[3 3]	0.9487	0.9998	-	0.9764
Median filtering	[1 1]	1.0000	1.0000	0.9985	0.9995
	[3 3]	0.9565	0.9999	-	0.9948
Sharpening		0.9903	0.9854	-	0.9968
Cropping	1/2	0.9884	0.9873	-	-
	1/4	0.9980	0.9975	-	-
	1/8	0.9998	0.9996	-	-

Upon average and median filtering, the NC score slightly decreases when we increase the noise density. In a JPEG attack, the NC score is recorded as more than 0.9953 against different quality factors (QFs). In cropping attacks, our method delivers better results, and it outperforms the mentioned technique [20] at various cropping sizes. The proposed method offers high resistance compared to previous methods [9], [20], [25] against cropping, histogram equalization, JPEG, and average and median filtering. However, the NC score of the proposed method is less than reported in a previous study [20] against listed noises such as salt and pepper, speckle, and Gaussian noise. Furthermore, the robustness score of the proposed method against the listed attacks is mentioned in Fig. 4.11.

4.5.3 Security analysis

Different types of images should be encrypted into unrecognizable cypher images via a competent image encryption technique. As shown in Fig. 4.12, encrypted images differ significantly from corresponding plain images. It is also clear from the figure that the proposed scheme decrypts the encrypted image without loss of any information. Therefore, the proposed scheme is visually secure and prevents anyone from learning anything about the plaintext image just by glancing at the encrypted image. Further, the security performance of the proposed encryption scheme is evaluated by statistical, differential, and key security analysis.

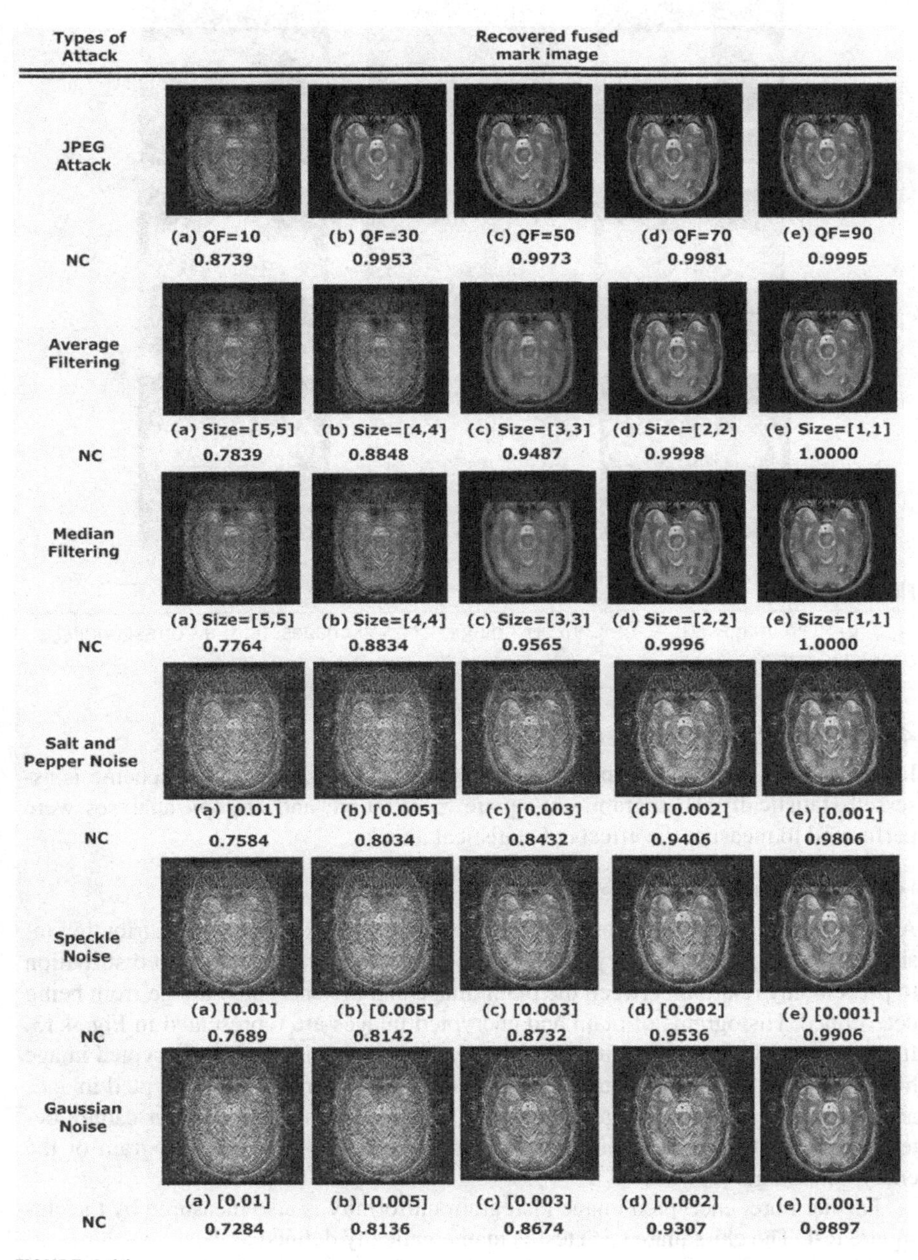

FIGURE 4.11

Robustness analysis of the proposed method against listed attacks.

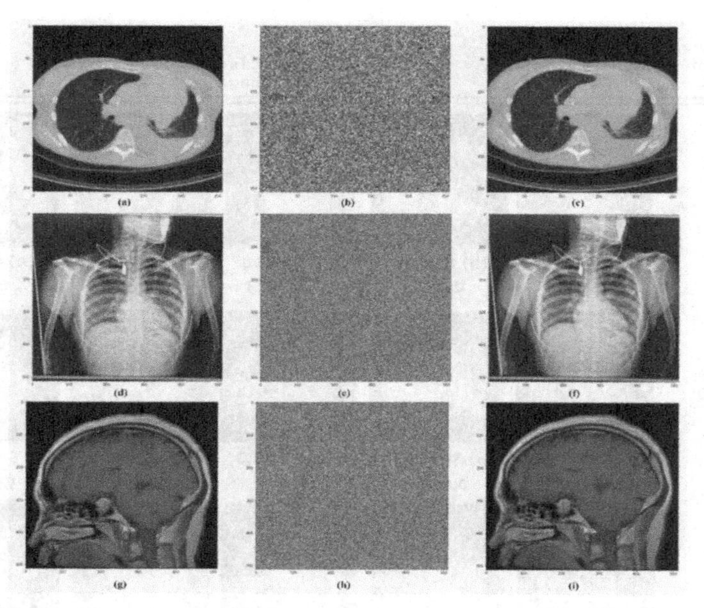

FIGURE 4.12

(a, d, g) Plain images. (b, e, h). Corresponding encrypted images. (c, f, i) Corresponding decrypted images.

4.5.3.1 Statistical analysis

In this section, the security performance of the proposed encryption scheme is assessed statistically. Histogram, chi-square, correlation, and entropy analyses were performed to measure the effects of statistical attacks.

Histogram and chi-square analysis

An image histogram is a pictorial representation of the pixel intensity distribution inside an image. An encrypted picture's histogram should have a uniform distribution to prevent any relation between the plain image and the encrypted image from being determined. Histograms of plain and encrypted images are represented in Fig. 4.13. It can be observed that plain image histograms are not uniform, but encrypted image histograms are uniformly distributed. Moreover, histograms of all encrypted images are different from plain image histograms. Because of this, the attacker cannot determine any information about the plain image by analyzing the histogram of the encrypted image.

Furthermore, encrypted image histogram uniformity is also measured by the chi-square test. The chi-square (χ^2) test is mathematically defined as

$$\chi^2 = \sum_{i=0}^{255} \left(\frac{(OB_i - EX_i)^2}{EX_i} \right), \tag{4.27}$$

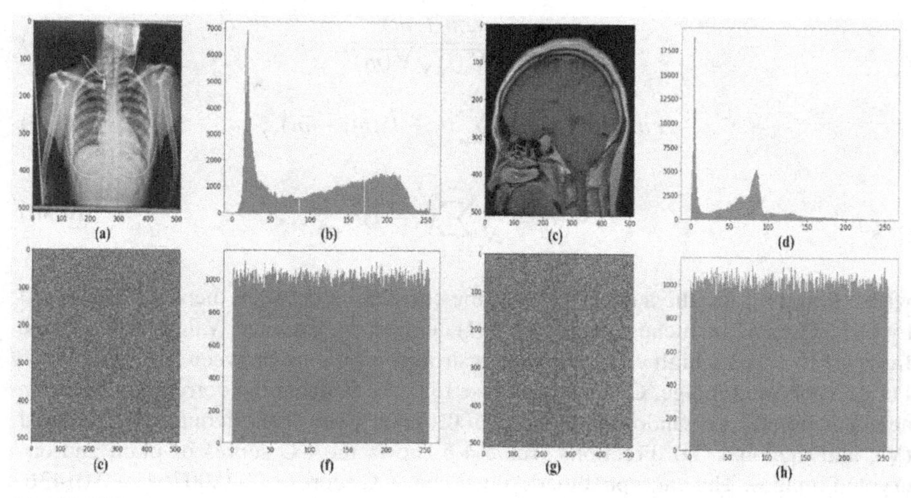

FIGURE 4.13

(a, c) Plain images and (b, d) their corresponding histograms. (e, g) Corresponding encrypted images and (f, h) their histograms.

where the observed frequency and expected frequency for every gray level are represented as OB_i and EX_i, respectively. For qualified encryption schemes, the χ^2 test value for encrypted images should be less than the ideal value 293.2478 [26]. The χ^2 test scores of encrypted images by the proposed scheme are shown in Table 4.5. The χ^2 score of all the encrypted images is much lower than the theoretical score, and the average χ^2 score of 100 images of the Kaggle dataset is 253.9154. This observation proves the uniformity of histograms of images encrypted with our scheme.

Table 4.5 χ^2 test analysis.

Tested image	χ^2 score	Result	Tested image	χ^2 score	Result
med1	265.5625	Pass	med6	242.8896	Pass
med2	228.9688	Pass	med7	249.9577	Pass
med3	287.2676	Pass	med8	277.0401	Pass
med4	243.7324	Pass	med9	260.8112	Pass
med5	279.0469	Pass	Avg. of Kaggle dataset	253.9154	Pass

Correlation analysis

Correlation analysis is conducted to determine the correlation among the image pixels since plain image pixels are highly correlated to their adjacent pixels. To minimize the effects of statistical attacks, an effective encryption method must minimize correlation. The correction metric termed correlation coefficient (CC) [27] is defined as

$$C_{l,m} = \frac{con(l,m)}{\sqrt{V(l)}\sqrt{V(m)}}, \tag{4.28}$$

$$con(l,m) = \frac{1}{N}\sum_{i=1}^{N}(l_i - \bar{l})(m_i - \bar{m}), \tag{4.29}$$

$$V(l) = \frac{1}{N}\sum_{i=1}^{N}(l_i - \bar{l})^2, \tag{4.30}$$

where l_i and m_i are the adjacent pixels, the correlation between them is represented by $C_{l,m}$, $\bar{l}$ and $\bar{m}$ are mean values, and $V(l)$ denotes the variance. Values of CC range from -1 to 1, and a high value indicates a strong correlation between adjacent pixels. For an encrypted image, CC must be close to 0. To evaluate the correlation between adjacent pixels, we randomly selected 5000-pixel pairs in horizontal (H), vertical (V), and diagonal (D) directions. Table 4.6 shows the CC scores of plain and encrypted images. The encrypted med6 image has CC scores of −0.00764, −0.01436, and −0.00025 in the H, V, and D directions, respectively. The average CC scores of 100 encrypted images of the Kaggle dataset are −0.00068 −0.00162, and 0.00072.

Table 4.6 Correlation analysis.

	Tested image	Plain image CC			Encrypted image CC		
		H	V	D	H	V	D
Proposed scheme	med1	0.95391	0.96290	0.93741	0.01930	0.00206	−0.01791
	med2	0.99376	0.98023	0.97494	0.01004	0.00194	−0.01440
	med3	0.98082	0.98816	0.96986	−0.01521	0.01025	−0.01052
	med4	0.98563	0.98800	0.97235	−0.01805	−0.02572	−0.00291
	med5	0.99595	0.99747	0.99167	−0.01295	−0.00456	−0.00168
	med6	0.96563	0.97274	0.94616	−0.00764	−0.01436	−0.00025
	med7	0.94412	0.95624	0.91255	−0.00820	0.01248	0.00314
	med8	0.96498	0.97703	0.94372	−0.00926	0.01071	−0.00817
	med9	0.90137	0.93506	0.85697	−0.00375	−0.02095	−0.00234
	Avg. of Kaggle dataset	0.94724	0.95796	0.91814	−0.00068	−0.00162	0.00072
[28]	OPENi4	0.9803	0.9803	0.9688	0.0048	−0.0012	−0.00023
[29]	X-ray1	0.8922	0.7470	0.7598	0.0021	0.0029	0.0016
[30]		-	-	-	−0.0027	0.0031	0.0011
[31]	MRI image	0.9563	0.9575	0.9131	0.0067	−0.0020	0.0081

Table 4.6 indicates that the CC score of all the encrypted images is significantly reduced compared to their corresponding plain images and approaches 0. To assess the performance of the proposed scheme, we also compared the results with recent reported schemes [28–31], which showed a better or comparable performance. Fig. 4.14 shows the pixel distributions of the plain and encrypted images in each direction. We can observe that the plain image pixel distribution along the diagonal

indicates a higher correlation, while the encrypted image pixel distribution shows minimal or no correlation. Therefore, correlation analysis proves that the proposed scheme can withstand statistical attacks.

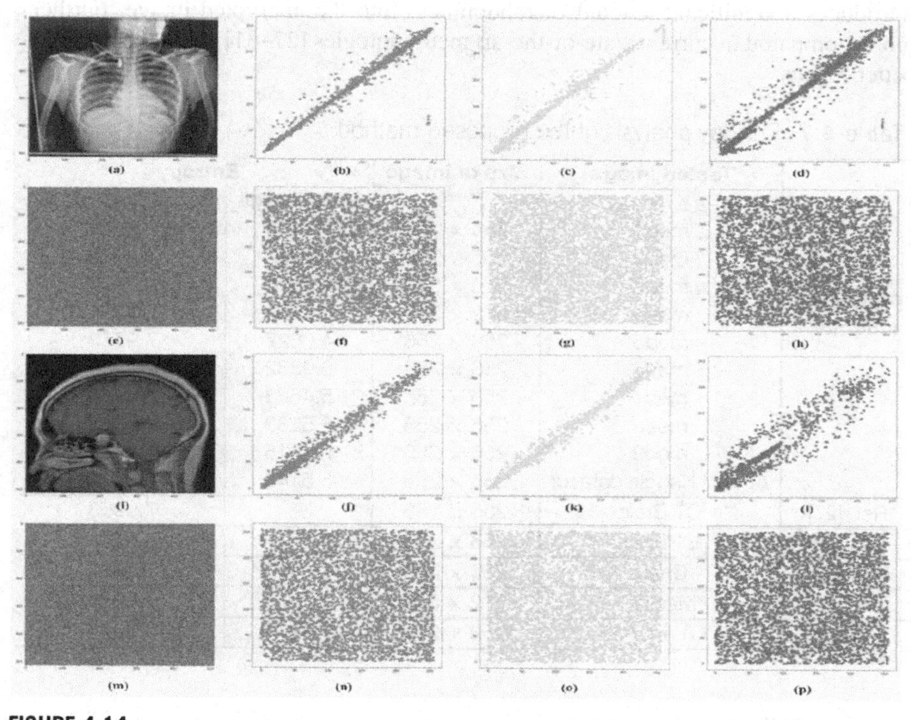

FIGURE 4.14

(a, i) Plain images and (e, m) corresponding encrypted images. (b–d, j–l) Plain image correlation in the H, V, and D directions and (f–h, n–p) encrypted image correlation in the H, V, and D directions.

Entropy analysis

Randomness in an image is measured by information entropy. The entropy score ranges from 0 to 8 for grayscale images. In general, for a plain image, a low entropy score indicates less randomness in the image information. A qualified encryption scheme increases the randomness in the image so that the entropy score should be close to the theoretical score, i.e., 8. Entropy [21] is defined as

$$En\,(l) = -\sum_{i=0}^{255}(P(l_i) \times log_2 P(l_i)), \tag{4.31}$$

where $P(l_i)$ indicates the probability of pixel l_i. Table 4.7 illustrates the entropy score of encrypted and plain images. The maximum and minimum entropy scores for encrypted images are 7.9993 and 7.9962, respectively. The average entropy of 100 images of a Kaggle dataset is 7.9974, and every encrypted image entropy approaches the ideal value of 8. These observations indicate that the proposed encryption scheme introduces a significant amount of randomness into the encrypted image. Furthermore, compared to current state-of-the-art methodologies [27–31], our scheme yields better results.

Table 4.7 Entropy analysis of the proposed method.

	Tested image	Size of image	Entropy	
			Plain image	Encrypted image
Proposed scheme	med1	256 × 256	3.7148	7.9973
	med2	256 × 256	6.1936	7.9970
	med3	512 × 512	7.5920	7.9993
	med4	512 × 512	6.7829	7.9991
	med5	256 × 256	4.0627	7.9972
	med6	256 × 256	6.9382	7.9963
	med7	256 × 256	5.4683	7.9964
	med8	256 × 256	6.5239	7.9962
	med9	256 × 256	7.1515	7.9963
	Avg. of Kaggle dataset	256 × 256	5.5140	7.9974
Ref. [27]	CT Brain	256 × 256	-	7.9953
Ref. [28]	OPENi4	256 × 256	-	7.9971
Ref. [29]	Chest	256 × 256	-	7.9935
Ref. [30]	Medical	512 × 512	-	7.9998
Ref. [31]	CT image	256 × 256	4.8274	7.9970

4.5.3.2 Differential analysis

A strong encryption method must be extremely responsive to even the slightest alteration in the plain image. An ideal encryption method, for instance, should generate totally different cypher images whenever there is even a minor variation between two plain images. Otherwise, a differential attack on the encryption mechanism is feasible. Differential analysis is performed by altering one bit in a plain image. The number of pixels change rate (NPCR) and the unified average changing intensity (UACI) are metrics to access robustness against differential attacks [26]. These are defined as follows:

$$NPCR = \frac{1}{M \times N} \sum_{i,j} Diff(i,j) \times 100\%, \tag{4.32}$$

$$UACI = \frac{1}{M \times N} \sum_{i,j} \frac{|X(i,j) - X'(i,j)|}{255} \times 100\%, \tag{4.33}$$

$$D(i,j) = \begin{cases} 0, & if\ X(i,j) = X'(i,j), \\ 1, & if\ X(i,j) \neq X'(i,j), \end{cases} \tag{4.34}$$

where $X(i, j)$ is the encrypted plain image and $X'(i, j)$ is the encrypted 1-bit-changed plain image. The ideal values of NPCR and UACI are 0.99609 and 0.33464, respectively [30]. The differential analysis of our proposed scheme is depicted in Table 4.8. The NPCR and UACI scores of our proposed scheme are very close to the ideal values. The average NPCR and UACI scores for 100 images of the Kaggle dataset are 99.6103 and 33.4589, respectively. Furthermore, comparison with recent techniques [27–31] also indicates that our scheme is superior. Therefore, the proposed scheme efficiently withstands differential attacks.

Table 4.8 Differential analysis.

	Tested image	NPCR	UACI		Tested image	NPCR	UACI
Proposed scheme	med1	99.5868	33.5119	**Proposed scheme**	med6	99.6023	33.4330
	med2	99.6066	33.4488		med7	99.6222	33.4775
	med3	99.6023	33.4766		med8	99.6274	33.5179
	med4	99.6066	33.4683		med9	99.6285	33.4291
	med5	99.5963	33.3896		Avg. of Kaggle dataset	99.6103	33.4589
[27]	CT Brain	99.6310	33.871	**[30]**	Image-01L	99.6103	33.4691
[28]	OPENi4	99.6155	33.5477	**[31]**	MRI	99.79	33.447
[29]	Chest	99.6132	33.5012				

4.5.3.3 *Key space and sensitivity analysis*

A strong cryptosystem should have a large key space and be highly sensitive to an encryption key. The key space should be greater than 2^{100} to reduce the effect of a brute-force attack [26]. In the proposed scheme, a 256-bit secret key is utilized to generate the initial point of the Lorenz map; therefore, the key space of our scheme is 2^{256}, which is sufficiently large to mitigate the effect of a brute-force attack. The key sensitivity analysis of our scheme is depicted in Fig. 4.15. To test key sensitivity, we chose encryption keys K1 and K2, which differ by only one bit. Fig. 4.15(b) and Fig. 4.15(c) are the images encrypted by K1 and K2, respectively. Fig. 4.15(d) is the result of decryption with K1, which was encrypted by K1. Fig. 4.15(e) is the result of decryption with key K2, which was encrypted by K1. This analysis shows that the image cannot be decrypted if there is a single bit difference in the key. Furthermore, the difference between images encrypted with K1 and those encrypted with K2 is 99.61%. These two encrypted images are totally different. This analysis indicates that our scheme is highly sensitive to the encryption key.

4.5.3.4 *Time cost analysis*

A secure cryptosystem should also be efficient at encrypting and decrypting images in a low amount of time. To analyze the efficiency of our proposed scheme, we calculated the time taken in the encryption and decryption process of different-sized medical images (see Table 4.9).

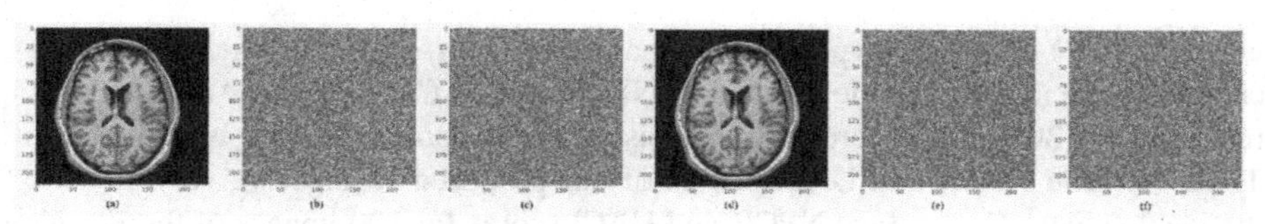

FIGURE 4.15

(a) Plain image. (b) Image encrypted with key K1. (c) Image encrypted with K2. (d) Decrypted image (encrypted and decrypted with K1). (e) Decrypted image (encrypted with K1 and decrypted with K2). (f) Difference between (b) and (c).

Table 4.9 Encryption (Enc.)/decryption (Dec.) time for different-sized images.

	Image	Size	Enc. time (sec)	Dec. time (sec)
Proposed scheme	med1	256 × 256	0.080305	0.128037
	med2	256 × 256	0.153087	0.177899
	med3	512 × 512	0.597068	0.458286
	med4	512 × 512	0.641664	0.598827
	med5	256 × 256	0.126391	0.126086
	med6	217 × 232	0.117328	0.131709
	med7	217 × 232	0.115786	0.090888
	med8	214 × 235	0.094718	0.09157
	med9	242 × 208	0.096603	0.092471
	Avg. of Kaggle dataset	Variable	0.348566	0.342713
[27]	-	256 × 256	4.8	-
[28]	OPENi4	256 × 256	3.9	-
[29]	-	256 × 256	2.12	-
[30]	COVID-00048	224 × 224	0.0140	-

The encryption times for image sizes of 256 × 256 and 512 × 512 are 0.080305 and 0.597068 seconds, respectively. The decryption times for image sizes of 256 × 256 and 512 × 512 are 0.128037 and 0.458286 seconds, respectively. The 100 different-sized images of the Kaggle dataset require average encryption and decryption times of 0.348566 and 0.342713 seconds, respectively. The average encryption time of an image size of 256 × 256 is 0.119928 seconds, so our method takes 98%, 97%, and 94% less time than the methods in [27–29], respectively.

4.6 Conclusion

In this chapter, we introduced a robust and secure watermarking algorithm based on multimodal medical image fusion. Initially, an NSST-based fusion approach was employed to fuse MRI and CT scan images and obtain the fused mark image. This fused mark image contains rich information, which is better suited for diagnosis and assessment than an individual image. Further, the integration of multiple decompositions (IWT-QR-SVD) was utilized to perform an imperceptible marking of the fused image within the cover media. Furthermore, an efficient encryption algorithm was performed by utilizing a 3D chaotic map on a marked image to ensure better security. Experimental outcomes on Kaggle and Open-i datasets offered better resistance against a wide range of attacks. Finally, the results of our method indicated that the proposed algorithm outperformed various state-of-the-art techniques in terms of robustness, visual quality, and security. The proposed work is suited for grayscale images only. In future work, we aim to utilize a deep learning model to enhance the embedding capacity while maintaining high robustness.

Acknowledgments

This work was supported by research project order no. IES212111 - International Exchanges 2021 Round 2, dt. 28 February 2022, under Royal Society, UK.

References

[1] S. Mahato, D. Yadav, D. Khan, A novel information hiding scheme based on social networking site viewers' public comments, Journal of Information Security and Applications 47 (2019) 275–283, https://doi.org/10.1016/j.jisa.2019.05.013.

[2] A. Singh, A. Anand, Z. Lv, H. Ko, A. Mohan, A survey on healthcare data: a security perspective, ACM Transactions on Multimedia Computing Communications and Applications 17 (2) (2021) 1–26, https://doi.org/10.1145/3422816.

[3] O. Singh, A. Singh, G. Srivastava, N. Kumar, Image watermarking using soft computing techniques: a comprehensive survey, Multimedia Tools and Applications 80 (20) (2020) 30367–30398, https://doi.org/10.1007/s11042-020-09606-x.

[4] Z. Xia, X. Wang, L. Zhang, Z. Qin, X. Sun, K. Ren, A privacy-preserving and copy-deterrence content-based image retrieval scheme in cloud computing, IEEE Transactions on Information Forensics and Security 11 (11) (2016) 2594–2608, https://doi.org/10.1109/TIFS.2016.2590944.

[5] K. Amine, K. Fares, K. Redouane, E. Salah, Medical image watermarking for telemedicine application security, Journal of Circuits, Systems, and Computers 31 (5) (2021) 1–12, https://doi.org/10.1142/s0218126622500979.

[6] N. Sharma, O. Singh, A. Anand, A. Singh, Improved method of optimization-based ECG signal watermarking, Journal of Electronic Imaging 31 (04) (2021) 1–12, https://doi.org/10.1117/1.JEI.31.4.041207.

[7] B. Wang, S. Jiawei, W. Wang, P. Zhao, Image copyright protection based on blockchain and zero-watermark, IEEE Transactions on Network Science and Engineering 9 (4) (2022) 2188–2199, https://doi.org/10.1109/TNSE.2022.3157867.

[8] O.P. Singh, A.K. Singh, Data hiding in encryption–compression domain, Complex & Intelligent Systems (2021) 1–14, https://doi.org/10.1007/s40747-021-00309-w.

[9] O.P. Singh, A.K. Singh, Image fusion-based watermarking in IWT-SVD domain, Lecture Notes in Electrical Engineering (2022) 163–175, https://doi.org/10.1007/978-981-19-0840-8_12.
[10] K.N. Singh, O.P. Singh, A.K. Singh, A.K. Agrawal, Watmif: multimodal medical image fusion-based watermarking for telehealth applications, Cognitive Computation (2022) 1–17, https://doi.org/10.1007/s12559-022-10040-4.
[11] R. Bhardwaj, Hiding patient information in medical images: an encrypted dual image reversible and secure patient data hiding algorithm for E-healthcare, Multimedia Tools and Applications 81 (1) (2021) 1125–1152, https://doi.org/10.1007/s11042-021-11445-3.
[12] S.P. Vaidya, Fingerprint-based robust medical image watermarking in hybrid transform, The Visual Computer (2022) 1–16, https://doi.org/10.1007/s00371-022-02406-4.
[13] M.S. Moad, M.R. Kafi, A. Khaldi, A wavelet based medical image watermarking scheme for secure transmission in telemedicine applications, Microprocessors and Microsystems 90 (2022) 1–14, https://doi.org/10.1016/j.micpro.2022.104490.
[14] P. Singh, K.J. Devi, H.K. Thakkar, K. Kotecha, Region-based hybrid medical image watermarking scheme for robust and secured transmission in IOMT, IEEE Access 10 (2022) 8974–8993, https://doi.org/10.1109/ACCESS.2022.3143801.
[15] O.P. Singh, A. Kumar Singh, H. Zhou, Multimodal fusion-based image hiding algorithm for secure healthcare system, IEEE Intelligent Systems (2022) 1–7, https://doi.org/10.1109/MIS.2022.3210331.
[16] H. Zhang, W. Yan, C. Zhang, L. Wang, Research on image fusion algorithm based on NSST frequency division and improved LSCN, Mobile Networks and Applications 26 (5) (2021) 1960–1970, https://doi.org/10.1007/s11036-020-01728-8.
[17] O. Singh, C. Kumar, A. Singh, M. Singh, H. Ko, Fuzzy-based secure exchange of digital data using watermarking in NSCT-RDWT-SVD domain, Concurrency and Computation: Practice and Experience (2021) 1–11, https://doi.org/10.1002/cpe.6251.
[18] S. Naaz, E. Sana, I. Ansari, Comparative analysis of digital image watermarking based on lifting wavelet transform and singular value decomposition, Advances in Intelligent Systems and Computing (2019) 65–81, https://doi.org/10.1007/978-981-15-0339-9_7.
[19] Q. Su, Y. Niu, G. Wang, S. Jia, J. Yue, Color image blind watermarking scheme based on QR decomposition, Signal Processing 94 (2014) 219–235, https://doi.org/10.1016/j.sigpro.2013.06.025.
[20] O.P. Singh, A.K. Singh, A.K. Agrawal, H. Zhou, SecDH: security of Covid-19 images based on data hiding with PCA, Computer Communications 191 (2022) 368–377, https://doi.org/10.1016/j.comcom.2022.05.010.
[21] K.N. Singh, O.P. Singh, A.K. Singh, A.K. Agrawal, EiMOL: a secure medical image encryption algorithm based on optimization and the Lorenz system, ACM Transactions on Multimedia Computing Communications and Applications (2022) 1–19, https://doi.org/10.1145/3561513.
[22] https://kaggle.com/tawsifurrahman/covid19-radiography-database. (Accessed 5 January 2023).
[23] https://openi.nlm.nih.gov/gridquery. (Accessed 5 January 2023).
[24] D.K. Mahto, O.P. Singh, A.K. Singh, FuSIW: fusion-based secure RGB image watermarking using hashing, Multimedia Tools and Applications (2022) 1–17, https://doi.org/10.1007/s11042-022-13454-2.
[25] O. Singh, A. Singh, A robust information hiding algorithm based on lossless encryption and NSCT-HD-SVD, Machine Vision and Applications 32 (4) (2021) 1–13, https://doi.org/10.1007/s00138-021-01227-0.

[26] G. Ye, C. Pan, X. Huang, Q. Mei, An efficient pixel-level chaotic image encryption algorithm, Nonlinear Dynamics 94 (1) (2018) 745–756, https://doi.org/10.1007/s11071-018-4391-y.

[27] K. Jain, A. Aji, P. Krishnan, Medical image encryption scheme using multiple chaotic maps, Pattern Recognition Letters 152 (2021) 356–364, https://doi.org/10.1016/j.patrec.2021.10.033.

[28] P. Sarosh, S.A. Parah, G.M. Bhat, An efficient image encryption scheme for Healthcare Applications, Multimedia Tools and Applications 81 (5) (2022) 7253–7270, https://doi.org/10.1007/s11042-021-11812-0.

[29] Y. Wu, L. Zhang, S. Berretti, S. Wan, Medical image encryption by content-aware DNA computing for Secure Healthcare, IEEE Transactions on Industrial Informatics 19 (2) (2023) 2089–2098, https://doi.org/10.1109/TII.2022.3194590.

[30] Q. Lai, G. Hu, U. Erkan, A. Toktas, High-efficiency medical image encryption method based on 2D logistic-Gaussian hyperchaotic map, Applied Mathematics and Computation 442 (2023) 1–12, https://doi.org/10.1016/j.amc.2022.127738.

[31] R. Ismail, A. Fattah, H.M. Saqr, M.E. Nasr, An efficient medical image encryption scheme for (WBAN) based on adaptive DNA and modern multi chaotic map, Multimedia Tools and Applications (2022) 1–15, https://doi.org/10.1007/s11042-022-13343-8.

CHAPTER

Fusion-based robust and secure watermarking method for e-healthcare applications

5

Ashima Anand[a], Amit Kumar Singh[b], and Huiyu Zhou[c]

[a]*Department of Computer Science and Engineering, Thapar Institute of Engineering and Technology, Patiala, Punjab, India*

[b]*Department of Computer Science and Engineering, National Institute of Technology Patna, Bihar, India*

[c]*School of Computing and Mathematical Sciences, University of Leicester, Leicester, United Kingdom*

5.1 Introduction

With the recent developments in the cloud environment, wearable devices, and Internet of Medical Things (IoMT), the storage and distribution of medical records have increased [1]. Medical devices are now connected to the internet, which provides a better clinical diagnosis. Patient digital records and related images are generated through these smart and intelligent devices and transmitted online for statistical accretion and precise diagnosis [2].

Further, the world recently suffered from the outbreak of COVID-19, which is a highly contagious disease [3]. Important patient digital records called electronic health data (EHD) were assembled, compiled, and communicated among different professionals using smart devices for accurate diagnosis. Also, storing large volumes of patient records in cloud-based healthcare applications has now become an important practice [4]. However, EHD exchange brings the issues of data exposure, ownership conflicts, and security, which have attracted more attention from the healthcare research community [2].

Therefore, the study of authenticity and copy protection of medical records and related information is becoming popular among researchers.

Among the different data-hiding schemes, watermarking is an effective tool that is intended as a solution to the issues of copyright protection and authentication of medical images [2]. It conceals the information called watermark(s) into the medical image, which can be recovered for ownership verification of the image [5,6]. Further, encryption-based watermarking techniques are widely used since they offer a high level of security [7,8]. Fusion of unimodal digital images to obtain a distinct multi-

Data Fusion Techniques and Applications for Smart Healthcare. https://doi.org/10.1016/B978-0-44-313233-9.00011-4

modal image is needed to ensure better information acquisition [1,9,10]. Robustness, visual quality, and embedding capacity are the mutually exclusive attributes of watermarking [5]. These three essentials need to be balanced to ensure high performance. Of late, many researchers have opted for optimization-based schemes to achieve high visual quality with resistance against attacks [11–13,4].

Motivated by these observations, this chapter introduces a secure watermarking algorithm using encryption and optimization using fused medical images as mark carriers, which can offer an excellent balance among robustness, invisibility, and capacity. The major novelties of this work are listed below.

- **Influence of infusing multimodal medical image fusion with watermarking:** This work implements nonsubsampled shearlet transform (NSST)-based fusion to generate fused images which hold more sophisticated and detailed information [14]. These fused images are later considered as watermark images for the watermarking process.
- **Superiority of the NSST-MSVD-based watermarking method:** The generated mark is concealed within the NSST–multiresolution singular value decomposition (MSVD) coefficients of the carrier image. The combination of these transforms offers better robustness and visual quality along with high embedding capacity at low cost.
- **Use of histogram of oriented gradients (HOG) to compute the optimized gain factors:** HOG is used to efficiently compute the values of optimal gain factors to maintain high visual quality and robustness during the embedding and removal of hidden marks.
- **Reliable communication of marked media in cloud environment:** The final marked image is ciphered using a step space filling curve (SSFC), redundant discrete wavelet transform (RDWT), and a nonlinear chaotic map, ensuring additional security of the media data. Moreover, the suggested framework is tested in a cloud environment to verify the suitability for e-healthcare applications.
- **Improved robustness and visual quality:** The results are extensively evaluated; the proposed method is not only robust but also imperceptible and secure with low cost. Its performance also exceeds that of state-of-the-art techniques when tested against various signal processing attacks.

The remainder of the chapter is ordered as follows. Section 5.2 provides a detailed explanation of the proposed framework, followed by the result analysis in Section 5.3. The performance evaluation of the improved technique is presented in Section 5.4 and a summary is provided in Section 5.5.

5.2 Literature review

Some of the related watermarking techniques using image fusion, optimization, and encryption are summarized in this section.

Singh offered a nonblind watermarking technique of color images in the discrete cosine transform (DCT)–lifting wavelet transform (LWT) domain [15]. MD5 encryption is applied on the signature mark and the patient's record is encoded with the BCH encoding method. The experimental results indicated high robustness and confidentiality of the scheme, while maintaining low computational cost. However, this technique requires manual selection of the gain factor to balance visual quality and robustness, which is not efficient. Further, a wavelet-based watermarking scheme was developed by Abdulazeez et al. [11]. For better security, the image mark is first encrypted using the chaotic map. This encrypted image is concealed in a singular matrix of third-level LWT coefficients of the cover image using a multiple gain factor which is calculated using artificial bee colony (ABC) optimization. The performance of the proposed work is compared with existing work [16] and the results confirm high robustness and visual quality, but the resistance against Gaussian filtering needs to be improved.

Another ABC optimization-based watermarking technique is presented by Amrit et al. [13]. In this article, a discrete wavelet transform (DWT)–singular value decomposition (SVD)-based dual watermarking technique is developed for healthcare applications. Hiding multiple marks ensured security and better identity authentication. The text mark is encoded using Hamming code before embedding. Further, the gain factor is optimized using ABC optimization for better robustness and visual quality simultaneously. The performance can be further improved with the concepts of neural networks and block. Hemdan developed a robust watermarking approach using the fused image as the watermark [17]. A fused image is scrambled using an Arnold map and a chaotic method and then concealed via a DWT-SVD-based marking technique.

Another wavelet-based blind watermarking technique is offered by Sharma et al. using LWT, DCT, and quantization [12]. Arnold map-based scrambling is applied on the color mark for extra security. Further, an ABC technique is used to optimize the value of the gain factor for robust embedding and extraction of the scrambled mark with high imperceptibility. Robustness, imperceptibility, and execution time of this work are better than those of traditional watermarking techniques [18,19]. Mahto et al. [20] introduced a fusion-based data-hiding scheme in a transform domain. First, nonsubsampled contourlet transform (NSCT) is employed to obtain the fused image, and then it is encrypted using SIE encryption to ensure high security. Further, the hash value of cover media along with an encrypted fused mark is concealed inside the multimedia object with the help of an optimal embedding factor, which ensures high robustness and better visual quality.

In a similar direction, Singh et al. [21] proposed a data-hiding framework in the healthcare domain. Initially, NSST is employed to determine the fused image, and then it is scrambled using Arnold transform. Further, a scrambled mark is hidden inside selected components of the multimedia object using fractional dual tree complex wavelet transform (Fr-DTCWT) and randomized SVD (RSVD) to ensure high security and better visual quality. This scheme delivers better resistance against hybrid attacks.

Singh et al. developed a secure fusion-based watermarking method for identity verification of medical images [10]. An NSCT-based fusion technique is used to fuse

MRI and CT images to generate a distinct image, which is hidden in the encrypted cover image. The authors used a hybrid version of RDWT and RSVD to robustly embed the fused image. Further, the quality of the extracted mark is enhanced by using a DnCNN-based denoising method. Anand and Singh presented a cloud-based robust dual watermarking scheme for secure communication of the fused multimodality images for smart healthcare applications [22]. A fused image is generated using an NSCT-based medical image fusion technique, which is treated as the cover image. It conceals the image watermark using NSCT, QR, and Schur decomposition. Further, the text watermark including the hash value of the cover image and the EPR text is marked via a magic cube-based text watermarking procedure for high payload. The final marked image is encrypted using an efficient encryption technique. The authors of [23] proposed a robust and secure watermarking solution for cloud-based healthcare applications. The fused image generated from NSST-based fusion is used as host image. Multiple marks, including MAC address, image, and EPR, are concealed within the host using DTCWT-SVD and pseudomagic cube-based watermarking. In addition, encryption of marked images is performed to provide better security.

Though many researchers have provided a fusion-based watermarking solution, optimization of the prime watermarking parameters has not been implemented in combination with fusion and watermarking. Some of the watermarking schemes are based on a simple encryption algorithm, which is ineffective for the secure healthcare systems.

5.3 Proposed method

This section provides a detailed discussion of the proposed fusion-based watermarking technique in the NSST-MSVD transform domain, as shown in Fig. 5.1. The entire proposed algorithm is divided into three phases, namely, **(a) watermark generation using image fusion, (b) imperceptible and robust embedding of the fused image, and (c) encryption of the final watermarked image**. These three phases are illustrated in Algorithms 5.1–5.3, respectively. Also, Table 5.1 presents a detailed description of the notations used in the algorithms.

5.3.1 Watermark generation using image fusion

In the initial phase, NSST-based fusion of medical images is implemented to generate the watermark using two input images, "*img_CT*" and "*img_MRI*." This image fusion algorithm uses two different rule sets to fuse the high- and low-frequency NSST coefficients of the input images. CNP systems are used to fuse the resultant low-frequency coefficients (LFCs), while WSE- and ISNML-based rules contribute to generating the fused high-frequency coefficients (HFCs).

Let δ_1 and δ_2 be two CNP systems with local topologies, which accepts the LFC of input image 1 and image 2 as external inputs, respectively. These two CNP systems iterate until the maximum iterations produce excitation number matrices Π_1 and Π_2

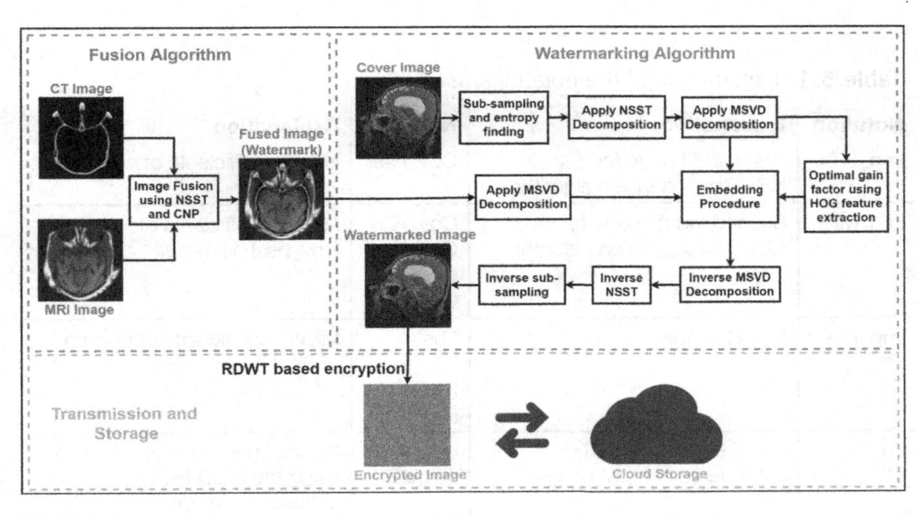

FIGURE 5.1

Framework of proposed NSST-MSVD-based robust and secure watermarking method.

as outputs, where $\Pi_1 = (e^1_{xy})_{m\times n}$ and $\Pi_2 = (e^2_{xy})_{m\times n}$. Here, e_{xy} represents the count of neurons firing in the excitation number matrix. Based on this information, the fusion rules for the fusion of LFC are defined as

$$Fus\_L(x, y) = \begin{cases} CT\_L(x, y) & \text{if } e^1_{xy} \geq e^2_{xy}, \\ MRI\_L(x, y) & \text{if } e^1_{xy} < e^2_{xy}. \end{cases} \tag{5.1}$$

Further, the edge and contours of the input images are preserved by fusing the HFC of input images using a hybrid version of WLE and INSML. The activity level measure, WLE, is mathematically calculated as

$$\text{WLE}_{ab}(x, y) = \sum_i \sum_j Wm'(i, j) H_{ab} \times (x + i, y + j)^2, \tag{5.2}$$

where $H_{ab}(x, y)$ denotes the high-frequency NSST coefficient at position (x, y) of direction b at layer a. Also, Wm' is the weighted matrix which is defined as

$$Wm' = \frac{1}{16} \begin{vmatrix} 1 & 2 & 1 \\ 2 & 4 & 2 \\ 1 & 2 & 1 \end{vmatrix}. \tag{5.3}$$

Further, INSML is used to extract the details of the input images using the following equation:

$$\text{INSML}_{ab}(x, y) = \sum_i \sum_j Wm'(i, j) \text{IML}_{ab} \quad \times (x + i, y + j). \tag{5.4}$$

Table 5.1 Explanation of the notations used.

Notation	Explanation	Notation	Explanation
img_CT	First input image for NSST-based fusion algorithm	Cov_nsst	NSST coefficients of max_s
img_MRI	Second input image for NSST-based fusion algorithm	Cov_A, Cov_H, Cov_V, Cov_D	MSVD coefficients of Cov_nsst{1, 3}
img_Fus	Fused image	Fus_A, Fus_H, Fus_V, Fus_V	MSVD coefficients of fw_img
CT_L, MRI_L	Low-frequency NSST coefficients of input images	opt_α1, opt_α2	Optimal gain factors calculated using the HOG feature extraction method
CT_H, MRI_H	High-frequency NSST coefficients of input images	Wat_V, Wat_nsst	Watermarked Cov_V and Cov_nsst
CT_m, MRI_m	Excitation number matrices of CNP system of CT_L and MRI_L, respectively	Wat_s	Watermarked max_s
Fus_L	Fused low-frequency coefficient	enc_WC	Encrypted watermarked cover image
CT_W, MRI_W	WLE associated with CT_H and MRI_H, respectively	scr_WC	Scrambled img_WC
CT_IN, MRI_IN	INSML of CT_H and MRI_H, respectively	src_A, src_H, src_V, src_D	RDWT components of scr_WC
Fus_H	Fused high-frequency coefficient	[r, c]	Count of rows and columns of src_A
img_C	Input cover image	Ch_seq	Chaotic sequence
img_WC	Watermarked cover image	Key_diff	Diffusion key
s1, s2, s3, s4	Subsampled components of cover image	K	Final key
es1, es2, es3, es4	Entropy of subsampled components of cover image	con_A, con_H, con_V, con_D	RDWT coefficients after applying the confusion method
max_s	Subsampled component with maximum entropy	enc_A, enc_H, enc_V, enc_D	Chaotic encrypted RDWT coefficients

Here, IML can mathematically be explained as

$$
\begin{aligned}
\mathrm{IML}_{ab}(x,y) = {} & |\,2H_{ab}(x,y) - H_{ab}(x-1,y) - H_{ab}(x+1,y)| + |2H_{ab}(x,y) \\
& - H_{ab}(x,y-1) - H_{ab}(x,y+1) + \frac{1}{\sqrt{2}}\,|\,2H_{ab}(x,y) \\
& - H_{ab}(x-1,y-1) - H_{ab}(x+1,y+1)| + +|\frac{1}{\sqrt{2}}|2H_{ab}(x,y) \\
& - H_{ab}(x-1,y+1) - H_{ab}(x+1,y-1)\,|\,.
\end{aligned}
\tag{5.5}
$$

The final rule for computing the fused high-frequency components can be defined as

$$
FH_{ab}(x,y) = \begin{cases} H^{1}_{ab}(x,y) & \text{if } \mathrm{WLE{-}INSML}^{1}_{ab}(x,y) \geq \mathrm{WLE}^{1} - \mathrm{INSML}^{2}_{ab}(x,y), \\ H^{2}_{ab}(x,y) & \text{otherwise,} \end{cases}
\tag{5.6}
$$

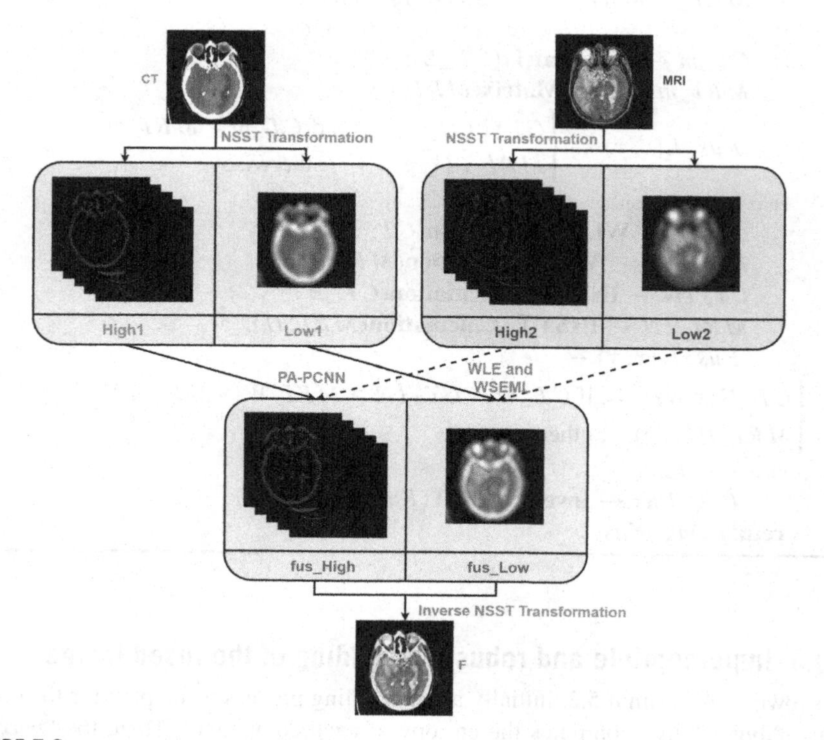

FIGURE 5.2

Framework of the NSST-CNP-based multimodal medical image fusion method.

where $H^1_{ab}(x, y)$ and $H^2_{ab}(x, y)$ are the high-frequency NSST coefficients of the input images "*img_CT*" and "*img_MRI*," respectively. Also, the fused high-frequency coefficient is denoted by $FH^1_{ab}(x, y)$.

The flowchart for generating the fused image is shown in Fig. 5.2. Input images "*img_CT*" and "*img_MRI*" are decomposed into high-pass and low-pass components using NSST. The resultant low-pass components "*CT_L*" and "*MRI_L*" are merged using equation (1), resulting in "*Fus_L*." Further, a hybrid version of WLE and INSML, shown in equation (6), is applied on the high-pass components "*CT_H*" and "*MRI_H*," generating the fused high-pass component "*Fus_H*." Finally, inverse NSST is applied to obtain the fused image "*img_Fus*," which is treated as a mark image in the later sections. The procedure of this process is summarized in Algorithm 5.1.

Algorithm 5.1: NSST-based medical image fusion

Input : img_CT, img_MRI
Output: img_Fus

```
// Phase 1: Transforming input images using NSST decomposition
```
1 $CT\_L, CT\_H \leftarrow$ **NSST**(img_CT);
2 $MRI\_L, MRI\_H \leftarrow$ **NSST**(img_MRI);
```
// Phase 2: Fusion of low-frequency coefficients
```
3 $CT\_m \leftarrow$ **Exc_Matrix**(CT_L);
4 $MRI\_m \leftarrow$ **Exc_Matrix**(MRI_L);
5 $Fus\_L(x, y) \leftarrow \begin{cases} CT\_L(x, y) & \text{, if } CT\_m \geq MRI\_m \\ MRI\_L(x, y) & \text{, otherwise} \end{cases}$
```
// Phase 3: Fusion of high-frequency coefficients
```
6 $CT\_W \leftarrow$ **WLE_Calculation**(CT_H);
7 $MRI\_W \leftarrow$ **WLE_Calculation**(MRI_H);
8 $CT\_IN \leftarrow$ **INSML_Calculation**(CT_H);
9 $MRI\_IN \leftarrow$ **INSML_Calculation**(MRI_H);
10 $Fus\_H(x, y) \leftarrow$
$\begin{cases} CT\_H(x, y) & \text{, if } CT\_W - CT\_IN \geq MRI\_W - MRI\_IN \\ MRI\_H(x, y) & \text{, otherwise} \end{cases}$
```
// Phase 4: Applying inverse NSST decomposition
```
11 $Img\_Fus \leftarrow$ **Inverse_NSST**(Fus_L, Fus_H) ;
12 **return** img_Fus

5.3.2 Imperceptible and robust embedding of the fused image

As shown in Algorithm 5.2, initially a subsampling process is adopted for the cover image "*img_C*" that computes the entropy of each component. Then, the "*max_s*" component is decomposed with maximum entropy by applying NSST, and MSVD is used to decompose the NSST high-frequency coefficient of the subimage. Finally, the

Algorithm 5.2: Watermarking process to hide fused image

Input : img_Fus, img_C
Output: img_WC

```
// Phase 1: Subsampling of cover image
```
1 $[s1, s2, s3, s4] \leftarrow$ **Sub-sampling**(img_C);
2 $[es1, es2, es3, es4] \leftarrow$ **Entropy**$(s1, s2, s3, s4)$;
3 $max\_s \leftarrow$ **Maximum**$(es1, es2, es3, es4)$;
```
// Phase 2: Transforming max_s and img_Fus with NSST and MSVD
   transforms
```
4 $Cov\_nsst \leftarrow$ **NSST**(max_s);
5 $[Cov\_A, Cov\_H, Cov\_V, Cov\_D] \leftarrow$ **MSVD**$(Cov\_nsst1, 3)$;
6 $[Fus\_A, Fus\_H, Fus\_V, Fus\_V] \leftarrow$ **MSVD**(img_Fus) ;
```
// Phase 3: HOG-based calculation of optimal gain factors
```
7 $[opt\_\alpha1, opt\_\alpha2] \leftarrow$ **HOG_FeatureExtraction**(Cov_V);
```
// Phase 4: Embedding using opt_α1 and opt_α2
```
8 $Wat\_V \leftarrow opt\_\alpha1 \times Cov\_V + opt\_\alpha2 \times Fus\_V$;
```
// Phase 5: Transformation using inverse MSVD and inverse NSST
```
9 $Wat\_nsst \leftarrow$ **iMSVD**$(Cov\_A, Cov\_H, Wat\_V, Cov\_D)$;
10 $Wat\_s \leftarrow$ **Inverse_NSST**(Wat_nsst) ;
```
// Phase 6: Inverse subsampling procedure
```
11 $img\_WC \leftarrow$ **iSub-sampling**$(s1, s2, s3, Wat\_s)$;
12 **return** img_WC

optimized gain factors "*opt_α1*" and "*opt_α2*" are used to conceal the generated mark "*img_Fus*" in the cover image. These numerical values are calculated using HOG feature extraction to optimize the robustness and visual quality. The final marked image "*img_WC*" is then generated by applying inverse MSVD and inverse NSST. To recover the final mark, reverse embedding steps are followed.

5.3.3 Encryption of the final watermarked image

In the next phase, the security of the proposed work is enhanced by encrypting the marked image using a chaotic map and RDWT-based encryption, as shown in Fig. 5.3. First, the final mark "*F_wat*" is scrambled using SSFC scrambling [20] to improve the security and enhance the authentication process. This scrambling performs a continuous scan on the image, traveling through each pixel exactly once. The traversal is performed in a stair format. Further, the scrambled image is decomposed using RDWT, producing the coefficient matrices "*src_A*," "*src_H*," "*src_V*," and "*src_D*." Each component is independently encrypted using nonlinear chaotic encryption. Finally, the encrypted image is generated by applying inverse RDWT on the encrypted coefficients. The detailed procedure is indicated in Algorithm 5.3.

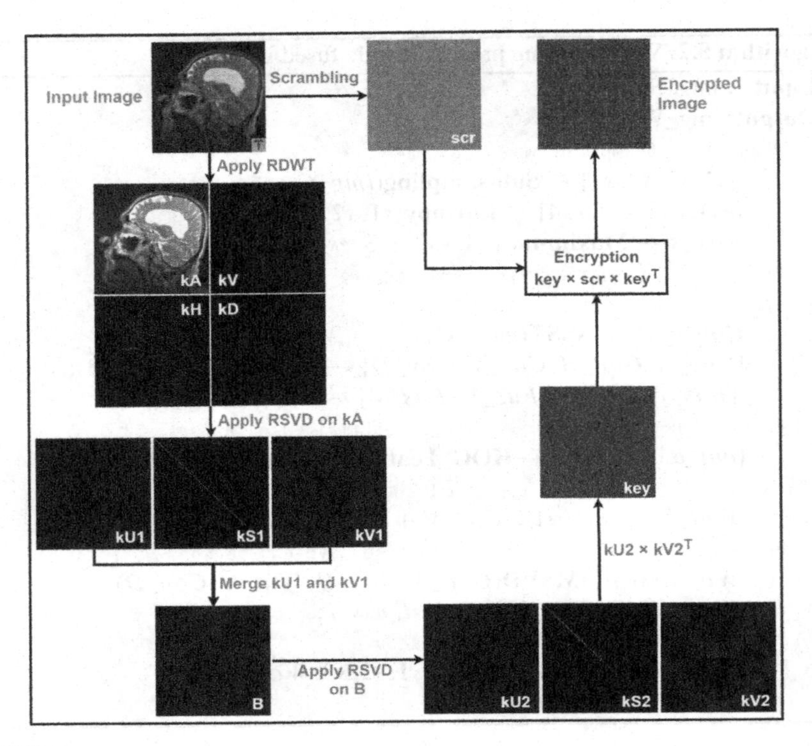

FIGURE 5.3

Framework of the proposed RDWT-chaotic map-based encryption method.

5.4 Results and analysis

This section presents the experimental setup and details of the results obtained from the implementation of the fusion-based watermarking method in the encryption domain. The performance evaluation is conducted in MATLAB® R2021b running on an Intel Xeon(R) Gold processor with 256 GB RAM, by fusing a CT scan and an MRI image to generate the mark image of size 256×256. This is followed by imperceptible embedding inside the medical image, producing the final marked image of size 512×512. Performance parameters including peak signal-to-noise ratio (PSNR), structural index similarity (SSIM), and normalized correlation (NC) values verify the imperceptibility and robustness of the watermarking technique [25]. We utilized Piella's structure similarity-based metric (PW) to compute the performance of the fusion scheme. Finally, the number of pixel changing rate (NPCR) and unified averaged changed intensity (UACI) are important indicators to evaluate the performance of the updated encryption scheme [26]. Also, suitable gain factors are calculated using HOG to optimize the robustness and invisibility.

Algorithm 5.3: RDWT-based image encryption

```
Input  : img_WC
Output: enc_WC
// Phase 1: Scrambling of img_WC
1     scr_WC ← SSFC(img_WC);
// Phase 2: Transforming src_WC using RDWT decomposition
2     [src_A, src_H, src_V, src_D] ← RDWT(src_WC);
// Phase 3: Key generation
3     [r, c] ← size(src_A);
4     r ← 3.62;
5     Ch_seq(1) ← 0.7;
6     for i ← 1 to (r × c) do
7         Ch_seq(i + 1) ← r × Ch_seq(i) × (1Ch_seq(i));
8     end for
9     p ← 3.628;
10    k(1) ← 0.632;
11    Key_diff ← Diffusion_Key(p, k(1));
12    K ← bitxor(Key_diff, CircularShift(Key_diff));
// Phase 4: Confusion of RDWT coefficients
13    con_A ← Img_Confusion(src_A, [r, c], Ch_seq) ;
14    con_H ← Img_Confusion(src_H, [r, c], Ch_seq) ;
15    con_V ← Img_Confusion(src_V, [r, c], Ch_seq) ;
16    con_D ← Img_Confusion(src_D, [r, c], Ch_seq) ;
// Phase 5: Encryption of confused RDWT coefficients
17    enc_A ← bitxor(con_A, K) ;
18    enc_H ← bitxor(con_H, K) ;
19    enc_V ← bitxor(con_V, K) ;
20    enc_D ← bitxor(con_D, K) ;
// Phase 6: Encryption of confused RDWT coefficients
21    enc_WC ← iRDWT(enc_A, enc_H, enc_V, enc_D) ;
22  return enc_WC
```

The proposed scheme is tested using different cover images; the results are displayed in Table 5.2. The achieved NC score is greater than or equal to 0.9921 for all considered images. Also, maximum scores of PSNR = 78.0413 dB and SSIM = 1 are obtained for the considered cover images. Further, NPCR $\geq$ 0.9808 and UACI $\geq$ 0.3164 indicate high security of the marked image when commuted through an online network.

The objective evaluation of the fusion method when implemented on 50 pairs of medical images is referred to in Table 5.3. Average OABF, FMI, SSIM, SF, and STD scores are 0.6893, 0.8978, 0.9463, 7.6622, and 63.7709, respectively. These results

Table 5.2 Objective evaluation of the proposed work on different images.

Cover image	PSNR (dB)	SSIM	NC	NPCR	UACI
MRI	73.5926	1	0.9955	0.9931	0.3692
Kidney stones	72.6452	1	0.9959	0.9932	0.3304
Colon MRI	73.4413	1	0.9921	0.9957	0.3615
Head CT scan	71.484	1	0.997	0.9808	0.3336
Barbara	76.3012	1	0.9968	0.9966	0.3164
Camera man	75.2708	1	0.9971	0.9967	0.3351
Cell	78.0413	1	0.9921	0.9953	0.3256
Rice	77.4283	1	0.9972	0.9966	0.3406
Lena	78.0031	1	0.9937	0.9963	0.3164
300 COVID-19 images [24]	77.2941	1	0.9968	0.9914	0.3383

Table 5.3 Performance evaluation of the NSST-CNP-based fusion method.

Image 1	Image 2	Fused image	OABF	FMI	SSIM	SF	STD
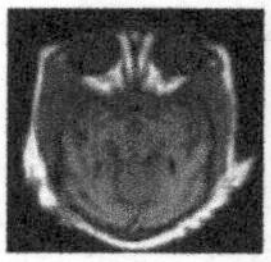	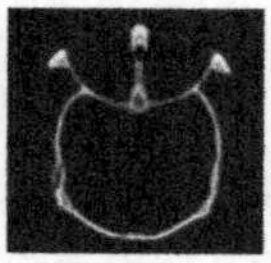	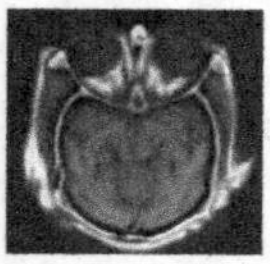	0.8115	0.8673	0.8765	7.5346	50.2588
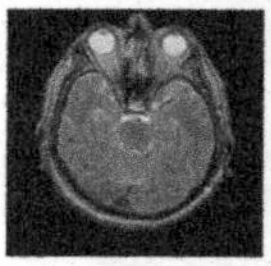	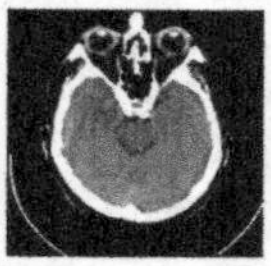	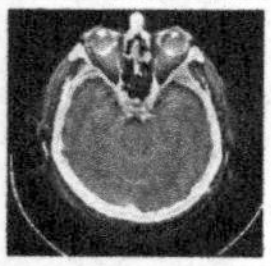	0.8075	0.9064	0.9526	7.8331	63.7699
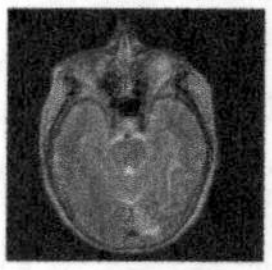	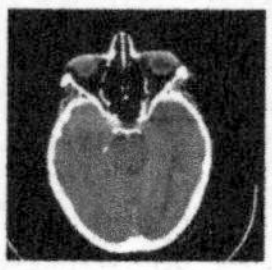	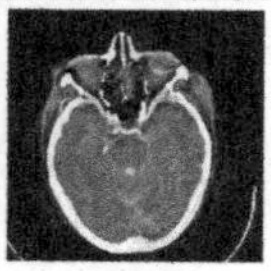	0.7162	0.9053	0.8963	7.5173	53.8248
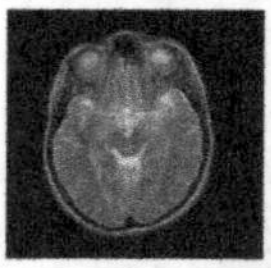	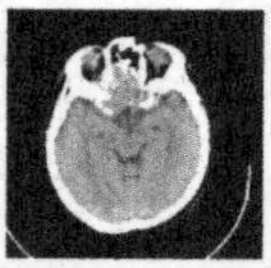	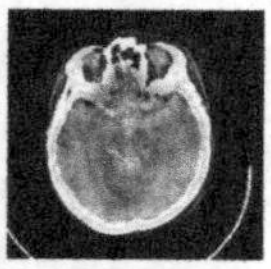	0.6659	0.9171	0.8701	7.1851	56.9732
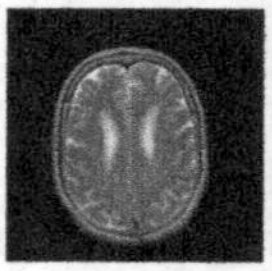	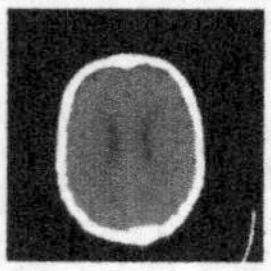	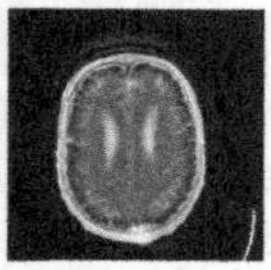	0.7944	0.9192	0.9028	6.7492	56.338
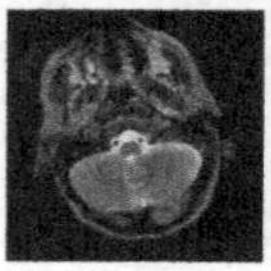	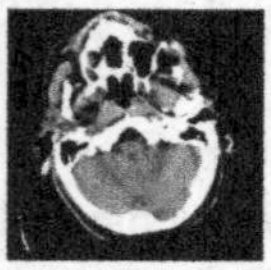	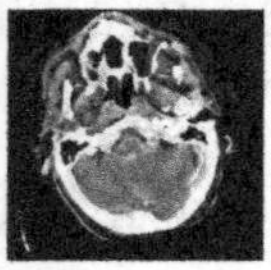	0.7216	0.9192	0.8744	7.1811	62.7901
Average of 50 pairs of images			0.6893	0.8978	0.9463	7.6622	63.7709

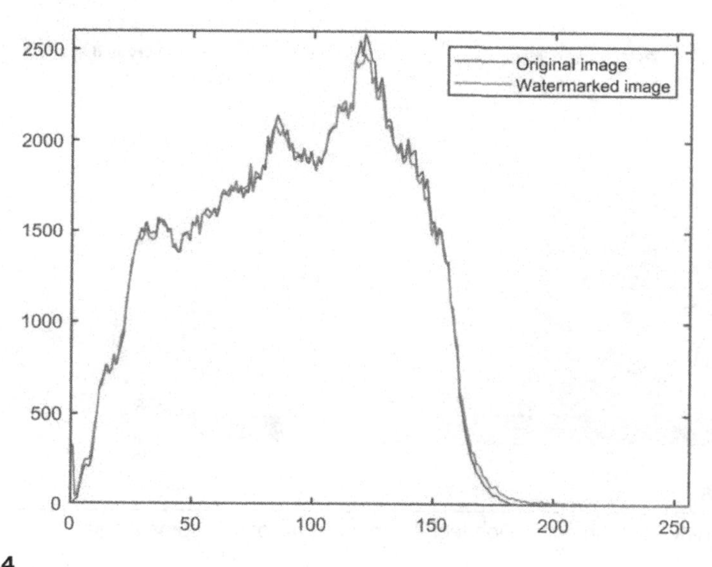

FIGURE 5.4

Subjective histogram analysis of the proposed work.

verify the satisfactory performance of the multimodal fusion method for healthcare images.

To subjectively verify the performance of the proposed work, a comparative histogram of the original cover image and marked images is presented in Fig. 5.4. This graphical comparison shows high similarity between the visual quality of the two images.

Another histogram-based analysis is shown in Fig. 5.5. It provides the frequency distribution of marked images and encrypted images. The histogram of marked images shows stiff peaks, providing statistical information. On the contrary, the histogram of ciphered images is uniform, which makes the method robust against statistical attacks.

Further, the robustness analysis of the proposed watermarking scheme is summarized in Table 5.4. Referring to the obtained results, NC score $\geq$ 0.9876 indicates high robustness for all the considered attacks. Compared to existing watermarking schemes [4,27–29], our proposed scheme has better robustness. The comparative NC results are shown in Fig. 5.6. Notably, the optimal NC score is 47.49% higher than that of existing schemes proposed in [4,27–29]. Comparative analysis suggests that the proposed method is superior due to the use of a hybrid version of NSST and MSVD along with HOG-based optimization for concealing the fused image in the carrier image.

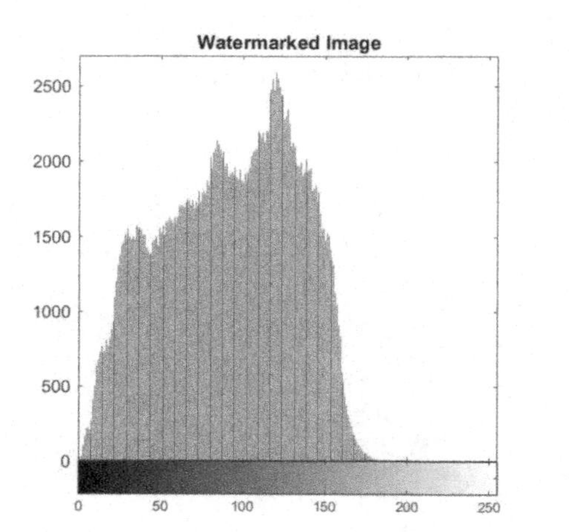

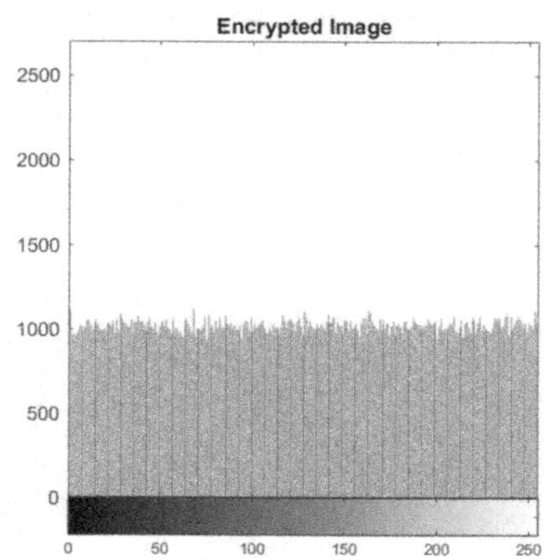

FIGURE 5.5

Security analysis of the encryption technique using histogram comparison.

Table 5.4 Robustness evaluation of the proposed work.

Attack	Noise density	NSST-MSVD
		NC
Salt and pepper noise	0.001	0.9967
	0.1	0.9948
Gaussian noise	0.001	0.9954
	0.1	0.9925
Rotation	45°	0.9942
	90°	0.9921
JPEG compression	QF = 10	0.9897
	QF = 90	0.9955
Gaussian low-pass filter	Var = 0.4	0.9955
	Var = 0.6	0.9955
Speckle noise	0.05	0.9931
	0.5	0.9908
Cropping	[20 20 400 480]	0.9876
Median filter	[2 2]	0.9913
	[3 3]	0.9886
Sharpening mask attack	0.01	0.9948
	0.1	0.9932
Histogram equalization		0.9901
Image scaling	0.5	0.9922
	2	0.9935
Translation	[5 5]	0.9919
	[7 7]	0.9891

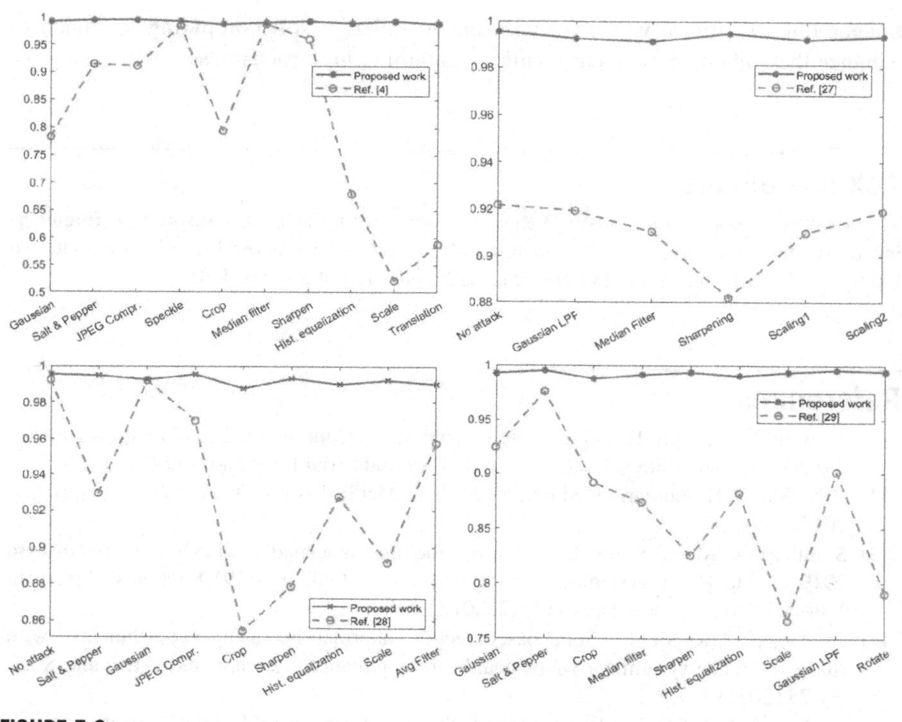

FIGURE 5.6

Comparative analysis of proposed work with Ref. [4], Ref. [27], Ref. [28], and Ref. [29].

5.5 Conclusion

A robust fusion-based watermarking solution for secure commuting of digital medical records is presented in this chapter. We utilized NSST along with MSVD to develop a better watermarking technique at a low cost. Also, NSST-based fusion is applied on CT scans and MRI images to generate a fused image, which is used as a watermark. Further, HOG is used to compute the values of the gain factor to maintain the balance between the robustness and visual quality of the proposed marking technique. Finally, the marked image is encrypted with a chaotic map and RDWT for better security in a cloud-based environment. PSNR $\leq$ 78.0413 dB and NC $\leq$ 0.9972 confirm the high visual quality and the ability to extract high-quality marks. Also, the high security of the encryption technique is indicated by the highest NPCR and UACI scores of 0.9966 and 0.3692, respectively. Better robustness is observed in comparison to traditional work, with a maximum improvement of 47.49%. The combination of NSST-based multimodality fusion, marking using NSST-MSVD, HOG-based optimization, and suitable encryption resulted in simultaneous improvement of robustness, security, and visual quality. This work is designed for grayscale images only. In future work, we will aim to utilize a soft computing approach to enhance the embedding capacity while maintaining high robustness.

Acknowledgment

This work was supported by a SEED Research Grant from the Thapar Institute of Engineering & Technology, Patiala, India, and research project order no. IES212111 - International Exchanges 2021 Round 2, dt. 28 February 2022, under Royal Society, UK.

References

[1] A. Anand, A. Singh, H. Zhou, Vimdh: visible-imperceptible medical data hiding for Internet of medical things, IEEE Transactions on Industrial Informatics (2022) 1–8.
[2] A.K. Singh, B. Kumar, G. Singh, A. Mohan, Medical Image Watermarking, Springer, 2017.
[3] S. Salehi, A. Abedi, S. Balakrishnan, A. Gholamrezanezhad, et al., Coronavirus disease 2019 (Covid-19): a systematic review of imaging findings in 919 patients, American Journal of Roentgenology 215 (1) (2020) 87–93.
[4] A. Anand, A.K. Singh, Cloud based secure watermarking using iwt-Schur-rsvd with fuzzy inference system for smart healthcare applications, Sustainable Cities and Society 75 (2021) 1–7.
[5] A. Anand, A.K. Singh, Watermarking techniques for medical data authentication: a survey, Multimedia Tools and Applications 80 (20) (2021) 30165–30197.
[6] N. Sharma, A. Anand, A.K. Singh, Bio-signal data sharing security through watermarking: a technical survey, Computing 103 (9) (2021) 1883–1917.
[7] S. Thakur, A.K. Singh, S.P. Ghrera, Nsct domain–based secure multiple-watermarking technique through lightweight encryption for medical images, Concurrency and Computation: Practice and Experience 33 (2) (2021) e5108.
[8] O. Singh, A. Singh, Data hiding in encryption–compression domain, Complex & Intelligent Systems (2021) 1–14.
[9] D.K. Mahto, O.P. Singh, A.K. Singh, Fusiw: fusion-based secure rgb image watermarking using hashing, Multimedia Tools and Applications (2022) 1–17.
[10] K.N. Singh, O.P. Singh, A.K. Singh, A.K. Agrawal, Watmif: multimodal medical image fusion-based watermarking for telehealth applications, Cognitive Computation (2022) 1–17.
[11] A.M. Abdulazeez, D.M. Hajy, D.Q. Zeebaree, D.A. Zebari, Robust watermarking scheme based lwt and svd using artificial bee colony optimization, Indonesian Journal of Electrical Engineering and Computer Science 21 (2) (2021) 1218–1229.
[12] S. Sharma, H. Sharma, J.B. Sharma, Artificial bee colony based perceptually tuned blind color image watermarking in hybrid lwt-dct domain, Multimedia Tools and Applications 80 (12) (2021) 18753–18785.
[13] P. Amrit, A. Anand, S. Kumar, A. Singh, Robust transmission of medical records using dual watermarking and optimization algorithm, Journal of Physics. Conference Series 1767 (2021) 1–10.

[14] B. Li, H. Peng, X. Luo, J. Wang, X. Song, M.J. Pérez-Jiménez, A. Riscos-Núñez, Medical image fusion method based on coupled neural p systems in nonsubsampled shearlet transform domain, International Journal of Neural Systems 31 (01) (2021) 1–17.
[15] A.K. Singh, Robust and distortion control dual watermarking in lwt domain using dct and error correction code for color medical image, Multimedia Tools and Applications 78 (21) (2019) 30523–30533.
[16] N. Harish, B. Kumar, A. Kusagur, Hybrid robust watermarking techniques based on dwt, dct, and svd, International Journal of Advanced Electrical and Electronics Engineering 2 (5) (2013) 137–143.
[17] E.E.-D. Hemdan, An efficient and robust watermarking approach based on single value decompression, multi-level dwt, and wavelet fusion with scrambled medical images, Multimedia Tools and Applications 80 (2) (2021) 1749–1777.
[18] T. Huynh-The, C.-H. Hua, N.A. Tu, T. Hur, J. Bang, D. Kim, M.B. Amin, B.H. Kang, H. Seung, S. Lee, Selective bit embedding scheme for robust blind color image watermarking, Information Sciences 426 (2018) 1–18.
[19] Q. Su, Y. Liu, D. Liu, Z. Yuan, H. Ning, A new watermarking scheme for colour image using qr decomposition and ternary coding, Multimedia Tools and Applications 78 (7) (2019) 8113–8132.
[20] D. Mahto, A. Singh, K. Singh, O. Singh, A. Agrawal, Robust copyright protection technique with high-embedding capacity for color images, ACM Transactions on Multimedia Computing Communications and Applications (2023) 1–12.
[21] O.P. Singh, A.K. Singh, H. Zhou, Multimodal fusion-based image hiding algorithm for secure healthcare system, IEEE Intelligent Systems (2022) 1–7.
[22] A. Anand, A.K. Singh, Sdh: secure data hiding in fused medical image for smart healthcare, IEEE Transactions on Computational Social Systems (2021) 1–8.
[23] A. Anand, A.K. Singh, Health record security through multiple watermarking on fused medical images, IEEE Transactions on Computational Social Systems (2021) 1–10.
[24] J.P. Cohen, P. Morrison, L. Dao, Covid-19 image data collection, arXiv:2003.11597, 2020, https://github.com/ieee8023/covid-chestxray-dataset.
[25] O.P. Singh, A.K. Singh, G. Srivastava, N. Kumar, Image watermarking using soft computing techniques: a comprehensive survey, Multimedia Tools and Applications 80 (2021) 30367–30398.
[26] K.N. Singh, N. Baranwal, O.P. Singh, A.K. Singh, Sielnet: 3d chaotic-map-based secure image encryption using customized residual dense spatial network, IEEE Transactions on Consumer Electronics (2022) 1–8.
[27] H.S. Alshanbari, Medical image watermarking for ownership & tamper detection, Multimedia Tools and Applications 80 (11) (2021) 16549–16564.
[28] F. Kahlessenane, A. Khaldi, R. Kafi, S. Euschi, A robust blind medical image watermarking approach for telemedicine applications, Cluster Computing 24 (3) (2021) 2069–2082.
[29] M.S. Moad, M.R. Kafi, A. Khaldi, Medical image watermarking for secure e-healthcare applications, Multimedia Tools and Applications (2022) 1–21.

CHAPTER

Recent advancements in deep learning-based remote photoplethysmography methods

Ankit Gupta[a,b], **Antonio G. Ravelo-García**[a,c], **and Fernando Morgado-Dias**[a,b]

[a]*Interactive Technologies Institute, ARDITI, LarSYS, Funchal, Portugal*

[b]*Universidade da Madeira, Engineering and Exact sciences, Funchal, Portugal*

[c]*Institute of Technological Development and Innovation in Communications, Universidad de Las Palmas de Gran Canaria, Las Palmas, Spain*

6.1 Introduction

An individual's physical and mental health or diseased state can be determined by monitoring cardiac activity based on a cardiac signal. Electrocardiography (ECG) and Plethysmography (PG) are the two conventional techniques for measuring a cardiac signal. ECG measures changes in the heart's electrical activity by estimating the change in the potential difference among the electrodes physically attached to the skin [1]. Therefore, ECG requires accurate placement of lead electrodes in contact with the skin using adhesive gel. On the other hand, PG works by detecting the changes in the blood volume due to vasomotor changes synchronized with the heartbeat. Among variants of PG, the one exploiting the optical properties of the skin tissues, also called Photoplethysmography (PPG), is widely used nowadays due to the existence of relatively few limitations, unlike its other variants [2]. Extracting PPG signals requires the placement of an oximeter probe over different organs such as a finger, earlobes, etc. Both of these techniques are contact-based approaches, which can estimate the physiological parameters accurately but have a limited scope in a few scenarios, such as prolonged and neonatal health monitoring, burned or sensitive skin, etc. Furthermore, due to the subject's constrained mobility in the case of contact-based approaches, accurate estimation could be infeasible for applications such as health monitoring during sleep analysis, driving, fitness exercise, etc. This motivates the need to develop noncontact approaches for extracting cardiac signals for accurate physiological parameter estimations and eventually health monitoring.

A noncontact version of ECG is feasible but requires sensor placement on things such as chairs [3], which limits the subject's mobility and, eventually, the above-

Data Fusion Techniques and Applications for Smart Healthcare. https://doi.org/10.1016/B978-0-44-313233-9.00012-6

mentioned applications. On the other hand, Remote photoplethysmography (rPPG) does not require any physical contact with the skin, which makes it applicable to all the abovementioned scenarios and applications. rPPG is based on reflectance mode PPG, wherein the illumination source and photodetector are placed on the same side. Therefore, since 2008, when the first study was conducted by Verkruysse et al. [4], numerous studies have been conducted focusing on devising noncontact methods for efficiently extracting PPG information. These methods primarily extract the PPG information from the subject's video by extracting the subtle color variations based on the reflected light intensities from the consecutive image frames of the face. Essentially, these methods follow a three-step procedure: (1) face detection followed by skin segmentation, (2) extraction of the rPPG signal, and (3) frequency identification of the corresponding physiological parameters to be estimated. Each step poses different challenges to ensure accurate extraction of rPPG signals, ultimately leading to correct physiological parameter estimations. For instance, inappropriate Region of interest (ROI) selection results in insufficient PPG information extraction, which may lead to false estimates. Comprehensive reviews discussing State-of-the-art (SOTA) face detection and skin segmentation have already been presented by Kumar, Kaur, and Kumar [5] and Juneja and Rana [6]. A detailed review of various time- and frequency-based methods for physiological parameter frequency identification has been presented by Wacker and Witte [7]. These topics do not need further explanations; hence, they are not discussed in this chapter. Thus, the scope of this chapter lies in exploring the merits and demerits of the existing SOTA Deep learning (DL)-based rPPG signal extraction algorithms.

The rPPG signal extraction methods typically suffer from motion, illumination variation artifacts, and camera quantization noise. Consequently, the proposed methods revolve around alleviating the effects of these artifacts to ensure accurate rPPG signal extraction. Existing SOTA rPPG signal extraction methods can be categorized into conventional and DL-based approaches. Traditional methods are based on assumptions, which may not hold in every real-world situation. For instance, it is assumed that there is constant illumination over the entire ROI, which is not possible due to the asymmetrical geometry of the face.

A detailed review of the conventional methods can be found in [8–11]. DL-based rPPG signal extraction methods can be further categorized into two types: considering the physiological parameter estimations as a regression problem by mapping the video to the physiological parameters [12–15] and mapping the videos with the ground truth signal or through handcrafted features such as spatiotemporal representations [12,16]. Several aspects have already been explored in existing review articles [17–19]. For example, these reviews have either tested a few DL methods for rPPG signal extraction based on the accurate estimation of the Heart rate (HR) or presented a comparative analysis of conventional and DL-based rPPG methods.

However, to the best of our knowledge, no review provides insight into factors such as type of Neural network (NN), datasets, etc., for designing a standardized study for rPPG signal extraction. Additionally, existing reviews serve as an essential starting point for experienced researchers in the domain. Still, they do not address the problems researchers encounter when analyzing different critical factors for designing a generalized DL-based rPPG estimation method. Therefore, this chapter provides answers related to various vital factors to effectively design and test DL-based rPPG signal extraction methods. Considering this, this chapter encompasses the following considerations:

- It provides insight into the fundamental principle of PPG with various operational configurations, PPG signal characteristics, and the associated physiological parameters.
- It summarizes and analyzes the architectural components of the rPPG signal extraction pipeline, i.e., ROI selection, different kinds of NN for signal extraction, and signal processing for physiological parameter estimations.
- It also presents the limitations and recommendations for designing and testing the proposed DL-based rPPG methods in the future.

The remaining sections are organized as follows. Section 6.2 presents a brief introduction to PPG, including its working principle, operational configurations, challenges, and noncontact PPG variants. The DL-based rPPG methods and their components are summarized and analyzed in Section 6.3. Finally, their limitations, future recommendations (Section 6.4), and conclusions (Section 6.5) highlight critical aspects of DL rPPG methods.

6.2 Photoplethysmography

PPG is a simple, noninvasive, inexpensive technique that exploits the optical properties of the skin to detect subtle variations in the blood volume synchronized with cardiac activity, resulting in a PPG signal. The resultant signal can be used to estimate physiological parameters such as HR, Oxygen saturation (SpO2), Breathing rate (BR), etc. The apparatus used for PPG analysis is shown in Fig. 6.1a, which consists of the illumination source(s), emitting constant light radiations, and a photodetector for capturing the resultant light.

This section provides insight into the working principle and measuring instrument (Section 6.2.1), followed by operational configurations, signal characteristics (Section 6.2.2), current challenges (Section 6.2.3), noncontact PPG, and potential physiological parameters to be estimated by rPPG (Section 6.2.4).

6.2.1 Working principle and measuring instrument

The working principle of PPG is relatively simple and easy to analyze. Light is absorbed by tissue, bones, and blood. Additionally, most of the light is predominantly absorbed by the blood, while small fractions are scattered due to the opaque properties of the biological tissues. The critical components affected by this light absorption are blood volume, blood vessel walls, and red blood cell orientation. Therefore, skin is used for PPG analysis. The skin is comprised of the following layers: epidermis,

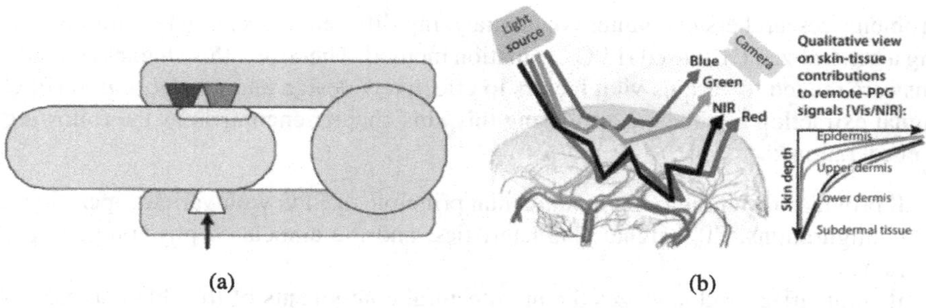

FIGURE 6.1

(a) Pulse oximeter [20]. (b) Working principle of PPG [21].

dermis (upper and lower), and hypodermis (subdermal), as shown in Fig. 6.1b. The epidermis is the visible part of the skin, followed by the upper and lower dermis and subdermal layers constituting the vascular bed (consisting of arteries and veins). Heart pumping allows the blood flow through multiple veins in the dermal and subdermal layers. The blood flow in arteries and arterioles can be used to detect the pulsatile component of the cardiac cycle [22].

Conventionally, optoelectronic oximeters are used for extracting the PPG signal. Oximeters consist of two Light-emitting diodes (LEDs) emitting red and infrared wavelengths and a photodetector for capturing the transmitted light after absorption through skin, bones, and blood. There are three reasons to use LEDs emitting visible red and Near infrared (NIR) light: (1) most of the light is absorbed by the water present in the blood, but only visible red and NIR wavelength radiations can penetrate deeper into the vascular bed to detect Blood volume change (BVP); (2) the 805 nm wavelength is an isosbestic wavelength which allows the signal to remain unaffected from the effect of SpO2; and (3) these wavelengths can penetrate through a volume of 1 cm^3 for transmission mode PPG [22]. Different penetration depths of visible and Infrared (IR) light are illustrated in Fig. 6.1b. If the skin is not compressed, the change in the blood volume resulting from the skin illumination by LED reflects the cardiovascular pressure wave. Therefore, the PPG signal is often obtained by optoelectronic pulse oximetry [23].

6.2.2 PPG operational configurations and signal characteristics

Based on the operational configuration, PPG can be classified into transmission mode (Fig. 6.2a) and reflection mode (Fig. 6.2b) PPG. In transmission mode PPG, the illumination source, i.e., LEDs, and the photodetector are placed opposite to each other. The light transmitted through the skin tissue is captured by a photodetector for PPG signal extraction. On the other hand, in reflectance mode PPG, the LEDs and the photodetector are placed on the same side, where the photodetector captures the reflected light from the skin tissue and is eventually used for PPG signal extraction. Transmission mode PPG has been used by pulse oximeters using organs such as the

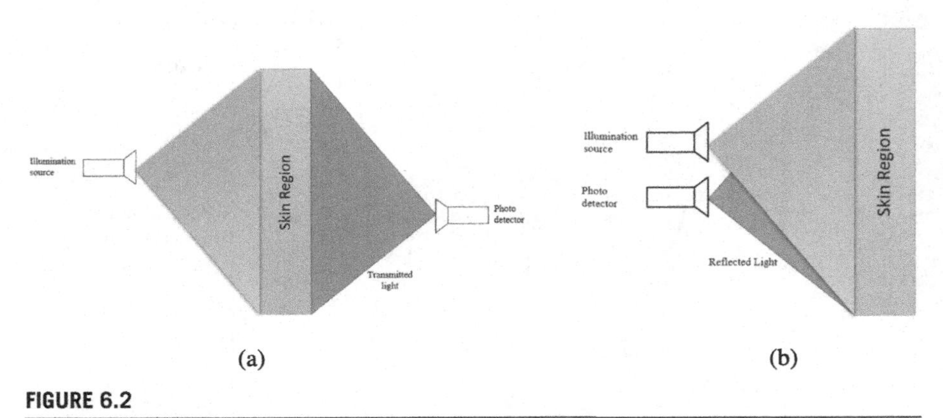

FIGURE 6.2

PPG operation configuration. Left, reflection mode PPG. Right, transmission mode PPG.

finger, earlobe, toe, etc. Reflectance mode PPG (noncontact) has been widely used for noncontact PPG using different organs such as the face, finger, palm, etc. The former imposes more restrictions than the latter regarding body movement and accurate placement of the oximeter probe, while these limitations are overcome in the latter due to its noncontact nature.

A PPG signal resulting from placing a finger in the pulse oximeter probe is shown in Fig. 6.3. It consists of a pulsatile component, also called the AC component, which is the source of cardiac information, and a slowly varying DC component due to multiple factors such as vasomotor activity, respiration, etc. Additionally, the AC component of the PPG signal consists of two peaks: the rising edge in the PPG signal, also called the anacrotic phase, which corresponds to the systole, and the falling edge, also called the catacrotic phase, corresponding to the diastole and wave reflection from the periphery. The PPG analysis eliminates the effect of a slowly varying dominant DC component by high-bandpass filtering [24].

6.2.3 Photoplethysmography challenges

The conventional way of obtaining the PPG signal is by optoelectronic pulse oximeters, which require the placement of an oximeter probe on organs such as the finger, toe, earlobe, etc. The disadvantages of using contact-based probes are multifold: prominent prevalence of artifacts due to skin compression, restricted movement of body locations, and limited applicability in a few scenarios. Precisely, if the skin is compressed while extracting the PPG signal, artifacts may be prominent, dominating the cardiac information. A minute deflection of the skin tissue from the actual position may result in a noisy signal. These probes can cause allergies, itching, or infection upon prolonged monitoring, in burned or sensitive skin, and in Neonatal intensive care units (NICUs). Fig. 6.4 illustrates the challenges associated with contact-based PPG approaches.

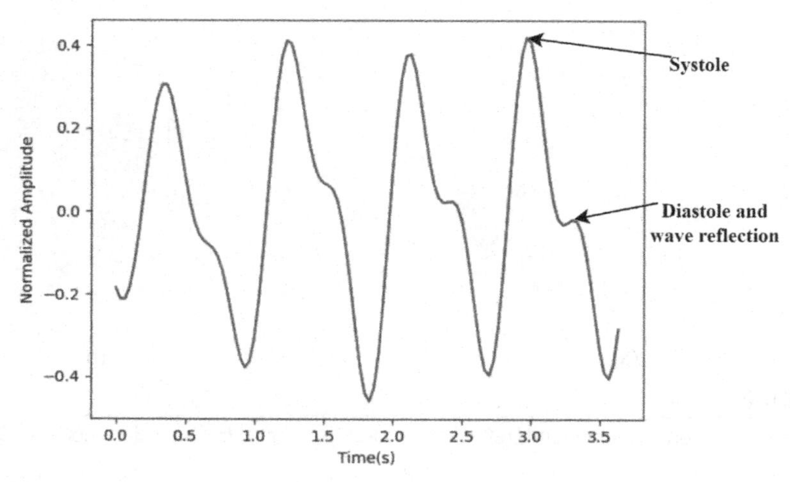

FIGURE 6.3

PPG signal.

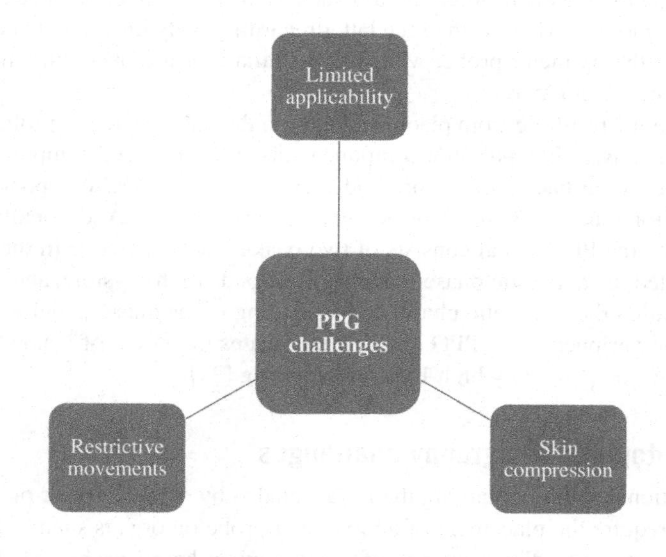

FIGURE 6.4

A flowchart depicting challenges associated with contact-based PPG approaches.

6.2.4 Noncontact photoplethysmography and potential physiological measurements

To overcome the problems associated with conventional PPG, the current research is primarily focused on noncontact PPG, based on reflection mode PPG, using visible light [16,25–39] or NIR cameras [35,39–42]. Although few studies have tried to explore several other modalities, such as microwave radars [43], due to the high cost of operating instruments and their high noise sensitivity, these studies are limited to studies utilizing visible light and NIR cameras. This variant eliminates the need to place the sensor in contact with the skin, avoiding the risk of skin compression, and can be used in the scenarios mentioned above. However, motion does degrade the accuracy of the signal. The signal extracted using this type of noncontact PPG is also called rPPG or Imaging photoplethysmography (iPPG) signal, and few studies have also reported the term BVP signal [44].

Like PPG, rPPG can be used to estimate various physiological parameters for the health monitoring of a subject. Physiological parameters, also called vital signs, are fundamental indicators of the subject's physiological state. There are five, which are listed below

Physiological parameters

Vital signs include Blood pressure (BP), Body temperature (BT), HR, BR, and SpO2 [45]. Although these parameters are sufficient to provide insight into the subject's health, their variabilities can help better diagnose disease. For example, atrial fibrillation and mental stress can be studied by exploring Heart rate variability (HRV) [46,47]. Similarly, HR and SpO2 variability can be used to detect sleep disorders like sleep apnea [48]. Although rPPG studies are predominantly focused on estimating BR, HR, and corresponding variabilities, limited studies have also explored the estimation of SpO2 [47] and BP [49]. Studies estimating several parameters are limited by the requirement of extreme conditions and the complexity of the estimation procedure [50].

6.3 Remote photoplethysmography methods

rPPG methods can be divided into conventional and DL-based approaches. Conventional methods are built upon certain assumptions, which may not hold in a few real-world scenarios. For instance, for Blind source separation (BSS)-based methods [51], the pulse signal is assumed to be periodic and independent of other artifacts, but this assumption does not hold when the subject is performing repetitive activities (i.e., periodic motion). Similarly, for projection methods such as CHROM [44] or POS [52], alpha tuning is used to separate motion artifacts and pulse from chrominance signals, which assumes comparatively different magnitudes of pulse and motion signals, which may not hold as in the abovementioned scenario. On the other hand, DL-based rPPG methods are independent of these impractical assumptions and provide better generalizability and accuracy than conventional methods. A detailed analysis of the conventional methods can be found in the study by Gupta, Ravelo-Garcia, and Morgado-Dias [50]. This section deals with analyzing and summarizing DL-based rPPG studies encompassing a brief description of existing publicly available datasets (Section 6.3.1), the basic principle of DL-based methods (Section 6.3.2), their architectural components (Section 6.3.3), and signal processing techniques (Section 6.3.4).

Table 6.1 List of RGB and NIR datasets used in rPPG studies.

Dataset image modality	Dataset name
RGB	VIPL-HR* [53]
	MAHNOB-HCI [54]
	MMSE-HR [55]
	COHFACE [56]
	PURE [57]
	UBFC-rPPG [58]
	DEAP [59]
	MMVS [60]
	OBF [61]
	VIPL-HR-V2 [62]
	AFRL [63]
NIR	VIPL-HR* [53]
	OBF* [61]
	MR-NIRP [64]

6.3.1 Publicly available datasets

The quality of rPPG signals obtained by conventional or DL-based extraction methods primarily depends on the video samples. Data acquisition for this task requires extraction of synchronized video and ground truth PPG signal samples. The data acquisition for this task is a complex procedure that requires proper handling of the measuring instruments and their synchronization. Therefore, various experts in the domain have released their datasets publicly, which can be accessed by signing a data protection agreement between the user and the owner of the dataset to avoid its illegal use for obvious purposes. The video samples of a few datasets, such as VIPL-HR [53], MAHNOB-HCI [54], MMSE-HR [55], COHFACE [56], etc., were released in compressed form, while the datasets like PURE [57], UBFC-rPPG [58], etc., were released in uncompressed format. Unfortunately, this information has been shared in a few studies. However, it was proven that uncompressed versions better suit rPPG signal extraction tasks than compressed versions because compressed versions suffer from data loss. Additionally, RGB image models have shown stronger pulsatile strength than other spectra, such as IR; consequently, a significant number of studies used RGB spectra. A list of the RGB and IR datasets used in DL-based rPPG signal extraction studies is presented in Table 6.1.

6.3.2 Skin reflection model

DL-based rPPG methods are primarily developed by following Shafer's skin dichromatic reflection model [52], which assumes that the illumination source has a fixed spectral composition, but a varying intensity depending on the subject–camera distance. The reflection of the color components at a particular point on the skin tissue varies with time due to motion-induced and intensity variations due to BVP, known as specular and diffuse reflection, respectively. Based on this, the reflection of each skin pixel at time t can be mathematically defined as

$$C_k(t) = I(t) \cdot (v_s(t) + v_d(t)) + v_n(t), \tag{6.1}$$

where $I(t)$ is the illuminance intensity level which absorbs the intensity changes due to the illumination source and $v_s(t)$, $v_d(t)$, and $v_n(t)$ depict the specular reflection caused by motion-induced intensity changes, diffuse reflection caused by BVP, and camera quantization noise, respectively. The specular and diffused reflection components can be further divided into their corresponding stationary and varying parts, denoted as

$$v_s(t) = u_s \cdot (s_0 + s(t)), \tag{6.2}$$

$$v_d(t) = u_d \cdot d_0 + u_p * p(t), \tag{6.3}$$

where u_s and u_d denote the unit color vectors of the light spectrum, s_0 and d_0 denote the stationary components of specular and diffuse reflection of skin tissue, respectively, u_p denotes the pulsatile strength in color channels, and $s(t)$ and $p(t)$ denote the continuously varying components due to motion and BVP, respectively. Substituting (6.2) and (6.3) in (6.1) and combining the stationary part of specular and diffuse reflection results in the following equation:

$$C_k(t) = I_0(1 + i(t)) \cdot (u_c \cdot c_0 + u_s \cdot s(t) + u_p \cdot p(t)) + v_n(t), \tag{6.4}$$

where $I(t)$ consists of the stationary and varying components I_0 and $I_0 * i(t)$ and $u_c * c_0$ is comprised of stationary components of specular and diffuse reflection ($u_s * s_0 + u_d * d_0$). DL models aim at extracting the pulsatile component by following this model. The effect of quantization noise is removed by resizing the image sizes of the video, followed by designing DL-based rPPG methods to extract the rPPG signal.

6.3.3 Architectural components of deep learning-based rPPG methods

Typically, an rPPG method consists of the following components: ROI selection, BVP or rPPG signal extraction, and a signal processing pipeline for identifying the frequency of interest based on physiological parameters. The rPPG signal extraction pipeline is shown in Fig. 6.5. For ROI selection, different Machine learning (ML) methods have been proposed for robust extraction of the face, followed by skin segmentation. Then the extracted region is fed to the DL network to extract a clean rPPG signal. Subsequently, standard signal processing techniques such as wavelet analysis or different variants of Fourier transform for frequency of interest extraction are applied. This section will focus on these steps.

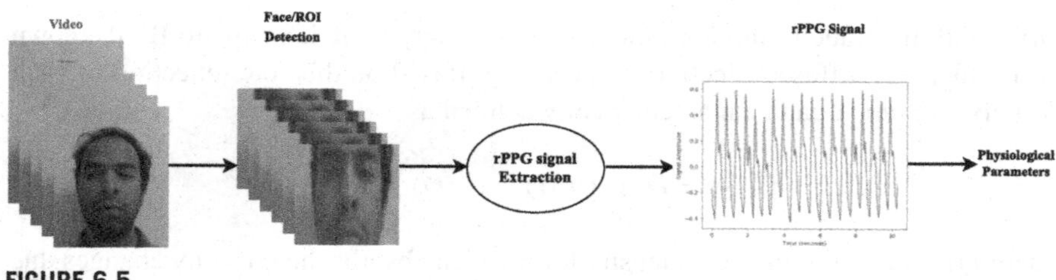

FIGURE 6.5

DL workflow for rPPG signal extraction followed by physiological parameters estimation.

Table 6.2 Characteristics of face detection methods used in the rPPG literature.

Detection method	Architecture	Occlusion	Motion	L/B	Facial points
Viola–Jones [65]	CC	No	No	B	NA
S3FD [66]	SSD	No	No	B	No
MTCNN [67]	CNN	Yes	Yes	L	5
Dlib (TS-DCN) [68]	CNN	Yes	Yes	L	68
Blazeface [69]	SSD	No	yes	L	6
OpenFace [70]	CLNF	No	Yes	L	68
CRF [25]	RF	No	Yes	L	10
RetinaFace [71]	SSD	No	No	Both	5

L and B stand for landmarks and bounding box, respectively.
CC, cascaded classifier; CNN, convolutional neural network; SSD, single-shot detector; CRF, conditional regression forest; RF, random forest.

6.3.3.1 Region of interest selection

Accurate ROI plays a critical role in identifying the potential regions that extract a clean rPPG signal, which consists of face selection methods followed by selection of skin regions rich in PPG information.

Face detection

Face detection for the rPPG signal extraction method can be broadly categorized based on its outputs, i.e., a face bounding box or facial landmarks extraction. The distribution of face detection methods in rPPG studies is presented in Fig. 6.6. Face bounding box methods include Viola–Jones [65] and Single-shot scale-invariant face detector (S3FD) [66]. In contrast, the facial landmarks methods include Convolutional neural networks (CNN)-based Multitask cascaded convolutional networks (MTCNN) [67], DLib (Task-constrained uncascaded deep convolution network (TC-DCN)) [68], blazeface [69], and other ML-based methods such as Conditional local neural field (CLNF)-NN (Openface) [70] and Conditional regression forests (CRF) [25]. On the other hand, RetinaFace [71] extracts both facial landmarks and the bounding box, providing enough flexibility based on the problem. The characteristic features of face detection methods are presented in Table 6.2.

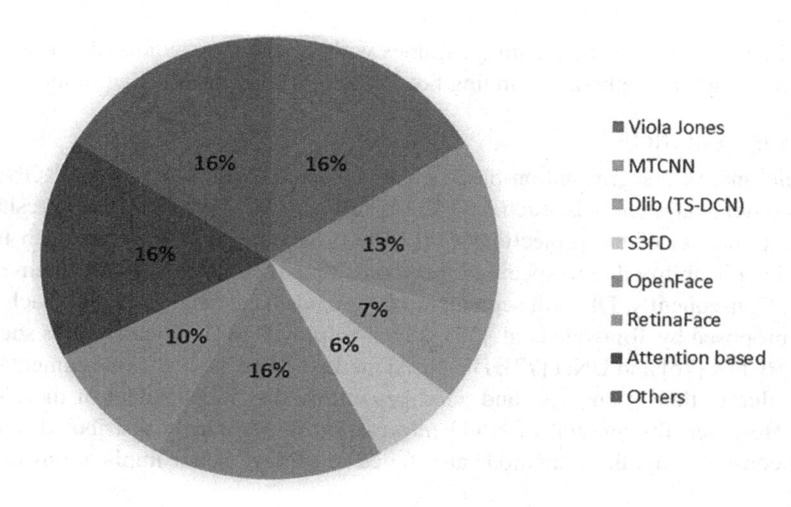

FIGURE 6.6

Distribution of ROI selection methods for rPPG signal estimation.

The Viola–Jones face detection method uses a Cascaded classifier (CC), employing Haar-like features, while S3FD can detect faces at various spatial scales using anchor tiling, matching, and max-ground false labeling.

The CNN-based methods, such as MTCNN, exploit the correlation between face alignment and detection by its cascaded architecture with three-stage convolution networks for face classification, bounding box, and facial landmarks detection. In contrast, the DLib face detection module uses TC-DCN, which is optimized for pose variations and occlusions. A mobilenetV2 [72] type Blazeface face detection method is based on a single-shot detector with a better tie resolution strategy, primarily designed for augmented reality and smartphone devices. Other methods, such as the Openface toolkit, use the CLNF model [73], which consists of a point distribution model and patch experts for landmark shape (e.g., based on eye region orientation) and appearance variations, respectively. CRF is a random forest-based motion robust algorithm which aims to explore the relationship between facial image patches and feature points from the set of faces, thereby learning conditional to global face properties.

On the other hand, the only face detection method providing both a bounding box and facial landmark points is RetinaFace. It is also a Single-shot detector (SSD) based on pixelwise localizations on various spatial scales using joint and self-supervised multitask learning.

The Viola–Jones face detection algorithm has been predominantly used in rPPG signal extraction methods due to its robustness and faster execution. At the same time, RetinaFace is the second-most used method due to its ability to detect faces under different spatial dimensions due to efficient pixelwise localizations. Besides, facial landmark point detection methods are preferable since these points can be tracked throughout all the consecutive image frames without any computational burden, unlike bounding box methods (bounding box extraction from each image frame).

Skin segmentation

Conventional skin segmentation-based methods were based on projecting RGB values to other image models, such as YCbCr, followed by identifying the thresholds for each channel of the projections [74]. The disadvantage of this approach is its limited applicability due to specified thresholds due to complex real-world environments. Consequently, DL skin segmentation methods have been proposed, such as a study proposed by Topiwala et al. [75], which employed SOTA architectures such as Mask-RCNN [76] and UNet [77]. DL-based methods have a better skin segmentation ability due to their robustness and good generalizability independent of the conditions. However, the amount of rPPG information is not evenly distributed among faces; consequently, these methods also failed to achieve better implications in this field.

Therefore, current studies rely on developing soft-attention mechanisms in conjunction with their rPPG signal extraction DL networks [27,30,39,41,42]. The benefit of using soft-attention mechanisms is twofold: First, these are very lightweight networks that can easily be incorporated into the architecture without the need for additional training and can be incorporated with the rPPG signal extraction framework; second, they can select the regions of skin containing PPG information and discard the regions with noise due to motion and illumination artifacts as proven by the study conducted by Nowara, McDuff, and Veeraraghavan [42]. Therefore, researchers have started designing soft-attention networks to improve the accuracy of the PPG signal. Face detection, in conjunction with the soft-attention mechanism, drastically reduces the computational complexity and training time of the rPPG signal extraction method due to the exclusion of an explicit skin segmentation module.

6.3.3.2 Deep learning networks for rPPG signal extraction methods

DL-based rPPG signal extraction methods are designed and developed based on Shafer's skin dichromatic model, as explained in Section 6.3.2. Additionally, these methods aim to extract spatial information from the regions selected by attention mechanisms and explore the temporal relationship from the consecutive image frames of the video. Literature suggests that the rPPG signal extraction task can be performed in two ways: simultaneously extracting spatial and temporal information or extracting spatial information followed by temporal information [28,34]. As an example of the former approach, 3D CNNs have been explored, whereas for the latter, 2D convolutions have been used for spatial feature extraction, followed by sequence models for modeling temporal relationships. Besides, several other sophisticated architectures, such as Generative adversarial network (GAN) and convolutional autoencoders, have been tested to extract a clean rPPG signal. Therefore, this subsection summarizes the SOTA DL-based rPPG methods based on the baselines presented in Table 6.3.

3D convolutional neural network-based rPPG methods

Most studies employing 3D CNNs used PhysNet, proposed by Yu, Li, and Zhao [38]. The original research suggested two variants (PhysNet-3DCNN and Physnet-2DCNN with Long short-term memory (LSTM)) of PhysNet, complying with the approaches mentioned in Section 6.3.3.2, i.e., simultaneous spatiotemporal extraction and extraction of spatial followed by temporal features. Speth et al. [34] explored the potential of a spatiotemporal network (RPNet [37]) and PhysNet in masked scenarios and found that PhysNet performed better. Similarly, Sun and Li [35] customized PhysNet to extract candidate rPPG samples and used contrastive learning to extract artifact-resistant rPPG signals. Later, Sahoo et al. [78] demonstrated the clinical implication of PhysNet-3D CNN in the NICU. To exploit the temporal variations between consecutive frames and spatial information simultaneously, Yu et al. [79] proposed a 3D CNN-based network similar to the spatiotemporal network proposed by Chen and McDuff [41] and optimized using Huber loss, which can use dual losses, namely, the Mean squared error (MSE) for smaller and the Mean absolute error (MAE) for larger error differences. The advantage of 3D CNN lies in its ability to extract spatial and temporal relationships in the video samples simultaneously, but it is computationally intensive and requires high-performance computing facilities for better performance.

Highlighting the problem of degraded performance in cross-database testing scenarios, Chung et al. [80] consider the rPPG signal extraction task as a domain generalization problem. Consequently, a DL architecture DG-rPPGNet is proposed, which consists of a 3D convolution-based architecture for domain-generalized rPPG features. This work also proposed domain permutation and augmentation to further ensure the domain-invariant rPPG features. Domain permutation aims at reducing the dependency of rPPG features on the domain, while domain augmentation using adversarial learning further refines this generalization. The architecture is trained using a combination of Negative Pearson correlation (NPC) and Cross-entropy (CE). The trained model aims at extracting the robust rPPG signal by calculating domain-invariant features. The abovementioned methods aim at extracting the rPPG signal from the single ROI. Zhao et al. [81] claimed that multiple ROIs improve accuracy and proposed a 3D CNN-based JAMSNet, which extracts rPPG features by creating Gaussian pyramids, followed by attention-based feature fusion. Subsequently, the fused rPPG features are fed to the rPPG extraction network consisting of Channel temporal joint attention (CTJA) followed by Spatiotemporal joint attention (STJA) for rPPG signal extraction. It is essential to mention that the order of these attention mechanisms (CTJA→CTJA) is vital for better performance.

Other works exploiting the advantages of 3D CNN include the study by Yue, she, and Ding [38] that used self-supervised contrastive learning to model rPPG signals from the videos. They proposed a DL architecture consisting of three modules: (1) Learnable frequency augmentation (LFA) for synthesizing samples with different frequency values of the same subject; (2) rPPG expert aggregation (REA) for extracting PPG information from different facial regions and aggregating them; and (3) optimizing the network using contrastive learning, using different spatial and frequency losses. Further, inspired by the Siamese network, Tsou et al. [36] proposed a 3D variant of the Siamese network and used aggregated features from the cheek and forehead to extract a clean rPPG signal. Addressing the computational burden due to 3D CNN training, Perepelkina et al. [32] proposed an rPPG signal extraction network employing (2+1)D convolutions, i.e., factorizing a single 3D convolution to two 3D convolutions (one in the spatial domain and one in the temporal domain) [82] for filtering out noise from the red, green, and blue signal traces. However, the disadvantage of this network is that it did not follow the characteristic pattern of the PPG signal.

2D CNN with sequence models or temporal shift module

As explained earlier, 3D CNNs are sophisticated and computationally intensive but are well suited for rPPG signal extraction, which seeks spatial and temporal relationships. Alternatively, 2D CNNs, in conjunction with the sequence models, can also be used for rPPG signal extraction, wherein spatial information extraction is followed by temporal information extraction. The spatial information is extracted using 2D convolutions, whereas sequence models such as LSTM [83] or Gated recurrent unit (GRU) [84] extract temporal relationships. Following this approach, Guo, Pen, and Chu [28] proposed a network consisting of ConvLSTM [85] and 2D CNN for extracting rPPG signals. The method could not perform well due to illumination variation.

Similarly, Chen and McDuff [41] proposed DeepPhys, a VGG-like [86] network that consists of an appearance and motion branch, which takes ROIs from the individual frames and their temporal differences (two consecutive frames) for rPPG signal extraction. Finally, the temporal information is extracted by GRU. To further denoise the signal and test the performance of the attention mechanism, Nowara, McDuff, and Veeraraghavan [42] proposed a concept of inverse attention and used the network presented by Chen and McDuff [41], followed by LSTM units to further denoise the rPPG and BR signal. Inspired by the capability of 3D CNN of simultaneous spatial and temporal feature extraction, Ren, Syrnyk, and Avadhanam [33] exploited the Temporal shift module (TSM) [87] with DeepPhys [41]. The advantage of TSM is that it not only provides the ability to extract spatial and temporal extraction simultaneously but also does not need any parameter for model training. Similarly, identifying the abilities of TSM, Liu et al. [30] replaced the appearance branch in [41] with TSM and proposed a network named EfficientPhys for the rPPG signal extraction framework. The TSM module ensured the similar accuracy and lower computational burden than 3D CNN for rPPG signal extraction. This observation is based on the results presented in the respective studies. Additionally, the literature lacks a comparative study for analyzing the performance of 3D CNN with 2D CNN with TSM.

DL architectures comprised of the combination of different convolutional variants based on spatial dimensions

A combination of 2D and 3D convolutions has been used in the literature for two purposes: creating soft-attention masks [88] and ensuring efficient spatial feature extraction [27,29]. The DL framework proposed by Zhang et al. [88] is a 3D spatiotemporal network consisting of three different modules: Low-level face feature generation (LFFG), Spatiotemporal stack convolutions (STSC), and Multihierarchical feature fusion (MHFF) (MHFF consisting of a skin map generator and a $C_{feature}$ extractor). The $C_{feature}$ module employs 2D convolution for channelwise feature extraction, while all other modules consist of 3D convolutions to extract spatiotemporal representations. On the other hand, the architecture proposed by Hu et al. [27] and Liu and Yuen [29] consists of 2D convolutions for better spatial representation and 3D convolutions for extracting spatiotemporal information.

Interestingly, Comas, Ruiz, and Sukano [40] preferred 1D convolution for temporal information extraction over sequential models. The architecture consists of 2D CNN and a Temporal derivative module (TDM) consisting of Differential temporal convolutions (DTC) satisfying the Taylor series. 2D CNN was used to extract the spatial information, while TDM consisting of DTC are 1D convolutions used to extract the temporal information. An ablation study carried out in this work proved that 2D CNN in conjunction with DTC constituted from 1D convolutions performed equally well as 3D CNN; however, by just using 2D CNN (without TDM), the performance degraded drastically.

Generative adversarial networks

GANs are designed to generate synthetic image samples to mitigate the problem of limited training data. GAN consists of two NNs: a generator and a discriminator. The generator is responsible for generating data, while the discriminator aims at identifying the fake and real data. Both generator and discriminator are trained in an adversarial manner, where the generator learns to produce counterfeit data mimicking the real data and the discriminator learns to identify the real and fake data [89] accurately. However, GANs have also been used to extract rPPG signals by designing efficient spatiotemporal representations [25,30] or by taking an input signal followed by denoising [90]. Dong, Yang, and Yin [26] and Lu, Han, and Zhou [31] used the spatiotemporal representation proposed by Niu et al. [12]. Both studies used a similar principle for creating the spatiotemporal representations, i.e., by noise modeling and its aggregation with the ground truth PPG signal to train the discriminator for categorizing real and fake data. The difference between the two studies lies in the number of inputs, ways of aggregation, and the discriminator.

On the one hand, DRNet [26] uses two input videos, extracts the noise maps from both videos, and aggregates them with the respective ground truth signals and two discriminators, while Dual-gan [31] uses a single video aggregation of a clean representation of the ground truth with modeled noise and a single discriminator for classifying the real and fake data samples. GANs are computationally intensive to train with high sensitivity to hyperparameters. Pulse-gan by Song et al. [90], which was inspired by the CHROM method by Haan et al. [44], uses GAN (a convolutional autoencoder-based generator) for rPPG signal denoising.

Autoencoders

Autoencoders consists of an encoder and decoder, where the encoder projects the input to a latent space and the decoder reverses the latent space to the desired output. The rPPG signal extraction task is the same, in which the encoder projects the video to a latent space where PPG components can be extracted from the noise component and the decoder translates the PPG information from the latent space to the noise-free PPG signal. Following the same approach, Monsalve, Benezeth, and Miteran [91] proposed a convolutional autoencoder rPPG signal extraction framework for which the encoder is inspired by the Physnet3D architecture [38], followed by a decoder network. Similarly, Li et al. [92] proposed an autoencoder-based rPPG signal extraction method which extracts relevant rPPG features using two attention mechanisms: 3D S/T and 3D-S-T. 3D S/T selects relevant rPPG features across channel dimensions, while 3D-S-T extracts spatiotemporal rPPG features. The decoder aims at extracting multiscale spatiotemporal features. Finally, global average pooling cascaded by two pointwise convolutions is used to extract the rPPG signal.

In contrast, Comas et al. [40] exploited a UNet architecture with GRU in the skip connection to extract spatial and temporal information simultaneously, similar to 3D CNN. This study utilized a spatiotemporal representation consisting of temporal information of 48 ROIs selected from the facial region. Both works employed 3D convolutions, which made these architectures more computationally intensive in terms of training time, with slightly better performance than the Physnet3D baseline.

Transformers

Transformers are a particular type of architecture that divides the input into different embeddings using a well-defined procedure, followed by processing each embedding separately, primarily for classification tasks. There are limited studies in this domain because efficient transformers are very deep and require enormous amounts of labeled data. On the other hand, shallow transformers are unable to perform well with limited data.

Park, Kim, and Dong [93] proposed a Video-based vision transformer (ViViT) to extract the rPPG signal using self-supervised contrastive learning. The embedding tokens for the RGB and NIR videos are separately created using tubulet embedding [93], which takes into account the temporal dimension, followed by calculating spatial and temporal self-attention using Multiheaded self-attention (MSA) [93]. Subsequently, average pooling and flattening are used to extract the feature vectors for contrastive learning using a loss function. Meanwhile, a transformer variant of EfficientPhys with TSM was proposed by Liu et al. [30], which performed more poorly than the respective CNN variant. Therefore, it was concluded that shallow transformers are infeasible for the rPPG signal extraction task due to limited labeled samples and insufficient depth to extract spatiotemporal information [30].

To solve the problem of large training samples, Yu et al. [94] proposed a transformer-based Physformer architecture consisting of a shallow stem, Etube, temporal difference-based MSA blocks, a spatiotemporal feedforward network, and a predictor head. The stemming aims to extract coarse local spatiotemporal features, followed by applying Etube to them for creating nonoverlapping tubes. Each tube is passed through temporal difference-based MSA blocks. Subsequently, the spatiotemporal feedforward network refines the feature map and passes it to the predictor head for rPPG signal extraction. Additionally, the network is trained using a loss function based on label distribution and curriculum learning. It is worth mentioning that Physformer also preserves long-term temporal dependencies from videos, unlike 3D CNNs. An improved version, Physformer++ [95], was also proposed, replacing a single MSA framework with two blocks, namely, slow and fast pathways, each taking different inputs. The slow pathway includes a Temporal difference multiheaded periodic self-attention module (TDMHPSA), which extracts rPPG clues for accurate positions, supervised by peak maps. In contrast, the fast pathway extracts fine-tuned rPPG features in interaction with slow pathways using a Temporal difference-based multiheaded cross-self-attention module (TDMHCSA) and lateral connections. The loss function of Physformer++ is similar to that of its previous version, with the addition of attention loss to penalize attention mechanisms for incorrect rPPG clues.

Transformers with other NN architectures

Gupta et al. proposed RADIANT [94], a video transformer-based rPPG signal method which takes multiple ROIs to construct rPPG signals using the CHROM method [44]. Subsequently, a fixed number of signals are selected using a quality score, calculated as the standard deviation of CHROM signals. The selected CHROM signals are passed to a Multilayered perceptron (MLP) to create noise-free signal embeddings. These embeddings are fed to multiheaded self-attention blocks followed by creating latent vectors. The rPPG signal is then derived from these latent vectors using another MLP. Furthermore, to overcome the problem of a large training dataset, RADIANT was first pretrained using an imagenet database, followed by synthetic PPG signals, before getting trained for extracting rPPG signals from videos.

Similarly, Revanur et al. [96] proposed a combination of DeepPhys and a vision transformer, emphasizing the significance of instantaneous HR and BR estimates. The architecture takes face videos and extracts signals using DeepPhys. Subsequently, the resultant signals are converted to corresponding signal embeddings using an MLP architecture. These embeddings are then fed to a vision transformer for rPPG signal extraction. The proposed architecture is trained using maximum cross-correlation loss [97].

Other methods

Liu et al. [88] proposed a different approach from the conventional DL approach, where the motion signal from the nose region and the color signal from the cheek region are extracted and undergo Fast Fourier transform (FFT). Then, both frequency spectra are fed to respective Bidirectional LSTM (BiLSTM), sharing weights. The motion signal output is used as an attention mask, which is used for denoising the color signal. The resultant denoised signal is again passed to LSTM to extract the frequency spectrum for the potential rPPG signal. Finally, the frequency spectrum is converted to the rPPG signal. Although this approach is computationally simple, it

Table 6.3 DL-based rPPG methods according to baseline architectures.

Baselines	Variants	Loss functions	Image modality	Limitations
3D CNN	3D CNN [29, 34–36,38, 39,78,79,81]	NPC [29,34,78,81], MSE [35,36,38], Huber loss (MSE and MAE) [79], frequency contrastive loss, frequency ratio consistency loss, cross-video frequency agreement loss [39].	RGB [29,34–36, 38,39,78,79,81], NIR [35,78]	Computationally intensive methods [29,34–36,38,39,78]. Limited applicability for periodic motion due to similar magnitudes of pulse and motion signals [35]. Computationally intensive method with susceptibility to rigorous motion and sharp decrease [81].
	(2+1)D CNN [32]	MSE [32], NPC	RGB	The characteristic pattern of the PPG signal is not considered; instead, the noisy components are filtered; may contain other information not relevant to the PPG signal [32].
2D CNN + sequence models	2D CNN + LSTM [42]	MSE	RGB, NIR	Their assumption of a precise attention mechanism may not be feasible in real-world scenarios such as face occlusion, variant illumination, etc.
	2D CNN + GRU [41]	MSE	RGB, NIR	The accuracy of the method is highly dependent on the effectiveness of the attention mechanism.
	2D CNN + ConvLSTM [28]	MSE and SNR loss	RGB	The method suffers in the case of variant illumination conditions.
2D CNN	2D CNN with TSM [30], TSDAN [33]	MSE	RGB	The methods depend on extracting the temporal differences, which may be challenging in real-world situations such as improper light conditions.
GAN	DRNet [26], DualGAN [31], PulseGAN [90]	Frequency and signal L1 distance [31,90], adversarial loss [90], NPC [31], PSD CE [26]	RGB [26,31], signal [90]	An attention mechanism is used after generating a spatiotemporal representation, which may prevent accurate estimations due to the limited inclusion of the spatial dimension [26,31]. Computationally intensive [90].

continued on next page

Table 6.3 *(continued)*

Baselines	Variants	Loss functions	Image modality	Limitations
Autoencoder	3D-conv autoencoders [40,91,92]	NPC, SNR [40,91,92], PSD CE, L1 distance, and MSE [40]	RGB [91], NIR [40]	The method works on monochromatic IR signals, which may be prone to noise for higher dimensions [40]. Ground truth signal is not employed; instead, the estimated signal for the POS algorithm is used; therefore, this method is likely to suffer more in real-world scenarios [91]. The method is computationally intensive due to 3D convolutions as building blocks of the autoencoder architecture; the effect of artifacts is also not addressed [92].
Transformer	SWIN [30], ViViT [99], ViT [94,95]	MSE [30], Softmax function type [99], Kullback divergence loss [94,95], CE [94,95], NPC [94,95]	RGB	Shallow transformers are infeasible for physiological measurement and require a larger labeled sample dataset [30]. The transformers are computationally intensive and require large labeled training samples [99]. Computationally intensive architectures [94,95].
Conv. fusion	2D and (2+1)D [27,29,88]	SNR and MSE [27], NPC [29,88], PSD CE, and skin/nonskin CE [88]	RGB	The proposed architecture is slightly overfitted, and performance degradation is seen during an abrupt change in physiological parameters due to emotion elicitation and insufficient spatial dimensions [27]. The performance degrades for smaller ROIs [29]. The method's performance depends heavily on its optical flow-based skin segmentation, which might hinder its robustness under challenging illumination conditions [88].
	(2+1)D CNN [24]	MSE, temporal shift probability	RGB	Better accuracy at the cost of higher execution time for training for learning parameters to identify the temporal offset.
LSTM	BiLSTM [100]	MSE	RGB	The method does not explore spatial dimensions.
TF with other NN architectures	2d conv + TF [96], MLP + Vision TF [101]	Maximum cross-correlation [96], MSE [101]	RGB	Combination of DeepPhys with video transfer results in a computationally intensive architecture [96]. Computationally intensive architecture, with intolerance to rigorous motion [101].
Other hybrid architectures	1D CNN + LSTM [98]	MSE	RGB	The proposed architecture depends on accurate extraction from ROI and is intolerant of head movement.

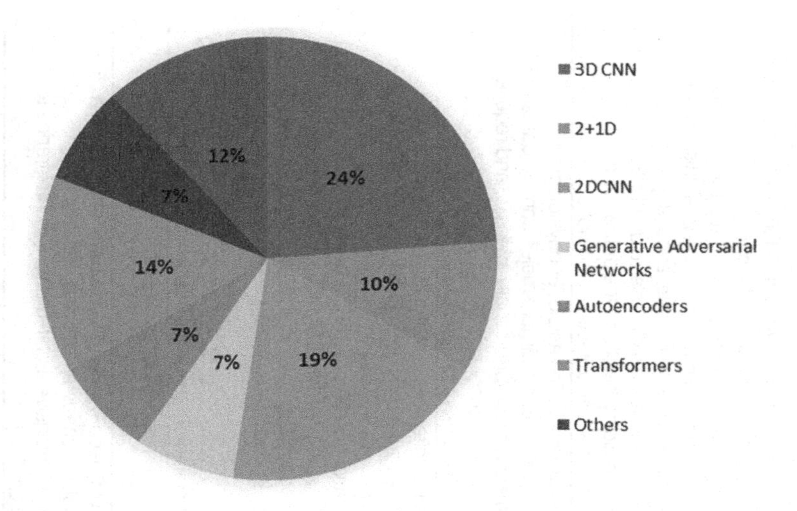

FIGURE 6.7

DL architectures for rPPG signal estimation.

does not consider spatial information extraction, which could be a concern regarding the accuracy of the rPPG signal.

Liu et al. [98] proposed an adaptive weighting scheme to extract robust rPPG signals from multiple ROIs. Specifically, the proposed method extracts the green channel temporal signal from multiple ROIs and uses 1D CNN (1D convolution with batch normalization and Rectified linear unit (ReLU)). The resultant rPPG signal is then constructed as the weighted combination of green temporal signals. In continuation, a three-layered LSTM-based architecture was also proposed to estimate HR. The entire architecture is trained using MSE.

Summarizing architectures

The distribution of DL architectures for rPPG signal extraction is depicted in Fig. 6.7. Overall, 3D CNNs have been predominant in the literature due to their ability to extract spatial and temporal information simultaneously. Although spatiotemporal networks could be used as an alternative for reducing computational complexity and faster execution, various studies have shown that 3D CNNs perform better than spatiotemporal networks [34,38].

Furthermore, a TSM can be used as an alternative, somewhat decreasing the computational burden without extra trainable parameters accuracy. Other architectures, such as GANs and autoencoders, are computationally expensive, but GANs may prove beneficial when generated rPPG signals are compared with the ground truth PPG signals using an additional neural network called discriminator. Transformers do not seem to be a reasonable choice for the rPPG signal extraction task due to the requirement of a sophisticated architecture and larger training samples, but this observation applies to currently available transformer architectures.

6.3.4 Signal processing techniques for physiological parameters estimation

Accurate rPPG signal extraction is vital for correct physiological parameter estimations, which eventually results in better health monitoring. Once a cleaner rPPG signal is extracted using one of the methods presented in Section 6.3.3.2, bandpass filtering is applied, wherein the low- and high-pass frequencies depend on the estimated physiological parameters. For HR estimation, the rPPG signal is low and high-pass-filtered with frequencies of 0.7 and 4.0 Hz, respectively. Then, techniques like wavelet analysis or Fourier transforms can be used to extract the frequency of interest by transforming the rPPG signal into its frequency domain. Since standard signal processing techniques are prevalent in the rPPG signal extraction literature, these techniques are beyond the scope of this chapter.

Literature suggests that the commonly estimated parameters after extracting the rPPG signal are HR and BR [33] due to the similar extraction procedure except for their respective high- and low-pass threshold frequencies. A few studies have also reported their variabilities since these provide more profound insights than values [38,41]. However, limited studies have reported the estimation of parameters such as SpO2 and BP due to the complexity involved in measuring the ground truth and parameter extraction. For example, designing scenarios to cover broader SpO2 value ranges may cause discomfort to the participant. Additionally, few DL methods also estimate parameters by learning a mapping between video and physiological parameters as a regression problem [13]. HR, BR, their variability, SpO2, and BP have been measured and reported in the respective studies.

6.4 Limitations and recommendations for future research

Although a large number of studies have been conducted in the domain, it is still in the proof-of-concept stage due to various limitations which need to be addressed in the future. Fig. 6.8 presents a graphical representation of these limitations. First, most rPPG signal extraction studies focused on publicly available datasets. These datasets are very small and possess limited applicability regarding including real-world situations during data acquisition. Consequently, the generalizability and robustness of the DL methods in this domain are limited. Additionally, a few data sets, such as PURE [57] and VIPL-HR [53], did not consider several factors, such as skin type, color, ethnicity, etc., for video sample acquisition, although it depends on the availability of subjects. Several studies have used spatial [34,40,99] and frequency [39] augmentation techniques to overcome this limitation of limited data availability, but those techniques are physiological parameter-dependent. Second, although the cur-

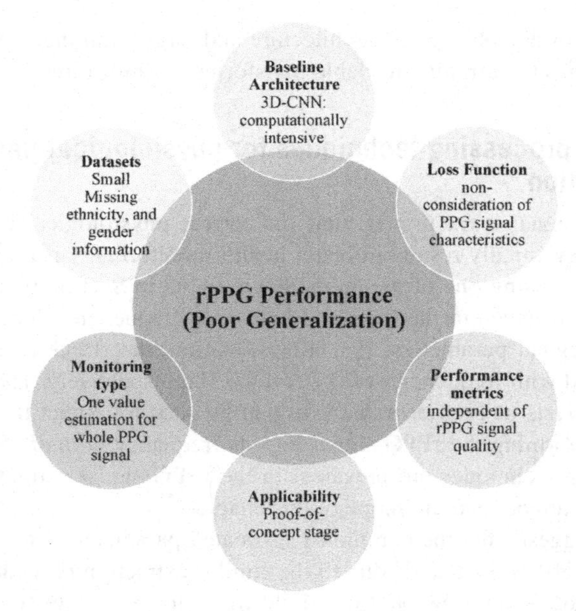

FIGURE 6.8

Limitations of the existing SOTA studies.

rent datasets are designed to match real-world scenarios, it is infeasible to consider them together in a dataset due to enormously challenging situations.

Third, most studies are based on 3D CNN architectures, which are computationally intensive to train and use in real-time. Although some modifications may reduce this intensity, further research is required to explore simpler, robust DL architectures. Fourth, various studies have investigated different SOTA deep NN architectures, and these architectures were optimized using a well-defined loss function. Most loss functions did not consider the PPG signal characteristics; instead, the loss functions are predominantly created using signal similarity without considering contextual information present in the PPG signal. Therefore, the loss functions preserving the PPG signal characteristics still need to be identified. Fifth, almost every rPPG signal extraction except two (Template matching [91], Irrelevant power ratio (IPR) [35]) has been tested based on the accurate estimation of physiological parameters. Therefore, the metrics for signal quality estimation of the rPPG signal need to be identified since the resultant rPPG signal may estimate a parameter correctly but fail to estimate other parameters accurately. Most SOTA studies focused on calculating the mean estimates of the physiological parameters, which are of limited applicability for real-time health monitoring. Overall, current methods require significant improvements to be deployable in real-world environments. As an instance, the trained DL models should be easily deployable to the portable gadgets with limited resources, which is almost not possible to date, except in a study by Nowara, McDuff, and Veeraraghavan [42].

6.5 Conclusion

This chapter discusses the basic principles of PPG, as well as its disadvantages in terms of its implications and challenges, and presents the noncontact variant of PPG. Furthermore, it also summarizes the SOTA ROI selection methods and their comparative analysis based on a few common characteristics. Also, existing rPPG signal extraction methods are categorized based on the standard DL baselines presented in Table 6.3, followed by summarizing them and highlighting their merits and demerits. Subsequently, the existing SOTA studies' limitations in the domain are presented and analyzed.

Addressing the limitations, specific recommendations for future studies are also presented in this chapter. To summarize, this domain suffers from limited data availability and diversification in terms of covering the real-world conditions for deploying methods in such situations. Additionally, the literature suggests the extensive use of 3D CNNs since they provide the ability to simultaneously extract accurate spatial and temporal information for rPPG signal extraction.

References

[1] D.B. Geselowitz, On the theory of the electrocardiogram, Proceedings of the IEEE 77 (6) (1989) 857–876.

[2] C.C. Brown, D.B. Giddbon, E.D. Dean, Techniques of plethysmography, Psychophysiology 1 (3) (1965) 253–266.

[3] E.M. Fong, W.Y. Chung, A hygroscopic sensor electrode for fast stabilized non-contact ecg signal acquisition, Sensors 15 (8) (2015) 19237–19250.

[4] W. Verkruysse, L.O. Svaasand, J.S. Nelson, Remote plethysmographic imaging using ambient light, Optics Express 16 (26) (2008) 21434–21445.

[5] A. Kumar, A. Kaur, M. Kumar, Face detection techniques: a review, Artificial Intelligence Review 52 (2) (2019) 927–948.

[6] K. Juneja, C. Rana, An extensive study on traditional-to-recent transformation on face recognition system, Wireless Personal Communications 118 (4) (2021) 3075–3128.

[7] M. Wacker, H. Witte, Time-frequency techniques in biomedical signal analysis, Methods of Information in Medicine 52 (04) (2013) 279–296.

[8] M.A. Hassan, A.S. Malik, D. Fofi, N. Saad, B. Karasfi, Y.S. Ali, et al., Heart rate estimation using facial video: a review, Biomedical Signal Processing and Control 38 (2017) 346–360.

[9] C. Wang, T. Pun, G. Chanel, A comparative survey of methods for remote heart rate detection from frontal face videos, Frontiers in Bioengineering and Biotechnology 6 (2018) 33.

[10] Y. Sun, N. Thakor, Photoplethysmography revisited: from contact to noncontact, from point to imaging, IEEE Transactions on Biomedical Engineering 63 (3) (2015) 463–477.

[11] M. Harford, J. Catherall, S. Gerry, J.D. Young, P. Watkinson, Availability and performance of image-based, non-contact methods of monitoring heart rate, blood pressure, respiratory rate, and oxygen saturation: a systematic review, Physiological Measurement 40 (6) (2019) 06TR01.
[12] X. Niu, S. Shan, H. Han, X. Chen, Rhythmnet: end-to-end heart rate estimation from face via spatial-temporal representation, IEEE Transactions on Image Processing 29 (2019) 2409–2423.
[13] Z. Yu, X. Li, X. Niu, J. Shi, G. Zhao, Autohr: a strong end-to-end baseline for remote heart rate measurement with neural searching, IEEE Signal Processing Letters 27 (2020) 1245–1249.
[14] X. Niu, H. Han, S. Shan, X. Chen, Continuous heart rate measurement from face: a robust rppg approach with distribution learning, in: 2017 IEEE International Joint Conference on Biometrics (IJCB), IEEE, ISBN 1538611244, 2017, pp. 642–650.
[15] J. Rumiński, A. Kwaśniewska, M. Szankin, T. Kocejko, M. Mazur-Milecka, Evaluation of facial pulse signals using deep neural net models, in: 2019 41st Annual International Conference of the IEEE Engineering in Medicine and Biology Society (EMBC), IEEE, ISBN 1538613115, 2019, pp. 3399–3403.
[16] R. Song, S. Zhang, C. Li, Y. Zhang, J. Cheng, X. Chen, Heart rate estimation from facial videos using a spatiotemporal representation with convolutional neural networks, IEEE Transactions on Instrumentation and Measurement 69 (10) (2020) 7411–7421.
[17] A. Ni, A. Azarang, N. Kehtarnavaz, A review of deep learning-based contactless heart rate measurement methods, Sensors 21 (11) (2021) 3719.
[18] C.H. Cheng, K.L. Wong, J.W. Chin, T.T. Chan, R.H. So, Deep learning methods for remote heart rate measurement: a review and future research agenda, Sensors 21 (18) (2021) 6296.
[19] Z. Yang, H. Wang, F. Lu, Assessment of deep learning-based heart rate estimation using remote photoplethysmography under different illuminations, IEEE Transactions on Human-Machine Systems 52 (6) (2022) 1236–1246.
[20] B. Anupama, K. Ravishankar, Working mechanism and utility of pulse oximeter, International Journal of Sport, Exercise and Health Research 2 (2) (2018) 111–113.
[21] A. Moço, W. Verkruysse, Pulse oximetry based on photoplethysmography imaging with red and green light, Journal of Clinical Monitoring and Computing 35 (1) (2021) 123–133.
[22] P. Sahindrakar, G. de Haan, I. Kirenko, Improving motion robustness of contact-less monitoring of heart rate using video analysis, Technische Universiteit Eindhoven, Department of Mathematics and Computer Science, 2011.
[23] J. Allen, Photoplethysmography and its application in clinical physiological measurement, Physiological Measurement 28 (3) (2007) R1.
[24] J. Comas, A. Ruiz, F. Sukno, Efficient remote photoplethysmography with temporal derivative modules and time-shift invariant loss, in: Proceedings of the IEEE/CVF Conference on Computer Vision and Pattern Recognition, 2022, pp. 2182–2191.
[25] M. Dantone, J. Gall, G. Fanelli, L. Van Gool, Real-time facial feature detection using conditional regression forests, in: 2012 IEEE Conference on Computer Vision and Pattern Recognition, IEEE, ISBN 1467312282, 2012, pp. 2578–2585.
[26] Y. Dong, G. Yang, Y. Yin, Drnet: decomposition and reconstruction network for remote physiological measurement, arXiv preprint, arXiv:2206.05687, 2022.
[27] M. Hu, D. Guo, M. Jiang, F. Qian, X. Wang, F. Ren, rppg-based heart rate estimation using spatial-temporal attention network, IEEE Transactions on Cognitive and Developmental Systems (2021).

[28] M. Hu, D. Guo, X. Wang, P. Ge, Q. Chu, A novel spatial-temporal convolutional neural network for remote photoplethysmography, in: 2019 12th International Congress on Image and Signal Processing, BioMedical Engineering and Informatics (CISP-BMEI), IEEE, ISBN 1728148529, 2019, pp. 1–6.

[29] S.Q. Liu, P.C. Yuen, A general remote photoplethysmography estimator with spatiotemporal convolutional network, in: 2020 15th IEEE International Conference on Automatic Face and Gesture Recognition (FG 2020), IEEE, ISBN 1728130794, 2020, pp. 481–488.

[30] X. Liu, B. Hill, Z. Jiang, S. Patel, D. McDuff, Efficientphys: enabling simple, fast and accurate camera-based cardiac measurement, in: Proceedings of the IEEE/CVF Winter Conference on Applications of Computer Vision, 2021, pp. 5008–5017.

[31] H. Lu, H. Han, S.K. Zhou, Dual-gan: joint bvp and noise modeling for remote physiological measurement, in: Proceedings of the IEEE/CVF Conference on Computer Vision and Pattern Recognition, 2021, pp. 12404–12413.

[32] O. Perepelkina, M. Artemyev, M. Churikova, M. Grinenko, Hearttrack: convolutional neural network for remote video-based heart rate monitoring, in: Proceedings of the IEEE/CVF Conference on Computer Vision and Pattern Recognition Workshops, 2020, pp. 288–289.

[33] Y. Ren, B. Syrnyk, N. Avadhanam, Dual attention network for heart rate and respiratory rate estimation, in: 2021 IEEE 23rd International Workshop on Multimedia Signal Processing (MMSP), IEEE, ISBN 1665432888, 2021, pp. 1–6.

[34] J. Speth, N. Vance, P. Flynn, K. Bowyer, A. Czajka, Remote pulse estimation in the presence of face masks, in: Proceedings of the IEEE/CVF Conference on Computer Vision and Pattern Recognition, 2022, pp. 2086–2095.

[35] Z. Sun, X. Li, Contrast-phys: unsupervised video-based remote physiological measurement via spatiotemporal contrast, in: European Conference on Computer Vision, Springer, 2022, pp. 492–510.

[36] Y.Y. Tsou, Y.A. Lee, C.T. Hsu, S.H. Chang, Siamese-rppg network: remote photoplethysmography signal estimation from face videos, in: Proceedings of the 35th Annual ACM Symposium on Applied Computing, 2020, pp. 2066–2073.

[37] S. Vijayarangan, R. Vignesh, B. Murugesan, S. Preejith, J. Joseph, M. Sivaprakasam, Rpnet: a deep learning approach for robust r peak detection in noisy ecg, in: 2020 42nd Annual International Conference of the IEEE Engineering in Medicine & Biology Society (EMBC), IEEE, ISBN 1728119901, 2020, pp. 345–348.

[38] Z. Yu, X. Li, G. Zhao, Remote photoplethysmograph signal measurement from facial videos using spatio-temporal networks, arXiv preprint, arXiv:1905.02419, 2019.

[39] Z. Yue, M. Shi, S. Ding, Video-based remote physiological measurement via self-supervised learning, arXiv preprint, arXiv:2210.15401, 2022.

[40] A. Comas, T.K. Marks, H. Mansour, S. Lohit, Y. Ma, X. Liu, Turnip: time-series with recurrence for nir imaging ppg, in: 2021 IEEE International Conference on Image Processing (ICIP), IEEE, ISBN 1665441151, 2021, pp. 309–313.

[41] W. Chen, D. McDuff, Deepphys: video-based physiological measurement using convolutional attention networks, in: Proceedings of the European Conference on Computer Vision (ECCV), 2018, pp. 349–365.

[42] E.M. Nowara, D. McDuff, A. Veeraraghavan, The benefit of distraction: denoising camera-based physiological measurements using inverse attention, in: Proceedings of the IEEE/CVF International Conference on Computer Vision, 2021, pp. 4955–4964.

[43] Y. Rong, P.C. Theofanopoulos, G.C. Trichopoulos, D.W. Bliss, A new principle of pulse detection based on terahertz wave plethysmography, Scientific Reports 12 (1) (2022) 1–15.
[44] G. De Haan, V. Jeanne, Robust pulse rate from chrominance-based rppg, IEEE Transactions on Biomedical Engineering 60 (10) (2013) 2878–2886.
[45] W.Q. Mok, W. Wang, S.Y. Liaw, Vital signs monitoring to detect patient deterioration: an integrative literature review, International Journal of Nursing Practice 21 (2015) 91–98.
[46] J. Taelman, S. Vandeput, A. Spaepen, S.V. Huffel, Influence of mental stress on heart rate and heart rate variability, in: 4th European Conference of the International Federation for Medical and Biological Engineering, Springer, 1999, pp. 1366–1369.
[47] M. Fioranelli, M. Piccoli, G. Mileto, F. Sgreccia, P. Azzolini, M. Risa, et al., Analysis of heart rate variability five minutes before the onset of paroxysmal atrial fibrillation, Pacing and Clinical Electrophysiology 22 (5) (1999) 743–749.
[48] A. Sabil, M. Blanchard, C. Annweiler, S. Bailly, F. Goupil, T. Pigeanne, et al., Overnight pulse rate variability and risk of major neurocognitive disorder in older patients with obstructive sleep apnea, Journal of the American Geriatrics Society 70 (11) (2022) 3127–3137.
[49] B.F. Wu, L.W. Chiu, Y.C. Wu, C.C. Lai, P.H. Chu, Contactless blood pressure measurement via remote photoplethysmography with synthetic data generation using generative adversarial network, in: Proceedings of the IEEE/CVF Conference on Computer Vision and Pattern Recognition, 2022, pp. 2130–2138.
[50] A. Gupta, A.G. Ravelo-García, F.M. Dias, Availability and performance of face based non-contact methods for heart rate and oxygen saturation estimations: a systematic review, Computer Methods and Programs in Biomedicine (2022) 106771.
[51] A. Gupta, A.G. Ravelo-García, F.M. Dias, A motion and illumination resistant non-contact method using undercomplete independent component analysis and Levenberg-Marquardt algorithm, IEEE Journal of Biomedical and Health Informatics 26 (10) (2022) 4837–4848.
[52] W. Wang, A.C. Den Brinker, S. Stuijk, G. De Haan, Algorithmic principles of remote ppg, IEEE Transactions on Biomedical Engineering 64 (7) (2016) 1479–1491.
[53] X. Niu, H. Han, S. Shan, X. Chen, Vipl-hr: a multi-modal database for pulse estimation from less-constrained face video, in: Asian Conference on Computer Vision, Springer, 2018, pp. 562–576.
[54] M. Soleymani, J. Lichtenauer, T. Pun, M. Pantic, A multimodal database for affect recognition and implicit tagging, IEEE Transactions on Affective Computing 3 (1) (2011) 42–55.
[55] Z. Zhang, J.M. Girard, Y. Wu, X. Zhang, P. Liu, U. Ciftci, et al., Multimodal spontaneous emotion corpus for human behavior analysis, in: Proceedings of the IEEE Conference on Computer Vision and Pattern Recognition, 2016, pp. 3438–3446.
[56] G. Heusch, A. Anjos, S. Marcel, A reproducible study on remote heart rate measurement, arXiv preprint, arXiv:1709.00962, 2017.
[57] R. Stricker, S. Müller, H.M. Gross, Non-contact video-based pulse rate measurement on a mobile service robot, in: The 23rd IEEE International Symposium on Robot and Human Interactive Communication, IEEE, ISBN 1479967653, 2014, pp. 1056–1062.
[58] S. Bobbia, R. Macwan, Y. Benezeth, A. Mansouri, J. Dubois, Unsupervised skin tissue segmentation for remote photoplethysmography, Pattern Recognition Letters 124 (2019) 82–90.

[59] S. Koelstra, C. Muhl, M. Soleymani, J.S. Lee, A. Yazdani, T. Ebrahimi, et al., Deap: a database for emotion analysis; using physiological signals, IEEE Transactions on Affective Computing 3 (1) (2011) 18–31.
[60] Z. Yue, S. Ding, S. Yang, H. Yang, Z. Li, Y. Zhang, et al., Deep super-resolution network for rppg information recovery and noncontact heart rate estimation, IEEE Transactions on Instrumentation and Measurement 70 (2021) 1–11.
[61] X. Li, I. Alikhani, J. Shi, T. Seppanen, J. Junttila, K. Majamaa-Voltti, et al., The obf database: a large face video database for remote physiological signal measurement and atrial fibrillation detection, in: 2018 13th IEEE International Conference on Automatic Face & Gesture Recognition (FG 2018), IEEE, ISBN 1538623358, 2018, pp. 242–249.
[62] X. Li, H. Han, H. Lu, X. Niu, Z. Yu, A. Dantcheva, et al., The 1st challenge on remote physiological signal sensing (repss), in: Proceedings of the IEEE/CVF Conference on Computer Vision and Pattern Recognition Workshops, 2020, pp. 314–315.
[63] J.R. Estepp, E.B. Blackford, C.M. Meier, Recovering pulse rate during motion artifact with a multi-imager array for non-contact imaging photoplethysmography, in: 2014 IEEE International Conference on Systems, Man, and Cybernetics (SMC), IEEE, ISBN 1479938408, 2014, pp. 1462–1469.
[64] E. Magdalena Nowara, T.K. Marks, H. Mansour, A. Veeraraghavan, Sparseppg: towards driver monitoring using camera-based vital signs estimation in near-infrared, in: Proceedings of the IEEE Conference on Computer Vision and Pattern Recognition Workshops, 2018, pp. 1272–1281.
[65] P. Viola, M. Jones, Rapid object detection using a boosted cascade of simple features, in: Proceedings of the 2001 IEEE Computer Society Conference on Computer Vision and Pattern Recognition. CVPR 2001, vol. 1, Ieee, ISBN 0769512720, 2001, I–I.
[66] S. Zhang, X. Zhu, Z. Lei, H. Shi, X. Wang, S.Z. Li, S3fd: single shot scale-invariant face detector, in: Proceedings of the IEEE International Conference on Computer Vision, 2017, pp. 192–201.
[67] K. Zhang, Z. Zhang, Z. Li, Y. Qiao, Joint face detection and alignment using multitask cascaded convolutional networks, IEEE Signal Processing Letters 23 (10) (2016) 1499–1503.
[68] Z. Zhang, P. Luo, C.C. Loy, X. Tang, Facial landmark detection by deep multi-task learning, in: European Conference on Computer Vision, Springer, 2014, pp. 94–108.
[69] V. Bazarevsky, Y. Kartynnik, A. Vakunov, K. Raveendran, M. Grundmann, Blazeface: sub-millisecond neural face detection on mobile gpus, arXiv preprint, arXiv:1907.05047, 2019.
[70] T. Baltrušaitis, P. Robinson, L.P. Morency, Openface: an open source facial behavior analysis toolkit, in: 2016 IEEE Winter Conference on Applications of Computer Vision (WACV), IEEE, ISBN 1509006419, 2016, pp. 1–10.
[71] J. Deng, J. Guo, E. Ververas, I. Kotsia, S. Zafeiriou, Retinaface: single-shot multi-level face localisation in the wild, in: Proceedings of the IEEE/CVF Conference on Computer Vision and Pattern Recognition, 2019, pp. 5203–5212.
[72] M. Sandler, A. Howard, M. Zhu, A. Zhmoginov, L.C. Chen, Mobilenetv2: inverted residuals and linear bottlenecks, in: Proceedings of the IEEE Conference on Computer Vision and Pattern Recognition, 2019, pp. 4510–4520.
[73] T. Baltrusaitis, P. Robinson, L.P. Morency, Constrained local neural fields for robust facial landmark detection in the wild, in: Proceedings of the IEEE International Conference on Computer Vision Workshops, 2013, pp. 354–361.
[74] T.M. Mahmoud, A new fast skin color detection technique, World Academy of Science, Engineering and Technology 43 (2008) 501–505.

[75] A. Topiwala, L. Al-Zogbi, T. Fleiter, A. Krieger, Adaptation and evaluation of deep learning techniques for skin segmentation on novel abdominal dataset, in: 2019 IEEE 19th International Conference on Bioinformatics and Bioengineering (BIBE), IEEE, ISBN 1728146178, 2019, pp. 752–759.

[76] K. He, G. Gkioxari, P. Dollár, R. Girshick, Mask r-cnn, in: Proceedings of the IEEE International Conference on Computer Vision, 2017, pp. 2961–2969.

[77] O. Ronneberger, P. Fischer, T. Brox, U-net: convolutional networks for biomedical image segmentation, in: International Conference on Medical Image Computing and Computer-Assisted Intervention, Springer, 2015, pp. 234–241.

[78] N.N. Sahoo, B. Murugesan, A. Das, S. Karthik, K. Ram, S. Leonhardt, et al., Deep learning based non-contact physiological monitoring in neonatal intensive care unit, in: 2022 44th Annual International Conference of the IEEE Engineering in Medicine & Biology Society (EMBC), IEEE, ISBN 1728127823, 2022, pp. 1327–1330.

[79] Y. Zhao, B. Zou, F. Yang, L. Lu, A.N. Belkacem, C. Chen, Video-based physiological measurement using 3d central difference convolution attention network, in: 2021 IEEE International Joint Conference on Biometrics (IJCB), IEEE, ISBN 1665437804, 2021, pp. 1–6.

[80] W.H. Chung, C.J. Hsieh, S.H. Liu, C.T. Hsu, Domain generalized rppg network: disentangled feature learning with domain permutation and domain augmentation, in: Proceedings of the Asian Conference on Computer Vision, 2022, pp. 807–823.

[81] C. Zhao, H. Wang, H. Chen, W. Shi, Y. Feng, Jamsnet: a remote pulse extraction network based on joint attention and multi-scale fusion, IEEE Transactions on Circuits and Systems for Video Technology (2022).

[82] D. Tran, H. Wang, L. Torresani, J. Ray, Y. LeCun, M. Paluri, A closer look at spatiotemporal convolutions for action recognition, in: Proceedings of the IEEE Conference on Computer Vision and Pattern Recognition, 2017, pp. 6450–6459.

[83] S. Hochreiter, J. Schmidhuber, Long short-term memory, Neural Computation 9 (8) (1997) 1735–1780.

[84] K. Cho, B. Van Merriënboer, C. Gulcehre, D. Bahdanau, F. Bougares, H. Schwenk, et al., Learning phrase representations using rnn encoder-decoder for statistical machine translation, arXiv preprint, arXiv:1406.1078, 2014.

[85] X. Shi, Z. Chen, H. Wang, D.Y. Yeung, W.K. Wong, Wc Woo, Convolutional lstm network: a machine learning approach for precipitation nowcasting, Advances in Neural Information Processing Systems (2015) 28.

[86] K. Simonyan, A. Zisserman, Very deep convolutional networks for large-scale image recognition, arXiv preprint, arXiv:1409.1556, 2014.

[87] J. Lin, C. Gan, S. Han, Tsm: temporal shift module for efficient video understanding, in: Proceedings of the IEEE/CVF International Conference on Computer Vision, 2019, pp. 7083–7093.

[88] P. Zhang, B. Li, J. Peng, W. Jiang, Multi-hierarchical convolutional network for efficient remote photoplethysmograph signal and heart rate estimation from face video clips, arXiv preprint, arXiv:2104.02260, 2021.

[89] I. Goodfellow, J. Pouget-Abadie, M. Mirza, B. Xu, D. Warde-Farley, S. Ozair, et al., Generative adversarial networks, Communications of the ACM 63 (11) (2020) 139–144.

[90] R. Song, H. Chen, J. Cheng, C. Li, Y. Liu, X. Chen, Pulsegan: learning to generate realistic pulse waveforms in remote photoplethysmography, IEEE Journal of Biomedical and Health Informatics 25 (5) (2021) 1373–1384.

[91] D. Botina-Monsalve, Y. Benezeth, J. Miteran, Rtrppg: an ultra light 3dcnn for real-time remote photoplethysmography, in: Proceedings of the IEEE/CVF Conference on Computer Vision and Pattern Recognition, 2022, pp. 2146–2154.

[92] B. Li, W. Jiang, J. Peng, X. Li, Deep learning-based remote-photoplethysmography measurement from short-time facial video, Physiological Measurement 43 (11) (2022) 115003.

[93] A. Arnab, M. Dehghani, G. Heigold, C. Sun, M. Lučić, C. Schmid, Vivit: a video vision transformer, in: Proceedings of the IEEE/CVF International Conference on Computer Vision, 2021, pp. 6836–6846.

[94] Z. Yu, Y. Shen, J. Shi, H. Zhao, P.H. Torr, G. Zhao, Physformer: facial video-based physiological measurement with temporal difference transformer, in: Proceedings of the IEEE/CVF Conference on Computer Vision and Pattern Recognition, 2022, pp. 4186–4196.

[95] Z. Yu, Y. Shen, J. Shi, H. Zhao, Y. Cui, J. Zhang, et al., Physformer++: facial video-based physiological measurement with slowfast temporal difference transformer, International Journal of Computer Vision 131 (6) (2023) 1307–1330.

[96] A. Revanur, A. Dasari, C.S. Tucker, L.A. Jeni, Instantaneous physiological estimation using video transformers, in: Multimodal AI in Healthcare: A Paradigm Shift in Health Intelligence, Springer, 2022, pp. 307–319.

[97] J. Gideon, S. Stent, The way to my heart is through contrastive learning: remote photoplethysmography from unlabelled video, in: Proceedings of the IEEE/CVF International Conference on Computer Vision, 2021, pp. 3995–4004.

[98] H. Liu, Y. Ding, M. Zhou, Q. Li, Adaptive-weight network for imaging photoplethysmography signal extraction and heart rate estimation, IEEE Transactions on Instrumentation and Measurement 71 (2022) 1–9.

[99] S. Park, B.K. Kim, S.Y. Dong, Self-supervised rgb-nir fusion video vision transformer framework for rppg estimation, IEEE Transactions on Instrumentation and Measurement 71 (2022) 1–10.

[100] X. Liu, X. Yang, Z. Meng, Y. Wang, J. Zhang, A. Wong, Manet: a motion-driven attention network for detecting the pulse from a facial video with drastic motions, in: Proceedings of the IEEE/CVF International Conference on Computer Vision, 2021, pp. 2385–2390.

[101] A.K. Gupta, R. Kumar, L. Birla, P. Gupta, Radiant: better rppg estimation using signal embeddings and transformer, in: Proceedings of the IEEE/CVF Winter Conference on Applications of Computer Vision, 2023, pp. 4976–4986.

CHAPTER

Federated learning in healthcare applications

7

Prasad Kanhegaonkar and Surya Prakash

Department of Computer Science & Engineering, Indian Institute of Technology Indore, Indore, India

7.1 Introduction

Recent developments in artificial intelligence (AI) have enabled computers to solve many complex problems which were very difficult to solve in the old days. AI techniques have enabled computers to deliver near-human-level performance in various real-life problems from different domains such as computer vision, natural language processing (NLP), audio/video processing, healthcare, etc. Machine learning (ML) and deep learning (DL) are the most popular and widely adopted techniques in AI. Some of the applications of ML/DL techniques in these domains are listed in Fig. 7.1. Due to some inherent challenges in the healthcare domain, a variety of ailments are treated by doctors in the physical presence of both patient and doctor. Prognosis, diagnosis, monitoring, treatment planning, and patient feedback need human-level efforts, knowledge, and experience. Medical systems are tightly regulated by the governments with some laws such as General Data Protection Regulation (**GDPR**) [1] and the Health Insurance Portability and Accountability Act (**HIPAA**) [2]. Medical data are treated as confidential, and the leakage of secret medical records or data is prohibitive and may invite stringent actions or penalties on concerned parties.

Medical problems can be addressed with ML techniques, although with certain limitations and restrictions. Models that aid medical practitioners at all phases of disease treatment can be designed using ML algorithms. The efficiency of ML-based algorithms is heavily influenced by the availability of a sufficient volume of high-quality, equally distributed input data. However, due to various hurdles, obtaining high-quality data is not always possible in real life. Due to this, we must deal with the trade-off between efficiency and accuracy in the actual world with a compromise on the performance of ML algorithms. Humans are now more equipped than ever to tackle the difficulties presented by ML approaches, and ML methods perform at levels that are very close to those of humans, thanks to recent advancements in data gathering, processing, handling, storage, and retrieval methods. Using additional data preprocessing methods can increase the quality of the input and the performance of the ML algorithms, hence resolving their shortcomings.

Federated learning (FL) [3] is a unique category of ML techniques in which the training process is carried out in a collaborative manner. In FL, the training process is

Data Fusion Techniques and Applications for Smart Healthcare. https://doi.org/10.1016/B978-0-44-313233-9.00013-8

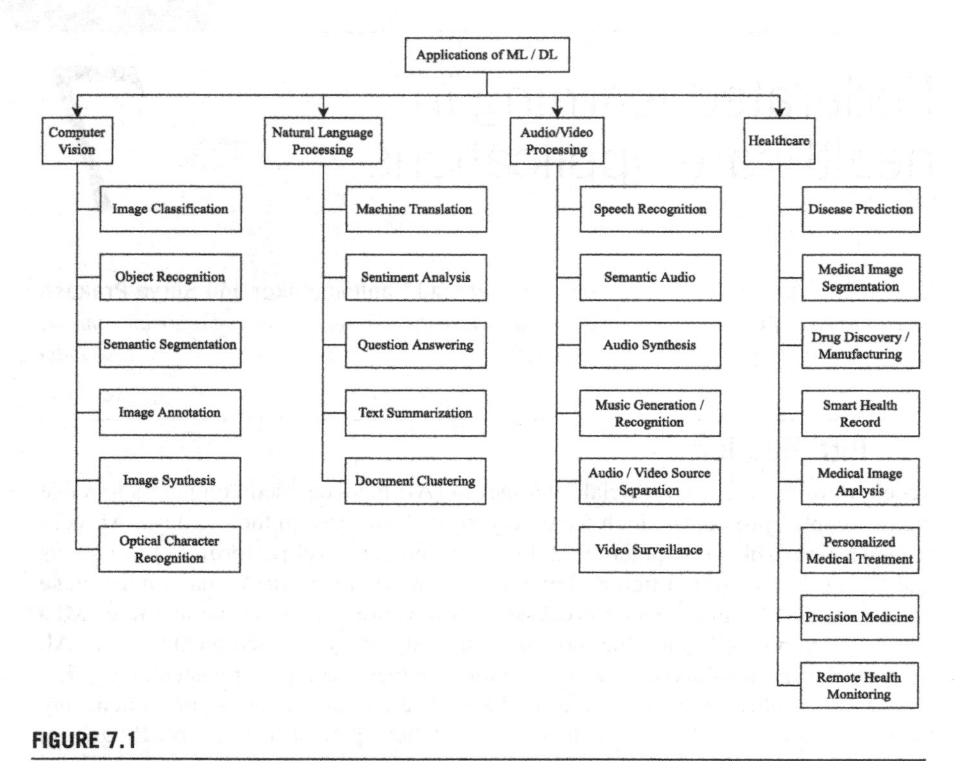

FIGURE 7.1

Applications of ML in various domains.

carried out on several computers, as opposed to centralized learning, where it is done on a single central server or workstation. FL lacks the condition, unlike decentralized learning, where it is assumed that the data samples follow a uniform distribution. In FL, there is no data interchange between various sites or places. The results of the local calculations made by the client workstations on their own local data are combined on the central server. By doing this, just the model parameters are sent across locations, leaving the actual data untouched. This guarantees the data's privacy and confidentiality, which is a fundamental necessity in the medical field.

This chapter contains a detailed discussion of FL and its applications in the medical and healthcare fields. The rest of the chapter is organized as follows. Section 7.2 describes the preliminaries of FL. Section 7.3 describes applications of FL in healthcare. Section 7.4 describes challenges and considerations in FL. Finally, Section 7.5 concludes the chapter.

7.2 Preliminaries

This section contains a discussion of some basic preliminaries which include centralized and decentralized learning, FL, algorithms for training FL models, and the impact of FL on stakeholders.

7.2.1 Centralized and decentralized learning

In centralized learning, an entire training dataset is kept on centralized computers or servers. These servers are often kept or managed in the cloud infrastructure's data centers. In centralized learning, the network infrastructures often use a star topology, allowing client and server computers to communicate directly with one another. However, in this, direct client–client contact is not made. The majority of the training data is kept on the central server, in which the central server makes the decisions about data transmission and task distribution. The tasks are completed by client computers locally, and they then provide the outputs to the central computer. Then, the learning results are created by this server once it has combined the client machines' outputs. Risks of confidentiality being compromised during client–server communication, central server failure, etc., are inherent in centralized learning. Additionally, server computers are vulnerable to network assaults. Due to an excess of communication and/or calculation, the throughput of server machines is also impacted, and latency increases. The system's potential to scale is limited as the cost of increasing the learning environment's total capacity grows. These dangers and drawbacks are amplified or increased enormously in the medical field.

In decentralized learning, data reside in multiple machines. There is no central server or controller machine to orchestrate the learning activity. The training data generally follow a uniform distribution. Nodes in the network infrastructure follow tree topology. The communication between nodes generally involves transmitting data as well as model parameter updates. Any node can use any data which may or may not be actually located at that particular node. The learning procedure involves the exchange of data and/or model parameters. The learning objective is normally minimization of the global loss function. In decentralized learning, the risk of central node failure is not present. Similarly, other risks which are more prominent due to the central learning mechanism are reduced or mitigated. Model capacity and scaling can be increased in this configuration. In medical applications, this learning configuration is better than the centralized learning technique.

7.2.2 Federated learning

Comparing centralized and decentralized learning settings, FL is distinct from both of them. In a conceptual sense, distributed learning mechanisms and FL processes are the same. The FL environment contains a single aggregating server. Data do not need to be exchanged between servers and clients. The learning problem in FL can be formulated as follows.

Let $\mathcal{L}$ be the global loss function. It is calculated using a weighted combination of K local loss values $\{\mathcal{L}_k\}_{k=1}^{K}$, which are calculated on private dataset X_k, where X_k is located within local boundaries of individual entities participating in the FL training process. These entities never share these private local data with any other entity. The learning problem in FL is given by

$$\min_{\phi} \mathcal{L}(X;\phi) \quad \text{with} \quad \mathcal{L}(X;\phi) = \sum_{k=1}^{K} w_k \, \mathcal{L}_k(X_k;\phi), \tag{7.1}$$

where w_k represents respective weight coefficients.

FL training stages: FL training entails several rounds of local computations and parameter aggregation. A central server aggregates the locally computed model updates at client sites as it tries to minimize the global loss function. In some other configurations, the central server does not exist at all to train the FL model. In such cases, a peer-to-peer method is used to train the FL model. Some examples of these are gossip [4] or consensus-based approaches [5]. In FL, the training process is completed once the global loss function converges to global minima or certain criteria for the threshold values are met. The training process in FL involves three fundamental steps as mentioned below.

- **Initialization and client selection:** The aggregation server decides the problem to be solved. Accordingly, it defines the loss functions and model parameters. It randomly selects a sufficient number of clients from available clients. The selected clients take part in the FL process to solve the defined problem.
- **Training and updates:** The server offers clients an initial model that includes a value for the initial global loss. Each client generates a local model with its own dataset and modifies it during each communication cycle. The server aggregates the model updates from each client after receiving model parameter values from a subset of clients.
- **Model aggregation:** After receiving updates from the specified clients, the server aggregates all model parameter updates. The data size that is accessible at the client location determines the weighted sum of the model parameters. The server updates the global model and distributes it to all clients so that the local models can be updated in the next learning cycle. Until the required accuracy is obtained or the global loss function converges, the FL training is repeated.

The above training stages assume synchronized model updates. However, some recent FL training methods allow asynchronicity during the training process. There is a way for training using dynamically evolving models [6]. Unlike their synchronous counterparts, which interchange local models after all layers of the neural network (NN) have been calculated, asynchronous techniques leverage NN features to reciprocate model changes as soon as computations for a single layer are available. These techniques are commonly referred to as split learning [7], [8]. Regardless of whether

FL is decentralized or centralized, split learning can be utilized at both the training and inference stages.

Characteristics of non-iid datasets: Datasets in the FL settings reside at different client sites. The datasets mostly possess nonidentically and independently distributed (non-iid) data in contrast to their ML counterparts which use iid datasets. Moreover, the datasets are imbalanced, which affects the performance of FL models. One can exploit some more complex methods of data normalization, instead of batch normalization, to aid in limiting the accuracy loss induced because of non-iid data [9]. Kairouz et al. [10] described the important properties of non-iid data and their effects in the FL setting. Non-iid data are described by the combined probability of features and labels for each node. Analysis of this joint probability enables decoupling the contribution of features and labels from the particular distribution offered by the local clients. The following list describes the major non-iid data characteristics:

- **Covariate shift:** The statistical distributions of the instances stored by local nodes may be different from those of other nodes. NLP datasets reveal that people regularly write the same numbers or letters with different stroke lengths or slants. [10]
- **Prior probability shift:** Local nodes may have labels with different statistical distributions than other nodes. This may occur if datasets are divided based on area and/or demographics. For instance, there are large regional differences in animal photograph databases. [10]
- **Concept drift (same label, different features):** Despite the fact that local nodes may use the same labels, some of them are related with the different attributes of different local nodes. Images of a specific item, for instance, might differ depending on the weather at the time they were taken. [10]
- **Concept shift (same features, different labels):** Even while some of the traits of local nodes may be identical, others connect to various local node labels. In NLP, for example, even if the same text is seen, sentiment analysis may give different feelings. [10]
- **Unbalancedness:** There may be wide variations in the size of the data that are available at the local nodes. [10] [6]

Broad categories of FL include horizontal FL (**HFL**), vertical FL (**VFL**), and federated transfer learning (**FTL**). In HFL, the feature space is shared among learning participant nodes while sample space is not shared and is different for different nodes, for instance, training an NN model by local clients on their own datasets where the shared NN model is exactly the same with the same values of the hyperparameters. An example of HFL in healthcare can be diabetic retinopathy detection where local clients access their own private data containing retina images to train the shared NN. In this case, the feature space is the same as all images are of human eyes to detect retinal disorders in diabetic patients, but the sample space is different as the images are different in different client datasets. In VFL, the sample space is shared among learning participant nodes while feature space is not shared and is different for different nodes. A healthcare example of VFL is a situation where the patient's data in a hospital can be utilized by both medical researchers and insurance companies to train their individual ML-based models. In this case, the sample space is the same as both medical researchers and insurance companies use the same dataset, but the feature space is different as they use that dataset for their own purpose with different ML models. In FTL, the feature space and sample space are not shared among learning participant nodes and are different for individual nodes. Here, the transfer learning technique is used. The same representation is used to calculate the feature values from various feature spaces, which are then used to train local models. Healthcare example of FTL include assistance in disease diagnosis by partnering with different governments that have different hospitals with a variety of patients, which indicates different sample spaces, and treatment interventions, which indicates different feature spaces. By doing this, FTL can enhance the output of the shared AI model, increasing diagnostic precision.

7.2.3 FL training algorithms

To train FL models, researchers have designed many advanced algorithms. Some of the popular FL training algorithms are described below, which include FedAvg [3], FedBN [11], FedDyn [12], FedProto [13], FedProx [14], FedSGD [15], FML [16], HyFDCA [17], IDA [18], SiloBN [19], Sub-FedAvg [20], FedAdagrad, FedAdam, and FedYogi [21].

In Federated Averaging (FedAvg) [3], to train the global model, the central server coordinates with individual local clients using their own private data. The surrounding clients generate the trained weights and train the generic model. This task is completed concurrently. The weights are combined at the global server after being calculated by local clients. All weights accumulated from clients are calculated by the server as a weighted average. In the following learning cycle, this weighted average will be used as the new set of weights. The stochastic gradient descent (SGD) technique, which is often employed in ML, is used by local clients for training. The learning process is repeated until the global loss function converges or a specified threshold is met. FedBN [11] employs local batch normalization to overcome the feature shift issue, which arises due to differences in data distributions which become apparent due to the difference in the data-capturing techniques and the use of different types of sensors. The developers of FedBN argued that their algorithm is more robust, the global loss function can converge faster, and the algorithm can achieve better performance on feature shift non-iid datasets as compared to other alternative models in the FL setting.

FL with Dynamic Regularization (FedDyn) [12] handles heterogeneously distributed datasets. In this algorithm, each client's local loss is dynamically regularized, which helps the true global loss to converge faster. This technique is more robust and agnostic to different heterogeneity levels. There is a theoretical guarantee of achieving a global optimal convergence point by using the FedDyn algorithm. The authors have verified their claim by doing extensive experimentation on different datasets. FedDynOneGD, which is an extension of FedDyn, enables minimization of the computation levels by local clients and reduces communication between clients and the aggregation server during training. This algorithm is suitable in the case in which the client machines have less powerful computing resources. It updates the model by using a regularized version of the gradients while calculating only one gradient per device in each training round. The algorithm offers linear complexity with respect to local computations. The local computations can be performed parallelly at local client machines, which helps to decrease the overall training time. By offering the advantage of less local computation, FedDynOneGD theoretically achieves similar convergence guarantees as in the case of FedDyn. FedProto [13] offers increased tolerance to heterogeneity, which arises due to the different data distributions, network latency, input/output space, and model architectures. In FedProto, instead of communicating gradients, the clients and servers communicate prototypes of abstract classes. The local prototypes from local clients are gathered by the server. The final step is to send each client the final global prototype. In this manner, local model training is regulated. On the local dataset, the local models seek to reduce the classification error. The generated local prototypes are kept in suitably close proximity to the equivalent global prototypes. The authors also offered a theoretical study of this algorithm's rate of convergence for nonconvex goals. This method manages heterogeneity better than previous FL algorithms, and experiments on various datasets have proved this claim.

FedProx [14] manages statistical heterogeneity caused by data collection in a non-identical manner or differences in data-capturing devices. This technique employs a proximal term to reduce the influence of local updates and limits the local updates to be near to the global model. FedProx offers improved robustness and stability for learning in heterogeneous networks and provides an efficient way to tackle statistical heterogeneity in federated settings. Federated Stochastic Gradient Descent (FedSGD) [15] is a federated variant of SGD. In FedSGD, the aggregation server chooses a random fraction of the nodes C. The server computes a gradient average proportionate to the amount of training samples present on the specified nodes, which is then used to update the global loss value in the training iterations of the FedSGD. The local client machines train their local models independently using their own datasets. These local clients share a tiny fraction of the important model parameters during training. It enables these clients to protect the privacy of their own data while benefiting from the models of other clients. This increases the learning accuracy, which is difficult to obtain with simply their own inputs.

Federated Mutual Learning (FML) [16] handles data heterogeneity (DH), objective heterogeneity (OH), and model heterogeneity (MH). These three heterogeneities are collectively termed DOM by the developers of this algorithm. DH arises because of data non-iid-ness, OH arises due to the conflict between server and clients to train a generalized vs. personalized model, and MH arises due to the requirement of the clients to create their own tailored model for particular settings and jobs. FML allows training the personalized models to handle DH and OH. In the notion of OH, DH is good for clients but bad for servers. Here, the non-iid-ness of data is no longer considered as a drawback, but it is treated as an advantage to allow serving the clients in a better and personalized way. Working together to train FML on comparable but different tasks can be beneficial for the locally tailored models. FML enables users to create their own unique models for a variety of scenarios and jobs. Hybrid Federated Dual Coordinate Ascent (HyFDCA) [17] is a primal-dual algorithm that solves convex problems in the hybrid FL setting. It is an extension of CoCoA, a primal-dual distributed optimization algorithm developed by Jaggi et al. [22] and Smith et al. [23], which considers the case where both samples and characteristics are distributed across clients. It safeguards the client data by simply making it infeasible for the server to recover them. It offers convergent proofs for a variety of problem contexts, incorporating one-of-a-kind circumstances in which just a fraction of customers are available to engage in each cycle. This approach can be used in the context of distributed optimization, where samples and features are disseminated (doubly distributed) across computer nodes. Outside of Parikh and Boyd's block-splitting ADMM [24], HyFDCA is the only algorithm in the doubly distributed scenario that has guaranteed convergence. According to its developers, HyFDCA is the only hybrid FL technique that achieves convergence to the same solution as if the model was trained in a centralized context.

Inverse Distance Aggregation (IDA) [18] offers a promising solution to an important challenge in FL, particularly in the medical domain, to create a more robust shared model capable of providing improved performance and tolerance to noisy and out-of-distribution clients. In the FL environment, IDA solves the issue of statistical heterogeneity in data, which is very desirable in the medical arena, where data may be obtained at various places in the world and with varied scanner settings. IDA is a cutting-edge, metainformation-based adaptive weighting technique for clients that deals with imbalanced and non-iid data. The performance of IDA is confirmed with extensive analysis and evaluation against the FedAvg algorithm, which is considered a baseline for IDA. SiloBN [19] is based on the batch normalization (BN) concept and tries to handle multicentric data heterogeneity such as in the medical domain. This approach provides a domain-adaptive new FL method based on local-statistic BN layers, resulting in models that were jointly trained yet were center-specific. This strategy increases robustness to data heterogeneity while considerably minimizing the chance of information leaks by not exposing the center-specific layer activation statistics. Because local statistics ensure that intermediate activations are centered at the same value across centers, keeping BN statistics private enables federated training of a model resilient to the heterogeneity of the many centers. BN layers can be added to the base model in order to enhance both domain generalization and robustness to heterogeneity.

Sub-FedAvg [20] attempts to broaden the lottery ticket hypothesis [25]. It concurrently addresses the communication effectiveness, resource limitations, and precision of the tailored models. It presents two techniques, namely, Sub-FedAvg(HY) (hybrid pruning) and Sub-FedAvg(UN) (unstructured pruning), to efficiently design the customized subnetworks for each client. In this way, the algorithm efficiently calculates the averaging on the remaining parameters of each subnetwork of each client. Further, clients are not required to be aware of any underlying data distributions or labeling patterns shared by other clients. FedAdagrad, FedAdam, and FedYogi [21] are the federated variants of the Adagrad, Adam, and Yogi algorithms. ClientOpt seeks to minimize the local loss function based on the local data of each client. ServerOpt optimizes from a global standpoint. ServerOpt is an adaptive optimization approach that employs one of the Adagrad, Adam, or Yogi methods, whereas ClientOpt optimizes using the SGD method. These methods bear similar communication costs as FedAvg and ensure to perform well in cross-device settings.

7.2.4 Impact of FL on stakeholders

FL can primarily improve the operations and productivity of its stakeholders [26]. This section describes the influence of FL on clinicians, patients, hospitals, researchers, healthcare providers, and manufacturers.

- **Patients:** In general, healthcare choices may be made more quickly and accurately by using expert knowledge. Expert doctors who practice in remote regions may be able to provide patients with high-quality clinical care. Doctors can recognize and research some uncommon or rare ailments that are only seen in specific geographic regions, allowing patients to receive better care in such unusual circumstances. Patients can also be reassured about the confidentiality and anonymity of their data. They are free to revoke their permission to access their personal data, and doing so has no negative effects on the functionality of the FL model.
- **Clinicians:** FL possesses enormous potential to broaden the scope of illness diagnostics. In contrast to conventional treatment approaches, doctors can benefit from the reach and access to geographically scattered patient data. They can lessen the incidence of biased diagnostic choices brought on by using just locally accessible patient data. Another advantage is that physicians may employ FL-based solutions for illness diagnosis as supportive or supplementary techniques and access resources from distant regions, which is challenging in a conventional manner. The main difficulty with this procedure is storing high-quality raw input data across all locations while adhering to a uniform, widely accepted data representation format.
- **Hospitals:** For private patient data, complete custody and control may be promised. To prevent unauthorized parties from using these data, access to it may be fully tracked and managed. It is crucial to install the expensive on-site infrastructure or sign up for the expensive cloud platform and to manage it appropriately. Whether hospitals take part in the training process, the inference process, or both will determine the cost. Because it requires more infrastructure and support, the latter is more expensive for hospitals.
- **Researchers:** They have access to a huge dispersion of real-world data. Instead of wasting time on data collection, logistics, and preparation, the efforts may be focused on tackling more difficult clinical or technological challenges. To address the relevant issues, such as distribution shift, model performance deterioration, etc., researchers can create unique and reliable FL methods. They can look into, audit, or keep track of model failures, develop models that are failure-proof in response, and stop models from acting in a given way in the future.
- **Healthcare providers:** They may reorient their work such that it focuses on value-based services rather than volume-based services, which basically provide the

framework for precision medicine. By doing this, it is guaranteed that patients will receive more effective, less expensive medical care. The financial viability of healthcare providers is unaffected at the same time.

- **Manufacturers:** By utilizing integrated learning from several devices and apps, they may profit from ongoing integration, validation, and enhancement in the FL frameworks with the certainty that the patient-specific data will be kept private. It costs a lot to upgrade the computing infrastructure, data storage facilities, networking facilities, and related software purchases or subscriptions in order to access such capable models and resources. FL can nevertheless guarantee an improvement in the involved organizations' profitability.

7.3 FL applications in healthcare

This section describes FL applications in healthcare such as electronic health record (EHR) mining, remote health monitoring, medical imaging, and disease prediction.

7.3.1 Electronic health record mining

EHRs store patients' historical information, which includes patient's personal details, medical history, line of treatment, suggested diagnostic tests and their results, etc. Governments, hospitals, insurance companies, pharmaceutical companies, diagnostic labs, researchers, developers, etc., can exploit such comprehensive information stored in the patient database generated by the collection of EHRs of each patient for different purposes. This information can serve multiple requirements such as extracting meaningful information, identifying correlations among multiple factors, making decisions, planning future tasks, etc. In a traditional ML setting, it is difficult to ensure security and privacy as confidentiality and secrecy can be lost at any stage in handling EHRs. Anonymization or metadata deletion is not a foolproof solution to safeguard privacy and secrecy. FL can serve this requirement as the patient data are maintained in certain discrete locations only and data do not travel outside the boundaries of such locations. EHR mining facilitates multiple medical operations including but not limited to the above-described ones.

Computational phenotyping is a biomedical informatics technique to study/extract some meaningful characteristics or design some concise and meaningful concepts from the vast and noisy EHR data. It uses horizontally partitioned EHR data. It has the potential to improve precision medicine, accelerate biomedical discovery, and raise healthcare standards. The approach in [27] combines advanced privacy-preserving technologies with collaborative learning to provide an accurate and communication-efficient solution for computational phenotyping. Tensor factorization is a technique for reducing dimensionality. It can be used to construct a meaningful low-dimensional latent representation of high-dimensional data, in this example, EHR data. Each hospital in this method decomposes its local tensors containing EHRs and perturbs the intermediary outcomes. The differentially private intermediary results are then shared with a semitrusted local server, which generates phenotypes to aid in the learning of essential clinical concepts. A structured sparsity term deals with patient population heterogeneity. To overcome the optimization challenge, the proposed DPFact framework in this method employs an elastic averaging SGD (EASGD) strategy. Another method proposed in [28] performs computational phenotyping using federated tensor factorization. Tensors are updated iteratively in this manner, and summarized secure data are delivered to the central server. When compared to centralized training models, this strategy is more resistant to patient size and distribution and faster. Patient information is kept secret by summarizing it. To design a safe data harmonization and federated computation technique, the federated optimization problem is tackled utilizing the alternating direction method of multipliers (ADMM).

The study in [29] presents existing FL applications in EHRs and evaluates the performance of state-of-the-art FL training algorithms on two clinically significant problems, viz. in-hospital mortality prediction and acute kidney injury (AKI) prediction in the intensive care unit (ICU). In this work, the eICU dataset [30] is used for experimentation that contains a wide range of clinical data. The results confirm the excellent performance of FL algorithms, although the experiments lack diversity in the dataset to mimic the real-world data scenario. The authors state their plan to extend the experiments to more diverse real-world data and identify trade-offs between model security and performance. Another method by Pfohl et al. [31] predicts in-hospital mortality and prolonged length of stay in the hospital. In this work, the authors used FedAvg and differentially private SGD (DPSGD) to collaboratively train the FL model for clinical risk prediction with a formal privacy guarantee. To ensure differential privacy, Gaussian noise is added to the intermediate gradients. The ratio of batch size to the training set size is selected as the noise multiplier. The authors claim that introducing differential privacy in FL training worsens the model performance and tuning the model for certain datasets to improve model performance bears the risk of model overfitting. Federated autonomous DL (FADL) [32] predicts the patient mortality rate using in-silico patient data stored in EHRs. This approach attempts to balance global and local training to create a high-performance FL model by training a portion of the model using data from all global sources and the remaining portion of the model using local data. The authors show that the resultant model outperforms FL and delivers similar performance as compared to the centrally trained conventional models.

Recent studies have demonstrated that integrating ML with EHR may significantly speed up precision medicine. In the context of EHR, Boughorbel et al. [33] proposed a novel FL model called federated uncertainty-aware learning algorithm (FUALA). To limit the contribution of models with high levels of uncertainty to the aggregate model, FUALA embeds uncertainty information in two different methods. By retaining the last layers of each hospital from the final round, it also introduces model ensembling at prediction time. The exploratory study for the goal of predicting preterm births on a cohort of 87,000 deliveries revealed that the suggested technique performs better than FedAvg when tested on out-of-distribution data. The authors demonstrated how the suggested method may be used to quantify uncertainty.

The work in [34] suggests a novel decentralized healthcare architecture based on blockchain and mobile edge computing (MEC) for distributed electronic medical record (EMR) sharing across federated institutions. The authors concentrate on a fully decentralized access control solution utilizing smart contracts that allow EMR access verification at the edge of the network without needing any central authority, in contrast to existing systems that frequently rely on a private entity for clinical governance. Additionally, the MEC network also incorporates a decentralized interplanetary file system (IPFS) platform with smart contracts, which greatly lowers data retrieval time and improves security for EMR sharing. The experimental findings and analyses demonstrate the suggested method's improved performance over the competition in terms of faster data retrieval, a more efficient blockchain, and security assurances.

Using a distributed approach, Brisimi et al. [35] seek to resolve a binary supervised classification issue to forecast hospitalizations for cardiac events. Without explicitly transferring raw data, the goal is to create a universal decentralized optimization framework that would allow many data holders to work together and converge to a single prediction model. The authors concentrate on the sparse support vector machine (sSVM) classifier with soft-margin L1 regularization. They create the decentralized iterative cluster primal-dual splitting (cPDS) technique to solve the massive sSVM issue. Based on data from the patient's EHR from the year before, the authors evaluate cPDS on the challenge of forecasting hospitalizations for heart disorders within a calendar year. They pinpoint crucial algorithmic components that predict upcoming hospitalizations, enabling them to comprehend categorization results and direct preventive steps.

Liu et al. [36] create a two-stage federated NLP technique that allows for the use of clinical notes from several hospitals or clinics without relocating the data and shows its effectiveness using the medical job of obesity and comorbidity phenotyping. This method not only enhances the quality of a particular clinical activity but also makes it easier to advance information throughout the whole healthcare system, which is a crucial component of learning the healthcare system. The authors believe that this is the first time federated ML has been used in clinical NLP.

ML algorithms for forecasting illness occurrence, patient responses to therapy, and other healthcare events may be developed with the use of EMRs. Yet up until now, the majority of algorithms have been centralized, paying little attention to the decentralized, non-iid, and privacy-sensitive features of EMRs that might impede data gathering, sharing, and learning. Huang et al. [37] developed a community-based federated ML (CBFL) approach and tested it on non-iid ICU EMRs to address this problem. The proposed algorithm organizes the dispersed data into clinically relevant communities that included comparable geographical areas and illnesses, and it then learns a single model for each community. According to evaluation findings, CBFL outperforms the standard federated ML algorithms in terms of communication cost between hospitals and the server as well as AUC-ROC, AUC-PRC, and other metrics.

Furthermore, the degree to which one group differed from others might account for variations in performance among communities.

7.3.2 Remote health monitoring

Remote health monitoring has multiple use cases in human life to make health monitoring safer and more convenient than before. Monitoring the health of patients (or any person) remotely can produce timely alerts about potential health risks. These alerts certainly help in taking timely precautions and providing treatment to taper the disease penetration in the patient's body. This can also be helpful for doctors in case the patient's treatment is to be done by a doctor who is located at a geographically distant place. There are numerous applications in which FL facilitates remote health monitoring.

Xu et al. [38] present a privacy-preserving federated depression detection method that uses multisource mobile health data. This work uses mentally ill patients' behavioral patterns to analyze and diagnose depression. Depression patients typically type more slowly than other people do, which may be due to emotional instability at the first occurrence of depression. The authors believe that the information about keystroke patterns of depressed patients can be leveraged as a source of evidence to predict depression. For privacy preservation, the model converts raw data into multiview data. The proposed DeepMood architecture calculates Hamilton Depression Rating Scale (HDRS) scores for depression prediction. The experiments are performed in two cases, viz. iid data and non-iid data. Performance on iid data is about 10% to 15% higher than that of local training, whereas on non-iid data, accuracy drops by at most 13%, which is acceptable in the case in which patient privacy is tightly maintained.

FedHealth [39] presents a federated transfer learning framework for wearable healthcare. In this work, the UCI Smartphone dataset is used, which contains information about six activities collected using smartphones worn on the waist. The model is trained using regular FL training procedures, except that the parameters are encrypted using homomorphic encryption before aggregation to infuse security in the model architecture. Applying transfer learning makes the model more personalized according to the desired task. This extensibility of the model helps its deployment for any new disease prediction task. The authors tested the model for auxiliary diagnosis of Parkinson's disease (PD). For this, symptomatic data using acceleration and gyroscope signals which contain information about arm droop, balance, gait, postural tremor, and resting tremor are collected through the custom-built smartphone application. The results prove the efficiency of the model in real-world disease classification use cases.

FedHome [40] is a novel cloud edge-based personalized FL solution for in-home health monitoring. In this work, generative convolutional autoencoder (GCAE) is used to deal with non-iid and imbalanced datasets by synthesizing artificial samples to generate a class-balanced dataset. GCAE is used for dimensionality reduction and it is a lightweight module that can be added on top of any baseline architectures, thereby reducing communication costs in FL training. Security is provided using homomorphic encryption in the model. Personalization is provided using fine-tuning on the reconstructed class-balanced dataset. The method is tested on the publicly available MobiAct dataset [41], which contains human activity samples of daily living activities, fall-like activities, and actual falls. The model performance shows that FedHome is a robust and efficient solution and it can be used in multiple real-world use cases of in-home health monitoring activities in a privacy-preserving and communication-efficient way.

The work in [42] solves the human activity recognition problem in resource-constrained environments in the presence of label and model heterogeneity. The authors demonstrated the working of the proposed model on the Raspberry Pi 2 SOC platform and the Heterogeneity Human Activity Recognition (HHAR) dataset [43]. The proposed model leverages a knowledge distillation mechanism to assimilate model independence and heterogeneity with transfer learning. Two versions of model updates, namely, model distillation update and weighted-α update, are proposed. Label heterogeneities are modeled using overlapping pairs of four activities, viz. sitting, walking, standing, and going up the stairs. Model heterogeneities are modeled by using three different model architectures of convolutional NNs (CNNs) and artificial NNs. One distilled student architecture is also taken separately. The average accuracy values indicate a gain of about 11% in the model distillation update and about 9.15% gain in the weighted-α update. The developed model is shown to be robust and feasible for on-device FL problems.

7.3.3 Medical imaging

FL can be helpful in multiple medical image analyses, processing, or similar activities such as medical image reconstruction, segmentation, registration, synthesis, etc. The advantages of using FL in such activities include performance improvement, ensuring security and privacy, better generalizability, tailor-made personalized models, and so on.

An application of FL in medical imaging is presented in [44] for the reconstruction of MRIs. In this work, a robust method that is invariant to domain shift is proposed. The learned intermediate latent features from the source domain are aligned to those in the target domain to model a cross-site FL architecture. This work also overcomes the challenges due to undersampled and heterogeneous datasets. The given approach offers a privacy-preserving solution to leverage multiinstitutional data in the FL architecture. The model was tested on the fastMRI [45], HPKS [46], IXI [47], and BraTS [48] datasets. Results indicate the efficiency of the proposed domain-generalized MRI reconstruction framework with multiinstitution collaboration.

Another application is proposed in FL for computational pathology on gigapixel whole slide images (WSIs) [49]. The given approach incorporates weakly supervised attention multiple-instance learning. It offers a privacy-preserving, robust, and better generalized model. Differential privacy is provided through a randomized mechanism. Thousands of gigapixel histopathology WSIs from multiple institutions are used to test the model's performance. The model overcomes the challenges arising due to the lack of histopathology datasets with detailed annotations. The effectiveness of the model is demonstrated in breast cancer and renal cell cancer histological subtyping. Another demonstration includes an interpretable and weakly supervised framework for survival prediction in renal cell carcinoma patients. The proposed model can be deployed for computational pathology tasks to identify rare diseases which lack diversity in morphology. It can also be deployed and tailored at places where it is difficult to provide access to pathology and laboratory medicine services.

The work in [50] evaluates multiinstitutional collaborations and the generalizability of FL-based architectures. This paper uses different data-sharing approaches including collaborative data sharing (CDS), FL, institutional incremental learning (IIL), and cyclic institutional incremental learning (CIIL). The datasets used in this work include BraTS 2017, MD Anderson Cancer Center (MDACC), and Washington University School of Medicine (WashU), which contain brain tumor MRI scans. To simulate real-world data distribution, data are fragmented across 10 institutions. Model performance is evaluated using a dice score and results indicate the feasibility and effectiveness of the proposed model.

Fan et al. [51] present an FL framework for 3D brain MRI images. This work handles cross-site heterogeneity using guide-weighted gradient updates during model training. To test the model, an autism spectrum disorder (ASD) dataset [52] is used. The model is made secure through the differential privacy mechanism. For feature extraction, 3D Resnet-10, 3D Resnet-18, and 3D Densenet-121 are used. The proposed model shows about 0.92–4.02% performance improvement on diagnostic classification problems.

An FL framework is presented in [53] for securely accessing and metaanalyzing biomedical data. The brain structural relationships across diseases and clinical cohorts are investigated. The Enhancing NeuroImaging Genetics through Meta-Analysis (ENIGMA) network is utilized to metaanalyze neuroimaging data. The proposed method offers an end-to-end solution for data standardization, confounding factor correction, and multivariate analysis of variability of high-dimensional features. The model is based on the ADMM optimization approach. The model is trained initially on synthetic data, and after that it analyzes subcortical thickness and shape features across diseases including Alzheimer's disease (AD), progressive and nonprogressive mild cognitive impairment (MCI), and PD and in healthy individuals.

The method proposed in [54] performs microvasculature segmentation and referable diabetic retinopathy (RDR) classification using optical coherence tomography (OCT) and OCT angiography (OCTA) images with an FL-based deep NN (DNN) architecture. The performance of microvasculature segmentation is measured using the dice similarity coefficient (DSC), while that of RDR uses accuracy, AUROC, AUPRC, balanced accuracy, F1 score, sensitivity, and specificity. The claimed results prove the method is well suited for both segmentation and classification.

FedMix [55] is a mixed supervised label-agnostic FL model to segment medical images. It incorporates strong pixel-level labels, weak bounding box labels, and weakest image-level class labels. On top of this, it employs an adaptive weighting style across local clients and achieves rich and discriminative feature representations. The model is tested on breast tumor and skin tumor datasets, namely, BUS, BUSIS, UDIAT, and HAM10K. This method outperforms many state-of-the-art algorithms by a large margin and is suitable for multiclass problems and clinical environments.

The authors of [56] explore the application of FL to develop medical imaging classification models in a practical collaborative scenario. This FL initiative brought together seven clinical institutions from around the world to train a breast density classification model using the Breast Imaging Reporting and Data System (BIRADS). The authors demonstrate that one may successfully train AI models in the federation despite significant variances in the datasets from all locations (mammography system, class distribution, and dataset size) and without centralizing data. The findings demonstrate that FL-trained models outperform models trained only on local data from a single institute by an average of 6.3%. Additionally, when the models' generalizability is assessed using testing data from the other participating sites, they exhibit a relative improvement of 45.8%.

The multisite fMRI classification issue is addressed in [57] using a privacy-protecting method. The authors suggest an FL technique to address the issue, in which a shared local model's weights are modified via a randomization mechanism and using a decentralized iterative optimization algorithm. In this FL framework, they further present two domain adaptation algorithms taking into account the systemic heterogeneity of fMRI distributions from various locations. They look into some useful facets of federated model optimization and contrast FL with other training approaches. Overall, the findings show that using multisite data without data sharing to improve neuroimage analysis performance and identify trustworthy disease-related biomarkers is promising. Other privacy-sensitive medical data analysis issues can be generalized using the suggested methodology.

The authors of [58] examine the viability of using differential-privacy approaches to safeguard patient data in an FL scenario. On the BraTS dataset, they construct and assess useful FL algorithms for segmenting brain tumors. The experimental findings demonstrate a trade-off between model effectiveness and expenses associated with privacy protection.

The authors in [59] suggest a specificity-preserving FL method for MRI reconstruction (FedMRI). The basic concept is to split the MRI reconstruction model into two components: a globally shared encoder to get a generalized representation at the global level and a client-specific decoder to keep the domain-specific properties of each client, which is crucial for collaborative reconstruction when the clients have a unique distribution. A system of this kind is then put into practice in the frequency space and the picture space, enabling the investigation of a universal representation and client-specific features simultaneously in several areas. Furthermore, a weighted contrastive regularization is included to immediately rectify any discrepancy between the client and server during optimization, which will substantially improve the convergence of the globally shared encoder when a domain shift is present. Extensive trials show that FedMRI surpasses state-of-the-art FL approaches and that its reconstructed findings are the most accurate for multiinstitutional data.

A variation-aware FL (VAFL) framework is proposed by Yan et al. [60], in which client variations are reduced by combining all clients' photos into a single shared image space. To define the target picture space and create a collection of images using a privacy-preserving generative adversarial network termed PPWGAN-GP, the authors first choose one client with the least data complexity. Then, an automated decision is made to share a subset of those synthesized pictures with other clients that successfully capture the traits of the raw photos and are sufficiently distinct from any raw image. To translate each client's raw photos into the target image space specified by the shared synthesized images, a modified CycleGAN is implemented. The issue of cross-client variance is solved in this manner while maintaining anonymity. The authors utilize many sources of decentralized apparent diffusion coefficient (ADC) imaging data to apply the framework for automated categorization of clinically relevant prostate cancer and assess it. The suggested VAFL framework outperforms the existing horizontal FL framework, according to experimental data. The authors claim that the suggested framework may be used for a variety of additional medical image classification problems because VAFL is not dependent on DL architectures for classification.

A customized Federated Framework with Local Calibration (LC-Fed) is proposed by Wang et al. [61] to take use of intersite discrepancies at the feature and prediction levels in order to improve segmentation. Concretely, the authors first construct the contrastive site embedding along with channel selection operation to calibrate the encoded features since each local site has a different focus on the different characteristics. Additionally, they suggest using the understanding of prediction-level inconsistency to direct the customized modeling on unclear locations, such as anatomical borders. Calibrating the forecast is done by calculating a disagreement-aware map. The efficacy of this approach has been demonstrated on three medical picture segmentation tasks using various modalities, where the suggested approach consistently outperforms the most advanced tailored FL approaches.

Ucar et al. [62] show how an AI-based framework outperforms the investigations at hand. A Bayesian optimization technique is used to optimize SqueezeNet's light network architecture for COVID-19 diagnostics. The proposed network outperforms current network designs and achieves greater COVID-19 diagnostic accuracy thanks to fine-tuned hyperparameters and enhanced datasets.

The work in [63] suggests a method of attacking the federated GAN (FedGAN) by subjecting the discriminator to a backdoor attack classification model technique known as data poisoning. The authors show that the FL-GAN model can be damaged by introducing a tiny trigger with a size of less than 0.5% of the original picture size. The proposed method offers two strong protection tactics based on the suggested attack: global malicious detection and local training regularization. The authors demonstrate that combining the two defensive tactics results in reliable medical image synthesis.

7.3.4 Disease prediction

FL can be used for disease diagnosis or prediction. In the FL setting, the patient data across multiple locations become available to the global FL model. This helps to increase the model performance. Also, more security and privacy can be embedded in the model architecture. The computing infrastructure from multiple locations can be utilized for model training, which increases the scalability of the model. It can be personalized as per the problem requirements.

A method in [64] offers a privacy-preserving and autocrafted feature extraction-based solution to diagnose cerebellar ataxia (CA). CA is characterized by the incoordination of movement caused by cerebellar dysfunction due to which movements of the eyes, trunk, and limbs as well as speech are affected. Existing conventional methods perform well for diagnosing CA but are not scalable and require large amounts of data, thereby raising privacy concerns. The proposed method uses motion capture sensors during the performance of the standard neurological balance test. By incorporating the differential privacy mechanism, the proposed model employs security in the operation. The authors evaluated the Mel spectrogram, recurrence plot, and Poincaré plot to test model performance. The recurrence plot yields the highest validation accuracy of 86.69% with MobileNetV2 to diagnose CA.

Another method incorporates a federated multitask learning (MTL) mechanism to diagnose multiple mental disorders [65]. In this study, an effective federated MTL framework is designed by using fMRI scans. To extract high-level features, the authors designed a federated contrastive learning-based feature extractor. A federated multigate mixture of expert classifiers performs joint classification and eases optimization conflicts of updating shared parameters in MTL. Along with practical modules, the proposed system provides federated biomarker interpretation and customized model learning and safeguards privacy. This method was tested on real-world datasets and produced robust diagnosis accuracy on ASD, attention-deficit/hyperactivity disorder, and schizophrenia. This architecture eases domain shift across clients through federated MTL. It reveals how to enhance computer-aided detection systems' ability to generalize by taking advantage of the beneficial information shared in related psychiatric diseases.

FedNI [66] employs network inpainting for federated graph learning and offers a population-based disease prediction system. This method exploits graph convolutional NNs (GCNs) for disease prediction on a population graph. People are represented by graph nodes, while similarities between people are represented by graph edges. GCNs deliver better performance only upon having access to large amounts of input data and it is difficult to ensure in the medical domain. Additionally, dealing with disease prediction in isolation with limited data sources remains a significant obstacle. FL handles this situation by enabling isolated local institutions to jointly train a global model without data sharing. FedNI exploits a graph generative adversarial network (GAN) to fill in the gaps caused by the missing data of local networks. In FedNI, a global GCN node classifier is trained using the federated graph learning technique. The proposed model delivers good performance on the neurodisorder classification task. It is evident from the results on two real brain disease datasets: Autism

Brain Imaging Data Exchange (ABIDE) [52] and Alzheimer's Disease Neuroimaging Initiative (ADNI) [67].

A COVID-19 detection method called FedFocus is proposed in [68]. This method dynamically focuses on different areas of chest X-ray images. Each local model's training loss serves as the foundation for the parameter aggregation weights and this increases training efficiency and accuracy. A dynamic factor that is changed frequently is intended to stabilize the aggregation process as the training layer becomes more complex. Additionally, the training sets in the tests are segregated according to the population and infection of three different cities in order to correctly reproduce the original data. Numerous tests on the real-world chest X-ray images dataset prove that FedFocus performs better than its peers in terms of model training effectiveness, precision, and stability.

A breast cancer classification method [69] utilizes memory-aware curriculum learning in a federated setting. The method increases the local models' consistency by penalizing inconsistent predictions, often known as forgotten samples. To address domain shift and offer data privacy protection, this method utilizes unsupervised domain adaptation. The proposed model is tested on three clinical datasets from various vendors. The performance of the suggested method is assessed using two classification metrics: AUC-PRC and AUC-ROC. In comparison to the standard federated setup, AUC and PR-AUC are, on average, enhanced by 5% and 6%, respectively. The results prove the usefulness and suitability of the proposed memory-aware curriculum FL in multisite breast cancer categorization.

A scalable FL- and Internet of Medical Things (IoMT)-based model architecture [70] classifies chest CT images for COVID-19 prediction. The framework initially uses IoMT to get CT images from multiple nearby hospitals, and then it aggregates those photographs for storage in a Hadoop Distributed File System (HDFS) Spark big data framework. The model training is then completed independently using the proposed framework, and the trained parameters are then transferred to a centralized server for FL aggregation. To evaluate the effectiveness of the proposed study, extensive experimentation is carried out on three separate COVID-19 databases. The numerical analysis demonstrated that the suggested approach performs significantly better than the state-of-the-art methods. Additionally, the global server outperforms the local server by 7.57% in terms of classification accuracy and by 3.33% in terms of area under the curve.

In [71], by leveraging knowledge distillation to ensure data privacy and safety, the authors provide a wholly decentralized FL approach. Each node functions autonomously and does not require access to outside data. This method of AI accuracy has been proven to be on par with centralized training, and when nodes include low-quality data, which is prevalent in the healthcare industry, AI accuracy can outperform conventional centralized training.

The authors of [72] suggest a brand-new paradigm for cooperative training on such data that they name Flexible FL (FFL). The authors show how one large AI model can be trained using this approach using publicly available data of 695,000 chest radiographs from five institutions, each with a different set of labels. They discover that FFL-trained models outperform those that are only trained on matching annotations. This could make it possible to train really large-scale AI models that effectively utilize all available data.

The model under [73] examines Federated Semi-Supervised Learning (FSSL), which tries to develop a federated model by combining the data from labeled and unlabeled clients (i.e., hospitals). With a new interclient relation matching scheme, the authors' creative solution to this issue outperforms the conventional consistency regularization mechanism. In order to address the lack of task expertise among unlabeled customers and to encourage discriminative information from unlabeled samples, the suggested learning scheme directly connects the learning across labeled and unlabeled clients by aligning their extracted illness associations. On the basis of two sizable medical image categorization datasets, they validate this methodology. The method's efficiency has been shown by the observable advancements above the state-of-the-art as well as the exhaustive ablation study on both workloads.

In order to facilitate privacy-enhanced COVID-19 detection with GANs in edge cloud computing, Nguyen et al. [74] present a new FL technique, termed FedGAN. In particular, the authors initially suggest a GAN in which for each edge-based medical institution a discriminator and a generator based on CNNs are alternately trained to imitate the genuine COVID-19 data distribution. Then, in order to improve the global GAN model for producing realistic COVID-19 images without the requirement for actual data sharing, they provide a new FL approach that enables local GANs to cooperate and communicate learned parameters with a cloud server. The authors implement a differential privacy solution at each hospital institution in order to improve privacy in federated COVID-19 data analytics. In addition, they suggest a new FedGAN architecture built on blockchain for secure COVID-19 data analytics by decentralizing the FL process and utilizing a new mining technique for low operating latency. The superiority of this method for effective COVID-19 detection over cutting-edge methods is shown by simulation results.

In [75], a customized image dataset with four kinds of skin diseases was created, a CNN model was developed and evaluated against a number of benchmark CNN algorithms, and an experiment was conducted to test an FL strategy for data privacy. To increase the dataset and make the model more inclusive, an image augmentation approach was used. For acne, eczema, and psoriasis, the proposed model attained a precision of 86%, 43%, and 60% and a recall of 67%, 60%, and 60%, respectively. Following the distribution of the dataset across 1000, 1500, 2000, and 2500 clients using the FL technique, the model displayed an average accuracy of 81.21%, 86.57%, 91.15%, and 94.15%. The FL technique combined with the CNN-based skin disease classification is a novel method to categorize human skin disorders while maintaining data security.

The work in [76] is the first attempt to create a comprehensive collaborative FL framework that respects privacy, called FLDISCO, which combines GAN and GNN to produce molecular graphs. Results from experiments show that FLDISCO is effective for producing highly novel compounds with high druglikeness, uniqueness, and LogP scores compared to the baseline on (1) iid data for ESOL and QM9 datasets and (2) non-iid data for ESOL and QM9 datasets, where FLDISCO generates 100% novel compounds with high validity and LogP scores compared to the baseline. The authors also show how various client, generator, and discriminator architecture fractions influence the evaluation results.

The authors in [77] created a cutting-edge platform as part of the European Innovative Medicines Initiative project MELLODDY (grant no. 831472), a collaboration between 10 pharmaceutical firms, academic research centers, major industrial companies, and startups, to apply FL to drug discovery. The MELLODDY platform was the first platform at an industrial scale to make it possible to build a worldwide federated model for drug discovery without disclosing the private datasets of individual partners. After each training cycle, the platform's federated model was trained by securely and cryptographically averaging the gradients of all cooperating partners. The platform was set up using a Kubernetes cluster running in a private subnet on an Amazon Web Services (AWS) multiaccount architecture. Organizationally, the various partners' responsibilities were outlined as various rights and permissions on the platform, which were then decentralized and managed. A companion study to the MELLODDY platform describes the fresh scientific findings that were made.

With three real-world datasets – EHR, images of skin cancer, and ECG datasets – the authors in [78] present empirical benchmarks of FL approaches in this study that takes into account both performance and monetary cost. Additionally, using a straightforward combination of FedProx and FedBN, they suggest FL with Proximal regularization eXcept local Normalization (FedPxN), which beats all existing FL methods while using barely any more power than the most power-efficient approach.

In order to solve FL's non-iid data problem, the work in [79] offers a unique technique for dynamically balancing the data distributions of clients by enhancing images. For the purpose of detecting several chest diseases in chest X-ray images in extremely non-iid FL situations, the approach substantially stabilizes the model training and increases the model's test accuracy. The results of iid, non-iid, and non-iid with the suggested federated training show that the method encourages institutions or researchers to create better systems to derive values from data with regard to data privacy, not only for the healthcare industry but also for other fields.

7.4 Challenges and considerations

This section describes some crucial challenges and considerations of FL which include domain generalization, data and model heterogeneity, privacy and security, system architecture and resource sharing, and lack of dataset and training.

7.4.1 Domain generalization

Domain generalization refers to the adoption of FL models to new and diverse data distributions. FL enables dispersed healthcare organizations to cooperatively build a common prediction model while maintaining privacy. When used in clinical settings,

FL models may still perform poorly when used in wholly unknown hospitals outside the federation. This is a situation that could be dealt with by exploiting the benefits offered by FL. FL models generalize well compared to conventional models because of access to global datasets. Some recent works on federated domain generalization are highlighted in the following discussion.

FedDG [80] is a federated domain generalization-based approach for medical image segmentation. This approach tries to learn a federated model from a number of dispersed source domains so that it may instantaneously adapt to new target domains. A technique called Episodic Learning in Continuous Frequency Space (ELCFS) has been designed to offer domain generalization by allowing each client to use multisource data distributions while adhering to the difficult data decentralization requirement. Through an efficient continuous frequency space interpolation technique, this method transmits the distribution information among clients while ensuring privacy. A boundary-oriented episodic learning paradigm has been developed to use the transferred multisource distributions in order to expose the local learning to domain distribution shifts and specifically address the issues of model generalization in medical image segmentation scenarios. The proposed model outperforms cutting-edge approaches. Segmentation performance is measured using the dice similarity coefficient and the Hausdorff distance. Thorough ablation studies on two medical image segmentation tasks, viz. Optic Disc/Cup segmentation and Prostrate MRI Segmentation, prove the usefulness of the proposed method.

Federated Knowledge Alignment (FedKA) [81] generalizes text and image classification tasks. The negative transfer problem reduces a model's generalizability to unforeseen tasks because of the distribution shift in features that are inherent to particular domains. In FL, learned model parameters are shared in order to train a global model that utilizes the underlying knowledge across client models trained on various data domains. However, standard domain adaption techniques that demand prior knowledge of various domain data are less effective due to FL's data secrecy. To enable the global model to learn domain-invariant client features while being constrained by unknown client data, FedKA uses feature distribution matching in a global workspace. To implement global model fine-tuning, FedKA uses a federated voting system that creates target domain pseudolabels based on client consensus. The performance of the suggested strategy is assessed on both imaging and text classification tasks using various model architectures and comprehensive experiments including an ablation study. Additionally, using a brand new metric named the group effect, it has been assessed how FedKA mitigates the negative FL transfer. According to the findings, FedKA is able to minimize negative transfer, which enhances the performance gain from model aggregation by about four times.

Federated Adversarial Domain Generalization (FedADG) proposed in [82] tries to increase the domain generalization capability of FL-based models. By comparing each distribution to a reference distribution, FedADG uses the federated adversarial learning approach to quantify and align the distributions among several source domains. By accommodating all source domains, the reference distribution is constructed adaptively in order to reduce the domain shift distance during alignment.

The learned feature representation, therefore, has a propensity for universality and exhibits high generalization performance over the unknown target domains while maintaining the privacy of local data. Analyses on numerous datasets show that FedADG performs on par with state-of-the-art algorithms.

In [83], a method to improve the generalizability of CNN-based segmentation is proposed to assess the structure and function of the heart using cardiac magnetic resonance (CMR) images. In this approach, data standardization and augmentation algorithms are deliberately designed to account for typical circumstances in multisite, multiscanner clinical imaging datasets to develop a straightforward yet effective method for enhancing the network generalization capacity. In this method, an NN is trained on a single-site, single-scanner dataset, for example, the UK Biobank (UKBB) dataset, to segment CMR images collected at several sites and using different scanners, without compromising the performance. Testing was done on a few different subjects from the UKBB dataset for intradomain testing. After that, for cross-domain testing, it was also tested on the Automated Cardiac Diagnosis Challenge (ACDC) dataset and the British Society of Cardiovascular Magnetic Resonance Aortic Stenosis (BSCMR-AS). On the UKBB test set, the proposed method achieves promising segmentation results that are equivalent to previously reported results. It also performs well on cross-domain test sets, attaining satisfactory accuracy on the ACDC and BSCMR-AS datasets when evaluated using the mean dice metric.

A blockchain-enabled federal domain generalization-based architecture [84] has been designed for dependable medical image segmentation. In this work, by utilizing FL, blockchain technology, and a Fourier-based domain generalization method, the model's segmentation performance is improved. The proposed approach is evaluated using the optic cup and optical disc segmentation job for retinal fundus images. As evident from the evaluation results, the proposed architecture offers reliability while outperforming traditional FL in terms of model generalization and accuracy.

7.4.2 Data and model heterogeneity

In FL, model and data heterogeneity create major issues. These issues are tackled using various innovative designs in model architectures as well as data handling techniques. For example, it has been shown [85] that self-attention-based architectures, such as transformers, are more resistant to distribution alterations, thereby enhancing the performance of FL with heterogeneous data. The lack of convergence and the risk of catastrophic forgetting across real-world heterogeneous devices remain key concerns, regardless of major developments. To put it more specifically, the authors carried out a thorough empirical assessment of several neural architectures using a variety of federated techniques, real-world baselines, and heterogeneous data partitions. Specifically, when working with heterogeneous input, experiments suggest that just substituting transformers for CNNs can significantly prevent catastrophic forgetting of prior devices, expedite convergence, and attain a superior global model.

SplitAVG [86] is a heterogeneity-aware federated DL method for medical imaging that overcomes the performance drops due to data heterogeneity. SplitAVG makes use of the straightforward network split and feature map fusion approaches to assist the federated model in training a neutral predictor of the intended data distribution. The SplitAVG method achieves results that are comparable to those of the baseline methods in all heterogeneous settings, achieving 96.2% accuracy and a mean absolute error of 110.4% in the case of a dataset for predicting bone age and classifying diabetic retinopathy, respectively. The SplitAVG approach is able to offset performance dips caused by differences in data distributions among institutions. Additionally, the findings of the experiments demonstrate that SplitAVG may be exploited for many kinds of medical imaging activities and modified to various basic CNN models.

A robust and label-efficient self-supervised FL framework to tackle data heterogeneity for medical image analysis is proposed in [87]. The goal of this method is to enable more robust representation learning on heterogeneous data and efficient knowledge transfer to downstream models. It does this by introducing a novel transformer-based self-supervised pretraining approach that pretrains models directly on decentralized target datasets using masked image modeling. The robustness of models against different levels of data heterogeneity is considerably improved by masked image modeling using transformers, according to thorough empirical results on synthetic and real-world medical imaging non-iid federated datasets. In particular, under extreme data heterogeneity, this method, without relying on any additional pretraining data, improves test accuracy by 5.06%, 1.53%, and 4.58% in retinal, dermatology, and chest X-ray classification, especially in comparison to the supervised baseline with ImageNet pretraining. Additionally, the authors demonstrate that in comparison to previous FL algorithms, these federated self-supervised pretraining methods produce models that generalize better to out-of-distribution data and operate more effectively when fine-tuning on sparsely labeled data.

HarmoFL [88] is a technique that correlates local and global drifts in FL on heterogeneous medical images. In order to create a consistent feature space among local clients, the authors first suggest reducing the local update drift by normalizing the amplitudes of images that have been converted into the frequency domain. Subsequently, the authors construct a client weight perturbation depending on harmonized features that direct each local model to a flat optimum, where a neighborhood of the local optimal solution has a uniformly low loss. By combining numerous local flat optima, the perturbation helps the global model optimize toward a converged optimal solution without incurring any additional communication costs. The proposed method has been thoroughly empirically tested on three medical imaging classification and segmentation tasks, and theoretical analysis by the authors has shown that HarmoFL outperforms a set of recently developed state-of-the-art algorithms with encouraging convergence behavior.

7.4.3 Privacy and security

In FL, the privacy and security of patient data are offered through different techniques such as differential privacy, homomorphic encryption, secure multiparty computation, etc. Due to the abundance and diversity of data, collaboration across various federated clients can speed up the training process and produce stronger ML models. But maintaining confidentiality while exchanging data is difficult due to privacy issues. With FL, collaborative training is made possible by exchanging model parameters instead of raw data. The majority of FL solutions currently in use operate under the presumption that participating clients are trustworthy and as a result are susceptible to poisoning attacks by bad actors whose objective is to degrade the performance of the global model.

In [89], the authors present a robust aggregation rule that is resistant to byzantine failures, called Distance-based Outlier Suppression (DOS), to suppress poisoning attacks. The suggested method uses Copula-based Outlier Detection (COPOD) to determine an outlier score for each client by calculating the distance between local parameter updates of various clients. A weighted average of the local parameters is utilized to update the global model after the resulting outlier scores are transformed into normalized weights using a softmax function. Even when the data distributions are diverse, DOS aggregation can successfully suppress parameter changes from fraudulent clients without the use of any hyperparameter selection. The DOS approach is more robust against different poisoning attacks than other state-of-the-art methods as revealed through the experimental results on the CheXpert and HAM10000 medical imaging datasets.

Shao et al. [90] propose an approach for medical data privacy-preserving stochastic channel-based FL with NN pruning. The suggested approach, known as stochastic channel-based FL (SCBFL), enables users to jointly and dispersedly train a high-performance model without revealing their inputs. The authors invented, constructed, and tested a channel-based update mechanism for a central server in a decentralized system. According to the most active features in a training loop, the update algorithm chooses the channels, uploads the learned data from nearby datasets, and then updates those channels. The algorithm developed using the validation set was then subjected to a pruning procedure, which acts as a model accelerator. The authors contend that in contrast to the FedAVG method, which divulges all of the parameters of local models to the server, the proposed model performs better and reaches saturation faster. Adding a pruning process can encourage the performance to reach saturation, and fine-tuning the pruning rate can result in even more improvement.

Hao et al. [91] propose Privacy-aware and Resource-saving Collaborative Learning (PRCL) for healthcare in a cloud computing environment. The authors develop a unique model-splitting technique that divides the NN into three pieces and offloads the computationally intensive middle half to cloud servers in order to reduce the local computational overhead. The privacy of the original data and labels, as well as the model's parameters, is protected by PRCL through the use of lightweight data perturbation and partly homomorphic encryption. Furthermore, the authors examine the proposed protocol's security and show that PRCL outperforms it in terms of accuracy and efficiency.

The authors in [92] show how differentially private gradient descent-based FL is used for the first time to complete the computational tomography job of semantic segmentation. According to the authors, good segmentation performance is feasible with a tolerable training time penalty given strict privacy protections. The adoption of DP, according to the authors, prevents the first effective gradient-based model inversion attack on a semantic segmentation model from revealing private picture information.

The authors in [93] offer an FL-based framework that may use dispersed health data stored locally at many sites to train a global model. Two degrees of privacy protection are provided by the framework. First, during the model training process, raw data are not sent between sites or shared with a central server. To further defend the model against potential privacy threats, it uses a differential privacy method. Using actual EHRs from 1 million patients, the authors carry out a thorough assessment of the proposed methodology on two healthcare applications. The authors present evidence for the viability and efficiency of the FL framework in providing a high level of privacy while upholding the usefulness of the global model.

Dopamine [94] is a system for training DNNs on distributed datasets that combines secure aggregation with FL with Differentially Private Stochastic Gradient Descent (DPSGD) and can achieve a better trade-off between differential privacy guarantee and DNN accuracy than other methods. The results of the Diabetic Retinopathy (DR) test demonstrate that Dopamine offers a differential privacy guarantee that is comparable to that of the centralized training counterpart while attaining a higher classification accuracy than FL with parallel DP where DPSGD is applied in a decentralized setting.

Guo et al. [95] propose to improve privacy-preserving logistic regression by stacking, driven by the success of enhancing prediction performance through ensemble learning. They demonstrate that this is possible using both sample-based and feature-based partitioning. However, they demonstrate that when privacy budgets are equal, feature-based partitioning requires fewer samples than sample-based partitioning and hence probably performs better empirically. The challenge of learning across businesses is further addressed by combining the suggested method with hypothesis transfer learning, which is challenging to integrate with a differential privacy guarantee. The method is tested on cross-organizational diabetes prediction from the RUIJIN dataset.

In [96], a publicly available dataset is used to build an in-hospital mortality prediction model and domain transfer is carried out. The study provides quantitative evidence that increasing the number of datasets from various hospitals enhances domain transfer's generalizability and effectiveness. This approach details a recent pilot project using FL to create a model of in-hospital mortality prediction. Empirically, it demonstrates that FL does reach a comparable level of performance with centralized training, with the added advantage of preventing the exchange of datasets among several hospitals. Additionally, it empirically analyzes the effectiveness of FedAvg and FedProx in an ICU environment.

The authors of [97] offer FedIPR, a unique ownership verification approach, by embedding watermarks in FL models to confirm FL model ownership and safeguard model intellectual property rights (IPR). They reaffirm the idea of FL and suggest Secure FL (SFL), with the ultimate objective being to construct reliable AI with strong privacy-preserving and IPR-preserving capabilities. They give a thorough overview of the efforts that are already out there, including threats, attacks, and countermeasures at each stage of SFL from a lifecycle viewpoint.

A key objective of precision medicine is the quick and accurate identification of patients with severe and diverse illnesses. Based on blood transcriptomes, ML can be used to identify leukemia patients. But because of privacy laws, there is a growing gap between what is technically possible and what is permitted. The authors of [98] introduce swarm learning, a decentralized ML approach that combines edge computing, blockchain-based peer-to-peer networking, and coordination while maintaining confidentiality without the need for a central coordinator, thereby going beyond FL, to make it easier to integrate any medical data from any data owner around the world without breaking privacy laws. Four use cases of diverse diseases were selected to demonstrate the viability of utilizing swarm learning to create disease classifiers using distributed data (COVID-19, tuberculosis, leukemia, and lung pathologies). They demonstrate that swarm learning classifiers perform better than those created at individual sites using more than 16,400 blood transcriptomes obtained from 127 clinical studies with nonuniform case-control distributions and significant study biases. Additionally, swarm learning was made specifically to comply with all local confidentiality laws. They think that by using this strategy, precision medicine will be adopted far faster.

7.4.4 System architecture and resource sharing

System architecture and resource sharing decide the efficiency and performance of FL algorithms. The FL model architecture needs to possess certain important properties. The model architecture should be scalable. It should use available resources efficiently and effectively. The time and space complexity of the underlying algorithms should be as low as possible. One should be able to customize the models as per the situations and requirements. Also, it should be feasible to deploy the resulting models on available resources. This section describes related concepts regarding designing and deploying effective FL model architectures.

SplitNN [7] suggests unique configurations of a distributed DL methodology. The proposed configurations take into account the real situations of entities holding various patient data modalities, centralized and local health entities working together on various tasks, and learning without label sharing. SplitNN is adaptable in that it supports a variety of plug-and-play configurations according to the necessary application. It can leverage any cutting-edge DL architecture and is scalable to large-scale environments. Additionally, by integrating splitNN with NN compression techniques for smooth distributed learning with edge devices, the limits of resource efficiency in distributed DL can be pushed even further. Given that there have been recent developments in all these areas, combining split learning and differential privacy with secure multiparty computation is an intriguing field for future studies. The authors compare the trade-offs between splitNN's performance and resource efficiency with those of other distributed DL techniques, like FL and big batch synchronous SGD, and present findings on the proposed architectures.

In [99], the authors present Adaptive FL (FedAP) as a means of dealing with domain transformations and obtaining customized models for local clients. FedAP maintains the uniqueness of each client using distinct local batch normalization while learning the similarity across clients based on the characteristics of the batch normalization layers. FedAP uses a pretrained model to identify client commonalities. The distances between the data distributions, which may be estimated using the statistical values of the pretrained network's layer outputs, indicate the similarities. After determining the commonalities, the server personally averages the model parameters to create a distinct model for each client. Each client maintains its own batch normalization and applies a momentum mechanism to update the model. FedAP is deployable in a variety of healthcare applications and is expandable. Extensive tests on five healthcare benchmarks show that FedAP outperforms state-of-the-art techniques with higher convergence speeds in terms of accuracy.

The authors of [100] suggest and investigate Collapsing Bandits, a novel restless multiarmed bandit (RMAB) setting in which each arm follows a binary-state Markovian process with a unique structure: when an arm is played, the state is fully observed, "collapsing" any uncertainty; however, when an arm is passive, no observation is made, allowing uncertainty to develop. By designing a constrained budget of actions per round, the objective is to keep as many arms in the "good" state as is feasible. Health professionals must simultaneously monitor patients and administer interventions in a way that maximizes the health of their patient cohort in various healthcare areas, and thus Collapsing Bandits models are natural models for such situations. Based on the Whittle index technique for RMABs, the authors determine the circumstances in which the Collapsing Bandits problem is indexable. The basis of the derivation is a novel set of conditions that define the circumstances in which "forward" or "reverse" threshold policies are the most advantageous. In order to create quick algorithms for computing the Whittle index, including a closed version, the authors use the optimality of threshold rules. The authors test the method using a variety of data distributions, including data from a real-world healthcare task where a worker must monitor and administer treatments to ensure that patients take their tuberculosis medication as prescribed. In comparison to cutting-edge RMAB approaches, the proposed algorithm accelerates computation by a factor of three while achieving comparable performance.

Different federations rarely cooperate in practical applications because of data heterogeneity and mistrust of or lack of a central server. MetaFed framework [101] enables reliable FL between various federations. Through the suggested Cyclic Knowledge Distillation, MetaFed generates a personalized model for each federation without the need for a central server. In particular, MetaFed gathers knowledge from each federation in a circular fashion by treating each as a metadistribution. The two components of the training are general knowledge acquisition and personalization. Comprehensive tests on three benchmarks show that MetaFed without a server delivers greater accuracy than cutting-edge techniques with fewer communication costs.

In [102], the authors proposed a model based on graph NNs (GNNs) that have made remarkable strides in a variety of sectors, such as network neuroscience and medical imaging, where they demonstrated a high degree of accuracy in the diagnosis of difficult neurological illnesses like autism. The most discriminative biomarkers (i.e., features) chosen by the GNN models inside an FL paradigm must be investigated for their repeatability; however, state-of-the-art GNNs and FL approaches ignore this crucial unsolved challenge. One of the largest challenges in creating translational clinical applications is quantifying the repeatability of a predictive medical model against disruptions in the distribution of training and testing data. The classification of medical imaging and brain connectivity datasets using federated GNN models is presented in this method as the first study examining the reproducibility of such models. Various GNN models trained on datasets from connectomics and medical imaging were used by the authors to test the framework. More significantly, they demonstrated how FL improves GNN models' accuracy and reproducibility in several medical learning tasks.

Kumar et al. [103] suggest an architecture that gathers a modest quantity of data from diverse sources (hospitals) and uses blockchain-based FL to train a global deep learning model for diagnosing COVID-19. While maintaining the organization's privacy, blockchain technology uses FL to train the model worldwide and authenticate the data. The authors first provide a data normalization method that addresses the heterogeneity of the data since the information was acquired from hospitals with various types of CT scanners. Second, they identify COVID-19 patients using segmentation and classification based on capsule networks. Thirdly, they develop a technique that can use blockchain technology and FL to jointly train a global model while maintaining anonymity. Additionally, they gather publicly available data on actual COVID-19 patients. The suggested framework may make use of current information, which enhances CT image recognition. Finally, in order to validate the suggested strategy, they carry out extensive trials. The findings show improved performance in identifying COVID-19 patients.

In order to reduce training latency, the work in [104] presents a multiarmed bandit-based framework for online client scheduling in FL without requiring knowledge of wireless channel status information or statistical client characteristics. First, the authors provide a Client Scheduling Upper Confidence Bound (CS-UCB) based on the upper confidence bound strategy under ideal conditions where local datasets of clients are evenly distributed and independently distributed. The suggested CS-UCB method is given with an upper bound of the predicted performance regret, which shows that the regret increases logarithmically with the number of communication rounds. Then, they further present a CS algorithm based on the UCB policy and virtual queue technique to solve nonideal cases with non-iid and imbalanced features of local datasets and changing client availability. The suggested algorithm's expected performance regret can, under certain circumstances, grow sublinearly over the course of communication rounds, according to an upper bound that is also determined. Additionally, the FL training's convergence performance is examined. The outcomes of the simulations confirm the effectiveness of the suggested algorithms.

The authors in [105] address the cumulative reward maximization problem in a secure FL setting, where multiple data owners keep their data stored locally and collaborate under the coordination of a central orchestration server. They rely on cryptographic schemes and propose Samba, a generic framework for Secure federAted Multiarmed BAndits. Each data owner has data associated with a bandit arm and the bandit algorithm has to sequentially select which data owner is solicited at each time step. They instantiate Samba for five bandit algorithms and show that Samba returns the same cumulative reward as the nonsecure versions of bandit algorithms while satisfying formally proven security properties. They also show that the overhead due to cryptographic primitives is linear in the size of the input, which is confirmed by the proof-of-concept implementation.

In [106], the authors first introduce reputation as the metric to gauge the dependability and trustworthiness of mobile devices in order to address the aforementioned issues. Then, using a multiweight subjective logic model, they design a reputation-based worker selection system for trustworthy FL. They also use the blockchain to achieve safe reputation management for workers with decentralized nonrepudiation and tamper-resistance features. In addition, they provide a strong incentive system integrating reputation and contract theory to encourage high-reputation mobile devices with high-quality data to take part in model learning. The proposed systems are effective for trustworthy FL, as demonstrated by the numerical findings, which also show a considerable improvement in learning accuracy.

7.4.5 Lack of dataset and training

The lack of sufficient, proper, and well-distributed datasets significantly affects the performance of FL algorithms. Due to data scarcity, the FL models tend to underfit. Due to the nonuniform distribution of the data, the FL models tend to learn wrong concepts such as learning only majority classes and neglecting minority classes. Due to the wrong format or low quality of data, the FL models tend to diverge or do not reach global minima during training. Much deliberation is required to handle such issues. Designing a robust and accurate model architecture in the presence of such issues needs a lot of customization and innovation.

The method in [107] develops and explores the federated partially supervised learning (FPSL) problem for restricted decentralized medical images with partial labels. By using FPSL multilabel classification as an example, the authors investigate the effects of decentralized partially labeled data on DL-based models. The authors formulate and analyze two key FPSL challenges through the analysis of FedAVG, a seminal FL system, and present FedPSL, a straightforward yet reliable FPSL framework, to overcome these issues. FedPSL, in particular, has two modules: task-dependent model aggregation and task-agnostic decoupling learning, where the first module deals with weight assignment and the second module enhances the feature extractor's generalization capabilities. The authors use simulated trials to provide a thorough empirical understanding of FPSL in the presence of data scarcity. The empirical results demonstrate that the proposed FedPSL can outperform baseline methods on data challenges such as data scarcity and domain shifts, in addition to demonstrating that FPSL is an understudied subject with application value. The results of this study also provide a fresh line of research into label-effective label learning on medical imagery.

In order to address performance degradation (i.e., retrogression) following each communication and the unavoidable class imbalance, the authors of [108] suggest a revolutionary tailored FL framework. They initially create progressive Fourier aggregation (PFA) at the server side to gradually incorporate client model parameters in the frequency domain in order to solve the retrogression problem. They then create a deputy-enhanced transfer (DET) on the client side to seamlessly transfer global knowledge to the local model that is tailored to each customer. They suggest the conjoint prototype-aligned (CPA) loss help with the balanced FL framework optimization for the class imbalance problem. The CPA loss determines the global conjoint objective based on global imbalance and then modifies the client-side local training through the prototype-aligned refinement to eliminate the imbalance gap with such a balanced goal, taking into account the inaccessibility of private local data to other participants in FL. The findings on real-world dermoscopic and prostate MRI FL datasets reveal that this method significantly outperforms state-of-the-art FL approaches, highlighting the benefits of the proposed FL framework in practical medical settings.

The novel compression architecture suggested in [109] named sparse ternary compression (STC) satisfies the needs of the FL environment. STC adds a novel mechanism to the top-k gradient sparsification compression method that enables downstream compression, ternarization, and the best Golomb encoding of the weight updates. Tests on four different learning tasks show that STC performs noticeably better than federated averaging in typical FL situations. These findings support a paradigm shift in federated optimization in favor of high-frequency low-bit-width communication, especially in bandwidth-restricted educational settings.

Konecny et al. [110] examine learning techniques for an FL scenario. Communication effectiveness is of utmost importance in this situation because the typical clients are mobile phones. The authors propose two strategies for lowering uplink communication costs: structured updates, in which we learn an update directly from a constrained space parametrized with fewer variables, such as a low-rank or random mask, and sketched updates, in which we learn a full model update and then condense it using a combination of quantization, random rotations, and subsampling before sending it to the server.

The efficiency of split learning and FL is examined by Singh et al. [111]. They compare the two strategies in each of a number of actual cases for distributed learning setups. They demonstrate that growing the client or model size benefits split learning configurations over federated ones, although growing the amount of data samples while maintaining a small client or model size boosts FL communication effectiveness.

7.5 Conclusions and future scope

In this chapter, we have discussed FL and its applications in the healthcare domain. More specifically, we have discussed the preliminaries of FL, its working principle, types, training algorithms, impact on stakeholders, applications, challenges, considerations, etc. We have seen the applications of FL in EHR mining, remote health monitoring, medical imaging, and disease prediction. We have also discussed challenges and considerations which include domain generalization, data and model heterogeneity, privacy and security, system architecture and resource sharing, lack of dataset and training, etc. We have also summarized multiple use cases in which the global community has addressed various core challenges and their most probable solutions by offering novel solutions. The detailed discussion of these issues shows that FL can be used as a very powerful tool to overcome multiple challenges in the healthcare sector.

The future scope of research in FL spans multiple domains. In healthcare, more research is required to address certain fundamental challenges which include data and model heterogeneity, scalability, privacy and security, complexity, feasibility, efficiency, readiness to deployment on edge platforms, etc. Also, the resultant models should conform to regulatory standards, especially in the medical domain.

References

[1] Reform of EU Data Protection Rules 2018, https://eur-lex.europa.eu/eli/reg/2016/679, Last Accessed: 01 February 2023.

[2] Centers for Medicare and Medicaid Services, 104th Congress, Public Law 104–191, The health insurance portability and accountability act of 1996 (HIPAA), https://www.cms.gov/regulations-and-guidance/administrative-simplification/hipaa-aca, Last Accessed: 01 February 2023.

[3] B. McMahan, E. Moore, D. Ramage, S. Hampson, Arcas BAy, Communication-efficient learning of deep networks from decentralized data, in: Proc. of the 20th International Conference on Artificial Intelligence and Statistics (PMLR 2017), vol. 54, 2017, pp. 1273–1282.

[4] P. Vanhaesebrouck, A. Bellet, M. Tommasi, Decentralized collaborative learning of personalized models over networks, https://doi.org/10.48550/ARXIV.1610.05202, 2016.

[5] S. Savazzi, M. Nicoli, V. Rampa, Federated learning with cooperating devices: a consensus approach for massive IoT networks, IEEE Internet of Things Journal 7 (5) (2020) 4641–4654, https://doi.org/10.1109/JIOT.2020.2964162.

[6] E. Diao, J. Ding, V. Tarokh, HeteroFL: computation and communication efficient federated learning for heterogeneous clients, in: Proc. of International Conference on Learning Representations (ICLR 2021), 2021.

[7] P. Vepakomma, O. Gupta, T. Swedish, R. Raskar, Split learning for health: distributed deep learning without sharing raw patient data, https://doi.org/10.48550/ARXIV.1812.00564, 2018.

[8] O. Gupta, R. Raskar, Distributed learning of deep neural network over multiple agents, Journal of Network and Computer Applications (2018) 116, https://doi.org/10.1016/j.jnca.2018.05.003.

[9] K. Hsieh, A. Phanishayee, O. Mutlu, P. Gibbons, The non-IID data quagmire of decentralized machine learning, in: Proc. of the 37th International Conference on Machine Learning (ICML 2020), in: Proc. of Machine Learning Research, vol. 119, 2020, pp. 4387–4398.
[10] P. Kairouz, H.B. McMahan, B. Avent, A. Bellet, M. Bennis, A.N. Bhagoji, et al., Advances and open problems in federated learning, Foundations and Trends in Machine Learning 14 (1–2) (2021), https://doi.org/10.1561/2200000083.
[11] X. Li, M. Jiang, X. Zhang, M. Kamp, Q. Dou, FedBN: federated learning on non-IID features via local batch normalization, in: Proc. of International Conference on Learning Representations (ICLR 2021), 2021.
[12] D.A.E. Acar, Y. Zhao, R. Matas, M. Mattina, P. Whatmough, V. Saligrama, Federated learning based on dynamic regularization, in: Proc. of International Conference on Learning Representations (ICLR 2021), 2021.
[13] Y. Tan, G. Long, L. Liu, T. Zhou, Q. Lu, J. Jiang, et al., FedProto: federated prototype learning across heterogeneous clients, in: Proc. of the AAAI Conference on Artificial Intelligence (AAAI-22), 2022, pp. 8432–8440.
[14] T. Li, A.K. Sahu, M. Zaheer, M. Sanjabi, A. Talwalkar, V. Smith, Federated optimization in heterogeneous networks, in: Proc. of Machine Learning and Systems (MLSys 2020), vol. 2, 2020, pp. 429–450.
[15] R. Shokri, V. Shmatikov, Privacy-preserving deep learning, in: Proc. of the 22nd ACM SIGSAC Conference on Computer and Communications Security (CCS 2015), 2015, pp. 1310–1321.
[16] T. Shen, J. Zhang, X. Jia, F. Zhang, G. Huang, P. Zhou, et al., Federated mutual learning, https://doi.org/10.48550/ARXIV.2006.16765, 2020.
[17] T. Overman, G. Blum, D. Klabjan, A primal-dual algorithm for hybrid federated learning, https://doi.org/10.48550/ARXIV.2210.08106, 2022.
[18] Y. Yeganeh, A. Farshad, N. Navab, S. Albarqouni, Inverse distance aggregation for federated learning with non-IID data, in: Proc. of Domain Adaptation and Representation Transfer, and Distributed and Collaborative Learning (DART DCL 2020), 2020, pp. 150–159.
[19] M. Andreux, J.O. du Terrail, C. Beguier, E.W. Tramel, Siloed federated learning for multi-centric histopathology datasets, in: Proc. of Domain Adaptation and Representation Transfer and Distributed and Collaborative Learning (DART DCL 2020), 2020, pp. 129–139.
[20] S. Vahidian, M. Morafah, B. Lin, Personalized federated learning by structured and unstructured pruning under data heterogeneity, in: Proc. of 41st International Conference on Distributed Computing Systems Workshops (ICDCSW 2021), 2021, pp. 27–34.
[21] S.J. Reddi, Z. Charles, M. Zaheer, Z. Garrett, K. Rush, J. Konečný, et al., Adaptive federated optimization, in: Proc. of International Conference on Learning Representations (ICLR 2021), 2021.
[22] M. Jaggi, V. Smith, M. Takáč, J. Terhorst, S. Krishnan, T. Hofmann, et al., Communication-efficient distributed dual coordinate ascent, in: Proc. of the 27th International Conference on Neural Information Processing Systems - Volume 2 (NIPS 2014), 2014, pp. 3068–3076.
[23] V. Smith, S. Forte, C. Ma, M. Takáč, M.I. Jordan, M. Jaggi, CoCoA: a general framework for communication-efficient distributed optimization, Journal of Machine Learning Research 18 (1) (2017) 8590–8638.
[24] N. Parikh, S. Boyd, Block splitting for distributed optimization, Mathematical Programming Computation 6 (1) (2014) 77–102, https://doi.org/10.1007/s12532-013-0061-8.

[25] J. Frankle, M. Carbin, The lottery ticket hypothesis: finding sparse, trainable neural networks, in: Proc. of International Conference on Learning Representations (ICLR 2019), 2019.
[26] N. Rieke, J. Hancox, W. Li, F. Milletarì, H.R. Roth, S. Albarqouni, et al., The future of digital health with federated learning, npj Digital Medicine 3 (1) (2020) 119, https://doi.org/10.1038/s41746-020-00323-1.
[27] J. Ma, Q. Zhang, J. Lou, J.C. Ho, L. Xiong, X. Jiang, Privacy-preserving tensor factorization for collaborative health data analysis, in: Proc. of the 28th ACM International Conference on Information and Knowledge Management (CIKM 2019), ISBN 9781450369763, 2019, pp. 1291–1300.
[28] Y. Kim, J. Sun, H. Yu, X. Jiang, Federated tensor factorization for computational phenotyping, in: Proc. of the 23rd ACM SIGKDD International Conference on Knowledge Discovery and Data Mining (KDD 2017), 2017, pp. 887–895.
[29] T.K. Dang, X. Lan, J. Weng, M. Feng, Federated learning for electronic health records, ACM Transactions on Intelligent Systems and Technology 13 (5) (2022), https://doi.org/10.1145/3514500.
[30] T.J. Pollard, A.E.W. Johnson, J.D. Raffa, L.A. Celi, R.G. Mark, O. Badawi, The eICU collaborative research database, a freely available multi-center database for critical care research, Scientific Data 5 (1) (2018) 180178, https://doi.org/10.1038/sdata.2018.178.
[31] S.R. Pfohl, A.M. Dai, K. Heller, Federated and differentially private learning for electronic health records, https://doi.org/10.48550/ARXIV.1911.05861, 2019.
[32] D. Liu, T. Miller, R. Sayeed, K.D. Mandl, FADL: federated-autonomous deep learning for distributed electronic health record, Machine Learning for Health (ML4H) Workshop at NeurIPS 2018, https://doi.org/10.48550/ARXIV.1811.11400, 2018.
[33] S. Boughorbel, F. Jarray, N. Venugopal, S. Moosa, H. Elhadi, M. Makhlouf, Federated uncertainty-aware learning for distributed hospital EHR data, Machine Learning for Health (ML4H) at NeurIPS 2019, https://doi.org/10.48550/ARXIV.1910.12191, 2019.
[34] D.C. Nguyen, P.N. Pathirana, M. Ding, A. Seneviratne, Blockchain and edge computing for decentralized EMRs sharing in federated healthcare, in: Proc. of IEEE Global Communications Conference (GLOBECOM 2020), 2020.
[35] T.S. Brisimi, R. Chen, T. Mela, A. Olshevsky, I.C. Paschalidis, W. Shi, Federated learning of predictive models from federated electronic health records, International Journal of Medical Informatics 112 (2018) 59–67, https://doi.org/10.1016/j.ijmedinf.2018.01.007.
[36] D. Liu, D. Dligach, T. Miller, Two-stage federated phenotyping and patient representation learning, in: Proc. of the Conference Association for Computational Linguistics Meeting, 2019, pp. 283–291.
[37] L. Huang, A.L. Shea, H. Qian, A. Masurkar, H. Deng, D. Liu, Patient clustering improves efficiency of federated machine learning to predict mortality and hospital stay time using distributed electronic medical records, Journal of Biomedical Informatics 99 (2019) 103291, https://doi.org/10.1016/j.jbi.2019.103291.
[38] X. Xu, H. Peng, M.Z.A. Bhuiyan, Z. Hao, L. Liu, L. Sun, et al., Privacy-preserving federated depression detection from multisource mobile health data, IEEE Transactions on Industrial Informatics 18 (7) (2022) 4788–4797, https://doi.org/10.1109/TII.2021.3113708.
[39] Y. Chen, X. Qin, J. Wang, C. Yu, W. Gao, FedHealth: a federated transfer learning framework for wearable healthcare, IEEE Intelligent Systems 35 (4) (2020) 83–93, https://doi.org/10.1109/MIS.2020.2988604.

[40] Q. Wu, X. Chen, Z. Zhou, J. Zhang, FedHome: cloud-edge based personalized federated learning for in-home health monitoring, IEEE Transactions on Mobile Computing 21 (8) (2022) 2818–2832, https://doi.org/10.1109/TMC.2020.3045266.

[41] G. Vavoulas, C. Chatzaki, T. Malliotakis, M. Pediaditis, M. Tsiknakis, The MobiAct dataset: recognition of activities of daily living using smartphones, in: Proc. of the International Conference on Information and Communication Technologies for Ageing Well and e-Health (ICT4AGEINGWELL 2016), ISBN 978-989-758-180-9, 2016, pp. 143–151.

[42] G.K. Gudur, S.K. Perepu, Resource-constrained federated learning with heterogeneous labels and models for human activity recognition, in: Proc. of Deep Learning for Human Activity Recognition (DL-HAR 2021), ISBN 978-981-16-0575-8, 2021, pp. 57–69.

[43] A. Stisen, H. Blunck, S. Bhattacharya, T.S. Prentow, M.B. Kjærgaard, A. Dey, et al., Smart devices are different: assessing and mitigating mobile sensing heterogeneities for activity recognition, in: Proc. of the 13th ACM Conference on Embedded Networked Sensor Systems (SenSys 2015), ISBN 9781450336314, 2015, pp. 127–140.

[44] P. Guo, P. Wang, J. Zhou, S. Jiang, V.M. Patel, Multi-institutional collaborations for improving deep learning-based magnetic resonance image reconstruction using federated learning, in: Proc. of IEEE/CVF Conference on Computer Vision and Pattern Recognition (CVPR 2021), 2021, pp. 2423–2432.

[45] F. Knoll, J. Zbontar, A. Sriram, M.J. Muckley, M. Bruno, A. Defazio, et al., FastMRI: a publicly available raw k-space and DICOM dataset of knee images for accelerated MR image reconstruction using machine learning, Radiology: Artificial Intelligence 2 (1) (2020) e190007, https://doi.org/10.1148/ryai.2020190007, PMID: 32076662.

[46] S. Jiang, C.G. Eberhart, M. Lim, H.Y. Heo, Y. Zhang, L. Blair, et al., Identifying recurrent malignant glioma after treatment using amide proton transfer-weighted MR imaging: a validation study with image-guided stereotactic biopsy, Clinical Cancer Research 25 (2) (2019) 552–561, https://doi.org/10.1158/1078-0432.CCR-18-1233.

[47] IXI dataset, https://brain-development.org/ixi-dataset/, Last Accessed: 01 February 2023.

[48] B.H. Menze, A. Jakab, S. Bauer, J. Kalpathy-Cramer, K. Farahani, J. Kirby, et al., The multimodal brain tumor image segmentation benchmark (BRATS), IEEE Transactions on Medical Imaging 34 (10) (2015) 1993–2024, https://doi.org/10.1109/TMI.2014.2377694.

[49] M.Y. Lu, R.J. Chen, D. Kong, J. Lipkova, R. Singh, D.F. Williamson, et al., Federated learning for computational pathology on gigapixel whole slide images, Medical Image Analysis 76 (2022) 102298, https://doi.org/10.1016/j.media.2021.102298.

[50] M.J. Sheller, B. Edwards, G.A. Reina, J. Martin, S. Pati, A. Kotrotsou, et al., Federated learning in medicine: facilitating multi-institutional collaborations without sharing patient data, Scientific Reports 10 (1) (2020), https://doi.org/10.1038/s41598-020-69250-1.

[51] Z. Fan, J. Su, K. Gao, D. Hu, L.L. Zeng, A federated deep learning framework for 3D brain MRI images, in: Proc. of International Joint Conference on Neural Networks (IJCNN 2021), 2021.

[52] A. Di Martino, C.G. Yan, Q. Li, E. Denio, F.X. Castellanos, K. Alaerts, et al., The autism brain imaging data exchange: towards a large-scale evaluation of the intrinsic brain architecture in autism, Molecular Psychiatry 19 (6) (2014) 659–667, https://doi.org/10.1038/mp.2013.78.

[53] S. Silva, B. Gutman, E. Romero, P.M. Thompson, A. Altmann, M. Lorenzi, Federated learning in distributed medical databases: meta-analysis of large-scale subcortical brain data, https://doi.org/10.48550/ARXIV.1810.08553, 2018.

[54] J. Lo, T.T. Yu, D. Ma, P. Zang, J.P. Owen, Q. Zhang, et al., Federated learning for microvasculature segmentation and diabetic retinopathy classification of OCT data, Ophthalmology Science 1 (4) (2021) 100069, https://doi.org/10.1016/j.xops.2021.100069.

[55] J. Wicaksana, Z. Yan, D. Zhang, X. Huang, H. Wu, X. Yang, et al., FedMix: mixed supervised federated learning for medical image segmentation, IEEE Transactions on Medical Imaging (2022), https://doi.org/10.1109/TMI.2022.3233405.

[56] H.R. Roth, K. Chang, P. Singh, N. Neumark, W. Li, V. Gupta, et al., Federated learning for breast density classification: a real-world implementation, in: Proc. of Domain Adaptation and Representation Transfer, and Distributed and Collaborative Learning (DART 2020, DCL 2020), ISBN 978-3-030-60548-3, 2020, pp. 181–191.

[57] X. Li, Y. Gu, N. Dvornek, L.H. Staib, P. Ventola, J.S. Duncan, Multi-site fMRI analysis using privacy-preserving federated learning and domain adaptation: ABIDE results, Medical Image Analysis 65 (2020) 101765, https://doi.org/10.1016/j.media.2020.101765.

[58] W. Li, F. Milletarì, D. Xu, N. Rieke, J. Hancox, W. Zhu, et al., Privacy-preserving federated brain tumour segmentation, in: Proc. of Machine Learning in Medical Imaging (MLMI 2019), ISBN 978-3-030-32692-0, 2019, pp. 133–141.

[59] C.M. Feng, Y. Yan, S. Wang, Y. Xu, L. Shao, H. Fu, Specificity-preserving federated learning for MR image reconstruction, IEEE Transactions on Medical Imaging (2022), https://doi.org/10.1109/TMI.2022.3202106.

[60] Z. Yan, J. Wicaksana, Z. Wang, X. Yang, K.T. Cheng, Variation-aware federated learning with multi-source decentralized medical image data, IEEE Journal of Biomedical and Health Informatics 25 (7) (2021) 2615–2628, https://doi.org/10.1109/JBHI.2020.3040015.

[61] J. Wang, Y. Jin, L. Wang, Personalizing federated medical image segmentation via local calibration, in: Proc. of European Conference on Computer Vision (ECCV 2022), ISBN 978-3-031-19803-8, 2022, pp. 456–472.

[62] F. Ucar, D. Korkmaz, COVIDiagnosis-Net: deep Bayes-SqueezeNet based diagnosis of the coronavirus disease 2019 (COVID-19) from X-ray images, Medical Hypotheses 140 (2020) 109761, https://doi.org/10.1016/j.mehy.2020.109761.

[63] R. Jin, X. Li, Backdoor attack is a devil in federated GAN-based medical image synthesis, in: Proc. of Simulation and Synthesis in Medical Imaging (SASHIMI 2022), ISBN 978-3-031-16980-9, 2022, pp. 154–165.

[64] T. Ngo, D.C. Nguyen, P.N. Pathirana, L.A. Corben, M.B. Delatycki, M. Horne, et al., Federated deep learning for the diagnosis of cerebellar ataxia: privacy preservation and auto-crafted feature extractor, IEEE Transactions on Neural Systems and Rehabilitation Engineering 30 (2022) 803–811, https://doi.org/10.1109/TNSRE.2022.3161272.

[65] Z.A. Huang, Y. Hu, R. Liu, X. Xue, Z. Zhu, L. Song, et al., Federated multi-task learning for joint diagnosis of multiple mental disorders on MRI scans, IEEE Transactions on Biomedical Engineering (2022), https://doi.org/10.1109/TBME.2022.3210940.

[66] L. Peng, N. Wang, N. Dvornek, X. Zhu, X. Li, FedNI: federated graph learning with network inpainting for population-based disease prediction, IEEE Transactions on Medical Imaging (2022), https://doi.org/10.1109/TMI.2022.3188728.

[67] Alzheimer's disease neuroimaging initiative (ADNI), https://adni.loni.usc.edu, 2003.

[68] Z. Li, X. Xu, X. Cao, W. Liu, Y. Zhang, D. Chen, et al., Integrated CNN and federated learning for COVID-19 detection on chest X-ray images, IEEE/ACM Transactions on Computational Biology and Bioinformatics (2022), https://doi.org/10.1109/TCBB.2022.3184319.

[69] A. Jiménez-Sánchez, M. Tardy, M.A. González Ballester, D. Mateus, G. Piella, Memory-aware curriculum federated learning for breast cancer classification, Computer Methods and Programs in Biomedicine 229 (2023) 107318, https://doi.org/10.1016/j.cmpb.2022.107318.

[70] S. Dara, A. Kanapala, A.R. Babu, S. Dhamercherala, A. Vidyarthi, R. Agarwal, Scalable federated-learning and Internet-of-things enabled architecture for chest computer tomography image classification, Computers & Electrical Engineering 102 (2022) 108266, https://doi.org/10.1016/j.compeleceng.2022.108266.

[71] T.V. Nguyen, M.A. Dakka, S.M. Diakiw, M.D. VerMilyea, M. Perugini, J.M.M. Hall, et al., A novel decentralized federated learning approach to train on globally distributed, poor quality, and protected private medical data, Scientific Reports 12 (1) (2022), https://doi.org/10.1038/s41598-022-12833-x.

[72] S.T. Arasteh, P. Isfort, M. Saehn, G. Mueller-Franzes, F. Khader, J.N. Kather, et al., Collaborative training of medical artificial intelligence models with non-uniform labels, https://doi.org/10.48550/ARXIV.2211.13606, 2022.

[73] Q. Liu, H. Yang, Q. Dou, P.A. Heng, Federated semi-supervised medical image classification via inter-client relation matching, in: Proc. of 24th International Conference on Medical Image Computing and Computer Assisted Intervention (MICCAI 2021), ISBN 978-3-030-87198-7, 2021, pp. 325–335.

[74] D.C. Nguyen, M. Ding, P.N. Pathirana, A. Seneviratne, A.Y. Zomaya, Federated learning for COVID-19 detection with generative adversarial networks in edge cloud computing, IEEE Internet of Things Journal 9 (12) (2022) 10257–10271, https://doi.org/10.1109/JIOT.2021.3120998.

[75] M.N. Hossen, V. Panneerselvam, D. Koundal, K. Ahmed, F.M. Bui, S.M. Ibrahim, Federated machine learning for detection of skin diseases and enhancement of Internet of medical things (IoMT) security, IEEE Journal of Biomedical and Health Informatics (2022), https://doi.org/10.1109/JBHI.2022.3149288.

[76] D. Manu, Y. Sheng, J. Yang, J. Deng, T. Geng, A. Li, et al., FL-DISCO: federated generative adversarial network for graph-based molecule drug discovery, in: Proc. of IEEE/ACM International Conference on Computer Aided Design (ICCAD 2021), 2021.

[77] M. Oldenhof, G. Ács, B. Pejó, A. Schuffenhauer, N. Holway, N. Sturm, et al., Industry-scale orchestrated federated learning for drug discovery, https://doi.org/10.48550/ARXIV.2210.08871, 2022.

[78] S. Yang, H. Hwang, D. Kim, R. Dua, J.Y. Kim, E. Yang, et al., Towards the practical utility of federated learning in the medical domain, https://doi.org/10.48550/ARXIV.2207.03075, 2022.

[79] A.E. Cetinkaya, M. Akin, S. Sagiroglu, Improving performance of federated learning-based medical image analysis in non-IID settings using image augmentation, in: Proc. of International Conference on Information Security and Cryptology (ISCTURKEY 2021), 2021, pp. 69–74.

[80] Q. Liu, C. Chen, J. Qin, Q. Dou, P.A. Heng, FedDG: federated domain generalization on medical image segmentation via episodic learning in continuous frequency space, in: Proc. of 2021 IEEE/CVF Conference on Computer Vision and Pattern Recognition (CVPR 2021), 2021, pp. 1013–1023.

[81] Y. Sun, N. Chong, H. Ochiai, Feature distribution matching for federated domain generalization, https://doi.org/10.48550/ARXIV.2203.11635, 2022.

[82] L. Zhang, X. Lei, Y. Shi, H. Huang, C. Chen, Federated learning for IoT devices with domain generalization, IEEE Internet of Things Journal (2023), https://doi.org/10.1109/JIOT.2023.3234977.

[83] C. Chen, W. Bai, R.H. Davies, A.N. Bhuva, C.H. Manisty, J.B. Augusto, et al., Improving the generalizability of convolutional neural network-based segmentation on CMR images, Frontiers in Cardiovascular Medicine 7 (2020), https://doi.org/10.3389/fcvm.2020.00105.

[84] X. Liao, J. Zhou, J. Shu, A blockchain-enabled federal domain generalization based architecture for dependable medical image segmentation, in: Proc. of IEEE 6th Advanced Information Technology, Electronic and Automation Control Conference (IAEAC 2022), 2022, pp. 1655–1658.

[85] L. Qu, Y. Zhou, P.P. Liang, Y. Xia, F. Wang, E. Adeli, et al., Rethinking architecture design for tackling data heterogeneity in federated learning, in: Proc. of IEEE/CVF Conference on Computer Vision and Pattern Recognition (CVPR 2022), 2022, pp. 10051–10061.

[86] M. Zhang, L. Qu, P. Singh, J. Kalpathy-Cramer, D.L. Rubin, SplitAVG: a heterogeneity-aware federated deep learning method for medical imaging, IEEE Journal of Biomedical and Health Informatics 26 (9) (2022) 4635–4644, https://doi.org/10.1109/JBHI.2022.3185956.

[87] R. Yan, L. Qu, Q. Wei, S.C. Huang, L. Shen, D. Rubin, et al., Label-efficient self-supervised federated learning for tackling data heterogeneity in medical imaging, IEEE Transactions on Medical Imaging (2022), https://doi.org/10.1109/TMI.2022.3233574.

[88] M. Jiang, Z. Wang, Q. Dou, HarmoFL: harmonizing local and global drifts in federated learning on heterogeneous medical images, in: Proc. of 36th AAAI Conference on Artificial Intelligence (AAAI 2022), 2022, pp. 1087–1095.

[89] N. Alkhunaizi, D. Kamzolov, M. Takáč, K. Nandakumar, Suppressing poisoning attacks on federated learning for medical imaging, in: Proc. of Medical Image Computing and Computer Assisted Intervention (MICCAI 2022), ISBN 978-3-031-16452-1, 2022, pp. 673–683.

[90] R. Shao, H. He, Z. Chen, H. Liu, D. Liu, Stochastic channel-based federated learning with neural network pruning for medical data privacy preservation: model development and experimental validation, JMIR Formative Research 4 (12) (2020), https://doi.org/10.2196/17265.

[91] M. Hao, H. Li, G. Xu, Z. Liu, Z. Chen, Privacy-aware and resource-saving collaborative learning for healthcare in cloud computing, in: Proc. of IEEE International Conference on Communications (ICC 2020), 2020.

[92] A. Ziller, D. Usynin, N. Remerscheid, M. Knolle, M. Makowski, R. Braren, et al., Differentially private federated deep learning for multi-site medical image segmentation, https://doi.org/10.48550/ARXIV.2107.02586, 2021.

[93] O. Choudhury, A. Gkoulalas-Divanis, T. Salonidis, I. Sylla, Y. Park, G. Hsu, et al., Differential privacy-enabled federated learning for sensitive health data, https://doi.org/10.48550/ARXIV.1910.02578, 2019.

[94] M. Malekzadeh, B. Hasircioglu, N. Mital, K. Katarya, M.E. Ozfatura, D. Gündüz, Dopamine: differentially private federated learning on medical data, https://doi.org/10.48550/ARXIV.2101.11693, 2021.

[95] X. Guo, Q. Yao, J. Kwok, W. Tu, Y. Chen, W. Dai, et al., Privacy-Preserving Stacking with Application to Cross-Organizational Diabetes Prediction, ISBN 978-3-030-63076-8, 2020, pp. 269–283.

[96] T.K. Dang, K.C. Tan, M. Choo, N. Lim, J. Weng, M. Feng, Building ICU In-Hospital Mortality Prediction Model with Federated Learning, ISBN 978-3-030-63076-8, 2020, pp. 255–268.

[97] Q. Yang, A. Huang, L. Fan, C.S. Chan, J.H. Lim, K.W. Ng, et al., Federated learning with privacy-preserving and model IP-right-protection, Machine Intelligence Research 20 (1) (2023) 19–37, https://doi.org/10.1007/s11633-022-1343-2.

[98] S. Warnat-Herresthal, H. Schultze, K.L. Shastry, S. Manamohan, S. Mukherjee, V. Garg, et al., Swarm learning for decentralized and confidential clinical machine learning, Nature 594 (7862) (2021) 265–270, https://doi.org/10.1038/s41586-021-03583-3.

[99] W. Lu, J. Wang, Y. Chen, X. Qin, R. Xu, D. Dimitriadis, et al., Personalized federated learning with adaptive batchnorm for healthcare, IEEE Transactions on Big Data (2022), https://doi.org/10.1109/TBDATA.2022.3177197.

[100] A. Mate, J. Killian, H. Xu, A. Perrault, M. Tambe, Collapsing bandits and their application to public health intervention, in: Proc. of Advances in Neural Information Processing Systems (NeurIPS 2020), vol. 33, 2020, pp. 15639–15650.

[101] Y. Chen, W. Lu, X. Qin, J. Wang, X. Xie, MetaFed: federated learning among federations with cyclic knowledge distillation for personalized healthcare, https://doi.org/10.48550/ARXIV.2206.08516, 2022.

[102] M.Y. Balık, A. Rekik, I. Rekik, Investigating the predictive reproducibility of federated graph neural networks using medical datasets, in: Proc. of Predictive Intelligence in Medicine (PRIME 2022), ISBN 978-3-031-16919-9, 2022, pp. 160–171.

[103] R. Kumar, A.A. Khan, J. Kumar, Zakria, N.A. Golilarz, S. Zhang, et al., Blockchain-federated-learning and deep learning models for COVID-19 detection using CT imaging, IEEE Sensors Journal 21 (14) (2021) 16301–16314, https://doi.org/10.1109/JSEN.2021.3076767.

[104] W. Xia, T.Q.S. Quek, K. Guo, W. Wen, H.H. Yang, H. Zhu, Multi-armed bandit-based client scheduling for federated learning, IEEE Transactions on Wireless Communications 19 (11) (2020) 7108–7123, https://doi.org/10.1109/TWC.2020.3008091.

[105] R. Ciucanu, P. Lafourcade, G. Marcadet, M. Soare, SAMBA: a generic framework for secure federated multi-armed bandits, Journal of Artificial Intelligence Research (2022) 73, https://doi.org/10.1613/jair.1.13163.

[106] J. Kang, Z. Xiong, D. Niyato, S. Xie, J. Zhang, Incentive mechanism for reliable federated learning: a joint optimization approach to combining reputation and contract theory, IEEE Internet of Things Journal 6 (6) (2019) 10700–10714, https://doi.org/10.1109/JIOT.2019.2940820.

[107] N. Dong, M. Kampffmeyer, I. Voiculescu, E. Xing, Federated partially supervised learning with limited decentralized medical images, IEEE Transactions on Medical Imaging (2022), https://doi.org/10.1109/TMI.2022.3231017.

[108] Z. Chen, C. Yang, M. Zhu, Z. Peng, Y. Yuan, Personalized retrogress-resilient federated learning toward imbalanced medical data, IEEE Transactions on Medical Imaging 41 (12) (2022) 3663–3674, https://doi.org/10.1109/TMI.2022.3192483.

[109] F. Sattler, S. Wiedemann, K.R. Müller, W. Samek, Robust and communication-efficient federated learning from non-i.i.d. data, IEEE Transactions on Neural Networks and Learning Systems 31 (9) (2020) 3400–3413, https://doi.org/10.1109/TNNLS.2019.2944481.

[110] J. Konečný, H.B. McMahan, F.X. Yu, P. Richtarik, A.T. Suresh, D. Bacon, Federated learning: strategies for improving communication efficiency, in: Proc. of NIPS Workshop on Private Multi-Party Machine Learning, 2016.
[111] A. Singh, P. Vepakomma, O. Gupta, R. Raskar, Detailed comparison of communication efficiency of split learning and federated learning, https://doi.org/10.48550/ARXIV.1909.09145, 2019.

CHAPTER

Riemannian deep feature fusion with autoencoder for MEG depression classification in smart healthcare applications

Srikireddy Dhanunjay Reddy[a], Shubhangi Goyal[a], Tharun Kumar Reddy[a], Ramana Vinjamuri[b], and Javier Andreu-Perez[c]

[a]*Department of Electronics and Communication Engineering, Indian Institute of Technology, Roorkee, India*

[b]*Department of Computer Science and Electrical Engineering, University of Maryland, College Park, MD, United States*

[c]*Centre for Computational Intelligence, University of Essex, Colchester, United Kingdom*

8.1 Introduction and motivation

Depression is a challenging mental disorder that is common among people irrespective of their age. People who suffer from depression are in an interminable state of sadness and have an uninterested mood. Clinically, this state is called major depression disorder (MDD). In several instances, the subject himself/herself is unable to recognize that he/she is depressed [1]. The severity of the problem and its impending social impact motivated us to work on a methodology to determine the mental state of a person (depressed or healthy). Early assessment of an MDD patient's mental state will help us to provide proper care and medication. Compared to other modalities, magnetoencephalography (MEG) provides a better temporal resolution and requires less computational complexity with high resolution in brain source localization [2]. Preprocessing of MEG signals typically involves several steps to clean and prepare the data for analysis. Artifacts in MEG signals are mainly induced due to activities like eye blinks, rapid eye movements, and cardiac activity [3]. MEG data are high-dimensional and can be effectively handled in non-Euclidean spaces.

Riemannian geometry is a branch of differential geometry that studies the geometry of smooth manifolds. A manifold is a topological space that locally resembles Euclidean space, and Riemannian geometry provides a way to define notions such as length, angle, and curvature on these spaces [4][5]. In Riemannian geometry, the notions of length and angle are defined using a Riemannian metric, which is a mathematical object that assigns a length to every smooth curve on the manifold. The

Data Fusion Techniques and Applications for Smart Healthcare. https://doi.org/10.1016/B978-0-44-313233-9.00014-X

curvature of a Riemannian manifold is then defined using the Riemann curvature tensor, which describes how the manifold deviates from being flat [4][5]. Through this technique, we can use the covariance matrices and the tangent space vectors as the input parameters of the classifier models [6].

Markov chain correlation matrices can be useful for modeling the relationships between different variables and predicting how they may change over time. However, it is important to note that the probabilities in the matrix are based on historical data and may not always accurately reflect future relationships between the variables [7][8][9]. In this chapter, as a novel approach, we use the feature fusion technique to use Markov chain and Riemannian correlation matrices as the input parameters of the classifiers. After extracting both Riemannian and Markov chain features, feature fusion techniques are applied using autoencoders. Feature fusion is a technique used in machine learning to combine multiple sets of features, or characteristics, extracted from different sources into a single set of features. This can be done in order to improve the performance of a model or to combine the strengths of different feature extraction methods [10][11]. An autoencoder is a type of unsupervised learning algorithm that learns to encode input data into a lower-dimensional representation and then decode that representation back into the original data [12]. Autoencoders belong to the class of neural network models and methods useful for feature fusion.

In this work, we focused on distinguishing between MDD patients and healthy subjects. This dataset was part of a data challenge held at BIOMAG 2022. The challenge comprised a dataset with no labels for patients. As the dataset provided by BIOMAG is comparatively small, we implemented a transfer learning technique with the help of the OpenNeuro dataset. Transfer learning is a machine learning technique in which a model trained on one task is repurposed on a second related task. The idea behind transfer learning is that models that were pretrained on large datasets can be fine-tuned on smaller and unsupervised datasets in order to improve their performance on a specific task. It can be especially useful in cases where there is a limited amount of data available for the new task, as it allows the model to leverage its existing knowledge to learn more effectively [13]. Using this transfer learning technique, features obtained from the autoencoder were fed to well-known classifiers like support vector machine (SVM) with radial basis function kernel, minimum distance mean square (MDM), K-nearest neighbor (KNN), and linear discriminant analysis (LDA). For all these classifiers, covariance matrices or tangent space vectors are given as input parameters to predict the labels.

This chapter presents a deep learning (DL) approach that combines multiple sources of information to improve the accuracy of depression classification in the context of smart healthcare. By leveraging autoencoders and Riemannian geometry, the framework extracts discriminative features from MEG data to enhance classification performance. This contribution aligns with the broader goal of exploring data fusion techniques and their applications in the healthcare domain, particularly in the context of smart healthcare.

8.2 Literature review and state-of-the-art

In this section, recent works and methodologies for depression classification are discussed. Lu et al. [14] proposed a matching pursuit (MP) methodology using the mean power spectrum (MPS) on MEG signals for discriminative oscillation pattern detection in depression. Lu et al. [15] conducted Granger analysis (GA) by extracting minimum redundancy–maximum relevance (mRMR) features from MEG signals to predict depression-based dynamic regional connectivity. Kahkonen et al. [2] worked on peak latencies (PLs) and conducted statistical analysis (SA) on both MEG and electroencephalography (EEG) data to improve their statistical understanding of the proposed methodology for depression classification. Isomura et al. [16] analyzed the auditory steady-state response (ASSR) using Gamma band neural oscillations (GNOs) to distinguish between depressed subjects and healthy subjects using MEG signals. Hosseinifard et al. [17] performed nonlinear analysis (NA) on EEG signals using detrended fluctuation analysis (DFA) and correlation dimensional (CA) features for depression classification. Huang et al. [18] proposed generalized gradient-based visual explanations for a deep convolutional network (Grad-cam) visualization technique for MDD classification of bipolar disorder (BD). Caglar et al. [19] implemented differential convolutional neural network (DCNN) models to distinguish between MDD patients and healthy subjects through images using topography maps. Shuming et al. [20] conducted a study to understand the cognitive impairments of brain subnetworks among unmedicated MDD subjects and healthy controls using network connectivity (NC) by implementing group representative tensor construction. Shuting et al. [21] proposed a clustering method for identifying MDD subjects using resting-state EEG signals with the help of functional network connectivity (FNC) features. Ahmad et al. [22] implemented an automated MDD diagnosis technique on EEG signals using FNC features through dual-input DL models. Based on these state-of-the-art methods, MDD classification is difficult due to its requirements for large amounts of data and high computational cost. Here, we present a Riemannian feature selection methodology and fusion techniques as a means to overcome this challenge and put into practice an effective classification method. We were able to lessen the computational complexity by efficiently selecting channels using these methods. As a result, we went from 275 channels of data to only 40 channels of data, which are nonredundant. In Table 8.1, * indicates the performance analysis conducted by authors using statistical metrics.

8.3 Problem definition

The data analysis competition of BIOMAG 2022 provided 250 seconds of eyes closed unsupervised MEG signal data of 36 subjects (both male and female), including 22 MDD patients and 14 healthy subjects. Resting MEG data were taken using the CTF Omega 275-channel system. As BIOMAG data are unsupervised, we implemented the transfer learning technique using the OpenNeuro multimodal MEG dataset [23]

Table 8.1 Performance comparison of recent works on depression classification.

No.	Authors	Method	Features	Source	Performance (accuracy)
1	Lu et al. [14]	MP	MPS	MEG	86.40%
2	Lu et al. [15]	GA	mRMR	MEG	87.50%
3	Seppo et al. [2]	SA	PL	MEG, EEG	*
4	Shuichi et al. [16]	ASSR	GNO	MEG	*
5	Behshad et al. [17]	NA	DFA, CD	EEG	90%
6	Huang et al. [18]	Grad-cam	-	MEG	95.71%
7	Caglar et al. [19]	DCNN	Topography maps	EEG	90.55%
8	Shuming et al. [20]	NC	Tensor graphs	MEG	*
9	Shuting et al. [21]	Clustering	FNC	EEG	78.04%
10	Ahmad et al. [22]	Dual-input DL	FNC	EEG	95.24%

of labeled MEG data with almost similar features. This process is clearly explained in the proposed solution section.

In this chapter, we propose a novel Riemannian channel selection, transfer learning, and autoencoder feature fusion technique to reduce the complexity and distinguish between MDD patients and healthy subjects. The proposed approach is novel as it combines Riemannian geometry and DL techniques to address the problem of depression classification using MEG data in the context of smart healthcare applications. By modeling the covariance matrices of the MEG data using Riemannian geometry, the approach provides a more robust representation of the underlying data structure. Autoencoders are employed to extract high-level, discriminative features from the MEG data, which are then fused to improve classification performance. This unique combination of techniques from the fields of machine learning, signal processing, and neuroimaging presents a distinctive promising solution due to its innovative approach. The proposed workflow starts with preprocessing of the data and comes up with label prediction through feature extraction, feature fusion, and classification.

8.4 Proposed solution

In this section, the complete proposed methodology is explained in detail with supporting flow graphs and architectures and possible mathematical and analytical proofs. Throughout the methodology, we send both BIOMAG and OpenNeuro datasets through similar blocks except in the classification section. Unlabeled BIOMAG data were used as the test set and the labeled OpenNeuro data were used as the train and validation sets.

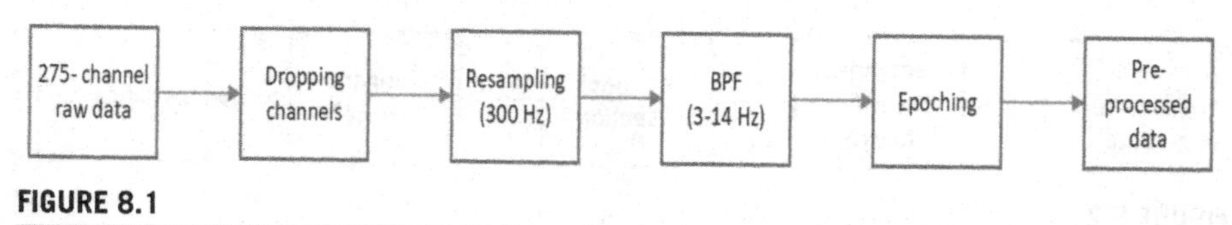

FIGURE 8.1

Preprocessing stage workflow.

8.4.1 Preprocessing

In the preprocessing stage, 275-channel resting raw data for 250 seconds of 36 subjects were taken and unnecessary channels like reference and improperly recorded data channels were excluded. In our particular problem solution, to maintain similar characteristics between the MEG and OpenNeuro datasets, three channels were excluded (two reference channels and one improperly collected data channel). After excluding the channels, both datasets were resampled to a common frequency of 300 Hz. After resampling the data, the overall data size became (36*25, 271, 300*10), which is in this structure (subjects*n_trials, n_channels, sampling frequency*time). Now the resampled data of both datasets were sent through a bandpass filter to remove the basic artifacts of MEG signals. In general, all those artifacts lie in the range of 3–14 Hz; therefore, a fourth-order Butterworth filter was also designed to remove the artifacts present in that range. Now the filtered signal was passed through the epoching block, where we split our long data into small data streams. In our work, we epoched the 250-second filtered data into 8-second data blocks. These preprocessing steps are shown in Fig. 8.1. Now, these preprocessed data were driven to the feature extraction stage.

8.4.2 Feature extraction

In this step, features of the MEG signals were extracted using two different concepts: Riemannian geometry and Markov chain. Then both features were fused and forwarded to the classification stage to predict the labels of the targeted BIOMAG dataset.

8.4.2.1 Riemannian features

Analyzing electrode placement and the correlation between the MEG channels, the Riemannian manifold is much easier and less complex when compared to Euclidean space. Another reason for analyzing MEG electrode placement using the Riemannian manifold is that it is more efficient in analyzing curved surface models. In this chapter, Riemannian covariance matrices are extracted using the Ledoit–Wolf covariance estimator [24][6] given as

$$A = (1-\alpha)K + \alpha trace(K)I/\eta_a, \tag{8.1}$$

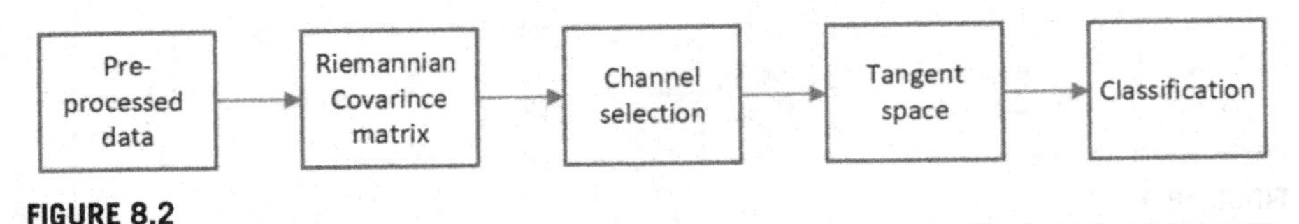

FIGURE 8.2

Riemannian geometry feature extraction flow graph.

where $\alpha \in [0, 1]$ is the shrinkage factor, I is the $\eta_a \times \eta_a$ identity matrix, and K is the simple covariance matrix calculated as

$$K = \frac{XX^T}{trace(XX^T)}. \tag{8.2}$$

The covariance matrix corresponding to the ith channel a_i is mapped to a tangent space as follows:

$$t_i = upper(A_\theta^{\frac{-1}{2}} log_\theta(A_i) A_\theta^{\frac{-1}{2}}). \tag{8.3}$$

In this way, covariance matrices are calculated as one of the input features to the classification block or for the feature fusion block. Now, we perform channel selection, where the channels with redundant data or irrelevant data are removed and the channels with important information are selected. In this stage, using Riemannian distance criteria, we selected 40 out of 275 data channels. Through this selection, the average threshold Riemannian distance is measured among A [25] using

$$A_{tres} = \sum_{i=1}^{N} \sum_{j>n}^{N} \delta_R(\overline{A}^{(i)}, \overline{A}^{(j)}), \tag{8.4}$$

where $K = 2$ is the number of categories and δ_R denotes the pairwise Riemannian distance calculated as

$$\delta_R = ||Log(A_1^{-1} A_2)||_F = \sqrt{\sum_{n=1}^{N_e} log^2 \lambda_n}, \tag{8.5}$$

where Log(.) is the log-matrix operator and $||.||_F$ is the Frobenius norm of a matrix. After channel selection, we have to feed these features to the classifier as input parameters. In order to do that, these features are fed to tangent space blocks to convert them from Riemannian space to Euclidean space. The whole procedure of Riemannian feature selection is shown in Fig. 8.2.

8.4.2.2 Markov chain features

A Markov chain is a mathematical system that undergoes transitions from one state to another according to certain probabilistic rules. In the context of a correlation matrix, a Markov chain can be used to model the relationship between different variables and how they may change over time [26][27].

A correlation matrix is a table of values that measures the linear relationship between two variables. It is typically used to assess the strength of the relationship between different variables, with a value of 1 indicating a perfect positive correlation, a value of -1 indicating a perfect negative correlation, and a value of 0 indicating no correlation. In a Markov chain correlation matrix, the elements in the matrix represent the probabilities of transitioning from one state (i.e., one variable) to another [26][27].

Let the random variables $X_1, X_2, ..., X_n$ represent the states of the Markov chain, where X_1 is the current state and $X_2, ..., X_n$ are the future states. P_{ij} represents the transition probability from state X_j to state X_i, where $i, j = 1, 2, ..., n$. $E(X_i)$ represents the expected value of the random variable X_i, and $Var(X_i)$ represents its variance. The vectorized Markov chain covariance matrix is a matrix whose elements are given by $Cov(X_i, X_j)$ for all pairs of states $i, j = 1, 2, ..., n$. The covariance between two random variables X_i and X_j can be computed as

$$Cov(X_i, X_j) = E[X_i X_j] - E[X_i]E[X_j]. \tag{8.6}$$

To compute $Cov[X_i, X_j]$, we can use the transition probabilities of the Markov chain as follows:

$$cov(X_i, X_j) = \sum_{i=1}^{n} P_{ik} P_{jk} (E[X_k] - E[X_i])(E[X_k] - E[X_j]). \tag{8.7}$$

Now we will see how this Markov concept will be useful for classification problems. Consider our dataset which has two classes, "Depressed" and "Healthy," and we want to build a classifier to predict the class label of a new sample. We can represent the class labels as a Markov chain, with the states "Depressed" and "Healthy" corresponding to the two classes. In this case, the transition probabilities of the Markov chain correspond to the probabilities of transitioning from one class to another. For example, the transition probability from the "Depressed" class to the "Healthy" class would represent the probability that a sample originally classified as "Depressed" is later reclassified as "Healthy."

We can use the vectorized Markov chain covariance matrix to quantify the covariance between the two classes and identify the states with the highest potential for risk or reward. For example, if the covariance between the "Depressed" and "Healthy" classes is high, it may indicate that the classifier is not very confident in its predictions and there is a high risk of misclassification. On the other hand, if the covariance is low, it may indicate that the classifier is more confident in its predictions and there is a lower risk of misclassification. By analyzing the vectorized Markov chain covariance matrix, we can gain insight into the stability of the classifier and identify potential areas for improvement. For example, if the classifier is frequently misclassifying samples, we may want to investigate the underlying cause and take steps to improve the performance of the classifier.

8.4.3 Feature fusion

Feature fusion is a technique in machine learning that involves combining multiple features or sets of features to create a more comprehensive and potentially more informative representation of the data. This can be done for a variety of purposes, such as improving the accuracy of a machine learning model, reducing the dimensionality of the data, or simply making the data more interpretable.

There are many different approaches to feature fusion, and the specific approach used will depend on the characteristics of the data and the goals of the machine learning task. Some common techniques include:

1. Concatenation: This involves simply combining the different sets of features into a single, longer feature vector.
2. Weighted sum: This involves combining the features using a weighted sum, where the weights reflect the relative importance of the different features.
3. Feature extraction: This involves using techniques like principal component analysis (PCA) or independent component analysis (ICA) to extract a smaller number of features that capture the most important information in the data.
4. Feature selection: This involves selecting a subset of the most important features from the original data, based on some criterion (e.g., mutual information, correlation, etc.).

In general, feature fusion can be a powerful tool for improving the performance of machine learning models, but it is important to carefully consider the appropriate approach for the specific task at hand.

Autoencoders are a type of neural network that is commonly used for feature fusion and feature extraction. They are particularly well suited for this purpose because they are designed to compress the input data into a lower-dimensional representation (called the latent space or bottleneck) and then reconstruct the original data from this compressed representation. This allows autoencoders to learn meaningful, compact representations of the data that can be used for a variety of purposes, including feature fusion.

There are several ways in which autoencoders can be used for feature fusion:

1. Concatenation: One approach is to simply concatenate the different sets of features and feed them into the autoencoder as a single input. The autoencoder can then learn to compress and reconstruct this input, effectively combining the different features into a single, more informative representation.
2. Multiple inputs: Another approach is to use multiple inputs for the autoencoder, with each input representing a different set of features. The autoencoder can then learn to compress and reconstruct these inputs separately and combine them in the latent space to create a fused representation.
3. Transfer learning: A third approach is to use pretrained autoencoders that have already been trained on a large dataset. These pretrained models can then be fine-tuned on a new dataset, using the learned representations from the original dataset

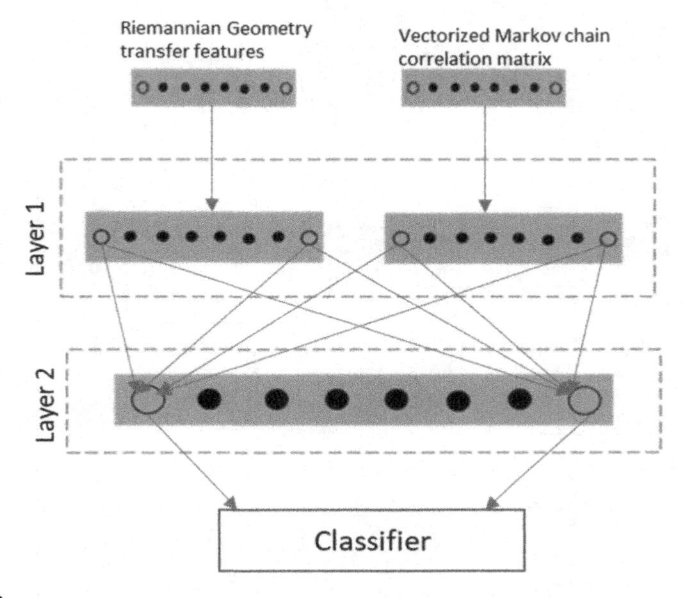

FIGURE 8.3

Deep feature fusion architecture.

as a starting point. This can be particularly useful when the new dataset is small or has a different structure than the original dataset.

Overall, autoencoders can be a powerful tool for feature fusion, allowing you to learn compact and meaningful representations of the data that can be used for various machine learning tasks. In this chapter, an autoencoder is used for feature fusion. Its architecture is shown in Fig. 8.3. Here both Riemannian geometry features and Markov chain features are given as inputs to the autoencoder to reconstruct them into latent space. The fused features are fed to classifiers for training, validation, or testing processes.

8.4.4 Classification

Classification is aimed at learning a model that can accurately predict the class label for a new, unseen data sample. As shown in Fig. 8.4, the model was trained with the OpenNeuro MEG dataset and then the BIOMAG competition dataset was sent to the model for prediction. SVM, MDM, KNN, and LDA models were used for predicting the labels and for comparing the performance of each model. SVM and LDA performed well with validation accuracies of 0.924 and 0.992, respectively. Our predicted labels of proposed models are displayed in Tables 8.2 and 8.3. Out of 36 subjects, our proposed LDA model classified 31 subjects correctly.

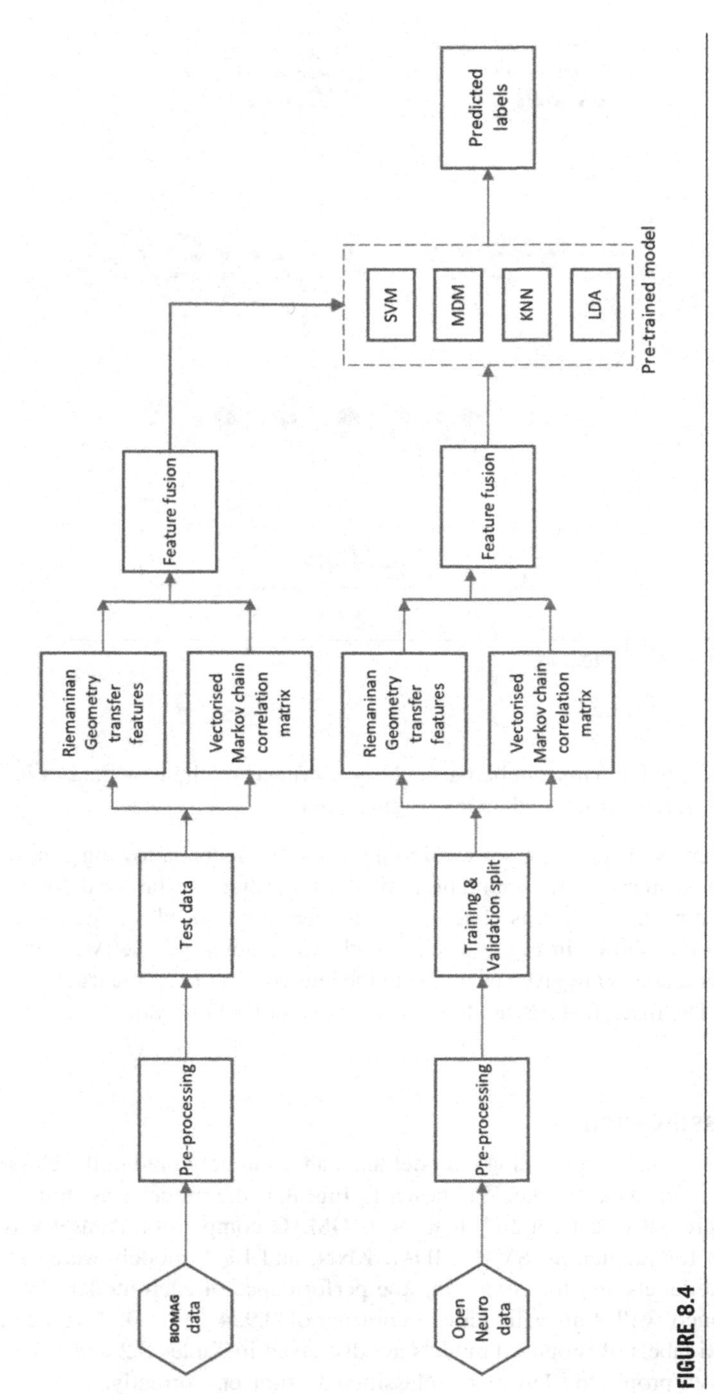

FIGURE 8.4

Proposed work flow.

Table 8.2 Performance comparison between SVM (0.924) and LDA (0.992).

No	Subject ID	SVM	LDA
1	BQBBKEBX	0	0
2	BYADLMJH	1	1
3	CECMHHYP	1	1
4	CEYAXQYB	1	1
5	CKTSVVFM	0	0
6	DXFTLCJA	1	1
7	EGQFCISZ	0	0
8	EVZQADGL	0	0
9	GFRLKCRT	1	1
10	IEKAAXXL	0	0
11	IRGXULEZ	1	1
12	JBGAZIEO	1	1
13	LUXESHXT	1	1
14	MIMGQLOB	1	1
15	MIVYRJFO	0	0
16	OLEGXGZO	0	0
17	PHMOFC IB	1	1
18	QGFMDSZY	1	1
19	RASMDGZN	0	0
20	SJHBZREE	1	1
21	SPIVRRSB	0	0
22	TCTIPUPB	1	1
23	TISSNJCK	1	1
24	TIUJUWXV	0	0
25	UPWNKWXJ	0	1
26	VATJUZCI	1	1
27	VWRRHGZE	1	1
28	WKQIVMBY	0	0
29	WRODLVBY	1	1
30	WWIZIWSJ	1	1
31	YDCCVVKE	1	0
32	YFLOQXEZ	0	0
33	YGSUCZFD	1	1
34	YIWMSUPM	1	1
35	YWRNZYI R	1	1
36	ZDIAXRUW	0	0

The specifications of the classifiers used in the present work are mentioned below:

SVM: It was implemented using a radial basis function kernel and it was optimized by balancing class weights ($0.1 \leq C \leq 1$). Regularization parameters were auto-tuned and the tangent space vectors were given as feature inputs to the classifier.

Table 8.3 Proposed predicted labels of the given data (LDA).

No	Subject	Predicted label
1	BQBBKEBX	0
2	BYADLMJH	1
3	CECMHHYP	1
4	CEYAXQYB	1
5	CKTSVVFM	0
6	DXFTLCJA	1
7	EGQFCISZ	0
8	EYZQADGL	0
9	GFRLKCRT	1
10	IEKAAXXL	0
11	IRGXULEZ	1
12	JBGAZIEO	1
13	LUXESHXT	1
14	MIMGQLOB	1
15	MIVYRJFO	0
16	OLEGXGZO	0
17	PHM OFCIB	1
18	QGFMDSZY	1
19	RASMDGZN	0
20	SJHBZREE	1
21	SPIVRRSB	0
22	TCTIPUPB	1
2.3	TISSNJCK	1
24	TTUJUWXV	0
25	UPWNKWXJ	1
26	VATJUZCI	1
27	VWRRHGZE	1
28	WKQIVMBY	0
29	WRODLVBY	1
30	WWIZIWSJ	1
31	YDCCVVKE	0
32	YFLOQXEZ	0
33	YGSUCZFD	1
34	YIWMSUPM	1
35	YWRNZYIR	1
36	ZDIAXRUW	0

MDM: Classification was done based on the shortest Riemannian distance between the test covariance matrix and intraclass covariance matrix means. Covariance matrices were given as the input parameters to the classifier.

KNN: Class weights and the nearest neighbors were given as hyperparameters for this model and tangent space vectors were given as feature inputs.

LDA: Here classification was done by reducing the dimensionality of input by projecting it to the most discriminative directions. A standard singular value decomposition solver was used and tangent space vectors were given as classifier input parameters.

Algorithm 8.1: Riemannian deep feature fusion technique for MDD classification

---

Input: $\{X^k, Y^k\} \in \mathbb{R}^{CXT}; k = \{1, 2, 3, ..., N\}$
X_k, Y_k: training and testing datasets;
C: number of channels;
T: length of time sequence;
N: number of subjects;
Output: Predicted labels $\{x_k\}$
PREP → Drop the reference and improper data channels;
→ Resample the datasets to f_s : 300 Hz and reshape the data as (k*C,C,f_s*T) for proper covariance matrices preparation;
→ Reshaped data are bandpass filtered to remove artifacts;
→ Compute covariance matrices A and $Cov(X_i, X_j)$ using (1), (2), (6), and (7);
→ Using (4), measure threshold Riemannian distance to remove redundant data channels;
→ As discussed in Section 8.4.3, extracted A and $Cov(X_i, X_j)$ are normalized and fused using the autoencoder;
→ We will get labels based on predicted labels x_k after ML/DL models are fed with fused features.
return: $\{x_k\}$

8.5 Experiment

The main objective of this section is to compare the effectiveness of the proposed method with that of similar, recently published methods and to compare the effectiveness of the method on the BIOMAG dataset with few publicly available datasets. Fig. 8.5 shows the performance comparison of the proposed methodology with the published works on MDD classification using MEG and EEG modalities.

Dataset 1: Public MEG data provided by Huang et al. [18] comprising 55 patients diagnosed with MDD and 74 healthy participants serving as controls. We were able to identify MDD patients with these data and got 94% accuracy using five-fold cross-validation.

Dataset 2: Public MEG data provided by Lu et al. [14] consisting of 22 individuals diagnosed with MDD and 22 individuals serving as healthy controls. Using these

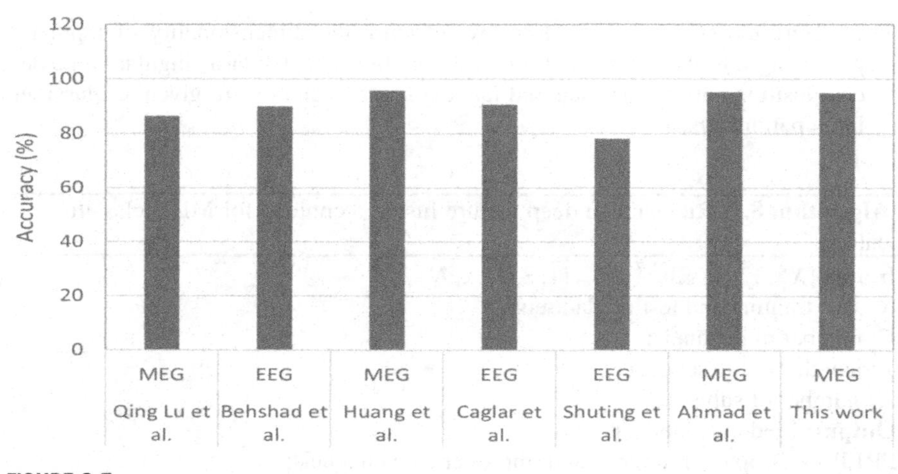

FIGURE 8.5

Comparison of the proposed method with other state-of-the-art methods.

data, we obtained 89.5% accuracy in identifying MDD patients by fivefold cross-validation.

Dataset 3: Along with the classification of MDD, we also used our proposed methodology on dementia screening using the data provided by the BIOMAG 2022 challenge for multiclass classification with an accuracy of 92.11%. The data include 100 healthy controls, 29 subjects with dementia, and 15 subjects with mild cognitive impairment.

Fig. 8.6 shows a performance comparison of the proposed method with other public datasets that are available, as well as a separate application, that is, dementia screening.

8.6 Conclusion and future work

In this chapter, we proposed a novel approach based on feature selection using Riemannian geometry, feature fusion with autoencoders, and transfer learning techniques to distinguish between MDD patients and healthy subjects from unlabeled data. Out of 36 subjects, the proposed methodology classified 31 subjects correctly and won the competition with high accuracy. Covariance matrices of Riemannian geometry were required in addition to tangent space vectors to improve the performance of certain classifiers. Even though this increases the amount of computational work, it is necessary to improve the performance of these particular classifiers. To address this issue in the future, we are looking toward the implementation of DL models with adaptive Riemannian methods. For the validation of the proposed methodology, we will check the performance of other modalities like EEG and resting-state fMRI.

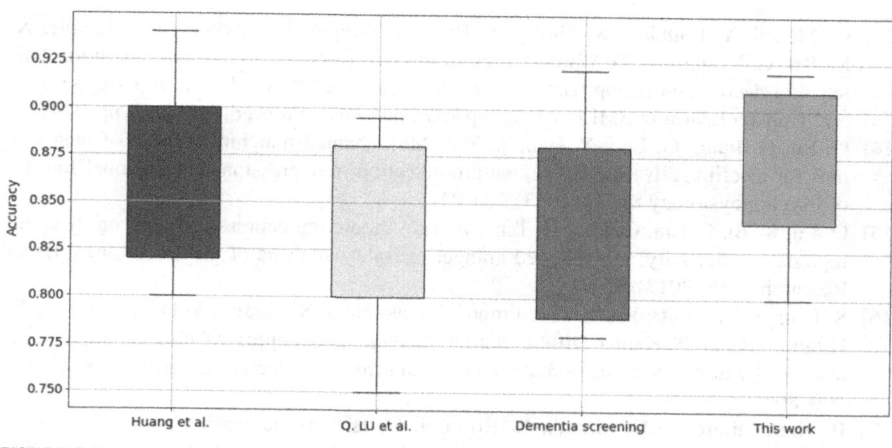

FIGURE 8.6

Comparison of the proposed methodology with other datasets and application in dementia screening.

References

[1] R.H. Belmaker, G. Agam, Major depressive disorder, The New England Journal of Medicine 358 (1) (2008) 55–68.

[2] S. Kähkönen, H. Yamashita, H. Rytsälä, K. Suominen, J. Ahveninen, E. Isometsä, Dysfunction in early auditory processing in major depressive disorder revealed by combined meg and eeg, Journal of Psychiatry and Neuroscience 32 (5) (2007) 316–322.

[3] A.H. Treacher, P. Garg, E. Davenport, R. Godwin, A. Proskovec, L.G. Bezerra, G. Murugesan, B. Wagner, C.T. Whitlow, J.D. Stitzel, et al., Megnet: automatic ica-based artifact removal for meg using spatiotemporal convolutional neural networks, NeuroImage 241 (2021) 118402.

[4] S. Gallot, D. Hulin, J. Lafontaine, Riemannian Geometry, vol. 2, Springer, 1990.

[5] M.P. Do Carmo, J. Flaherty Francis, Riemannian Geometry, vol. 6, Springer, 1992.

[6] A. Barachant, S. Bonnet, M. Congedo, C. Jutten, Classification of covariance matrices using a Riemannian-based kernel for bci applications, Neurocomputing 112 (2013) 172–178.

[7] J.R. Norris, J.R. Norris, Markov Chains, No. 2, Cambridge University Press, 1998.

[8] C.J. Geyer, Practical Markov chain Monte Carlo, Statistical Science (1992) 473–483.

[9] K.L. Chung, Markov Chains, Springer-Verlag, New York, 1967.

[10] N. Antropova, B.Q. Huynh, M.L. Giger, A deep feature fusion methodology for breast cancer diagnosis demonstrated on three imaging modality datasets, Medical Physics 44 (10) (2017) 5162–5171.

[11] F. Li, Y. Fan, X. Zhang, C. Wang, F. Hu, W. Jia, H. Hui, Multi-feature fusion method based on eeg signal and its application in stroke classification, Journal of Medical Systems 44 (2) (2020) 1–11.

[12] X. Yuan, L. Feng, Y. Wang, K. Wang, Stacked attention-based autoencoder with feature fusion and its application for quality prediction, in: 2021 IEEE 10th Data Driven Control and Learning Systems Conference (DDCLS), IEEE, 2021, pp. 1368–1373.

[13] G. Mesnil, Y. Dauphin, X. Glorot, S. Rifai, Y. Bengio, I. Goodfellow, E. Lavoie, X. Muller, G. Desjardins, D. Warde-Farley, et al., Unsupervised and transfer learning challenge: a deep learning approach, in: Proceedings of ICML Workshop on Unsupervised and Transfer Learning, JMLR Workshop and Conference Proceedings, 2012, pp. 97–110.

[14] Q. Lu, H. Jiang, G. Luo, Y. Han, Z. Yao, Multichannel matching pursuit of meg signals for discriminative oscillation pattern detection in depression, International Journal of Psychophysiology 88 (2) (2013) 206–212.

[15] Q. Lu, K. Bi, C. Liu, G. Luo, H. Tang, Z. Yao, Predicting depression based on dynamic regional connectivity: a windowed granger causality analysis of meg recordings, Brain Research 1535 (2013) 52–60.

[16] S. Isomura, T. Onitsuka, R. Tsuchimoto, I. Nakamura, S. Hirano, Y. Oda, N. Oribe, Y. Hirano, T. Ueno, S. Kanba, Differentiation between major depressive disorder and bipolar disorder by auditory steady-state responses, Journal of Affective Disorders 190 (2016) 800–806.

[17] B. Hosseinifard, M.H. Moradi, R. Rostami, Classifying depression patients and normal subjects using machine learning techniques and nonlinear features from eeg signal, Computer Methods and Programs in Biomedicine 109 (3) (2013) 339–345, https://doi.org/10.1016/j.cmpb.2012.10.008, https://www.sciencedirect.com/science/article/pii/S0169260712002507.

[18] C.-C. Huang, I. Low, C.-H. Kao, C.-Y. Yu, T.-P. Su, J.-C. Hsieh, Y.-S. Chen, L.-F. Chen, Meg-based classification and grad-cam visualization for major depressive and bipolar disorders with semi-cnn, in: 2022 44th Annual International Conference of the IEEE Engineering in Medicine & Biology Society (EMBC), IEEE, 2022, pp. 1823–1826.

[19] C. Uyulan, T.T. Ergüzel, H. Unubol, M. Cebi, G.H. Sayar, M. Nezhad Asad, N. Tarhan, Major depressive disorder classification based on different convolutional neural network models: deep learning approach, Clinical EEG and Neuroscience 52 (1) (2021) 38–51.

[20] S. Zhong, N. Chen, S. Lai, Y. Shan, Z. Li, J. Chen, A. Luo, Y. Zhang, S. Lv, J. He, et al., Association between cognitive impairments and aberrant dynamism of overlapping brain sub-networks in unmedicated major depressive disorder: a resting-state meg study, Journal of Affective Disorders 320 (2023) 576–589.

[21] S. Sun, H. Chen, G. Luo, C. Yan, Q. Dong, X. Shao, X. Li, B. Hu, Clustering-fusion feature selection method in identifying major depressive disorder based on resting state eeg signals, IEEE Journal of Biomedical and Health Informatics (2023).

[22] A. Afzali, A. Khaleghi, B. Hatef, R. Akbari Movahed, G. Pirzad Jahromi, Automated major depressive disorder diagnosis using a dual-input deep learning model and image generation from eeg signals, Waves Random Complex Media (2023) 1–16.

[23] L. Liuzzi, K. Chang, H. Keren, C. Zheng, D. Saha, D. Nielson, A. Stringaris, Mood induction in mdd and healthy adolescents, https://doi.org/10.18112/openneuro.ds003568.v1.0.3, 2022.

[24] O. Ledoit, M. Wolf, A well-conditioned estimator for large-dimensional covariance matrices, Journal of Multivariate Analysis 88 (2) (2004) 365–411.

[25] A. Barachant, S. Bonnet, Channel selection procedure using Riemannian distance for bci applications, in: 2011 5th International IEEE/EMBS Conference on Neural Engineering, IEEE, 2011, pp. 348–351.

[26] A. Papoulis, S.U. Pillai, Probability, Random Variables, and Stochastic Processes, Tata McGraw-Hill Education, 2002.

[27] R.W. Wolff, Stochastic Modeling and the Theory of Queues, Pearson College Division, 1989.

CHAPTER

Source localization of epileptiform MEG activity towards intelligent smart healthcare: a retrospective study

Sanjeev Kumar Varun[a], Tharun Kumar Reddy[a], Marios Antonakakis[b], and Michelis Zervakis[b]

[a]*Department of Electronics and Communication Engineering, Indian Institute of Technology, Roorkee, India*

[b]*School of Electrical and Computer Engineering, Technical University of Crete, Chania, Greece*

9.1 Introduction and motivation

Epilepsy is a neurological disorder that is common among people in developing countries irrespective of their age, sex, and color. People undergoing daily epileptic seizures suffer a lot personally and professionally. As per the International League Against Epilepsy (ILAE) guidelines, drug-resistant epilepsy (DRE) is defined as the persistence of seizures despite the use of at least two syndrome-adapted antiseizure drugs (ASDs) at efficacious daily doses [1]. For DRE patients, resective seizure surgery is one of the prominent solutions. However, accurate localization of the epileptogenic zone remains challenging [2]. In this regard, the localization of the irritative zone using the temporal content of the recorded interictal signal is laborious due to the multiparametric complexity of source analysis (modality type, noncerebral interference, volume conduction effects, etc.).

A combination of different modalities, e.g., EEG/MRI, MEG/MRI, or EEG/MEG/MRI, depending on the hypothesis about the orientation of the epilepsy origins as well as the available modalities [3,4], can be utilized to model the underlying epileptiform activity best. The combination of patient-specific EEG/MEG/MRI leads to high accuracy when all crucial parameters have been tuned appropriately [5]. In this regard, EEG suffers mainly from volume conduction effects, for which realistic head modeling is crucial due to the high intersubject variability [6]. MEG is less sensitive to those effects, making it a proper solution for this purpose [7].

Source localization using MEG and beamforming techniques has become increasingly popular over the last decades [7] mainly because of the lower complexity of MEG compared to EEG during the modeling of the brain source dynamics and the

Data Fusion Techniques and Applications for Smart Healthcare. https://doi.org/10.1016/B978-0-44-313233-9.00015-1

nonlaborious parametrization of the beamforming techniques during source localization. However, beamforming techniques can easily lead to suboptimal localizations when interpreting their raw outputs. Kurtosis beamforming [8] has shown a marginal performance improvement for the accurate spatiotemporal localization of the interictal spikes compared to other approaches, such as current dipole or current density approaches.

In this work, we investigate the performance of kurtosis-beamforming source localization on MEG and MRI data from two patients diagnosed with DRE. The comparison shows the localization results per patient using their postsurgery MRIs and includes comparisons with the standard current density approaches sLORETA and eLORETA. The data used and the analysis conducted are from the BIOMAG-2022 epilepsy competition, in which we secured second position (Silver Medal).

9.2 Literature review and state-of-the-art

Source localization of epileptic interictal spikes is an integral and noninvasive approach for potentially indicating the "irritative zone" during presurgical evaluation for focal epileptic patients [9]. The most popular techniques for MEG source localization in clinical applications are current dipole fitting (linear and nonlinear) and current density reconstruction [10]. In continuous MEG recordings, neurophysiologists visually annotate interictal spikes, and inverse algorithms are applied on single or averaged spikes to provide a spatial estimation of the underlying epilepsy foci. Manual annotation of spikes from MEG recordings can take several hours per patient, which is impractical.

Beamforming is a spatial filtering technique for source localization that can also reconstruct the source activity of the brain from continuous MEG recording segments by scanning the entire given source space (i.e., the model cloud of points that follow the structure of the brain tissues) [11]. Kurtosis beamforming has been proposed for the source time series on the scanned grid (cloud of points) of beamforming [12,13]. Kurtosis can identify outliers in data, which is a crucial aspect of interictal spikes [14]. Kurtosis beamforming can be used to localize spikes in continuous MEG data, eliminating the need for human intervention for the time-consuming spike annotation task.

Beamforming works through "virtual sensors" (VSs), which are comparable to the implanted sensors at specific locations, so as to reconstruct brain activity at the source level. As source candidates, we consider VSs that fit the local maxima of kurtosis distributions as artifacts and other brain waves, e.g., alpha activity, can also result in high degrees of kurtosis. It has been widely established in practice that visual inspection of VSs for real interictal spikes can improve localization performance; as a result, to identify spikes in VSs, the authors of [15] and [16] have shown all the data on VSs that exceeded a threshold of the peak-to-root mean square ratio. To include VSs that displayed synchronized spikes with MEG/EEG sensors, the authors of [8,17] and [18] depicted VSs along with the accompanying MEG sensor data.

To reject muscular artifacts, the authors of [12] visualized VSs. In the paper [19], eye inspection was used to eliminate nonepileptiform MEG epochs from VSs. Aside from the accurate prediction, one of the reasons to look for new algorithms is that examining VSs can be as laborious as manually annotating spikes at the MEG sensor level. In this book chapter, we will see how this novel method (kurtosis with LCMV beamforming) resolves this issue.

9.3 MEG recordings

The signal measured by an MEG system typically involves several steps to clean and prepare the data for analysis. Mainly, artifacts (noncerebral activity) in the MEG signal will be induced due to eye blinks, rapid eye movements, and cardiac activity [9]. Misdiagnosis can also occur due to mimickers, as in this section a patient behaves as if he is going through an epileptic seizure. In contrast, the reasons might differ, for example, the person could be psychogenic, malinger, or syncopic. Thus, identifying error-free focal or generalized epileptic signals along with the point of origin is always a cumbersome task.

9.4 MRI

In MRI, the human brain is converted to voxels, which could be thought of as building blocks that have some signal value and coordinates attached to them, just as pixels in an image; the only difference is that voxels are 3D. These voxels are resegmented, resliced, and realigned based on different anatomical regions. The human head is divided into different compartments, namely, skin, skull, cerebrospinal fluid, gray matter, white matter, etc., as depicted in Fig. 9.1.

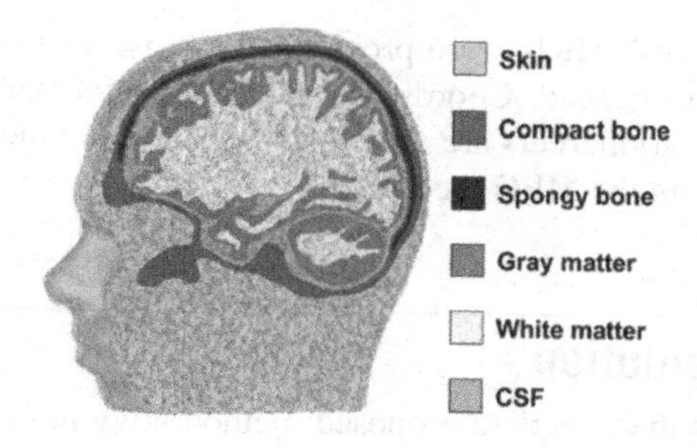

FIGURE 9.1

Sagittal cut through the segmented reference head model comprised of six tissue compartments [20].

Each compartment has different conductivity and resistivity; as a result, it is difficult to trace the path of electrochemical currents originating from the source and reaching the skull, which in turn is measured by the MRI device. Hence, localizing the source of an epileptic spike is a tedious task. Some methods use commercial software for source localization. In middle-income countries, it is required to find an intermediate solution that can not only give accurate localization but is also flexible in terms of tuning and cost. In the present study, we establish a benchmark for source localization on a novel dataset comprised of MEG and nonlesional MRI signals for neocortical epileptic patients and perform accurate localization of the hub node with sublobar resolution by overcoming the data recording (MEG-MRI modality) limitation.

9.5 Problem definition

The dataset is comprised of data collected from two medically refractory epilepsy patients, Subject 1 and Subject 2, from the BIOMAG-2022 Epilepsy Challenge [REF]. Each subject folder contains subfolders of MEG and MRI data. Resting-state MEG (4D neuroimaging data) and MRI (defaced) data are included. Our objective is to perform source localization of epileptic spikes with sublobar resolution. We used a novel linear constrained minimum variance (LCMV) approach with kurtosis beamforming on the MEG-MRI data.

9.6 Dataset description

MEG Data: The Marseille MEG Laboratory at the Institut de Neurosciences des Systèmes (INS) and the Timone Hospital Marseille (Assistance Publique - Hopitaux de Marseille) have recorded the data of two patients with epilepsy. The recording system is a 4D Neuroimaging 3600WH system. Bad channels had previously been identified and removed.

MRI data: The patients' MRIs were provided. To preserve the identity of the patients, they have been defaced. Coordinates of the fiducial (left preauricular, right preauricular, and nasion) markers are provided on the MRI volume in order to match the markers used during the MEG recording.

9.7 Proposed solution

This section explains the complete proposed methodology in detail with supporting flow graphs and architectures as well as mathematical and analytical proofs. Fig. 9.2 shows the overall flow diagram and a step-by-step process for epileptic spike localization. The flow diagram is divided into three parts; the first part is the MEG part, where the loading of MEG data, data preprocessing, component analysis, and artifact removal are performed. The second part is the MRI part, where loading of MRI data, realigning, reslicing, MRI segmentation, and preparation of the head model and source model are performed. After this, information from both parts is fused in the

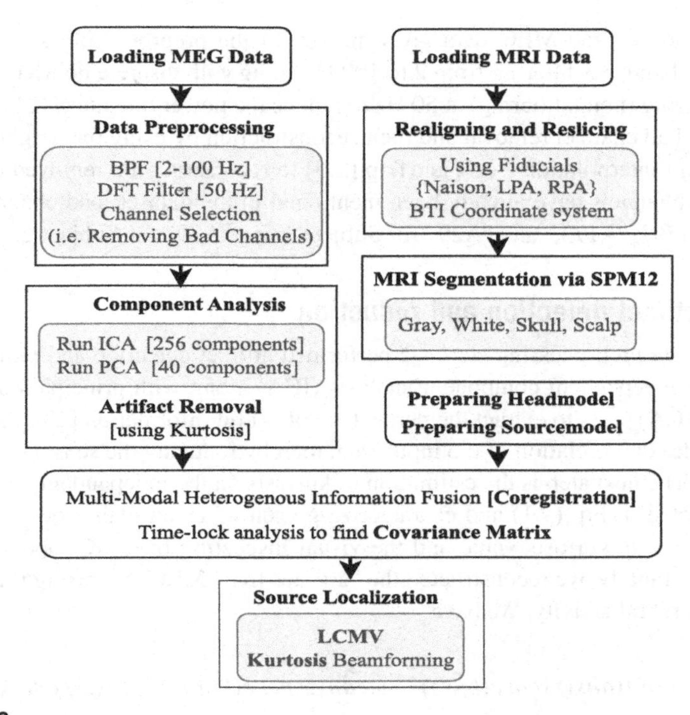

FIGURE 9.2

Architectural workflow of proposed methodology.

Algorithm 9.1: Preprocessing steps for source space (MEG signal)

Input: Raw MEG signals in c,rfDC format
Output: preprocessed source space (MEG signals)
MEG signal passed through $BPF \leftarrow [2 - 100]$ Hz and $DFT \leftarrow 50$ Hz
Removing bad channels, $Subj_1 \leftarrow 'A59', 'A173'$ and $Subj_2 \leftarrow 'A29'$;
Component analysis, $ICA \leftarrow 256$ and $PCA \leftarrow 40$;
Artifact removal using kurtosis through (1) ;
Return preprocessed source space (MEG signals)

---

third part, also known as coregistration, and then the covariance matrix is found using time-locked analysis. At last, we perform source localization using LCMV and kurtosis beamforming.

9.7.1 Data preprocessing

We used the FieldTrip toolbox to analyze MEG and MRI data in combination with our developments. The very first step in our analysis is the preprocessing step, for which we loaded the MEG data given in 4D. In the preprocessing step, we performed the bandpass filtering from 2 to 100 Hz along with discrete Fourier transform (DFT) filtering (notch filtering) at 50 Hz to reduce the power line noise [21]. We then performed bad channel removal and their reconstruction by using the neighbor channels through interpolation (see FieldTrip [22]) to reconstruct the removed channels. The noisy channels have already been mentioned in the dataset; bad channels were "A132," "A59," "A173," and "A29" for Subject 1 and "A29" for Subject 2.

9.7.2 Artifact detection and reduction

After the data preprocessing step, we performed artifact detection and reduction by employing independent component analysis (ICA) along with principal component analysis (PCA) so as to extract the percentage of useful information [23]. An a priori step includes decorrelation of the input data, thereby reducing the sensor dimensionality [23]. The next step is the estimation of kurtosis on the independent components (ICs) depicted in Eq. (9.1) and characterizing each IC as an ocular or cardiac artifact based on its kurtosis value and the visual inspection of its IC topography per signal [23]. Finally, we reconstructed the "artifact-free" MEG data using the ICs that preserve cerebral activity. We have

$$artK = find(abs(zscore(kur)) > \ddot{mean}(zscore(kur)) + std(zscore(kur))), \tag{9.1}$$

where $kur = kurtosis(X)$, $X = nsamples \times IC$, nsamples is the number of sample points of the MEG signal, abs is the absolute value, and IC is the independent component. The kurtosis beamformer, represented as $kur = kurtosis(X)$, works by reconstructing the source time series for each voxel in the source space grid and then computing the kurtosis value for each time series. This results in a volumetric map with a single kurtosis value depicting each voxel. Eq. (9.2) was applied to MEG data segments to estimate source activity at each coordinate of the brain. Here, r stands for the voxel's position, $S_r(t)$ for the dipole moment's strength at the voxel's location r, w_r^T for the transpose of a spatial filtering coefficient vector that applies to the data, and $m(t)$ for the data vector of magnetic field measurements at time t. We have

$$\widehat{S}_r(t) = \ddot{w}_r^T m(t). \tag{9.2}$$

For each coordinate, the excess kurtosis(X) is determined as the measure of spikiness. Kurtosis is computed from each source waveform as depicted in (9.3), where σ is the standard deviation and T is the number of time samples:

$$kur = \sum_T [S(t)(-\ddot{\overline{S}})]^4/(\sigma^4 T) - 3. \tag{9.3}$$

9.7.3 MRI

9.7.3.1 MRI data realigning, reslicing, and segmentation using SPM12

In this section, the very first step is loading MRI data, after which we process the MRI signal by orienting the MRI data in a BTi/4D coordinate system. For this, we have used fiducial markers such as Nasion, LPA (left preauricular), and RPA (right preauricular) (Table 9.1).

Table 9.1 BTi/4D coordinate points of the dataset.

	Nasion	LPA	RPA
Subject 1	[125 212 130]	[60 116 94]	[119 122 93]
Subject 2	[116 225 105]	[45 105 68]	[208 123 76]

Subsequently, we applied realigning along with reslicing and performed five-compartment segmentation (scalp, skull, cerebrospinal fluid, gray matter, and white matter) on the MRI data depicted in Figs. 9.3 and 9.4. Fig. 9.3 shows the brain MRI of Subject 1, obtained from the sagittal plane, axial plane, and coronal plane, while Fig. 9.4 shows the segmented brain MRI of Subject 1, obtained from the sagittal plane, the axial plane, and the coronal plane using the integrated version (v.12) of the SPM toolbox in FieldTrip.

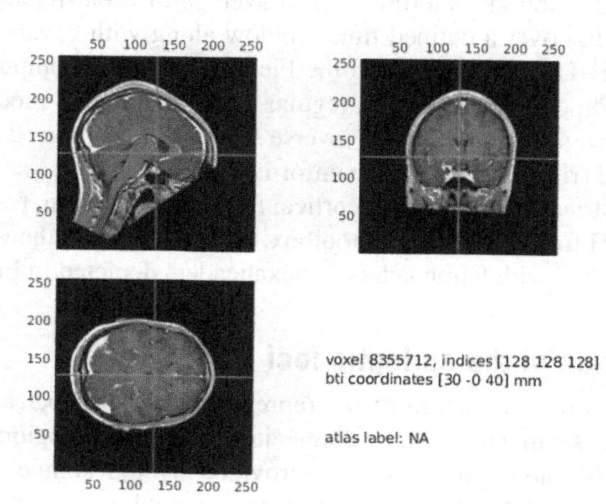

FIGURE 9.3

Brain MRI of Subject 1.

9.7.3.2 MRI head model and source model

We then created a head model and source model that followed the folding of the gray matter. We did not conduct detailed modeling for the forward solution as MEG is less affected by volume conduction effects than EEG, making it simpler to use in forward models. For this reason, we used the "singleshell" option offered by the

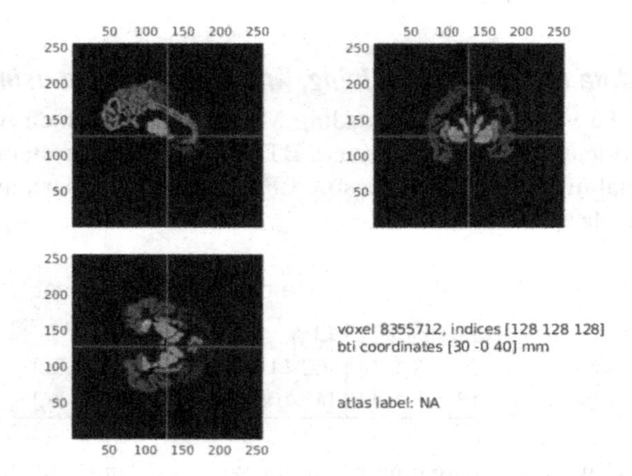

FIGURE 9.4

Segmented brain MRI of Subject 1.

FieldTrip toolbox. For forward modeling, we have to input the forward model with the data structure obtained from FT–TIMELOCKANALYSIS from the FieldTrip toolbox, as this function gives a time-locked average of event-related fields (ERFs) or potentials (ERPs) over a defined time window along with covariance over trials. So FT–PREPARE–LEADFIELD from the FieldTrip toolbox computes the forward model for many dipole locations on a regular 2D or 3D source model and stores it for efficient inverse modeling. For our inverse solution, we followed the approach as presented in FieldTrip using LCMV beamforming [12,22].

After this, we tried to visualize the cortical folding of the brain, for which we used FT–PLOT–MESH from the FieldTrip toolbox, which visualizes the volumetric mesh of the head described with tetraheaders or hexaheaders depicted in Fig. 9.5.

9.7.4 Estimation of the epileptic foci

The next step was the estimation of the representative signal per anatomic region. To parcellate the segmented brain volume into anatomical regions, we used the corresponding MRI atlas per patient as provided by the competition dataset. In our approach, we used the knnsearch (k-nearest neighbors search) function from MATLAB® to find the closest source space points to each one of the anatomical brain regions. By performing this, we had 246 anatomical regions with their corresponding source signals. We used the average value across all the source space points per signal, which were the closest points to each anatomical region according to knnsearch, in order to estimate each representative 3D centroid and source signal per anatomical region. From this procedure, we ended up with 246 signals and their centroids. Our next step was estimating the time-varying connectivity graph among all the anatomical regions to reveal the hub node. Our assumption is that the hub node will correspond to the anatomical region with the highest source activation (as

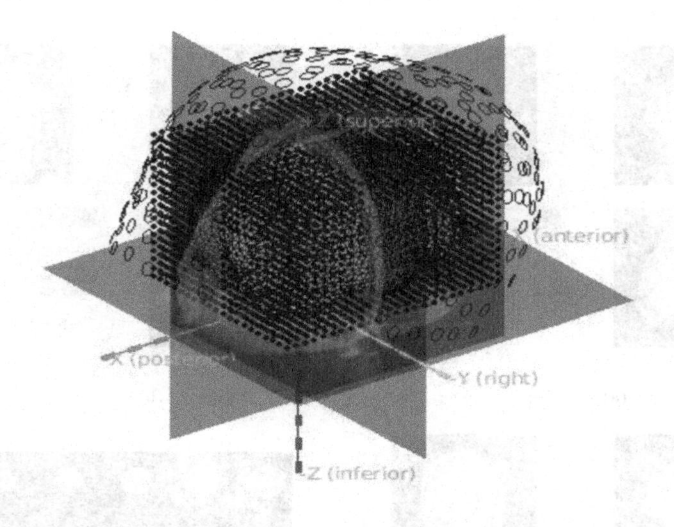

FIGURE 9.5

Mesh diagram for Subject 1.

provided by LCMV-kurtosis) and that this anatomical region will contain the brain malformation (e.g., focal cortical dysplasia [FCD]).

9.8 Results and discussion

In this section, Fig. 9.6 for Subject 1 and Fig. 9.7 for Subject 2 show the localization of epileptic sites in the sublobar resolution mainly if we consider the bright spots. In Subject 1 the epileptic node is present in the right temporal lobe of the brain; in Subject 2 the sites are in the right frontal and left temporal lobes of the brain.

In each figure, four plots are seen, namely, a, b, c, and d, where a, b, and c represent epilepsy originating at different epileptic sites at different time intervals and d represents the slicing done axially over the MRI signal depicting how the volume of the epileptic zone varies as we move axially. Each plot has three subplots in a, b, and c representing the localization of epileptic seizure using sagittal, coronal, and axial bisections of the brain, respectively. Here, the epileptic seizure is specified in terms of voxels, indices, BTi/4D coordinates, and their corresponding values. Here, the three viewpoints have been provided in a single figure. If we move from left to right, we get the first subpart as the sagittal bisection of the brain, the second subpart as the coronal bisection of the brain, and the third subpart as the axial bisection of the brain. This work develops a presurgical diagnostic assistive system for doctors or epileptologists.

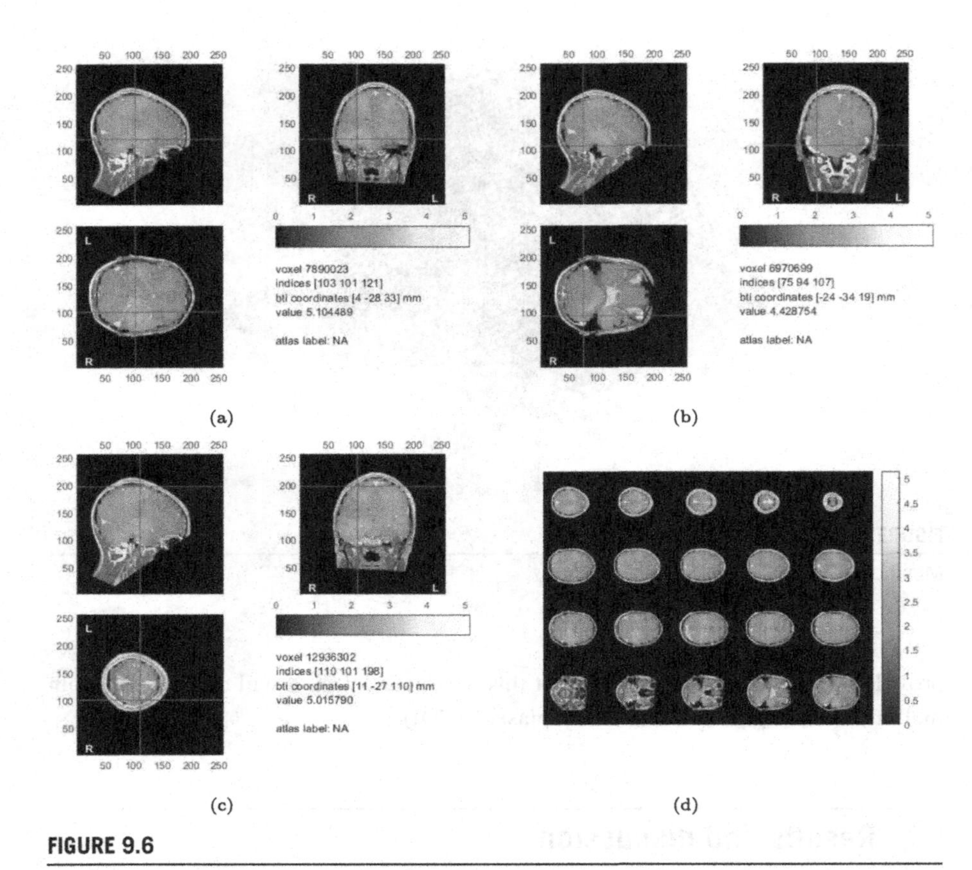

FIGURE 9.6

(a–c) Crosshair showing the epileptic seizure localization in sagittal, coronal, axial bisections of the brain for Subject 1 at different time instants. (d) The axial slicing of Subject 1.

9.9 Conclusion and future work

In the present study, we estimated the spatiotemporal localization of epileptic sites at sublobar resolution. We performed bandpass filtering and channel selection to mitigate artifacts. Then, we applied PCA and kurtosis beamforming over ICA to complete the preprocessing of MEG data. We oriented the MRI data in a BTi/4D coordinate system along with preparing the head model and source model by performing forward and inverse modeling and plotting the 3D model of the head represented as a mesh. Further, we have tried to estimate the time-varying connectivity graph and hub nodes using the LCMV-kurtosis approach to plot the high-source activation.

We conclude that the epileptic sites are mainly present in the right temporal lobe of the brain in Subject 1 and in the right frontal and left temporal lobes of the brain in Subject 2. Based on our previous studies and our results for Subject 1 and Subject 2 in the BIOMAG Epilepsy Challenge 2022, we received the Silver Medal. Also,

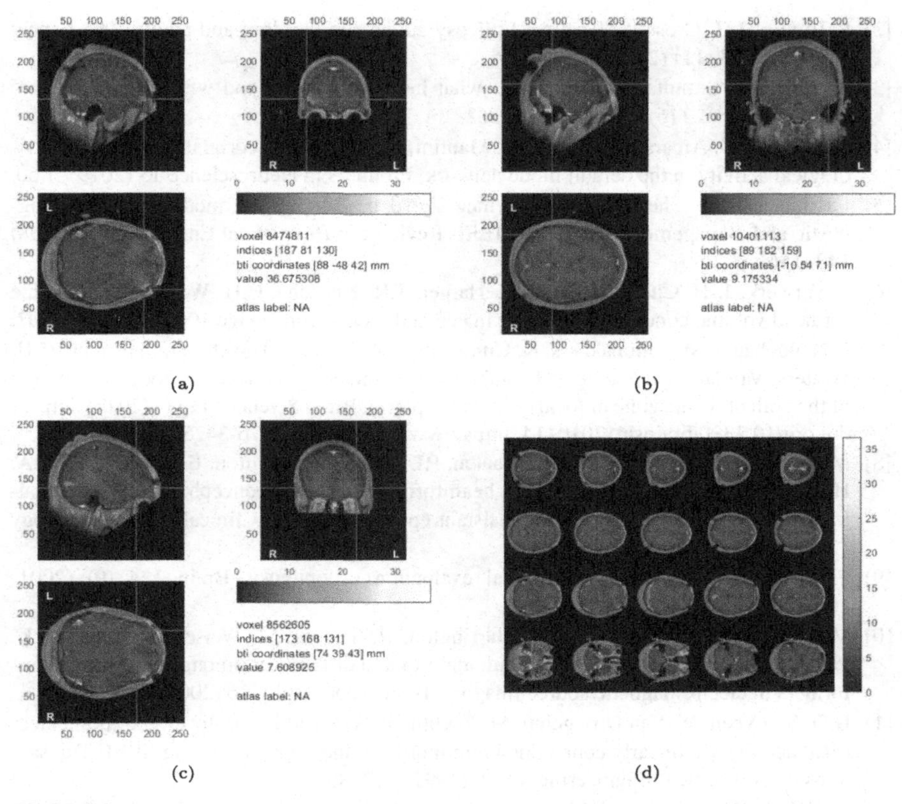

FIGURE 9.7

(a–c) Crosshair showing the epileptic seizure localization in sagittal, coronal, and axial bisections of the brain for Subject 2 at different time instants. (d) The axial slicing of Subject 2.

the organizers of BIOMAG 2022 acknowledged our approach as the most promising one. In addition, there is much room for improvement in this direction in the future, as researchers can intend to evaluate more detailed approaches such as finite element methods (FEMs), boundary element methods (BEMs), and additions of time-varying connectivity analysis to reveal the actual backbone of the underlying epilepsy network.

References

[1] P. Kwan, A. Arzimanoglou, A.T. Berg, M.J. Brodie, W. Allen Hauser, G. Mathern, S.L. Moshé, E. Perucca, S. Wiebe, J. French, Definition of drug resistant epilepsy: consensus proposal by the ad hoc task force of the ilae commission on therapeutic strategies, 2010.

[2] P. Ryvlin, J.H. Cross, S. Rheims, Epilepsy surgery in children and adults, The Lancet Neurology 13 (11) (2014) 1114–1126.

[3] T. Warbrick, Simultaneous eeg-fmri: what have we learned and what does the future hold?, Sensors 22 (6) (2022) 2231–2262.

[4] M. Marino, G. Arcara, C. Porcaro, D. Mantini, Hemodynamic correlates of electrophysiological activity in the default mode network, Frontiers in Neuroscience 13 (2019) 1060.

[5] F.H.L. da Silva, The impact of eeg/meg signal processing and modeling in the diagnostic and management of epilepsy, IEEE Reviews in Biomedical Engineering 1 (2008) 143–156.

[6] J. Vorwerk, J.-H. Cho, S. Rampp, H. Hamer, T.R. Knösche, C.H. Wolters, A guideline for head volume conductor modeling in eeg and meg, NeuroImage 100 (2014) 590–607.

[7] F. Neugebauer, M. Antonakakis, K. Unnwongse, Y. Parpaley, J. Wellmer, S. Rampp, C.H. Wolters, Validating eeg, meg and combined meg and eeg beamforming for an estimation of the epileptogenic zone in focal cortical dysplasia, Brain Sciences 12 (1) (2022), https://doi.org/10.3390/brainsci12010114, https://www.mdpi.com/2076-3425/12/1/114.

[8] M.B. Hall, I.A. Nissen, E.C. van Straaten, P.L. Furlong, C. Witton, E. Foley, S. Seri, A. Hillebrand, An evaluation of kurtosis beamforming in magnetoencephalography to localize the epileptogenic zone in drug resistant epilepsy patients, Clinical Neurophysiology 129 (6) (2018) 1221–1229.

[9] F. Rosenow, H. Lüders, Presurgical evaluation of epilepsy, Brain 124 (9) (2001) 1683–1700.

[10] M.X. Huang, J. Shih, R. Lee, D. Harrington, R. Thoma, M. Weisend, F. Hanlon, K. Paulson, T. Li, K. Martin, et al., Commonalities and differences among vectorized beamformers in electromagnetic source imaging, Brain Topography 16 (2004) 139–158.

[11] B.D. Van Veen, W. Van Drongelen, M. Yuchtman, A. Suzuki, Localization of brain electrical activity via linearly constrained minimum variance spatial filtering, IEEE Transactions on Biomedical Engineering 44 (9) (1997) 867–880.

[12] H. Kirsch, S. Robinson, M. Mantle, S. Nagarajan, Automated localization of magnetoencephalographic interictal spikes by adaptive spatial filtering, Clinical Neurophysiology 117 (10) (2006) 2264–2271.

[13] S. Robinson, S. Nagarajan, M. Mantle, V. Gibbons, H. Kirsch, Localization of interictal spikes using sam (g2) and dipole fit, Neurology & Clinical Neurophysiology 2004 (2004) 74.

[14] P.H. Westfall, Kurtosis as peakedness, 1905–2014. rip, American Statistician 68 (3) (2014) 191–195.

[15] Z. Agirre-Arrizubieta, N.J. Thai, A. Valentín, P.L. Furlong, S. Seri, R.P. Selway, R.D. Elwes, G. Alarcón, The value of magnetoencephalography to guide electrode implantation in epilepsy, Brain Topography 27 (2014) 197–207.

[16] K.L. de Gooijer-van de Groep, F.S. Leijten, C.H. Ferrier, G.J. Huiskamp, Inverse modeling in magnetic source imaging: comparison of music, sam (g2), and sloreta to interictal intracranial eeg, Human Brain Mapping 34 (9) (2013) 2032–2044.

[17] D.F. Rose, H. Fujiwara, K. Holland-Bouley, H.M. Greiner, T. Arthur, F.T. Mangano, Focal peak activities in spread of interictal-ictal discharges in epilepsy with beamformer meg: evidence for an epileptic network?, Frontiers in Neurology 4 (2013) 56.

[18] J.R. Tenney, H. Fujiwara, P.S. Horn, D.F. Rose, Comparison of magnetic source estimation to intracranial eeg, resection area, and seizure outcome, Epilepsia 55 (11) (2014) 1854–1863.

[19] M. Oishi, H. Otsubo, K. Iida, Y. Suyama, A. Ochi, S.K. Weiss, J. Xiang, W. Gaetz, D. Cheyne, S.H. Chuang, et al., Preoperative simulation of intracerebral epileptiform discharges: synthetic aperture magnetometry virtual sensor analysis of interictal magnetoencephalography data, Journal of Neurosurgery. Pediatrics 105 (1) (2006) 41–49.

[20] J.-H. Cho, V. Johannes, W. Carsten H, K. Thomas R, Influence of the head model on eeg and meg source connectivity analyses, NeuroImage 110 (1) (2015) 60–77.

[21] R. Oostenveld, Fries, Fieldtrip: MEG-MRI preprocessing, https://github.com/fieldtrip/fieldtrip/blob/release/ft_preprocessing.m, 2011.

[22] R. Oostenveld, Fries, analysis of epilepsy on MEG-MRI data, https://www.fieldtriptoolbox.org/tutorial/epilepsy/#virtual-channel-analysis-of-epilepsy-meg-data, 2011.

[23] M. Antonakakis, S.I. Dimitriadis, M. Zervakis, S. Micheloyannis, R. Rezaie, A. Babajani-Feremi, G. Zouridakis, A.C. Papanicolaou, Altered cross-frequency coupling in resting-state meg after mild traumatic brain injury, International Journal of Psychophysiology 102 (2016) 1–11.

CHAPTER

10 Early classification of time series data: overview, challenges, and opportunities

Anshul Sharma[a], Abhinav Kumar[b], and Sanjay Kumar Singh[b]

[a]*National Institute of Technology, Patna, Department of Computer Science & Engineering, Patna, Bihar, India*

[b]*Indian Institute of Technology (BHU), Department of Computer Science & Engineering, Varanasi, Uttar Pradesh, India*

10.1 Introduction

Classification is an important research topic in the field of data mining and machine learning. Basically, classification is a supervised learning task in which first a prediction model is trained from seen examples (training samples) and later it is used to classify unseen samples [1]. It has numerous real-world applications such as cancer detection [2,3], spam filtering [4], fault classification [5], and patient monitoring [6]. The classification task has been considered extensively on structural data, where training samples are represented in the form of feature vectors. Usually, these features are called attributes, which can be categorical or numerical. However, in many real-world applications, data are collected in a more complex form that may be structured, semistructured, or unstructured, such as text data, biological data, time series data, and multimedia data. This chapter addresses the classification problem in time series data.

A time series is an ordered sequence of measurements, called data points, recorded over time. Generally, the term time series refers to univariate time series (UTSs), where only one variable is measured, such as the temperature of a room, the electrical activity of a patient's heart (electrocardiography), etc. If two or more variables are measured, then the time series is called a multivariate time series (MTS). For example, when monitoring a patient, multiple variables such as temperature, pulse rate, blood pressure, and oxygen rate may be analyzed.

Time series classification (TSC) has gained popularity as a research field over the last few years, largely because of the wide range of real-world applications it has in industries, finance, healthcare, and other fields [7]. TSC's primary goal is to maximize prediction accuracy by leveraging full-length data. Since time series data are typically received over time, it is preferable to predict the class level of the time se-

Data Fusion Techniques and Applications for Smart Healthcare. https://doi.org/10.1016/B978-0-44-313233-9.00016-3

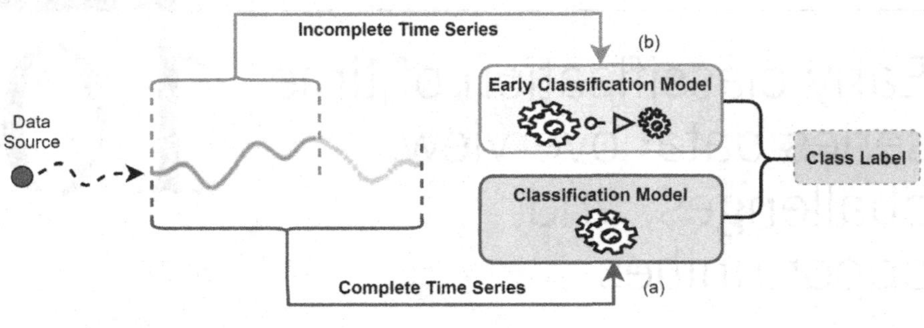

FIGURE 10.1

Illustration of traditional TSC and early classification.

ries before full-length series data become available. For instance, if a patient's disease is detected early from a series of medical observations, the cost of therapy and the length of the recovery period will be reduced. Additionally, an early diagnosis could save the patient's life by giving health practitioners more time for treatment. In agricultural monitoring, prediction of drought and shortage of resources ahead of time is highly useful to take necessary action to prevent food scarcity and to design sustainable policies [8]. A classification approach with the aim of classifying incomplete time series is referred to as early classification [9].

In recent times, early classification of time series has evoked great interest among researchers [10–18]. As a result, early classification of time series has shown promising applications such as early disease prediction [19,20], gas leakage detection [21], electricity demand prediction [22], malware detection [23], transportation mode detection [13], and crop mapping [18]. The fundamental difference between traditional and early classification approaches is shown in Fig. 10.1, where both models are learned from the same training set. In the figure, part (a) indicates the traditional TSC approach in which the classification model classifies the time series when its complete sequence becomes available, whereas part (b) demonstrates an early classification approach in which the classification model predicts the class label based on incomplete time series.

A general framework for early classification of time series is shown in Fig. 10.2. Let T be the length of the complete time series. The early classification framework processes the incoming time series at each time step t and gives it to the classifier. The classifier processes the observed data (incomplete time series) and predicts the class label if decision criteria are satisfied. Therefore, the earliness is defined as the number of data points in the incomplete time series at the time of decision making, that is, the number of data points used for making a decision. The early classification framework consists of two important components: an early classifier and decision criteria. In this regard, the classifier should be adaptable to incomplete time series so that it can produce classification results at different time step t. The decision criteria should be optimized well for achieving reliable class prediction and classify the time series when adequate information becomes available.

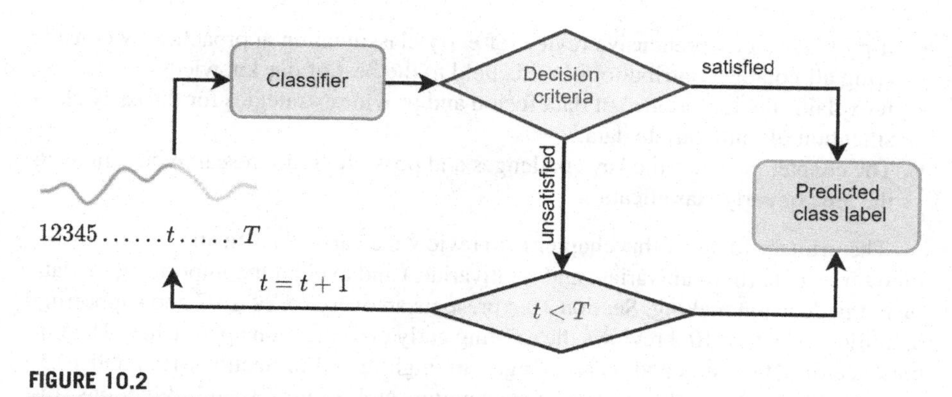

FIGURE 10.2

A general framework of an early classification model.

The objective of traditional TSC is to classify the time series accurately. Thus, it has only one objective, that is, to maximize accuracy. On the other hand, the aim of an early classification approach is to classify time series as early as possible with desirable accuracy. Hence, early classification has two objectives, i.e., accuracy and earliness [9]. The general intuition about an early classification problem is that the more data points a time series has, the more reliable its partition as it contains more information about the event or activity. Hence, earliness can be achieved only by compromising with accuracy. So, accuracy and earliness are two conflicting objectives of early classification approaches.

Data fusion is an another important area of interest, especially in the healthcare domain, that helps to improve the performance of the model. As discussed in the previous example of patient monitoring, single data information, such as the heart rate, blood pressure, oxygen level, or temperature of a patient, is sometimes not adequate to make reliable predictions. However, the collective information of such variables may be more helpful to make reliable predictions about the patient's conditions, e.g., during hospitalization. In this regard, the adoption of a suitable data fusion strategy will help to achieve desirable results. In [24], the authors present a review on data fusion strategies for multidimensional medical and biomedical data. Lin et al. [25] proposed a multisensor fusion strategy for body sensor networks by considering a human–robot interaction scenario in a medical setting. The basic aim of data fusion is to get more accurate, reliable, and comprehensive information from the model and the data. Data fusion strategies for early classification of time series are presented in Section 10.4.

A small survey of this area has been published in 2017 [26]. This review covers the basic structure of TSC approaches, including traditional methods, deep learning for time series, and early classification of time series. A nice and long literature review has been published in 2020 [27]. The major highlights of this chapter are as follows:

- The chapter explains the importance of early classification and how it is different from the traditional classification approach with a clear and precise definition.

- It provides a comprehensive review of early classification approaches by considering all notable contributions in this field to the best of our knowledge.
- It explains the importance of data fusion and provides strategies for the early classification of multivariate data.
- The chapter provides the key challenges and possible future research directions in the area of early classification.

The primary focus of this chapter is to review the early classification methods for time series data (both univariate and multivariate) and explain the importance of data fusion in decision making. Section 10.2 presents an overview of TSC and supporting definitions. Section 10.3 reviews the existing early classification approaches. The importance of data fusion and its challenges are highlighted in Sections 10.4 and 10.5, respectively. Section 10.6 presents opportunities and future research directions. Finally, conclusive remarks are provided in Section 10.7.

10.2 Overview

With the advancement in technology, especially in the Internet of Things (IoT), time series data are being generated in every field, including finance, healthcare, agriculture, industry, etc. Moreover, these time series data are being analyzed from different perspectives, including indexing, clustering, classification, and forecasting, to support various kinds of tasks using data mining and machine learning methods. In this chapter, we focus on early classification of time series, which is a special case of traditional TSC.

Definition 1 (Univariate time series (UTS)). A UTS is defined as an ordered sequence of values, typically recorded at equal time intervals. It is denoted as $S = \{s_1, s_2, \cdots, s_T\}$, where T is the length of the time series and $s_t \in R^v$ is the data point at time step t. In general, the term "time series" refers to UTSs.

Definition 2 (Multivariate time series (MTS)). An MTS is defined as a collection of at least two independent variables, where each variable is a UTS. It is formally defined as $\mathbf{S} = \{S^1, S^2, \ldots S^V\}$, where $S^v \in R^{T_v}$, $v \in [0..V]$, V is the number of variables, and T_v is the length of the vth variable in MTS.

Definition 3 (Incomplete time series). A time series is incomplete or partially observed if it has only t initial data points of the complete time series of length T. It is formally defined as $S_t = \{s_1, s_2, \ldots, s_t\}$, where $t \leq T$.

10.2.1 Traditional classification of time series

TSC has attracted much interest among researchers due to its applicability in various domains and the availability of labeled datasets in the public domain such as the UCI [28], UCR [29], and MTS datasets [30]. TSC is a supervised learning task in which a classifier $\mathcal{F}$ is built based on a given training set $\mathcal{D} = \{(S^i, y^i), 1 \leq i \leq N\}$, where

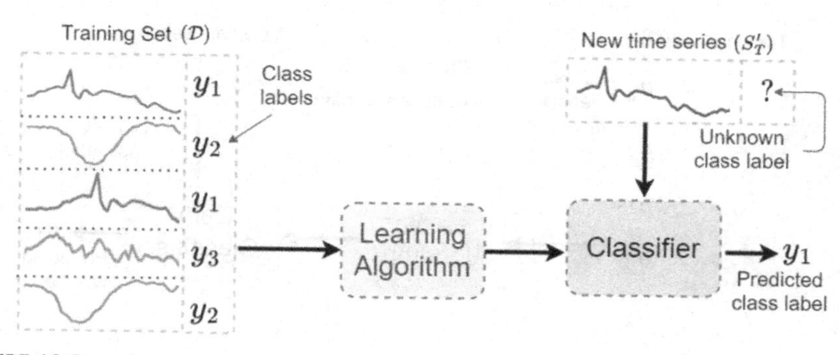

FIGURE 10.3

Traditional classification approach of time series.

S^i is the ith time series with corresponding class label $y^i \in \mathcal{Y}$ for some discrete set of class labels $\mathcal{Y}$ and N is the number of samples in the training set. The classifier learns the mapping function between time series and class label, which is formally defined as $\mathcal{F} : S_T \rightarrow y$, where $S_T \in \mathbb{R}^T$ denotes the complete time series. The main objective of classifier $\mathcal{F}$ is to classify the time series S'_T (new time series with an unknown class label, probably with the same domain) as accurately as possible. The TSC approach is depicted in Fig. 10.3.

10.2.2 Early classification of time series

Early prediction and forecasting are two distinct modeling issues with separate objectives and characteristics, despite the fact that the idea of early prediction can occasionally be imprecise in the literature and even be used interchangeably with forecasting by some researchers. Early classification on time series is a special case of the traditional TSC problem, where the goal is to optimize prediction quality with the additional characteristic of minimizing prediction time [9]. Formally, an early classifier $\mathcal{F}$ is able to provide the class prediction for incoming time series X' with t timestamps only, where $t \leq T$. It is formally defined as $\mathcal{F} : X'_t \rightarrow \hat{y}$, where X'_t is the incomplete time series and $\hat{y}$ is the predicted class label. Fig. 10.4 demonstrates the early classification of time series.

Early classification is highly advantageous when data points are generated sequentially over time. The aim of early classification is to predict the class label of incoming time series at the earliest with acceptable accuracy. The early classification approaches need to have two essential characteristics. The first one is the ability to classify the incomplete time series of any length and the second one is to have a decision policy that makes online decisions on incoming time series.

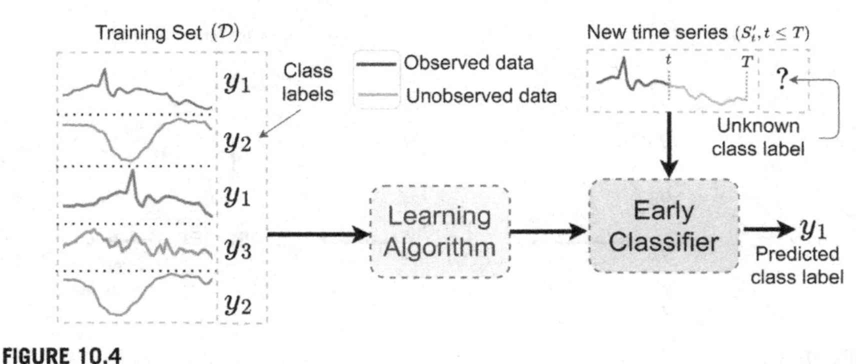

FIGURE 10.4

Early classification on time series.

10.3 Methods

TSC makes a significant contribution in numerous applications in data mining and machine learning. While several traditional TSC approaches have been investigated over the past few decades [7], early classification of time series has recently attracted a lot of research attention [31–35,12,11,36]. In the literature, early classification methods have been designed for UTSs as well as MTSs. Recently, early classification methods for MTSs have attracted more attention than early classification methods for UTSs. We group the early classification approaches into three categories, namely, instance-based [37–39,9,34], shapelet-based [40–42,32,43,44,10,36,45], and model-based [46,33,12,47–50,35,15,13]. *Instance-based* early classification methods basically learn the minimum predictive length (MPL) of time series in the training set and further use it to make reliable class predictions on incoming time series. The methods that fall in this category match the incomplete time series in the training dataset and classify the time series when the MPL condition is satisfied. *Shapelet-based* early classification methods mainly focus on extracting the key shapelets from the training set and using them to classify the new time series. The shapelets are the subsequences of the time series, having high discriminative power. Thus, they act as class representatives. Shapelet-based early classification approaches discover the local shapelet with the help of some utility measures to provide support in early decision making. *Model-based* early classification methods utilize discriminative or generative classifiers to provide classification results and predict the class label when the defined reliability threshold or decision function is satisfied. Table 10.1 provides a brief description as well as the advantages and limitations of these approaches.

10.3.1 Instance-based methods

Instance-based early classification methods predict the class level based on a best-matching instance in the training set. In this direction, the initial idea of early classification was published in [37]. Bregón et al. [37] presented an exploratory work

Table 10.1 Categorization of proposed early classification approaches.

Method	Description	Advantage	Limitation
Instance-based	These methods classify the new sample based on the closest sample in the training set.	The method is very easy and simple. A small training set is adequate for model training	These methods are slow if the dataset is large, and they do not work well for complex problems.
Shapelet-based	The class representation subsequence called shapelets is learned from the training set and the unknown time series is classified based on matching shapelets. Methods based on this approach have high discriminative power over classes in the dataset.	These methods are highly interpretable to the user.	They are computationally expensive and do not provide good results if classes do not have distinctive patterns.
Model-based	A mathematical model is learned from the data and the class is predicted based on learned features.	These methods are highly accurate and more effective when the training set is large.	These methods behave like a black box and lack interpretability.
Others	These methods do not fall in the above categories.	-	-

for fault classification at an early stage by considering a case-based reasoning (CBR) approach. CBR is an artificial intelligence approach that helps to solve the problem where domain knowledge of the problem is available. The proposed methodology implemented and analyzed the K-nearest neighbor classifier with Euclidean, Manhattan, and DTW distance measures. This method identifies the fixed set of points to classify the incoming time series. Thus, the decision policy is not adaptive in nature. In [38], the authors presented two interesting approaches. The first method mines the sequential classification rules, named SCR classifier. The second approach develops the classification model by utilizing a divide-and-conquer strategy, named a generalized sequential decision tree (GSDT). Moreover, both methodologies have been used to tackle the problem of early classification of symbolic sequences. These methods extract a large set of rules of variable length from a prefixed space and select the top rules for early classification with the help of support and prediction accuracy.

In [9], the author introduced a 1-nearest neighbor-based early classification strategy and developed an MPL concept by analyzing the nearest neighbor stability relationship in the training set. The proposed strategy determines the MPL for each time series in the training set and later uses it for early decision making. Xing et al. [9] formally define the early classification problem of time series and achieve decent early classification accuracy compared to the traditional 1-NN classifier for full-length time series. In [34], the authors extend this approach to MTSs by utilizing piecewise aggregate approximation (PAA) and named their method MTSECP. As a preprocessing step, this method generates a center sequence from a given MTS and then applies

PAA for dimensionality reduction. This method does not exploit the MTS properly and learns the MPL for each MTS in the training set by considering only accuracy as an evaluation measure. Thus, the proposed method does not consider the trade-off between accuracy and earliness during the MPL learning process.

10.3.2 Shapelet-based methods

Shapelets are defined as the subsequences of time series that have discriminating power over different classes in the dataset [51]. A shapelet is formally defined as $S = (s, l, \delta, c)$, where s is the subsequence of the time series, l is its length, δ is the threshold, and c is the class to which it belongs. A shapelet is considered an informative pattern/representative of a class in a training set. As a result, shapelet-based approaches for early classification are widely used, particularly in medical and health informatics. Early distinctive shapelet classification (EDSC) is the name of the initial baseline of this type of method and has been proposed for the early classification of UTSs [40]. EDSC has used two techniques to learn the class discriminative threshold, that is, kernel density estimation and Chebyshev's inequality. Moreover, they use well-defined utility measures to mine the best local shapelets. In [41], the authors extend this approach for early classification and utilize the concept of uncertainty measures to decide the best class prediction among multiple probable classes. In this proposed approach, the authors jointly address two practical requirements of the early classification of time series: (1) interpretability, which helps to understand the decision-making process, and (2) uncertainty estimation, which helps to measure the confidence in the prediction. They named this method Modified EDSC with Uncertainty estimates (MEDSC-U).

In [42], Ghalwash et al. developed an early classification method for MTSs and refined the idea of shapelets. Their method is called multimodal shapelet detection (MSD). MSD classifies the new incoming MTS based on the best matching key shapelet, and these local shapelets have been extracted from the N-dimensional MTS in the training set. The major drawback of this approach is that it uses a sliding window to mine the shapelets. As a result, a multivariate shapelet's subsequences have the same start and end points and fail to capture more realistic patterns from the MTS where informative patterns may lie in different parts of the segment. Also, it is very unlikely that all shapelet variates have the same start and end points. In [32], the author mitigated this problem by independently extracting the local shapelets for each variable in MTS and named this method Mining Core Feature for Early Classification (MCFEC). The authors also proposed a utility major that selects the best informative shapelet, called generalized extended F-measure. Finally, MCFCE offers two classification methodologies to classify the incoming MTS: rule-based and query-by-committee.

Ghalwash et al. [43] give a new method, named Interpretable Patterns for Early Diagnosis (IPED). This method formulates the problem as an optimization-based binary classification to extract the key multivariate shapelets from the training dataset having different start and end points compared to MSD. The IPED has been evaluated for human viral infection based on gene expression data. In [44], the authors developed a Reliable EArly ClassificaTion (REACT) method for heterogeneous MTS data, including categorical and numerical attributes. REACT generates the shapelets after discretizing the categorical time series and uses the concept of equivalence class mining to avoid redundant shapelets. The proposed early classifier ensures good accuracy and stability against the full-length time series classifier.

He et al. [10] proposed an ensemble-based early classification framework that is adaptable to data imbalance problems named Early Prediction on Imbalanced MTS (EPIMTS). EPIMTS learns the shapelets by considering correlations among multiple variables and helps to mitigate intraclass imbalance problems for early classification. Further, He et al. [36] extended this framework to improve prediction reliability by considering confidence estimation in shapelet learning. In [45], the authors present an early classification approach for patient monitoring in the intensive care unit. The proposed shapelet-based method can provide reliable class prediction on asynchronous MTSs. In [52], the author developed a shapelet-based early classification method that reduces shapelets by considering trend segmentation and learning the decision policy defining cost function that considers the trade-off between accuracy and earliness.

The above discussion demonstrates that the shapelet-based methods are highly interpretable during class prediction and especially more preferable medical data analysis. Despite good interpretability, shapelet-based methods have some critical issues. First, extracting informative shapelets is highly time consuming and complex. Second, it is difficult to learn the precise shapelet threshold if the classwise patterns are not well differentiable.

10.3.3 Model-based methods

Model-based early classification methods learn the mathematical model from the data and design decision criteria or rules to perform the classification task. Most of these methods define the decision criteria using conditional class probabilities. These probabilities are computed by either the generative model or the discriminative model. The authors in [46] present an early classification approach using a generative model which guarantees reliability in assigning class labels to incomplete time series. Basically, they develop a decision rule for early classification based on a quadratic discriminant analysis (QDA) classifier by assuming that training data follow a Gaussian distribution. Parish et al. in [53] further extend the approach and propose a more tractable decision rule by using different classifiers, including linear SVM and linear discriminant analysis (LDA).

In [33], the author develops a simple and straightforward method for early classification by adapting the approach of discriminating classes over time. The proposed method has a two-step learning process. First, a set of probabilistic classifiers is learned at different timestamps and then the reliability threshold is computed for each class. Second, a set of class-dependent safeguard points has been defined. Moreover, the model classifies the incoming time series when the reliability threshold (the difference between the two highest class probabilities) and designated safeguard points for the projected class are met. In [12], the authors present a relatively similar framework, in which the confidence threshold is defined by fusing the classifier's true prediction probabilities at successive time steps. This framework is adaptable for both probabilistic and discriminative classifiers. Yao et al. [47] propose an early classification framework by defining the distance-based feature transformation approach. In this framework, each time series is represented in distance space using informative local patterns and then the probabilistic classifiers are built. It defines the confidence area as a criterion for decision making based on time series. Li et al. [48] model the 3D action recognition problem as a stochastic process called a dynamic marked point process and develop an early classification approach for human activity recognition based on partially observed activity information. Hsu et al. [49] propose a deep learning-based early classification approach for MTSs through the attention mechanism, which helps identify best-performing segments in MTS variables.

Meanwhile, Dachraoui et al. in [31] propose a metaalgorithm for the early classification of time series. The proposed model is nonmyopic and uses the trade-off between accuracy and earliness in early decision making. To balance these two conflicting objectives, earliness and accuracy, the authors included misclassification and delaying decision costs in its optimization function. Thus, the proposed algorithm classifies the incoming time series only if the estimated cost at the current timestamp is less than that at all future timestamps. However, this method uses a clustering approach for evaluating future costs, which causes a lack of clarity in the overall process. Later, Tavenard et al. [50] eliminated this clustering step and proposed two new strategies (*NoCluster* and *2Step*) for further improvement. The authors in [35,54] present an early classification framework for time series data by defining the stopping rules as decision criteria and learning the rules by optimizing the accuracy and earliness simultaneously. However, in [54] a method for MTSs is proposed. Sharma et al. [13] developed a hybrid deep learning model for early transportation mode detection using smartphone sensing data. The proposed method first builds a model by adopting the capabilities of deep neural networks including CNN, RNN, and DNN and then defines the decision criteria as a confidence threshold by optimizing accuracy and earliness. Chen et al. [55] propose a deep learning-based end-to-end framework for early classification of time series. The proposed framework is composed of three subtasks, including feature extraction, a VTSC subnet, and gradient projection. Russwurm et al. [18] present an early classification model to predict crop cover maps captured through satellites. The model follows an end-to-end learning approach and produces a classification score as well as a scaler probability score to make early and accurate decisions.

10.3.4 Other methods

This section presents the methods that do not fall into the above three categories. Rodríguez and Alonso [56] came up with the concept of early classification as a natural finding. They presented the boosting interval-based literals method to classify variable-length time series. This method first partitions the time series into different time intervals and defines its states as increasing, decreasing, or constant. Later each interval is replaced by predicates that signal the presence or absence of the states. Finally, the predicate's output is used for classification at defined intervals. Thus, they did not attempt to optimize the trade-off between accuracy and earliness. In [57], the authors present a reinforcement learning-based early classification framework. They introduce an early classifier and reinforcement learning agent as a deep Q-network to perform the early classification task. The proposed framework defines the decision function as a learning agent that uses the reward function to balance accuracy and earliness. In [58], a divide-and-conquer-based early classification approach is proposed for MTSs. In this work, the authors consider an IoT problem where multiple sensors may generate data with different sampling rates. Gupta et al. [59] present an early as well as unseen fault classification approach for multivariate sensor data by using zero-shot learning.

A brief summary of early classification methods for UTSs and MTSs is provided in Table 10.2 and Table 10.3, respectively, based on three key factors:

- the type of classifier used,
- the type of decision strategy considered,
- the consideration of trade-off between accuracy and earliness.

Table 10.2 Comparative analysis of early classification approaches (UTSs).

ID	Year	Classifier used	Trade-off	Decision strategy	Method
[38]	2008	Decision tree and rule-based classifier	No	The decision function is triggered when the user-expected accuracy is achieved.	Instance-based
[39]	2009	1-NN	No	Minimum prediction length is defined as decision criterion.	Instance-based
[40]	2011	Closest shapelet using Euclidean	No	Early decision is performed when the defined shapelet is matched with the incoming time series.	Shapelet-based
[9]	2011	1-NN	No	Minimum prediction length is defined as decision criterion.	Instance-based
[60,46]	2012	Linear SVM and local QDA	No	The decision function is defined as the probability threshold.	Model-based
[46]	2012	QDA	No	The decision function has been defined as the probability threshold.	Model-based
[53]	2013	Linear and quadratic discriminants	No	Early decision criteria are defined as reliability thresholds.	Model-based
[41]	2014	Closest shapelet using ED	No	Decision function is defined based on shapelets matching with confidence threshold, computed using uncertainty in predictions.	Shapelet-based

continued on next page

Table 10.2 *(continued)*

ID	Year	Classifier used	Trade-off	Decision strategy	Method
[61]	2015	Linear SVM	No	Ensemble classifier developed with minimization of empirical risk and response time simultaneously.	Model-based
[31]	2015	Naïve Bayes and multilayer perceptrons	Yes	The decision policy is defined based on cost estimation at current as well as all future time steps. This approach is nonmyopic in nature.	Model-based
[50]	2016	SVM	Yes	The decision function is defined based on cost optimization between current and future predictions.	Model-based
[33]	2016	GP classifier	No	The decision criteria are defined based on the difference between the first two highest class probabilities, conditioned on user-defined parameters.	Model-based
[62]	2016	CNN	No	No explicit decision function is defined for early classification; hence, there is no trade-off optimization in decision learning.	Model-based
[35]	2018	GP and SVM	Yes	Stopping rules are defined to classify incomplete time series.	Model-based
[57]	2018	Reinforcement learning agent	Yes	A reinforcement learning agent is defined to make early decisions.	Other
[12]	2019	Dictionary classifier with WEASEL	Yes	Confidence threshold is defined to classify the incoming time series.	Model-based
[63]	2019	Combination of CNN and LSTM	Yes	The decision criteria are defined as probability threshold. Trade-off optimization is taken into consideration while learning the model.	Model-based
[11]	2019	GP and SVM	Yes	Rules are designed for making an early decision.	Model-based
[47]	2019	Closest shapelet using Euclidean	No	The decision criteria are defined as confidence areas. The framework considers the earliness in selecting the shapelets only.	Shapelet-based
[64]	2019	GP classifier	No	Confidence threshold is defined by considering uncertainty in prediction.	Model-based
[52]	2021	Shapelet and SVM	Yes	Learning decision policy as shapelet matching.	Shapelet-based

Table 10.3 Comparative analysis of early classification approaches (MTSs).

ID	Year	Classifier used	Trade-off	Decision strategy	Method
[56]	2004	Adaboost ensemble classifier	No	No explicit function was defined for decision making.	Other
[37]	2006	1-NN with Euclidean, Manhattan, and DTW distances	No	The decision function is defined as a user-defined threshold.	Instance-based
[42]	2012	Closest multivariate shapelet with Euclidean	No	A variatewise shapelet threshold is defined and a classification task is performed when variatewise thresholds are satisfied.	Shapelet-based
[21]	2013	SVM	No	The decision criterion is defined as a confidence score, considering accuracy only.	Model-based
[43]	2013	Closest multivariate shapelet with Euclidean	No	The problem is formulated as convex optimization and the key shapelets are extracted to classify the time series.	Shapelet-based
[65]	2013	Closest multivariate shapelet with Euclidean	No	Core shapelets are extracted using a defined utility measure that takes earliness to account.	Shapelet-based
[66]	2014	Rule-based and query-by-committee classifiers	No	Core shapelets are extracted using a defined utility measure.	Shapelet-based
[44]	2015	Decision tree	No	The decision criteria are defined as heterogeneous shapelets with a defined threshold. Earliness is not taken into consideration while learning the shapelets.	Shapelet-based
[20]	2015	Hybrid model using HMM and SVM	No	An early decision is made when the difference between two class probabilities is higher than a certain threshold value.	Model-based
[34]	2017	1-NN	No	A minimum required length is defined for making decisions.	Instance-based
[48]	2018	Stochastic process	No	No specific decision policy is defined for early classification.	Other
[67]	2018	Combination of CNN and LSTM	No	No decision policy is defined for making a reliable class prediction.	Model-based
[45]	2019	Decision tree and random forest	No	Shapelet matching.	Shapelet-based

continued on next page

Table 10.3 *(continued)*

ID	Year	Classifier used	Trade-off	Decision strategy	Method
[36]	2019	Multivariate shapelet with rule-based classifier	No	Decision criteria are defined as shapelet matching with cumulative confidence.	Shapelet-based
[58]	2020	GP classifier	No	MRL is defined for early stopping.	Other
[59]	2020	GP with similarity matrix	No	MRL is defined for early stopping.	Other
[54]	2020	GP classifier	No	Early stopping rules are defined.	Model-based
[13]	2020	CNN and RNN	Yes	Confidence threshold is defined for early stopping.	Model-based
[55]	2022	Deep learning with gradient projection	Yes	Stopping criteria are learned through gradient projection.	Model-based
[18]	2023	LSTM	Yes	Scalar probability is learned as stopping criteria.	Model-based

10.4 Data fusion

Nowadays, data fusion methods are very useful and essential for any kind of prediction activity where multimodal data are involved in the decision process. For example, the accurate prediction of a cardiovascular disease (related to the heart and blood vessels) requires various kinds of data/information that may include sensor data (blood pressure, body temperature, respiration rate, electrocardiograms, etc.), text data (medical history, general medical observations, etc.), and image data (X-ray images, MRIs, CT scans) [68]. A model developed by considering only a single kind of information may not provide reliable predictions due to insufficient information. Multimodal data help to capture the features/information about the disease from different perspectives, and the fusion of this captured information helps to make more reliable predictions.

Data fusion refers to the process of combining multiple sources of information to make a more accurate prediction or decision. In the context of early classification of time series data, data fusion can be used to improve the accuracy of class predictions by combining information from multiple time series or complementary data sources.

Generally, the model training follows three basic steps: data preprocessing, model training, and model prediction. Thus, in the case of MTS data, the fusion step can be applied in one of two ways: (a) the data fusion step is performed first and then complete data are considered as input for training the model or (b) an individual model is first developed for each variable of the MTS and then the data fusion step (combining the output) is performed for classifying the MTS. These two basic data fusion strategies for early classification of MTSs are shown in Fig. 10.5.

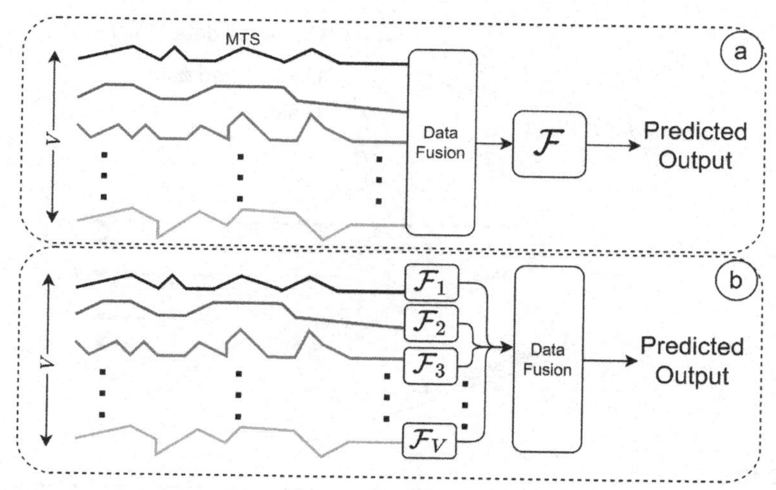

FIGURE 10.5

Data fusion strategies in early classification of MTSs.

In Section 10.3, we have discussed various early classification models developed for MTS data. Moreover, many of them have been tested for medical data. Some examples of scenario (a) (as given in Fig. 10.5) are [42], [32], [44], [34], [67], [16], etc. Some examples of scenario (b) (as given in Fig. 10.5) are [13], [10], [36], etc.

10.5 Challenges

Traditional classification models require complete time series to predict class labels. In contrast, early classification models process incomplete time series to predict class label at an early stage. This means the input of the classifier always has some missing values. Thus, various challenges arise in the design of early classification models.

- To define a *classifier strategy* so that early classifiers become adaptable to missing values.
- The implementation of a *decision policy* so that it can provide reliable class prediction at an early stage.
- Maintaining the trade-off between two *conflicting objectives.*
- Handling *multivariate signals.*

10.5.1 Handling missing values

Basically, three types of strategies have been suggested for early classifiers to handle incomplete time series [69].

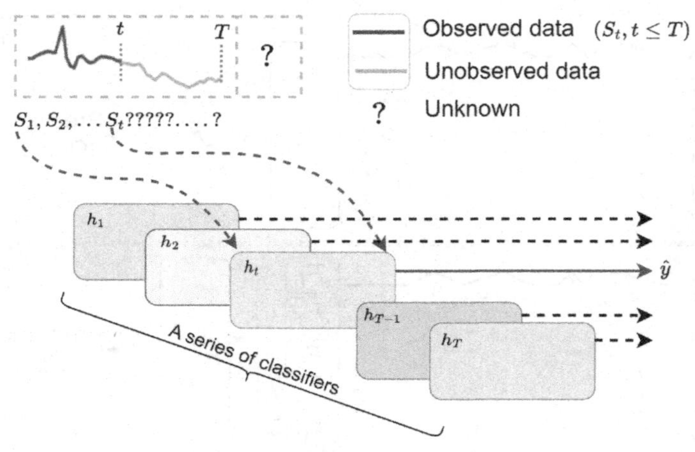

FIGURE 10.6

Adapting to missing values as a series of classifiers.

1. *Adapting to missing values:* The methods in this category do not use all the information contained in the complete training time series. These methods deal directly with an incoming time series and predict its class label without carrying out any operation to complete it. These methods are implemented as distance-based models or as a series of classifiers. The idea of adapting missing values by considering a series of classifiers is depicted in Fig. 10.6.
2. *Imputation of missing values:* The methods in this category utilize the full-length time series information to make a class prediction on incomplete time series. Basically, these methods perform implicit or explicit imputation to make the time series complete.

   - Implicit imputation methods leverage the information included in the complete series data to make a class prediction. For example, the cluster-based approach utilizes the closest cluster of incoming time series for imputing the missing values.
   - Explicit imputation methods first complete the incomplete time series with explicit imputation, for example, single imputation, machine learning imputation [70], imputation with forecasting [71], etc. The former method fills the missing values with zero or the conditional mean value or uses the last observation carry forward approach. The latter methods explicitly learn the model from training data for imputing the missing values of incomplete time series. An example of this kind is depicted in Fig. 10.7. The data imputation model (learned from training data) takes incomplete input $(S_1, S_2, \ldots S_t)$ and generates the missing data points $(U_{t+1} \ldots. U_T)$ to make the series complete. Now the complete time series is used to predict an unknown class label.

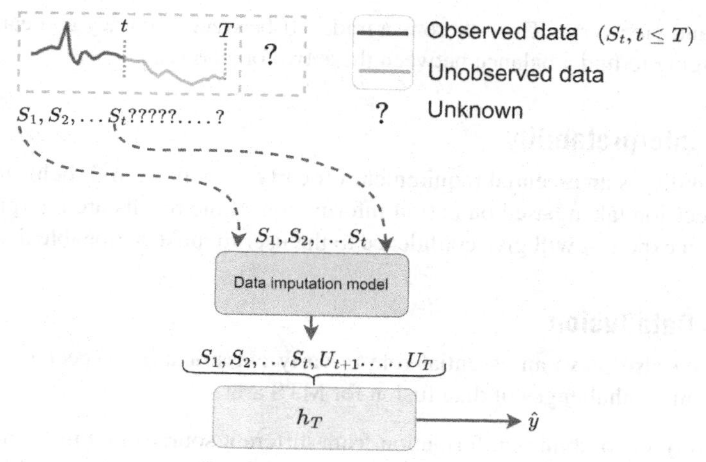

FIGURE 10.7

Explicit imputation of missing values.

3. *Representation of missing values:* The methods in this category implicitly use all the information contained in the complete time series. Basically, these methods change the representation of incomplete time series in another time-invariant domain in order to make it complete. Moreover, the prediction of the class label is performed on the transformed data. Let an incomplete time series $X_t \in \mathbb{R}^t$ be in the time domain. Then it is transformed in another domain by some function $\Phi(X_t) \in \mathbb{R}^K$.

10.5.2 Decision strategy

The decision strategy is the heart of the early classification model, which takes the following conditions [69].

- A decision criterion decides when to stop measuring additional information and predict the final output.
- Optimization between time and quality in prediction trade-off helps in finding the optimal balance between these two objectives, i.e., accuracy and earliness.
- Adaptability refers to the handling of incomplete time series X_t. An early classification approach is said to be adaptive if it is able to make decisions at any time point t while collecting the data points over time.

10.5.3 Conflicting objectives

The early classification of time series has two conflicting objectives: *accuracy* and *earliness*. In general *accuracy* is directly proportional to the number of observations (data points). If we have more observations about an activity, our prediction is more reliable, and vice versa. Thus, there is a trade-off between accuracy and earliness. It is challenging to find a balance between these two objectives.

10.5.4 Interpretability

Interpretability is an essential requirement for early classification. It helps in analyzing the decision taken based on partial information. If the results are interpretable to the domain expert, it will give confidence to the user to make actionable decisions.

10.5.5 Data fusion

Data fusion also plays an essential role in early classification, especially for MTS data. The main challenges of data fusion for MTS are:

- Difficulty in combining information from different sources in a meaningful way. The information from different sources may be in different formats or have different levels of relevance to the classification task, making it difficult to integrate the information effectively.
- Variability in time series data. Time series data can exhibit different patterns and behaviors across different time periods, making it difficult to generalize the learned patterns to new data. This can make it difficult to incorporate information from other sources that may have been learned during different time periods.

Despite these challenges, data fusion has the potential to significantly improve the accuracy of early classification of time series data. There are several approaches to data fusion, such as feature-level fusion, decision-level fusion, and hybrid fusion. Feature-level fusion combines information at the feature level by combining features from different sources to create a more comprehensive representation of the data. Decision-level fusion combines the predictions of multiple classifiers to make a final prediction by using techniques such as majority voting or weighting the predictions based on their confidence. Hybrid fusion combines elements of both feature-level and decision-level fusion by combining both the features and the predictions from multiple sources.

10.6 Opportunities and future directions

Early classification of time series is a vibrant research area. ECTS focuses on developing quality decision rules that need to provide reliable class prediction to make actionable decisions. The early classification approaches are beneficial in time-sensitive applications where either data collection is costly or there is a requirement for event prediction ahead of time with acceptable accuracy. It has numerous valuable applications in various domains, such as irrigation and crop monitoring in agriculture, health and disease prediction in medicine, quality and process monitoring in industry, traffic flow and user mode prediction in transportation, etc. Future research directions include the following:

- In the analysis of previous studies, early classification approaches have been classified into four categories, including instance-based, shapelet-based, model-based, and other. Shapelet-based methods are interpretable but computationally expensive. On the other end, model-based approaches are computationally effective but lack interpretability. So, the future recommendation is to develop more accurate and interpretable approaches.
- Early classification aims to predict the class as early as possible with acceptable accuracy. This leads to optimization problems facing a trade-off between accuracy and earliness. So, the recommendation is to develop decision policies that take a core balance between these two objectives.
- The existing approaches simulate their results by assuming that the length of the activity is known and calculating the earliness measure. However, in real scenarios, knowledge of the length of the activity is limited. Therefore, the future recommendation is to develop early classification methods by considering the open time series problem, which means predicting the activity at an early stage on infinite-length series.
- Maximally existing early classification approaches have been designed in general and have yet to be fully optimized for specific applications. Therefore, the recommendation is to develop application-specific approaches to benefit society in various domains such as agriculture monitoring, health monitoring, etc.

10.7 Conclusion

This chapter discusses the problem of early classification on time series data as well as its applications, challenges, and opportunities. It also covers the importance of data fusion and strategies for early classification of time series. Early classification of time series refers to the early prediction of class labels with acceptable accuracy. It is highly beneficial in time-sensitive applications such as disease prediction and patient monitoring. This chapter first described what is early classification and how it is different from traditional TSC approaches, followed by a discussion of categories of early classification methods, including instance-based, shapelet-based, and model-based approaches. Shapelet-based approaches are highly interpretable but computationally expensive, whereas model-based approaches are more accurate and robust but lack interpretability. When dealing with MTSs, data fusion can be an effective approach for improving the accuracy of early classification. However, it is important to carefully consider the challenges and opportunities in data fusion and to select an appropriate approach based on the specific requirements of the task. In conclusion, this chapter presents an overview, challenges, and opportunities of early classification of time series by highlighting the importance of data fusion and its strategies.

References

[1] T. Hastie, R. Tibshirani, J. Friedman, The Elements of Statistical Learning, Springer New York, 2009.

[2] A. Kumar, A. Sharma, V. Bharti, A.K. Singh, S.K. Singh, S. Saxena, Mobihisnet: a lightweight cnn in mobile edge computing for histopathological image classification, IEEE Internet of Things Journal 8 (24) (2021) 17778–17789, https://doi.org/10.1109/JIOT.2021.3119520.

[3] A. Kumar, S.K. Singh, S. Saxena, K. Lakshmanan, A.K. Sangaiah, H. Chauhan, et al., Deep feature learning for histopathological image classification of canine mammary tumors and human breast cancer, Information Sciences 508 (2020) 405–421, https://doi.org/10.1016/j.ins.2019.08.072, https://www.sciencedirect.com/science/article/pii/S0020025519308229.

[4] B.K. Dedeturk, B. Akay, Spam filtering using a logistic regression model trained by an artificial bee colony algorithm, Applied Soft Computing 91 (2020) 106229, https://doi.org/10.1016/j.asoc.2020.106229.

[5] S.S. Udmale, S.K. Singh, Application of spectral kurtosis and improved extreme learning machine for bearing fault classification, IEEE Transactions on Instrumentation and Measurement 68 (11) (2019) 4222–4233, https://doi.org/10.1109/tim.2018.2890329.

[6] M. Akkaş, R. Sokullu, H.E. Çetin, Healthcare and patient monitoring using IoT, Internet of Things 11 (2020) 100173, https://doi.org/10.1016/j.iot.2020.100173.

[7] A. Bagnall, J. Lines, A. Bostrom, J. Large, E. Keogh, The great time series classification bake off: a review and experimental evaluation of recent algorithmic advances, Data Mining and Knowledge Discovery 31 (3) (2016) 606–660, https://doi.org/10.1007/s10618-016-0483-9.

[8] M. Rußwurm, R. Tavenard, S. Lefèvre, M. Körner, Early classification for agricultural monitoring from satellite time series, http://arxiv.org/abs/1908.10283v1.

[9] Z. Xing, J. Pei, P.S. Yu, Early classification on time series, Knowledge and Information Systems 31 (1) (2011) 105–127.

[10] G. He, W. Zhao, X. Xia, R. Peng, X. Wu, An ensemble of shapelet-based classifiers on inter-class and intra-class imbalanced multivariate time series at the early stage, Soft Computing 23 (15) (2018) 6097–6114.

[11] U. Mori, A. Mendiburu, I. Miranda, J. Lozano, Early classification of time series using multi-objective optimization techniques, Information Sciences 492 (2019) 204–218.

[12] J. Lv, X. Hu, L. Li, P. Li, An effective confidence-based early classification of time series, IEEE Access 7 (2019) 96113–96124.

[13] A. Sharma, S.K. Singh, S.S. Udmale, A.K. Singh, R. Singh, Early transportation mode detection using smartphone sensing data, IEEE Sensors Journal (2020) 1, https://doi.org/10.1109/jsen.2020.3009312.

[14] G. Ottervanger, M. Baratchi, H.H. Hoos, MultiETSC: automated machine learning for early time series classification, Data Mining and Knowledge Discovery 35 (6) (2021) 2602–2654, https://doi.org/10.1007/s10618-021-00781-5.

[15] A.G. Nath, A. Sharma, S.S. Udmale, S.K. Singh, An early classification approach for improving structural rotor fault diagnosis, IEEE Transactions on Instrumentation and Measurement 70 (2021) 1–13, https://doi.org/10.1109/tim.2020.3043959.

[16] Y. Huang, G.G. Yen, V.S. Tseng, Snippet policy network for multi-class varied-length ECG early classification, IEEE Transactions on Knowledge and Data Engineering (2022) 1, https://doi.org/10.1109/tkde.2022.3160706.

[17] A. Sharma, S.K. Singh, A. Kumar, A.K. Singh, S.K. Singh, Adaptive early classification of time series using deep learning, in: Neural Information Processing, Springer International Publishing, 2023, pp. 533–542.
[18] M. Rußwurm, N. Courty, R. Emonet, S. Lefèvre, D. Tuia, R. Tavenard, End-to-end learned early classification of time series for in-season crop type mapping, ISPRS Journal of Photogrammetry and Remote Sensing 196 (2023) 445–456, https://doi.org/10.1016/j.isprsjprs.2022.12.016.
[19] M. Ghalwash, V. Radosavljevic, Z. Obradovic, Early diagnosis and its benefits in sepsis blood purification treatment, in: 2013 IEEE International Conference on Healthcare Informatics, IEEE, 2013.
[20] M.F. Ghalwash, D. Ramljak, Z. Obradović, Patient-specific early classification of multivariate observations, International Journal of Data Mining and Bioinformatics 11 (4) (2015) 392, https://doi.org/10.1504/ijdmb.2015.067955.
[21] N. Hatami, C. Chira, Classifiers with a reject option for early time-series classification, in: Computational Intelligence and Ensemble Learning (CIEL), 2013 IEEE Symposium on, IEEE, 2013, pp. 9–16.
[22] A. Dachraoui, A. Bondu, A. Cornuejols, Early classification of individual electricity consumptions, in: RealStream2013 (ECML), 2013, pp. 18–21.
[23] A. Sharma, S.K. Singh, A novel approach for early malware detection, Transactions on Emerging Telecommunications Technologies (2020), https://doi.org/10.1002/ett.3968.
[24] K.S.F. Azam, O. Ryabchykov, T. Bocklitz, A review on data fusion of multidimensional medical and biomedical data, Molecules 27 (21) (2022) 7448, https://doi.org/10.3390/molecules27217448.
[25] K. Lin, Y. Li, J. Sun, D. Zhou, Q. Zhang, Multi-sensor fusion for body sensor network in medical human–robot interaction scenario, Information Fusion 57 (2020) 15–26, https://doi.org/10.1016/j.inffus.2019.11.001.
[26] T. Santos, R. Kern, A literature survey of early time series classification and deep learning, in: Sami@ iknow, 2016.
[27] A. Gupta, H.P. Gupta, B. Biswas, T. Dutta, Approaches and applications of early classification of time series: a review, IEEE Transactions on Artificial Intelligence 1 (1) (2020) 47–61, https://doi.org/10.1109/TAI.2020.3027279.
[28] D. Dua, C. Graff, UCI machine learning repository, http://archive.ics.uci.edu/ml, 2017.
[29] Y. Chen, E. Keogh, B. Hu, N. Begum, A. Bagnall, A. Mueen, et al., The ucr time series classification archive, 2015.
[30] M.G. Baydogan, Multivariate time series classification datasets, http://www.mustafabaydogan.com, 2015.
[31] A. Dachraoui, A. Bondu, A. Cornuéjols, Early classification of time series as a non myopic sequential decision making problem, in: Machine Learning and Knowledge Discovery in Databases, Springer International Publishing, 2015, pp. 433–447.
[32] G. He, Y. Duan, R. Peng, X. Jing, T. Qian, L. Wang, Early classification on multivariate time series, Neurocomputing 149 (2015) 777–787.
[33] U. Mori, A. Mendiburu, E. Keogh, J.A. Lozano, Reliable early classification of time series based on discriminating the classes over time, Data Mining and Knowledge Discovery 31 (1) (2016) 233–263.
[34] C. Ma, X. Weng, Z. Shan, Early classification of multivariate time series based on piecewise aggregate approximation, in: Health Information Science, Springer International Publishing, 2017, pp. 81–88.

[35] U. Mori, A. Mendiburu, S. Dasgupta, J.A. Lozano, Early classification of time series by simultaneously optimizing the accuracy and earliness, IEEE Transactions on Neural Networks and Learning Systems 29 (10) (2018) 4569–4578.
[36] G. He, W. Zhao, X. Xia, Confidence-based early classification of multivariate time series with multiple interpretable rules, Pattern Analysis & Applications (2019).
[37] A. Bregón, M.A. Simón, J.J. Rodríguez, C. Alonso, B. Pulido, I. Moro, Early fault classification in dynamic systems using case-based reasoning, in: Current Topics in Artificial Intelligence, Springer Berlin Heidelberg, 2006, pp. 211–220.
[38] Z. Xing, J. Pei, G. Dong, P.S. Yu, Mining sequence classifiers for early prediction, in: Proceedings of the 2008 SIAM International Conference on Data Mining, SIAM, 2008, pp. 644–655.
[39] Z. Xing, J. Pei, S.Y. Philip, Early prediction on time series: a nearest neighbor approach, in: Twenty-First International Joint Conference on Artificial Intelligence, Citeseer, 2009.
[40] Z. Xing, J. Pei, P.S. Yu, K. Wang, Extracting interpretable features for early classification on time series, in: Proceedings of the 2011 SIAM International Conference on Data Mining, SIAM, 2011, pp. 247–258.
[41] M.F. Ghalwash, V. Radosavljevic, Z. Obradovic, Utilizing temporal patterns for estimating uncertainty in interpretable early decision making, in: Proceedings of the 20th ACM SIGKDD International Conference on Knowledge Discovery and Data Mining - KDD-14, ACM Press, 2014.
[42] M.F. Ghalwash, Z. Obradovic, Early classification of multivariate temporal observations by extraction of interpretable shapelets, BMC Bioinformatics 13 (1) (2012) 195.
[43] M.F. Ghalwash, V. Radosavljevic, Z. Obradovic, Extraction of interpretable multivariate patterns for early diagnostics, in: 2013 IEEE 13th International Conference on Data Mining, IEEE, 2013.
[44] Y.F. Lin, H.H. Chen, V.S. Tseng, J. Pei, Reliable early classification on multivariate time series with numerical and categorical attributes, in: Advances in Knowledge Discovery and Data Mining, Springer International Publishing, 2015, pp. 199–211.
[45] L. Zhao, H. Liang, D. Yu, X. Wang, G. Zhao, Asynchronous multivariate time series early prediction for ICU transfer, in: Proceedings of the 2019 International Conference on Intelligent Medicine and Health - ICIMH 2019, ACM Press, 2019.
[46] H.S. Anderson, N. Parrish, K. Tsukida, M.R. Gupta, Reliable early classification of time series, in: 2012 IEEE International Conference on Acoustics, Speech and Signal Processing (ICASSP), IEEE, 2012.
[47] L. Yao, Y. Li, Y. Li, H. Zhang, M. Huai, J. Gao, et al., DTEC: distance transformation based early time series classification, in: Proceedings of the 2019 SIAM International Conference on Data Mining, Society for Industrial and Applied Mathematics, 2019, pp. 486–494.
[48] S. Li, K. Li, Y. Fu, Early recognition of 3d human actions, ACM Transactions on Multimedia Computing Communications and Applications 14 (1s) (2018) 1–21.
[49] E.Y. Hsu, C.L. Liu, V.S. Tseng, Multivariate time series early classification with interpretability using deep learning and attention mechanism, in: Advances in Knowledge Discovery and Data Mining, Springer International Publishing, 2019, pp. 541–553.
[50] R. Tavenard, S. Malinowski, Cost-aware early classification of time series, in: Machine Learning and Knowledge Discovery in Databases, Springer International Publishing, 2016, pp. 632–647.
[51] L. Ye, E. Keogh, Time series shapelets: a new primitive for data mining, in: Proceedings of the 15th ACM SIGKDD International Conference on Knowledge Discovery and Data Mining - KDD '09, ACM Press, 2009.

[52] W. Zhang, Y. Wan, Early classification of time series based on trend segmentation and optimization cost function, Applied Intelligence 52 (6) (2021) 6782–6793, https://doi.org/10.1007/s10489-021-02788-3.

[53] N. Parrish, H.S. Anderson, M.R. Gupta, D.Y. Hsiao, Classifying with confidence from incomplete information, Journal of Machine Learning Research 14 (1) (2013) 3561–3589.

[54] A. Sharma, S.K. Singh, Early classification of multivariate data by learning optimal decision rules, Multimedia Tools and Applications 80 (28–29) (2020) 35081–35104, https://doi.org/10.1007/s11042-020-09366-8.

[55] H. Chen, Y. Zhang, A. Tian, Y. Hou, C. Ma, S. Zhou, Decoupled early time series classification using varied-length feature augmentation and gradient projection technique, Entropy 24 (10) (2022) 1477, https://doi.org/10.3390/e24101477.

[56] C.J.A. González, J.J.R. Diez, Boosting interval - based literals: variable length and early classification, in: Series in Machine Perception and Artificial Intelligence, WORLD SCIENTIFIC, 2004, pp. 149–171.

[57] M. Coralie, G. Perrin, E. Ramasso, M. Rombaut, A deep reinforcement learning approach for early classification of time series, in: 26th European Signal Processing Conference (EUSIPCO 2018), Rome, Italy, 2018, https://hal.archives-ouvertes.fr/hal-01825472.

[58] A. Gupta, H.P. Gupta, B. Biswas, T. Dutta, A divide-and-conquer–based early classification approach for multivariate time series with different sampling rate components in IoT, ACM Transactions on Internet of Things 1 (2) (2020) 1–21, https://doi.org/10.1145/3375877.

[59] A. Gupta, H.P. Gupta, B. Biswas, T. Dutta, An unseen fault classification approach for smart appliances using ongoing multivariate time series, IEEE Transactions on Industrial Informatics (2020) 1, https://doi.org/10.1109/tii.2020.3016590.

[60] H.S. Anderson, N. Parrish, M.R. Gupta, Early Time Series Classification with Reliability Guarantee, Tech. Rep., Sandia National Lab. (SNL-NM), Albuquerque, NM (United States), 2012.

[61] S. Ando, E. Suzuki, Minimizing response time in time series classification, Knowledge and Information Systems 46 (2) (2015) 449–476.

[62] W. Wang, C. Chen, W. Wang, P. Rai, L. Carin, Earliness-aware deep convolutional networks for early time series classification, CoRR, arXiv:1611.04578, 2016, http://arxiv.org/abs/1611.04578, arXiv:1611.04578.

[63] M. Rußwurm, S. Lefèvre, N. Courty, R. Emonet, M. Körner, R. Tavenard, End-to-end learning for early classification of time series, arXiv:1901.10681v1.

[64] A. Sharma, S.K. Singh, Early classification of time series based on uncertainty measure, in: 2019 IEEE Conference on Information and Communication Technology, IEEE, 2019.

[65] G. He, Y. Duan, T. Qian, X. Chen, Early prediction on imbalanced multivariate time series, in: Proceedings of the 22nd ACM International Conference on Conference on Information & Knowledge Management - CIKM '13, ACM Press, 2013.

[66] G. He, Y. Duan, G. Zhou, L. Wang, Early Classification on Multivariate Time Series with Core Features, Lecture Notes in Computer Science, Springer International Publishing, 2014, pp. 410–422.

[67] H.S. Huang, C.L. Liu, V.S. Tseng, Multivariate time series early classification using multi-domain deep neural network, in: 2018 IEEE 5th International Conference on Data Science and Advanced Analytics (DSAA), IEEE, 2018.

[68] S. Amal, L. Safarnejad, J.A. Omiye, I. Ghanzouri, J.H. Cabot, E.G. Ross, Use of multimodal data and machine learning to improve cardiovascular disease care, Frontiers in Cardiovascular Medicine (2022) 9, https://doi.org/10.3389/fcvm.2022.840262.

[69] A. Dachraoui, Cost-Sensitive Early classification of Time Series, Ph.D. thesis, 2017.
[70] M.B. Richman, T.B. Trafalis, I. Adrianto, Missing data imputation through machine learning algorithms, in: Artificial Intelligence Methods in the Environmental Sciences, Springer Netherlands, 2009, pp. 153–169.
[71] G.E. Box, G.M. Jenkins, G.C. Reinsel, G.M. Ljung, Time Series Analysis: Forecasting and Control, John Wiley & Sons, 2015.

CHAPTER

11 Deep learning-based multimodal medical image fusion

Aditya Kahol and Gaurav Bhatnagar
Indian Institute of Technology Jodhpur, Jodhpur, Rajasthan, India

11.1 Introduction

Medical imaging refers to the process of image acquisition using some *special* imaging devices that are multimodal in nature, which informs the viewer about the internal parts of a human body. With the increasing advancement in the field of radiography and in particular medical imaging, the healthcare industry has been adept at exploiting the various uses of those imaging devices for effective treatment strategies. The human body is considered to contain structural and functional information [1], where structural information, more commonly known as anatomical information, includes bones, soft tissues, cartilage, tendons, etc. Generally, these anatomical features are acquired through X-ray-based computed tomography (CT) and radio wave-based MRI. CT scans are used to view information having dense structures, i.e., bones, cartilage, tumors, etc., while the radio wave-based MRI scans are used to view lower-density features, for example soft tissues or fluidic components, inside the body. Functional information refers to the physiological and metabolic changes within the body, and this information is gathered by the means of positron emission tomography (PET) and single photon emission computed tomography (SPECT). Both PET and SPECT use radiotracers to appraise organ and tissue functions. PET and SPECT are different in the sense that they use different kinds of radiotracers, being positron- and gamma-ray-based, respectively.

With such diverse set of modalities, the information provided by these images is quite complementary, and hence multimodal image fusion has been identified as a decisive solution which aims to integrate information from these images to obtain a single and more complete image, which can aid medical practioners to identify the presence of lesions or any kind of anomaly within the patient's body with ease. More formally, *image fusion* is the process of merging or integrating complementary information of several source images such that the resultant image provides more detail and quality than any of the individual source images.

Data Fusion Techniques and Applications for Smart Healthcare. https://doi.org/10.1016/B978-0-44-313233-9.00017-5

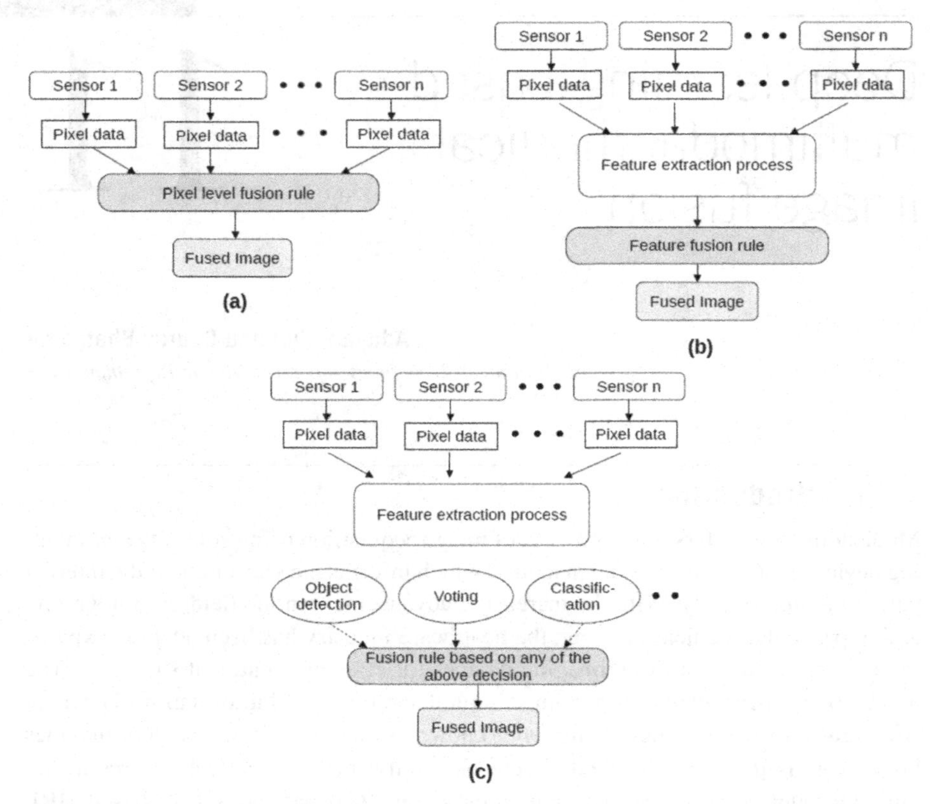

FIGURE 11.1

Image fusion levels. (a) Pixel level. (b) Feature level. (c) Decision level.

11.1.1 Fusion levels

On the basis of levels of abstraction, image fusion algorithms are categorized into three distinct fusion levels (Fig. 11.1), namely, (a) the pixel level, (b) the feature level, and (c) the decision level. Each level comes with its benefits and drawbacks when dealing with the complications put forward by multimodal information from each source image. Several factors can aid in the comparison of each fusion level, which are based on information loss, computational complexity, sensitivity to noise, and classification accuracy [2].

Pixel level

Also known as low-level image fusion, algorithms in this category work directly with the pixels of the input source images, and therefore amount to maximum information gain according to human perception. In terms of image processing jargon, fusion results produced by pixel-level algorithms have maximum energy. Based on the fusion rule, algorithms in this category are further split into two parts: (a) spatial domain methods and (b) transform domain methods.

Fusion rules in spatial domain methods are developed by smartly manipulating the input image pixels; methods such as weighted pixel averaging [3], min-max [3], and focus measure detection [4] fall in this category. Spatial domain techniques are quite effective when it comes to single sensor source images; however, medical images come from multiple sensors, and hence spatial domain algorithms fail to capture relevant details. Therefore, we transform the source images to some other domain where the relevant information from them can be easily captured. Fusion rules defined in this case fall under the transform domain methods. It must be noted that upon applying the fusion rule, inverse transformation should also be applied to return to the spatial domain in order to view the fused image. Pyramid-based [5], [6] and wavelet-based algorithms [7], [8], [9] fall in this category. Because of their effectiveness and ease of implementation, the majority of the image fusion literature is filled with pixel-level algorithms. Though intuitive to understand and easily implemented, the algorithms in this category are prone to errors such as presence of artifacts, shift-invariance, and blurriness [10].

Feature level

Also known as middle- or intermediate-level algorithms, feature-level algorithms happen to be a bit more complicated than their pixel-level counterparts. In hindsight, feature-level algorithms are divided into two parts. The first part consists of feature extraction from the input source images, and the second part consists of defining a fusion mapping which utilizes those extracted features to give a high-quality fused image. Feature-level algorithms aim at capturing the detailed parts of the source images, for example lines, edges, texture, corners, etc. Algorithms in this category are further divided into three classes, each of which has a different feature extraction procedure, namely, (a) region-based, (b) machine learning-based, and (c) similarity matching-based algorithms.

Region-based algorithms begins with region partitioning and semantic segmentation-based approaches to detect salient features from the input images, and then a fusion rule is devised in such a way that the salient features can be merged appropriately. A majority of the region-based algorithms have incorporated multiscale decompositions such as discrete wavelet transforms [7], contourlet transforms [8], shearlet transforms [9], and many more, most of which will be discussed in Section 11.2.

Machine learning and region-based algorithms are not too different. The only difference is that the fusion rule in the former is defined using a machine learning classifier; additionally, saliency detection and classification can be further improved using machine learning techniques. Fusion using genetic algorithms [11], support vector machines [12], and particle swarm optimization [13] falls in this category.

In similarity matching algorithms, the human visual system is incorporated in the fusion rule, that is, they utilize visual features such as lines, texture, shape, and other structural details of the input source images to design the fusion algorithm. Authors in [14], [15], and [16] have worked on similarity matching algorithms for multimodal and medical image fusion tasks.

Decision level

These are high-level information fusion algorithms, which are less explored in the literature. Feature-level fusion acts as a prerequisite for this level, which is followed by feature classification and building decision maps or indices for the final fusion operation. The fusion operation is carried out based on the best decision, or the decision which has the highest probability of giving better classification in the end. Hence, algorithms at this level are the most accurate of all; however, the downside of these algorithms is that they do not match well with the human visual system, that is, information loss is bound to happen. Decision-level fusion algorithms include voting, Bayes' inference [17,18], fuzzy integrals [19], and many other methods. Table 11.1 inspired from [2] gives a brief summary of the performances for each level discussed above.

Table 11.1 Attribute performance summary.

Attribute	Pixel level	Feature level	Decision level
Information loss	Minimum	Medium	Maximum
Information content	Highest	Medium	Lowest
Method complexity	Easiest	Moderate	Hardest
Classification performance	Worst	Moderate	Best
Noise sensitivity	Highest	Medium	Lowest

As illustrated in Fig. 11.2, the number of publications in the image fusion literature has been rising since the year 2000, reaching about 11000 in 2022; likewise, the medical image fusion literature has also increased since then (data source: https://www.scopus.com/). However, it must be noted that the majority of the publications on image fusion have been in the nonmedical domain, which motivated us to develop and further the research in this field. In Section 11.2, a comprehensive survey on traditional and recent multimodality image fusion techniques is provided, and a majority of them deal with medical images. Based on the strengths and weaknesses of each of the frameworks, in Section 11.3 a novel deep learning-based image fusion architecture is proposed in order to overcome the shortcomings of the recent deep learning-based frameworks.

11.1.2 Preprocessing pipeline

Medical images are acquired from multiple sensors, and it is possible that images obtained from any of the sensors have some kind of glitch or artifact in them. These glitches can be easily detected by a medical expert; however, once the multimodal images are fused, those glitches (if undetected) are also fused, and that would be rather undesirable. Hence, this stage is the most crucial step for any kind of image fusion task. Therefore, source images must be preprocessed to remove noise and other kinds

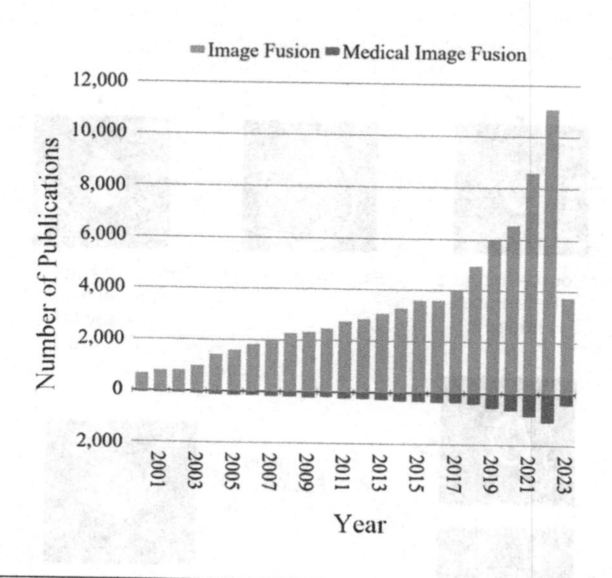

FIGURE 11.2

Yearwise publications in the image fusion literature.

of artifacts such as spatial blurring and nonuniform illumination. For illumination-related issues, image enhancement can also be carried out as a preprocessing step. A detailed survey on methods for noise reduction and image enhancement techniques is given in [20,21].

Another prerequisite for any kind of image fusion is *image registration*, in which source images are geometrically aligned with respect to size, orientation, and location of the reference source image. Fig. 11.3 illustrates image registration as a preprocessing step for image fusion. A large body of theory and many methods which deal with image registration are available in the image processing and computer vision literature, which is a separate field of research in its own right. A compresensive survey on traditional as well as modern deep learning-based image registration techniques can be found in [22,23].

11.2 Literature survey and state-of-the-art

Image fusion algorithms have been present for quite some time now; hence, it is imperative to first review the traditional approaches and then talk about the state-of-the-art deep learning-based methods in the field. Hence, this section is divided into two parts. The first is about nondeep learning-based techniques, and the second will be on deep learning-based advances for medical images.

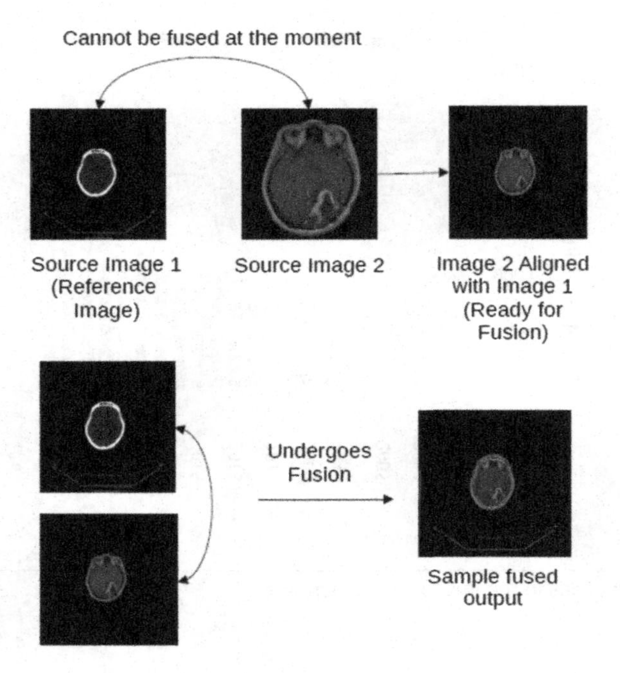

FIGURE 11.3

Image registration as a preprocessing step for CT-MRI fusion using pixel averaging criteria.

11.2.1 Traditional techniques

When it comes to traditional state-of-the-art techniques for medical image fusion, the majority of the techniques come from multiscale decomposition (MSD) methods, morphological methods, and fuzzy logic-based methods. Multiscale decomposition techniques are based on transforms such as pyramids [24], wavelets [25], contourlets [26], curvelets [27], and framelets [28]. In pyramid-based techniques for image fusion, the decomposition process involves creating a set of bandpass and low-pass versions of the original image by sequentially filtering and downsampling it, resulting in a pyramid-like structure.

The authors of [24] have incorporated a Laplacian filter for decomposition which uses the discrete cosine transform for data compression, and through an averaging fusion rule, the authors were able to fuse CT and MRI images. The authors of [5] used gradient pyramids incorporating Gaussian filters. Similarly, by varying the filters for decomposition, different pyramid-based fusion techniques can be developed, for example morphological pyramids [6], ratio pyramids [29], steerable pyramids [30], and many more. In [6], a nonlinear morphological approach is considered for decomposition, which shows promise when compared with linear approaches. The main issue with most of the pyramid-based schemes is that it suffers from blocking effects and edge distortion, and to overcome these problems variants of wavelet-based techniques were introduced. The authors of [25] introduced the gradient-based discrete wavelet transform to fuse MRI-T1, MRI-T2, and FLAIR images by incorporating two separate fusion rules which involve a max and an averaging operation, respectively. The authors of [31] propose a sparse representation-based method for medical image fusion in the tetrolum domain which is a new adaptive version of the Haar wavelet. The proposed scheme was able to preserve the color and contrast information, but had also introduced artifacts such as black dots. The authors of [32] propose a medical image fusion method which incorporates principal component analysis (PCA) and also use the intensity hue saturation color model to retain the color information in order to fuse MRIs and PET images. The proposed method is able to retain more spatial characteristics with no color and spatial distortion; however, the method is not robust enough to be tried on other types of modalities. In [33], the authors propose a new method for image fusion that uses the standard deviation and density function of the shift-invariant shearlet transform (SIST). The method is able to fuse images from different modalities, such as MRI, PET, SPECT, and CT, and it is able to capture both functional and spatial information well. The nonsubsampled contourlet transform (NSCT) domain is another quite popular domain transform technique in the medical image fusion literature. In [34], the authors propose a method for image denoising in the NSCT domain. The low-frequency subband coefficient is obtained by taking the square of the maximum entropy of the coefficients within a local window. The high-frequency subband coefficient is obtained using the maximum weighted sum modified Laplacian. The method was evaluated quantitatively and was found to outperform several existing methods in terms of noise reduction and contrast preservation. The authors of [26] introduce a fusion rule for CT and MRI brain images by exploiting the major properties of the NSCT by using maximum and average masking fusion rules that were devised for the approximate and directional coefficients, respectively. In [35], the authors propose a method for fusing medical images using the curvelet transform. The method first converts each source image into curvelet coefficients. These coefficients are then fused using a PCA fusion rule. Finally, the inverse curvelet transform is applied to produce the final fused image. The authors of [36] introduced a fuzzy transform-based method. The process begins by dividing the images into equal-sized blocks, which are then transformed into subblocks of varying sizes using the fuzzy transform. The maximum-entropy fusion rule is then applied to the subblocks, followed by the inverse fuzzy transform on the fused subblocks. The method's performance was evaluated through both subjective and objective means. Table 11.2 briefly summarizes the traditional nondeep learning-based methods discussed above.

11.2.2 Deep learning-based techniques

As Table 11.2 illustrates, traditional medical image fusion algorithms have major weaknesses, including a lack of robustness and the generation of artifacts.

This raises questions about the effectiveness of the feature extraction and feature fusion processes. To address these issues, researchers have turned to deep learning-based image fusion techniques. Deep learning-based techniques for image fusion

Table 11.2 Summary for traditional fusion techniques.

Work	Fusion method	Modalities used	Strengths	Weaknesses
Laplacian-based fusion using discrete cosine transform [24]	DCT & Laplacian pyramids	CT & MRI	Exhibits higher edge strength & good contrast with respect to human visual systems (HVS)	Introduces blocky artifacts
Gradient pyramids incorporating a Gaussian filter [5]	Fusion features calculated using local structure tensor in the gradient domain	CT & MRI	Shows superior results when compared with other pyramid methods	Unable to obtain a pristine noiseless image
Gradient-based discrete wavelet transform [25]	DWT & gradient pyramids	MRI-T1, MRI-T2, & FLAIR	Improved entropy, mutual information & fusion symmetry performance	Works well only for MRI brain scans, hence not robust
Sparse representation in Tetrolum domain [31]	Tetrolum transform	MRI, PET, & SPECT	Ensures the preservation of important details & minimal contrast loss	Black dotted artifacts upon fusion
Principal component analysis & IHS-based fusion [32]	PCA & IHS	MRI & PET	Retains more essential data & better spatial characteristics	Narrow application scope, with notable color inaccuracies
Fusion using shift-invariant shearlet transform [33]	Shearlet transform	MRI, PET, SPECT, & CT	Obtains both functional & spatial information	Does not include analysis of other multiscale geometric tools
Nonsubsampled contourlet transform [34], [26]	NSCT	MRI & CT	Outperforms multiple existing techniques while also producing strong contrast	Not robust, cannot be generalized to anatomical & functional medical images
Fusion using curvelet transform [35]	Curvelet & PCA	MRI & CT	Effectively captures the curves & edges of images in its representation	Requires more complex computation & parameter adjustment
Fusion using fuzzy transform [36]	Fuzzy transform	MRI-T1, MRI-T2, CT, & PET	Generates a fused image with improved contrast that conveys more information	More time consuming

are classified into four categories: autoencoder-based, convolutional neural network (CNN)-based, generative adversarial network (GAN)-based, and transformer-based architectures. Table 11.3 gives a brief overview of the image fusion process using the first three architectures mentioned.

Table 11.3 Different deep learning architectures for image fusion.

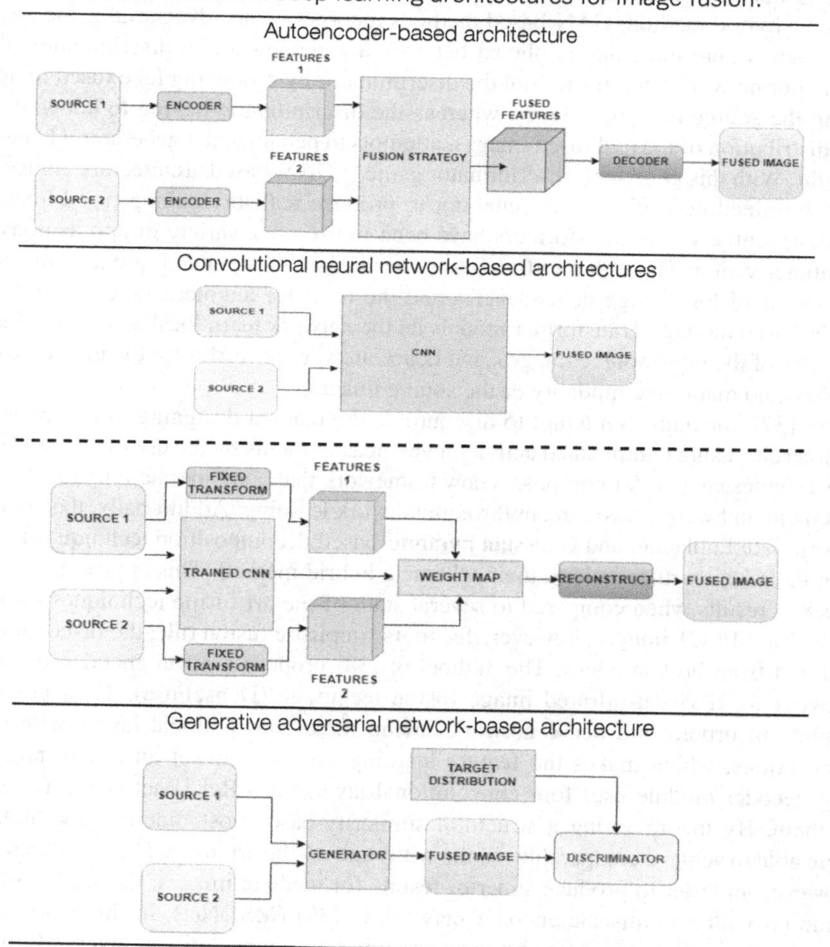

The autoencoder method usually pretrains an autoencoder on a different dataset in order to learn the most precise feature extraction and reconstruction process in order to generate a fused representation in a supervised or unsupervised setting; however, autoencoder-based methods face issues with edge distortion because they happen to lose important information while learning for latent representations of the source images. CNN-based architectures are quite flexible. As seen in Table 11.3, CNN-based methods can be applied in two ways. One way is to utilize just the convolutional layers to extract the useful features and then train another set of convolutional layers for the fusion process. A different approach suggests to use CNNs with a transform domain technique which can in turn bring in more flexibility towards the architecture and hence extract even deeper features without having to go deep and then use another set of convolutional layers for the reconstruction process; this technique is also called a hybrid method. GAN-based methods incorporate an adversarial game-based approach, where the game is played between a generator and a discriminator. The generator network attempts to fool the discriminator by producing fake fused images using the source images as input, whereas the discriminator, having to know about the distribution of the real fused images, attempts to penaltize the generator whenever caught. With this generator–discriminator game, a GAN-based architecture comes up with a trained network for the generator to produce realistic looking fused images. Since recently, vision transformers have been in use for a variety of applications in computer vision. Through a self-attention mechanism these models are able to solve the issues of long-range dependencies and the need for augmentation put forth by CNN-based models. Transformer models let the network learn local as well as global features of the input source images, and hence are very powerful for the image fusion process, no matter the modality of the source image.

In [37], the authors attempt to discontinue the manual designing of complicated fusion rules using complicated activity level measurements by the use of CNNs. They use a siamese network to propose a new framework that combines activity level measurement and weight assignment through network learning. Additionally, the authors incorporate Laplacian and Gaussian pyramid-based decomposition techniques to design their fusion rule, making their scheme a hybrid method. This approach yielded superior results when compared to several state-of-the-art fusion techniques for CT, MRI, and SPECT images; however, due to its simplistic fusion rule, the fused images suffered from broken edges. The authors of [38] propose a novel encoder–decoder network for a visual-infrared image fusion technique (DenseFuse). Their encoder module incorporates a dense network having three convolutional layers with skip connections, which makes the feature learning process quicker and more precise. The decoder module uses four convolutional layers with ReLU activation for each of them. By incorporating a structural similarity-based cost function, the authors were able to achieve comparable results with state-of-the art image fusion techniques. However, in order to produce superior results for medical images, the same authors came up with a multiscale encoder network in [39] (MSDNet). In this work, they improve the encoding process by incorporating three convolutional layers of different sizes simultaneously in order to learn more complex features. With this change the authors were able to design a fusion scheme for medical, visual-infrared, and multifocus images. However, the results were just slightly improved in comparison to their previous method, and hence the impact was not significant. The authors in [40] present a novel end-to-end image fusion framework (IFCNN) that utilizes the power of ResNet101 [41] pretrained over multifocus image datasets available online. The proposed framework is a general-purpose method that can handle different types of images. The authors demonstrated its ability to elegantly fuse multifocus, multimodal, and multiexposure images and overshadow a majority of the CNN-, autoencoder-, and GAN-based methods; however, in spite of its overwhelming performance when it came to multimodal medical images (CT-MRI), the fused results were rather dull and the method was not able to properly capture the texture information of the source images. On the same grounds as [40], the authors of [42] propose a unified unsupervised end-to-end fusion technique for multifocus, multimodal, and multiexposure images. The proposed technique utilizes the power of the VGG16 [43] architecture for its feature extraction process. Overall, the method produced decent results when compared with traditional fusion schemes but had issues majorly with intensity and contrast in the fused output. In [44], the authors propose EMFusion, an enhanced unsupervised image fusion framework designed specifically for medical images. The architecture of this framework ensures the enhancement by adding constraints to both the surface and deeper levels, thereby preserving relevant features of the source images. Another encoder–decoder-based fusion framework (MSENet) was proposed in [45] for medical images. The framework uses a multiscale feature extraction process using CNNs. This technique outperformed traditional state-of-the-art techniques; however, an extensive experiment with deep learning-based techniques was not provided. Using U-Net [46] as the backbone of the feature extraction process, the authors of [47] propose a self-supervised image fusion scheme for visible and infrared images. Their proposed architecture is able to capture relevant feature information uniformly; however, the technique was overshadowed by GAN- and autoencoder-based methods in terms of entropy and mutual information, respectively.

FusionGAN [48] was one of the first GAN-based image fusion frameworks designed to fuse multimodal images. Inspired from the DCGAN [49] architecture, its generator uses a CNN which takes infrared and visible images as its input and produces a fused result having the discriminator to penalize the fused output if its distribution is different from the true distribution. Similarly, the authors of [50] and [51] came up with their own GAN-based architectures for multimodality images and produced state-of-the-art results. However, the authors of [52] came up with a dual stream attention-based generator–discriminator architecture (DSAGAN) which outperformed most of the other generative models for medical image fusion. DSAGAN uses three CNN-based attention modules for its generator network in order to produce a high-quality fused result; having said that, a six-layer deep CNN-based discriminator is used to penalize the image if found fake. DSAGAN was able to outperform most of the traditional and GAN-based fusion schemes in terms of entropy and blind image quality metrics.

Though not extensively explored in the literature, transformer-based image fusion architectures are rising after their overwhelming performance in terms of image classification and object detection [53]. The authors of [54] propose a bimodal transformer-based visual-infrared image fusion framework with a complex three-level architecture, where in the first level it uses multiscale feature extraction using dense networks, in the second level it brings in two separate transformers which take in the dense features from visual and infrared inputs, respectively, and in the third level it prepares them for the fusion stage. Level three utilizes CNNs and fast Fourier transforms for the fusion to take place. With such a refined architecture, the framework was able to outperform many traditional and deep learning-based state-of-the-art techniques. Knowing this, transformer-based architectures [55], [56], [57] are quite heavy in terms of parameters and data required for training. Table 11.4 provides a brief summary of some of the methods discussed above.

Table 11.4 Summary of deep learning-based image fusion techniques.

Architecture/ method	Modalities	Strengths	Shortcomings
DenseFuse [38]	VI, IR	Results comparable to state-of-the-art techniques	Poor performance for medical images
MSDNet [39]	VI, IR, CT, MRI, SPECT	Improvement upon DenseFuse	Results not compared with enough deep learning-based methods
IFCNN [40]	VI, IR, CT, MRI, SPECT	Is able to capture textural features well	Unable to capture contrast details
EMFusion [44], MSENet [45]	CT, MRI, PET, SPECT	Unsupervised multiscale feature extraction	Extensive experiments not provided, poor contrast
FusionGAN [48], DDcGAN [50], DSAGAN [52]	VI, IR, CT, MRI, PET, SPECT	Outperforms most CNN-based and traditional methods	Sensitive to hyperparameters, cannot generalize well
TransFuse [56], THFuse [57]	CT, MRI, VI, IR	Maintains long-range dependencies and can capture local and global features well	Requires a lot of training data, time-inefficient

Fig. 11.4 showcases a pie chart illustrating the percentage difference of the publications on recent and traditional medical image fusion techniques since the year 2000 until 2023, whereas the yearwise difference is illustrated in Fig. 11.5 (data source: https://www.scopus.com/). Based on all the strengths and weaknesses of each of the

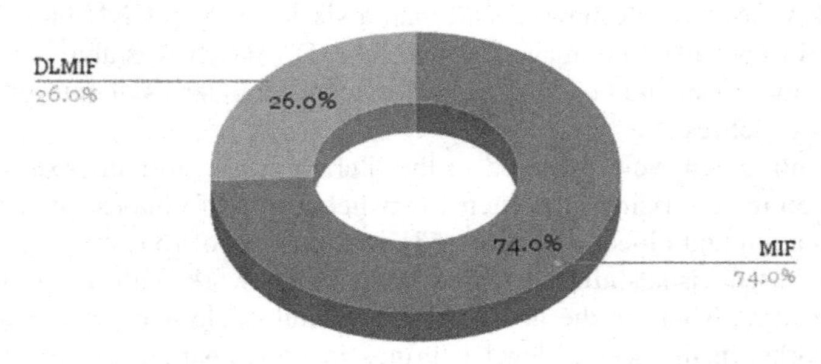

FIGURE 11.4

Percentages of traditional and recent (deep learning-based) medical image fusion techniques.

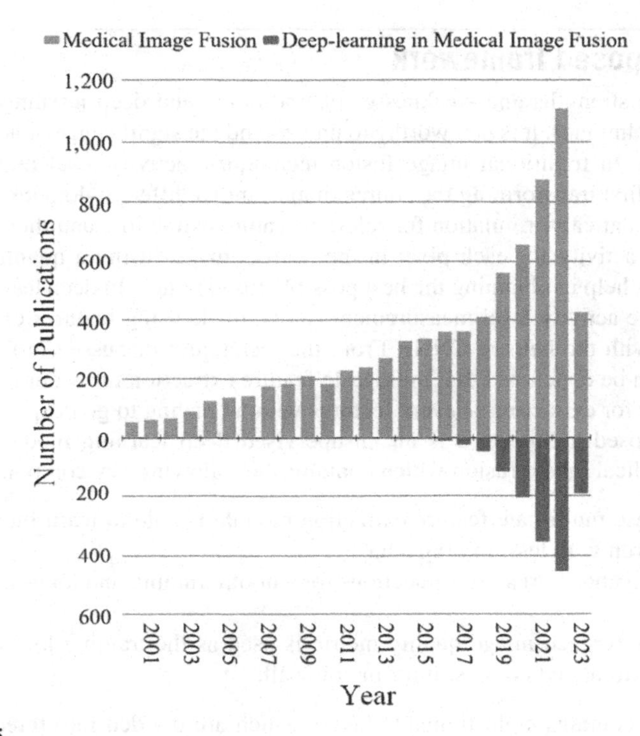

FIGURE 11.5

Yearwise publications in the medical image fusion literature.

deep learning-based frameworks for medical image fusion, the proposed architecture in this chapter aims to unify the strengths of most of the algorithms discussed so far and to mitigate the weaknesses put forth by each of the methods. Notable weaknesses of most of the methods are the following:

- Encoder–decoder-based architectures: Most results suffered from broken edges or were unable to capture the true intensity and contrast information from the source images.
- Purely CNN-based architectures: Having to train the network over databases other than medical images makes the network learn features which are not useful, and hence results in a fused outcome which is sometimes different from the original source images and leads to ambiguity.
- GAN-based architectures: These networks are sensitive to changes in hyperparameters. They also require abundant data for training, which becomes a challenge when just medical data are concerned.
- Transformer-based architectures: Though not explored in detail, their architectures are rather complex and the results obtained so far are comparable with encoder–decoder- and/or CNN-based architectures.

11.3 Proposed framework

Based on the strengths and weaknesses of traditional and deep learning-based image fusion techniques, it is noteworthy to understand the significance of activity level measurement. In traditional image fusion techniques, activity level measurements are taken by first transforming the source images into a different domain, coming up with a mathematical formulation for relevant feature extraction, and then measuring the levels of activity for each pixel in the source images using a quantifier. These activity levels help in obtaining the best possible fused image. In deep learning-based techniques the activity level measurement can be made using features extracted automatically with the help of CNNs. From the techniques discussed in the previous section, it can be concluded that multiscale feature extraction turns out to be the optimal method for extracting relevant features without having to go deep.

The proposed architecture is an unsupervised deep learning model for multimodality medical image fusion which contains the following key components:

- The siamese multiscale feature extraction module is able to learn more accurate features even with less training data.
- It uses long and short skip connections for smooth training and long-range dependencies.
- A partial reference image quality metric is used as the training loss function to capture textural and contrast information well.

The network contains eight trainable layers which are divided into three modules: (a) siamese multiscale feature extraction, (b) feature fusion, and (c) reconstruction. In order to solve the issues put forth by various deep learning-based techniques, it is worth mentioning that the architecture aims to incorporate the following things:

- long-range dependencies without having a complicated architecture,
- little dependence on hyperparameters,
- good generalization,
- the ability to learn significant features with fewer layers.

11.3.1 Feature extraction

The feature extraction module of the proposed architecture in Fig. 11.6 is inspired by the multiscale decomposition process used in the traditional medical image fusion algorithms. The module is comprised of a siamese CNN architecture having two identical subnetworks with shared weights that takes as an input two multimodality images. Instead of relying on a deeper architecture with numerous convolutional layers, the module relies heavily on learning multiscale feature representations, which in turn helps in capturing low- and high-level features simultaneously [58]. To capture these features, kernels of size 5×5 and 7×7 are used, respectively. All convolutional layers are followed by batch normalization with ReLU activation in the end. In order to initialize and remap the feature extraction process, the multiscale layers

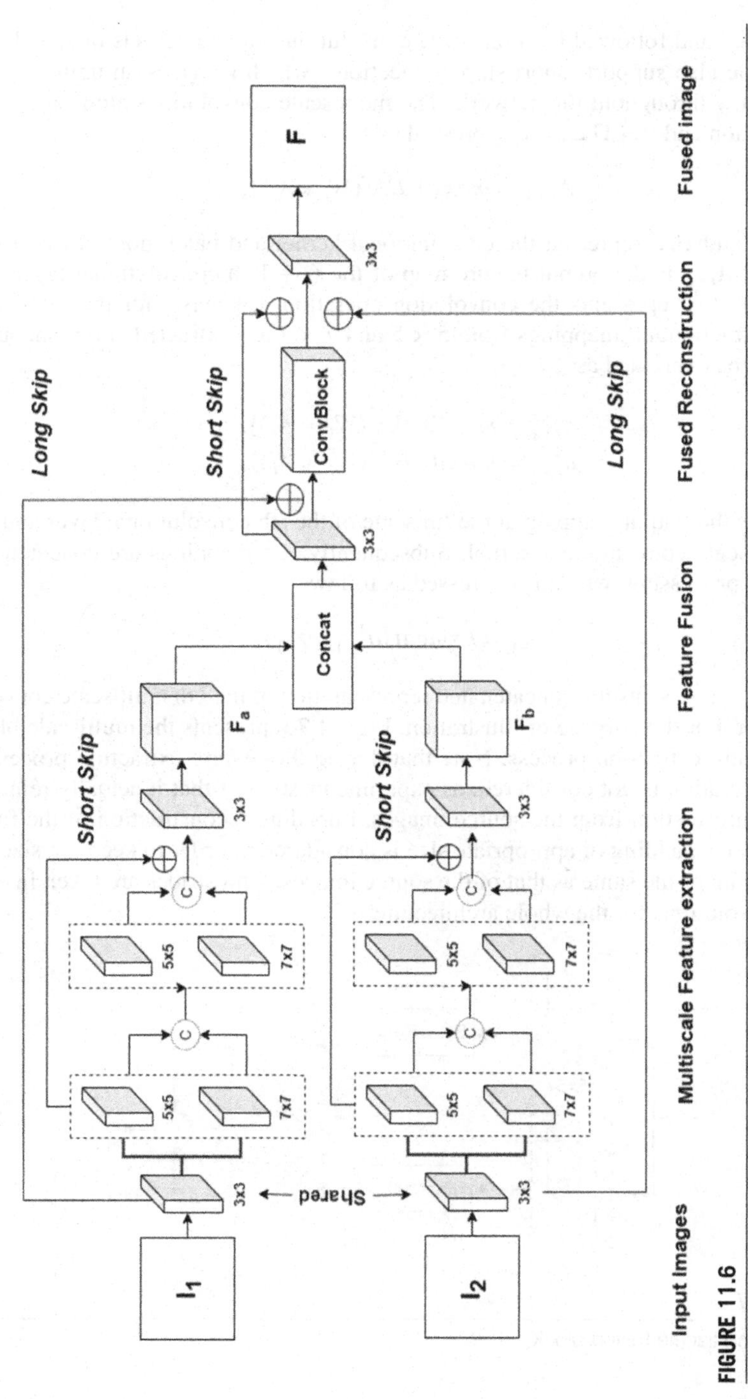

FIGURE 11.6

Proposed fusion architecture.

are preceded and followed by monoscale convolutions having kernels of size 3×3. The module also supports short skip connections, which provides an uninterrupted gradient flow throughout the network. The monoscale convolutions alongside batch normalization and ReLU can be expressed as

$$A_{i+1} = max\{0, \mathcal{BN}(W_i * A_i)\},$$

where W_i and $\mathcal{BN}$ represent the convolutional kernel and batch normalization, respectively, A_{i+1} is the output feature map of the $(i+1)$th convolutional layer, and the symbol "$*$" represents the convolution operation. Likewise, for the multiscale feature extraction, all mappings from 5×5 and 7×7 are extracted simultaneously, which can be expressed as

$$a^1_{i+1} = max\{0, \mathcal{BN}(W^1_i * A_i)\},$$
$$a^2_{i+1} = max\{0, \mathcal{BN}(W^2_i * A_i)\},$$

where a^l_i is the output mapping at the lth scale of the ith convolutional layer and W^l_i is the lth scale convolutional kernel. Subsequently, the mappings are concatenated for further processing, which is expressed as follows:

$$A^k_{i+1} = Concat(a^1_{i+1}, a^2_{i+1}),$$

where A^k_{i+1} represents the concatenated representation of the kth multiscale convolutional layer. For the purpose of illustration, Fig. 11.7 represents the multiscale block of the feature extraction process. Note that during the feature extraction process, a pooling operation is not considered, as experiments suggest that it actually removes essential information from the source images, impeding reconstruction of the fused image. Also, a padding of appropriate size is considered in order to keep the sizes of the feature maps the same as that of the source images. Unit strides are taken in each of the convolutions for the whole architecture.

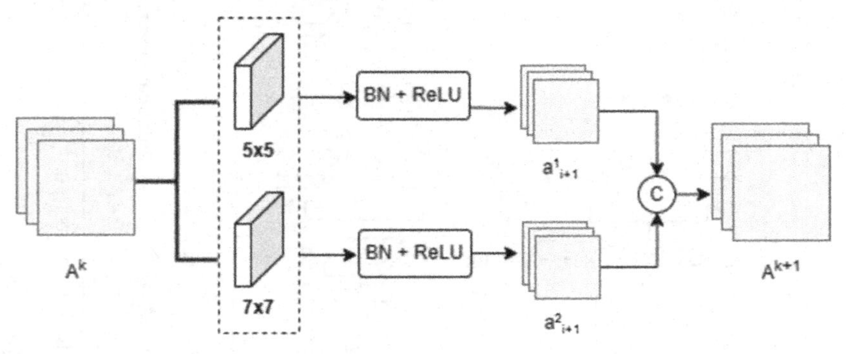

FIGURE 11.7

Proposed multiscale fusion block.

11.3.2 Feature fusion and reconstruction

The feature fusion process is fairly straightforward, similar to the concatenation step during multiscale extraction. The $Concat(F_a, F_b)$ operation completes the feature fusion process, where F_a and F_b are the final feature representations of the multiscale extraction process as illustrated in Fig. 11.6.

The reconstruction phase of the proposed architecture is comprised of a set of convolution operations having short and long skip connections in order to output the final fused image. It must be noted that deconvolutions or transposed convolutions are not used for this step since an appropriate amount of padding was performed in each step of the feature extraction phase, and hence applying transposed convolutions actually inhibits the model to reconstruct a well-defined fused image. The "ConvBlock" in the architecture consists of a sequence of two back-to-back convolution operations of size 3 × 3 each, which are preceded and followed by 3 × 3 kernels, having long and short skip connections in between to smoothen the reconstruction process. The long skip connections are the key to making the proposed model robust, as it takes care of the essential features brought forth by each of the imaging modalities.

11.3.3 Implementation details

The multiscale structural similarity index (MS-SSIM) is a modified version of the structural similarity index (SSIM) [59] that takes into account the structural information of an image at multiple scales. SSIM is a widely used image similarity metric that measures the similarity between two images by comparing their luminance, contrast, and structural information. While SSIM is effective in measuring the similarity between two images at a single scale, it can be sensitive to small changes in the images that may not be perceptually significant. MS-SSIM was developed to address this limitation by considering the structural information of the images at multiple scales, which allows it to better capture the perceived similarity between the images. In a patch P, SSIM for the center pixel $\tilde{p}$ can be expressed as

$$\mathrm{SSIM}(\tilde{p}) = \frac{(2\mu_x\mu_y + c_1)(2\sigma_{xy} + c_2)}{(\mu_x^2 + \mu_y^2 + c_1)(\sigma_x^2 + \sigma_y^2 c_2)} = l(\tilde{p}).cs(\tilde{p}),$$

where c_1 and c_2 are small constants that are used to stabilize the division and prevent the denominators from becoming too small and x and y are sliding windows in the reference image and source images, respectively, with mean μ_x and variance σ_x^2, and the covariance of x and y is denoted as σ_{xy}.

Given a dyadic pyramid of K levels, MS-SSIM can be defined as

$$\text{MS-SSIM}(\tilde{p}) = l_K^{\alpha}(\tilde{p}).\prod_{j=1}^{K} cs_j^{\beta_j}(\tilde{p}).$$

We explored the use of MS-SSIM as a loss function for the task of multimodal medical image fusion. Loss functions are an essential component of machine learning models, as they define the measure of how well the model is able to learn from the training data. The choice of loss function can have a significant impact on the performance and convergence of the model. For patch P and its center pixel $\tilde{p}$, the MS-SSIM loss is evaluated as

$$\mathcal{L}_{\text{MS-SSIM}}(P) = 1 - \text{MS-SSIM}(\tilde{p}).$$

Further details regarding MS-SSIM can be found in [60]. To evaluate the performance of the model, the researchers used MS-SSIM as the primary loss function, along with mean squared error loss, which can be termed as pixel loss, given by

$$\mathcal{L}_{\text{pixel}}(F) = \sum_{i=1}^{n} \frac{1}{n} \text{MSE}(I_i, F),$$

where I_i is the input source image at index i, n is the total number of images, and F is the fused image. The researchers found that using MS-SSIM as the primary loss function resulted in better fusion results. In comparison, using the pixel loss alone as the loss function resulted in intensity distortions and introduced artifacts as well. The final loss function is the sum of both of these losses:

$$\mathcal{L} = \lambda \mathcal{L}_{\text{MS-SSIM}}(F) + \mathcal{L}_{\text{pixel}}(F),$$

where the value of λ is chosen to be 500 for faster convergence. Overall, the findings suggest that MS-SSIM is a promising loss function for multimodal medical image fusion and warrants further investigation.

The major limitation for multimodal medical image fusion is the lack of training data and ground truth availability. The proposed architecture is therefore designed in such a way that even if it is trained on fewer images, the model is able to capture relevant features to perform fusion tasks well, which was possible due to the multiscale feature extraction and the long-short residual connections [61]. Data used for training and testing are openly available on Harvard Brain Atlas [62]. The model was trained on 180 image pairs of CT and MRI brain scans. Each image was resized to 200 × 200 for training. The one-cycle learning rate scheduler [63] is incorporated with a maximum learning rate of 0.01. The model was trained using the Adam optimizer [64] for 100 epochs with a batch size of 40. Implementation was done in the Google Colaboratory environment and the model was built using the pytorch framework [65].

11.4 Experimental results and discussion

11.4.1 Evaluation setup

The performance of image fusion techniques is evaluated using various criteria such as entropy, mutual information, and fusion symmetry. The choice of criteria depends on the fusion task and the type of fused image. Based on the mentioned criteria, discussed below are a few image quality metrics that are used for comparative analysis.

(a) Normalized mutual information (Q_{MI}): Mutual information (MI) is a statistical measure that quantifies the dependence between two variables. It is often used to measure the amount of information shared by two images. The mathematical definition of MI for two discrete random variables X and Y is as follows:

$$\text{MI}(X, Y) = \sum_{x \in X} \sum_{y \in Y} p(x, y) \log_2 \frac{p(x, y)}{p(x)p(y)}.$$

Note that $p(x, y)$ is the joint probability distribution of X and Y, with $p(x)$ and $p(y)$ being the marginal distributions of X and Y, respectively. Suppose the source images are given by I_1 and I_2, with the fused image as F. Then the normalized mutual information [66] is given by

$$Q_{\text{MI}} = 2\{\frac{\text{MI}(I_1, F)}{E(I_1) + E(F)} + \frac{\text{MI}(I_2, F)}{E(I_2) + E(F)}\},$$

where $E(.)$ is the entropy. The range of Q_{MI} is from 0 to 1, with 0 indicating poor fusion quality and 1 meaning perfect image fusion.

(b) Feature mutual information (Q_{FMI}): Feature mutual information is a metric used to measure the similarity between the fused image and a reference image. It is used to evaluate the quality of image fusion and ensures that important features of the original images are preserved in the fused image. Mathematical details regarding the metric can be found in [67]. Note that $Q_{\text{FMI}} \in [0, 1]$, with a higher score indicating better fusion results.

(c) Edge-based measure ($Q_{\text{AB/F}}$): This measures the total information transference of the source images to the fused image using edge information. It is defined as

$$Q_{\text{AB/F}} = \frac{\sum_i \sum_j [w^x_{i,j} Q_{\text{AF}} + w^y_{i,j} Q_{\text{BF}}]}{\sum_i \sum_j [w^x_{i,j} + w^y_{i,j}]},$$

where A, B, and F are the source and fused images, respectively. Note that Q_{AF} and Q_{BF} are edge-preserving values which are weighted by w^x and w^y. The range of $Q_{\text{AB/F}}$ is [0, 1], where values close to 0 indicate that little information is transferred and values close to 1 indicate that much information is transferred. Additional details about the metric can be obtained from [68].

(d) Structural similarity-based metric (Q_S): An alternative method for using SSIM in image fusion evaluation is presented in [69], which is based on the traditional definition of SSIM and is defined as

$$Q_S = \begin{cases} \lambda(w)\text{SSIM}(\text{A,F}|w) + (1 - \lambda(w))\text{SSIM}(\text{B,F}|w), \\ \quad \text{if SSIM}(\text{A,B}|w) \geq 0.75, \\ \\ \max[\text{SSIM}(\text{A,F}|w), \text{SSIM}(\text{B,F}|w)], \\ \quad \text{if SSIM}(\text{A,F}|w) < 0.75, \end{cases}$$

where w is a window of odd size that scans from top left to bottom right with unit steps and $\lambda(w)$ is a weight obtained from the local image features. A and B are the source images and F is the fused image. Similar to other metrics, $Q_S \in [0, 1]$, with 0 indicating poor fusion quality and 1 indicating perfect image fusion.

The indices/metrics mentioned above use the input and fused images in order to evaluate the performance of the fusion algorithm, and hence these metrics can be termed as partial reference image quality metrics because the ground truth is unavailable. Moreover, the literature is further extended to no-reference image quality metrics as well. These metrics incorporates the knowledge of human visual systems and use statistical methods to mathematically quantify the perceptual quality of the image. One such metric used for the analysis is Blind/Referenceless Image Spatial Quality Evaluator (BRISQUE) [70]. The goal of employing blind image quality metrics is to predict the naturalness of the fused images without using input or ground truth images as reference. The metric is designed to produce a positive numerical result, with a lower score indicating better visual quality. This allows for the comparison of the proposed result with other techniques.

11.4.2 Results and discussion

To reveal the strengths of the proposed medical image fusion framework, a variety of experiments were conducted using the evaluation metrics mentioned above, and the results obtained were compared to five state-of-the-art image fusion techniques, three of which are nondeep learning-based, which use the stationary wavelet transform [71], Laplacian pyramids [72] and the nonsubsampled contourlet transform [73], with the remaining two being deep learning-based techniques, one relying heavily on the VGG19 [74] architecture and the other being the IFCNN [40] method. The superiority of the results obtained is discussed both subjectively and objectively in the following sections. For the purpose of evaluation, CT and MRI brain scans are chosen. The image pairs were registered beforehand and are freely available online [62]. Table 11.5 represents the set of image pairs used for objective experimentation.

11.4.2.1 Qualitative evaluation

Evaluating the fused results through a qualitative approach is of paramount importance, as it ensures that the images meet not only the technical requirements but also the standards of the human visual system. The proposed architecture resulted in visually pleasing fused images upon training. As shown in Tables 11.6 and 11.7, the fused images not only retained texture details but also maintained contrast information from the input source images, which makes sure that if there exists an anomaly (tumor/lesion) in any of the source images, then it will be captured in the fused image as well. On the contrary, results produced by LRD- and NSCT-based approaches are overexposed, while SWT and IFCNN fused images have a rather dull appearance. Results of the VGG19-based methods are also visually appealing, and these methods appear to have captured most of the relevant information.

Table 11.5 Experimental dataset: CT (left)-MRI (right) image pairs used for comparison.

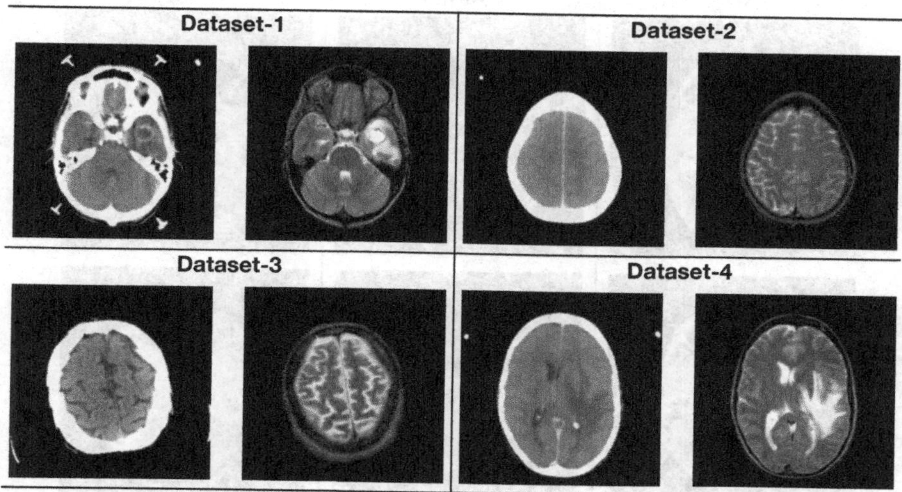

11.4.2.2 Quantitative evaluation

As can be seen from Table 11.8, the results obtained using the proposed architecture show dominance in terms of the blind image quality metric ($\mathcal{Q}_{\mathrm{BRISQUE}}$) and information transference ($\mathcal{Q}_{\mathrm{AB/F}}$), which further clarifies the claims of generalizability and long-range dependency. Additionally, it must also be noted that NSCT- and VGG19-based methods dominate in terms of mutual information ($\mathcal{Q}_{\mathrm{MI}}$, $\mathcal{Q}_{\mathrm{FMI}}$) and structural similarity ($\mathcal{Q}_{\mathrm{S}}$); however, after the VGG19-based model, the proposed results show overall better performance than the other four methods. Given that IFCNN had already outperformed several of the GAN- and CNN-based models, the proposed architecture gives very good results.

11.5 Conclusion

Image fusion is a process that combines multiple images with similar content but different information to create a single high-quality image. In this chapter, we provide an extensive review of various traditional and recent deep learning-based state-of-the-art techniques for multimodality image fusion. It was pointed out that deep learning-based techniques have outgrown a majority of the traditional image fusion methods. It was also pointed out that when it came to medical image fusion, deep learning-based techniques suffered from issues such as long-range dependency, sensitivity to hyperparameters, poor generalizability, and learning of insignificant features. All these issues were addressed by proposing a novel unsupervised deep learning architecture which incorporates a siamese multiscale feature extraction module, a feature fusion

Table 11.6 Results obtained using the proposed framework.

CT	MRI	Fused image

Table 11.7 Subjective quality evaluation.

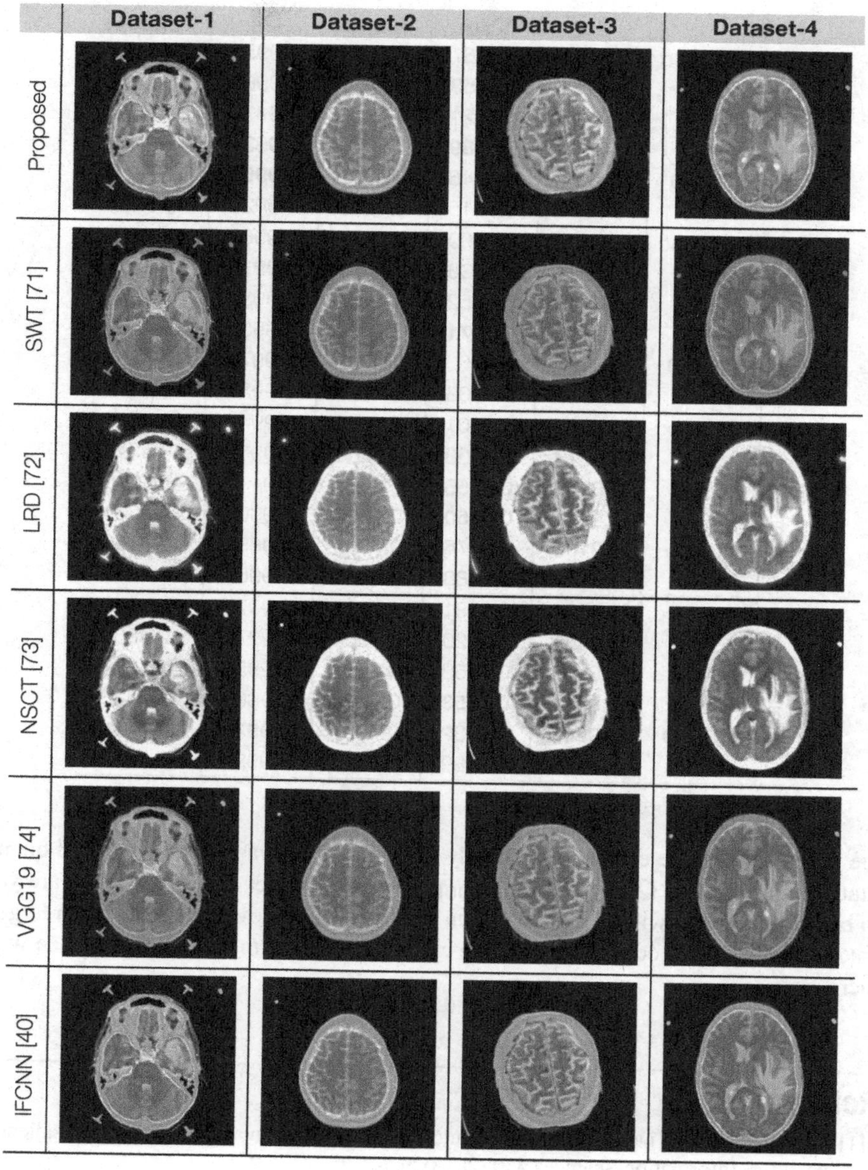

	Dataset-1	Dataset-2	Dataset-3	Dataset-4
Proposed				
SWT [71]				
LRD [72]				
NSCT [73]				
VGG19 [74]				
IFCNN [40]				

module, and a fused reconstruction module. The proposed framework relies heavily on CNNs in order to perform medical image fusion. Upon comparing the proposed framework with traditional and modern image fusion methods, it was concluded that

Table 11.8 Objective quality evaluation.

		$\mathcal{Q}_{MI}$	$\mathcal{Q}_{FMI}$	$\mathcal{Q}_{AB/F}$	$\mathcal{Q}_{S}$	$\mathcal{Q}_{BRISQUE}$
Dataset-1	Proposed	0.8118	0.8367	**0.7190**	0.9957	**34.5174**
	SWT	0.8157	0.8453	0.6780	0.9961	40.6390
	LRD	0.7860	0.8521	0.5880	0.9930	35.5579
	NSCT	0.8375	**0.8639**	0.6670	0.9936	38.8441
	VGG19	**0.8773**	0.8546	0.7040	**0.9962**	41.2680
	IFCNN	0.6385	0.7715	0.5519	0.9925	42.5134
Dataset-2	Proposed	0.9428	0.8909	**0.8149**	0.9963	**47.1250**
	SWT	0.9669	0.8933	0.8000	**0.9970**	50.7880
	LRD	0.9336	0.8952	0.7101	0.9933	48.6970
	NSCT	0.9725	**0.8971**	0.7552	0.9935	49.4562
	VGG19	**0.9990**	0.8960	0.8089	0.9970	49.5364
	IFCNN	0.7159	0.8238	0.6211	0.9924	50.9619
Dataset-3	Proposed	0.8730	0.8601	**0.7601**	0.9947	**43.4630**
	SWT	0.8710	0.8738	0.7265	0.9955	46.6320
	LRD	0.8840	0.8756	0.6398	0.9901	44.4210
	NSCT	0.9094	**0.8850**	0.6776	0.9908	44.5331
	VGG19	**0.9109**	0.8775	0.7378	**0.9956**	46.3750
	IFCNN	0.8999	0.8688	0.7668	0.9953	47.5811
Dataset-4	Proposed	0.8424	0.8781	**0.7131**	0.9970	**36.7379**
	SWT	0.8658	0.8820	0.7064	0.9973	42.4839
	LRD	0.8203	0.8852	0.6313	0.9955	41.7659
	NSCT	0.8644	**0.8953**	0.6903	0.9960	41.8448
	VGG19	**0.9245**	0.8882	0.7061	**0.9974**	43.0376
	IFCNN	0.6101	0.7889	0.5586	0.9921	43.8322

the proposed scheme excels qualitatively in terms of contrast and texture and quantitatively in terms of $\mathcal{Q}_{AB/F}$ and $\mathcal{Q}_{BRISQUE}$. The results over other modalities are yet to be evaluated and will be considered in the future. Moreover, tests on much larger datasets will also be considered; we are confident that the model's performance will increase.

References

[1] J. Du, W. Li, B. Xiao, Fusion of anatomical and functional images using parallel saliency features, Information Sciences 430 (2018) 567–576.

[2] H. Hermessi, O. Mourali, E. Zagrouba, Multimodal medical image fusion review: theoretical background and recent advances, Signal Processing 183 (2021) 108036.

[3] Gang Xiao, Durga Prasad Bavirisetti, Gang Liu, Xingchen Zhang, Image Fusion, Springer, Singapore, ISBN 978-981-15-4867-3, 2020.

[4] A. Kahol, G. Bhatnagar, A new multi-focus image fusion framework based on focus

measures, in: 2021 IEEE International Conference on Systems, Man, and Cybernetics (SMC), 2021, pp. 2083–2088.
[5] Z. Jin, Y. Wang, Z. Chen, S. Nie, Medical image fusion in gradient domain with structure tensor, Journal of Medical Imaging and Health Informatics 6 (5) (2016) 1314–1318.
[6] G.K. Matsopoulos, S. Marshall, J.N.H. Brunt, Multiresolution morphological fusion of MR and CT images of the human brain, IEE Proceedings. Vision, Image and Signal Processing 141 (3) (1994) 137–142.
[7] Yong Yang, A novel DWT based multi-focus image fusion method, Procedia Engineering 24 (2011) 177–181.
[8] T. Li, Y. Wang, Biological image fusion using a NSCT based variable-weight method, Information Fusion 12 (2011) 85–92.
[9] X. Jin, G. Chen, J. Hou, Q. Jiang, D. Zhou, S. Yao, Multimodal sensor medical image fusion based on nonsubsampled shearlet transform and S-PCNNs in HSV space, Signal Processing 153 (2018) 379–395.
[10] B. Meher, S. Agrawal, R. Panda, A. Abraham, A survey on region based image fusion methods, Information Fusion 48 (2019) 119–132.
[11] A. Mumtaz, A. Majid, A. Mumtaz, Genetic algorithms and its application to image fusion, in: 2008 4th International Conference on Emerging Technologies, Rawalpindi, Pakistan, 2008, pp. 6–10.
[12] C. Heng, L. Jie, Z. Weile, A novel support vector machine-based multifocus image fusion algorithm, in: 2006 International Conference on Communications, Circuits and Systems, Guilin, China, 2006, pp. 500–504.
[13] Yuan Gao, Shiwei Ma, Jingjing Liu, Yanyan Liu, Xianxia Zhang, Fusion of medical images based on salient features extraction by PSO optimized fuzzy logic in NSST domain, Biomedical Signal Processing and Control (ISSN 1746-8094) 69 (2021) 102852, https://doi.org/10.1016/j.bspc.2021.102852.
[14] Z.-S. Xiao, C.-X. Zheng, Medical image fusion based on the structure similarity match measure, in: 2009 International Conference on Measuring Technology and Mechatronics Automation, 2009, pp. 491–494.
[15] M.M. Rahman, B.C. Desai, P. Bhattacharya, A feature level fusion in similarity matching to content-based image retrieval, in: 2006 9th International Conference on Information Fusion, 2006, pp. 1–6.
[16] Zhizhong Fu, Yufei Zhao, Yuwei Xu, Lijuan Xu, Jin Xu, Gradient structural similarity based gradient filtering for multi-modal image fusion, Information Fusion (ISSN 1566-2535) 53 (2020) 251–268, https://doi.org/10.1016/j.inffus.2019.06.025.
[17] J. Kittler, Multi-Sensor Integration and Decision Level Fusion, The Institution of Electrical Engineers, London, 2001.
[18] B. Jeon, D.A. Landgrebe, Decision fusion approach for multitemporal classification, IEEE Transactions on Geoscience and Remote Sensing 37 (3) (1999) 1227–1233.
[19] Y. Yang, J. Wu, S. Huang, Y. Fang, P. Lin, Y. Que, Multimodal medical image fusion based on fuzzy discrimination with structural patch decomposition, IEEE Journal of Biomedical and Health Informatics 23 (4) (July 2019) 1647–1660, https://doi.org/10.1109/JBHI.2018.2869096.
[20] A. Ravishankar, S. Anusha, H.K. Akshatha, A. Raj, S. Jahnavi, J. Madhura, A survey on noise reduction techniques in medical images, in: 2017 International Conference of Electronics, Communication and Aerospace Technology (ICECA), Coimbatore, India, 2017, pp. 385–389.

[21] K. Koonsanit, S. Thongvigitmanee, N. Pongnapang, P. Thajchayapong, Image enhancement on digital x-ray images using N-CLAHE, in: 2017 10th Biomedical Engineering International Conference (BMEiCON), Hokkaido, Japan, 2017, pp. 1–4.
[22] A. Gholipour, N. Kehtarnavaz, R. Briggs, M. Devous, K. Gopinath, Brain functional localization: a survey of image registration techniques, IEEE Transactions on Medical Imaging 26 (4) (April 2007) 427–451, https://doi.org/10.1109/TMI.2007.892508.
[23] Y. Zhang, Z. Zhang, G. Ma, J. Wu, Multi-source remote sensing image registration based on local deep learning feature, in: 2021 IEEE International Geoscience and Remote Sensing Symposium IGARSS, Brussels, Belgium, 2021, pp. 3412–3415.
[24] A. Sahu, V. Bhateja, A. Krishn, Himanshi, Medical image fusion with Laplacian pyramids, in: 2014 International Conference on Medical Imaging, m-Health and Emerging Communication Systems (MedCom), 2014, pp. 448–453.
[25] B. Deepa, M.G. Sumithra, T.D. Bharathi, S. Rajesh, MRI medical image fusion using gradient based discrete wavelet transform, in: 2017 IEEE International Conference on Computational Intelligence and Computing Research (ICCIC), 2017, pp. 1–4.
[26] T.J. Reddy, S.N. Rao, A novel fusion approach for multimodal medical images using non-subsampled contourlet transform, in: 2016 International Conference on Advanced Communication Control and Computing Technologies (ICACCCT), 2016, pp. 838–841.
[27] Himanshi, V. Bhateja, A. Krishn, A. Sahu, Medical image fusion in curvelet domain employing PCA and maximum selection rule, in: S. Satapathy, K. Raju, J. Mandal, V. Bhateja (Eds.), Proceedings of the Second International Conference on Computer and Communication Technologies, in: Advances in Intelligent Systems and Computing, vol. 379, Springer, India, 2016, pp. 1–9.
[28] G. Bhatnagar, Q.M. Wu, An image fusion framework based on human visual system in framelet domain, International Journal of Wavelets, Multiresolution and Information Processing 10 (2012).
[29] M. Li, Y. Dong, Review of image fusion algorithm based on multiscale decomposition, in: Proceedings 2013 International Conference on Mechatronic Sciences, Electric Engineering and Computer (MEC), IEEE, 2013, pp. 1422–1425.
[30] H. Deng, Y. Ma, Image fusion based on steerable pyramid and PCNN, in: 2009 Second International Conference on the Applications of Digital Information and Web Technologies, London, UK, 2009, pp. 569–573.
[31] H.R. Shahdoosti, A. Mehrabi, Multimodal image fusion using sparse representation classification in tetrolet domain, Digital Signal Processing 79 (2018) 9–22.
[32] Q. Liu, C. He, H. Li, H. Wang, Multimodal medical image fusion based on IHS and PCA, Procedia Engineering 7 (2010) 280–285.
[33] L. Wang, B. Li, L. Tian, Multi-modal medical image fusion using the inter-scale and intra-scale dependencies between image shift-invariant shearlet coefficients, Information Fusion 19 (2014) 20–28.
[34] P. Ganasala, V. Kumar, CT and MR image fusion scheme in nonsubsampled contourlet transform domain, Journal of Digital Imaging 27 (2014) 407–418.
[35] P. Rn, U. Desai, V.B. Shetty, Medical image fusion analysis using curvelet transform, in: Int. Conf. on Adv. in Comp., Comm., and Inf. Sci. (ACCIS-14), 2014, pp. 1–8.
[36] M. Manchanda, R. Sharma, A novel method of multimodal medical image fusion using fuzzy transform, Journal of Visual Communication and Image Representation 40 (2016) 197–217.
[37] Y. Liu, X. Chen, J. Cheng, H. Peng, A medical image fusion method based on convolutional neural networks, in: 2017 20th International Conference on Information Fusion (Fusion), Xi'an, China, 2017, pp. 1–7.

[38] H. Li, X.-J. Wu, DenseFuse: a fusion approach to infrared and visible images, IEEE Transactions on Image Processing 28 (5) (May 2019) 2614–2623, https://doi.org/10.1109/TIP.2018.2887342.

[39] X. Song, X.J. Wu, H. Li, MSDNet for medical image fusion, in: Y. Zhao, N. Barnes, B. Chen, R. Westermann, X. Kong, C. Lin (Eds.), Image and Graphics, ICIG 2019, in: Lecture Notes in Computer Science, vol. 11902, Springer, Cham, 2019.

[40] Yu Zhang, Yu Liu, Peng Sun, Han Yan, Xiaolin Zhao, Li Zhang, IFCNN: a general image fusion framework based on convolutional neural network, Information Fusion (ISSN 1566-2535) 54 (2020) 99–118, https://doi.org/10.1016/j.inffus.2019.07.011.

[41] Kaiming He, Xiangyu Zhang, Shaoqing Ren, Jian Sun, Deep residual learning for image recognition, in: Proceedings of the IEEE Conference on Computer Vision and Pattern Recognition, 2016, pp. 770–778.

[42] H. Xu, J. Ma, J. Jiang, X. Guo, H. Ling, U2Fusion: a unified unsupervised image fusion network, IEEE Transactions on Pattern Analysis and Machine Intelligence 44 (1) (1 Jan. 2022) 502–518, https://doi.org/10.1109/TPAMI.2020.3012548.

[43] Karen Simonyan, Andrew Zisserman, Very deep convolutional networks for large-scale image recognition, CoRR, arXiv:1409.1556 [abs], 2014.

[44] Han Xu, Jiayi Ma, EMFusion: an unsupervised enhanced medical image fusion network, Information Fusion 76 (2021) 177–186.

[45] Weisheng Li, Ruyue Li, Jun Fu, Xiuxiu Peng, MSENet: a multi-scale enhanced network based on unique features guidance for medical image fusion, Biomedical Signal Processing and Control (ISSN 1746-8094) 74 (2022) 103534.

[46] Olaf Ronneberger, Philipp Fischer, Thomas Brox, U-Net: convolutional networks for biomedical image segmentation, arXiv, arXiv:1505.04597 [abs], 2015.

[47] X. Lin, G. Zhou, W. Zeng, X. Tu, Y. Huang, X. Ding, A self-supervised method for infrared and visible image fusion, in: 2022 IEEE International Conference on Image Processing (ICIP), Bordeaux, France, 2022, pp. 2376–2380.

[48] Jiayi Ma, Wei Yu, Pengwei Liang, Chang Li, Junjun Jiang, FusionGAN: a generative adversarial network for infrared and visible image fusion, Information Fusion (ISSN 1566-2535) 48 (2019) 11–26, https://doi.org/10.1016/j.inffus.2018.09.004.

[49] Alec Radford, Luke Metz, Soumith Chintala, Unsupervised representation learning with deep convolutional generative adversarial networks, CoRR, arXiv:1511.06434 [abs], 2015.

[50] J. Ma, H. Xu, J. Jiang, X. Mei, X.-P. Zhang, DDcGAN: a dual-discriminator conditional generative adversarial network for multi-resolution image fusion, IEEE Transactions on Image Processing 29 (2020) 4980–4995, https://doi.org/10.1109/TIP.2020.2977573.

[51] Yicheng Wang, Shuang Xu, Junmin Liu, Zixiang Zhao, Chunxia Zhang, Jiangshe Zhang, MFIF-GAN: a new generative adversarial network for multi-focus image fusion, Signal Processing. Image Communication (ISSN 0923-5965) 96 (2021) 116295, https://doi.org/10.1016/j.image.2021.116295.

[52] Jun Fu, Weisheng Li, Jiao Du, Liming Xu, DSAGAN: a generative adversarial network based on dual-stream attention mechanism for anatomical and functional image fusion, Information Sciences (ISSN 0020-0255) 576 (2021) 484–506, https://doi.org/10.1016/j.ins.2021.06.083.

[53] C.-F.R. Chen, Q. Fan, R. Panda, CrossViT: cross-attention multi-scale vision transformer for image classification, in: 2021 IEEE/CVF International Conference on Computer Vision (ICCV), Montreal, QC, Canada, 2021, pp. 347–356.

[54] S. Park, A.G. Vien, C. Lee, Infrared and visible image fusion using bimodal transformers, in: 2022 IEEE International Conference on Image Processing (ICIP), Bordeaux, France, 2022, pp. 1741–1745.

[55] W. Tang, F. He, Y. Liu, Y. Duan, MATR: multimodal medical image fusion via multiscale adaptive transformer, IEEE Transactions on Image Processing 31 (2022) 5134–5149, https://doi.org/10.1109/TIP.2022.3193288.

[56] Linhao Qu, Shaolei Liu, Manning Wang, Shiman Li, Siqi Yin, Qin Qiao, Zhijian Song, TransFuse: a unified transformer-based image fusion framework using self-supervised learning, arXiv, arXiv:2201.07451 [abs], 2022.

[57] Jun Chen, Jianfeng Ding, Yang Yu, Wenping Gong, THFuse: an infrared and visible image fusion network using transformer and hybrid feature extractor, Neurocomputing (ISSN 0925-2312) 527 (2023) 71–82, https://doi.org/10.1016/j.neucom.2023.01.033.

[58] Hafiz Tayyab Mustafa, Jie Yang, Masoumeh Zareapoor, Multi-scale convolutional neural network for multi-focus image fusion, Image and Vision Computing (ISSN 0262-8856) 85 (2019) 26–35, https://doi.org/10.1016/j.imavis.2019.03.001.

[59] W. Zhou, A.C. Bovik, H.R. Sheikh, E.P. Simoncelli, Image quality assessment: from error visibility to structural similarity, IEEE Transactions on Image Processing 13 (4) (April 2004) 600–612.

[60] Zhou Wang, Eero P. Simoncelli, Alan Conrad Bovik, Multiscale structural similarity for image quality assessment, in: The Thirty-Seventh Asilomar Conference on Signals, Systems & Computers, vol. 2, 2003, pp. 1398–1402.

[61] Kaiming He, Xiangyu Zhang, Shaoqing Ren, Jian Sun, in: Proceedings of the IEEE Conference on Computer Vision and Pattern Recognition (CVPR), 2016, pp. 770–778.

[62] M.E. Shenton, R. Kikinis, W. McCarley, P. Saiviroonporn, H.H. Hokama, A. Robatino, F.A. Jolesz, Harvard brain atlas: a teaching and visualization tool, in: Proceedings 1995 Biomedical Visualization, 1995, pp. 10–17.

[63] L.N. Smith, N. Topin, Super-convergence: very fast training of neural networks using large learning rates, in: Defense + Commercial Sensing, 2017.

[64] D.P. Kingma, J. Ba, Adam: a method for stochastic optimization, CoRR, arXiv:1412.6980 [abs], 2014.

[65] A. Paszke, S. Gross, F. Massa, A. Lerer, J. Bradbury, G. Chanan, T. Killeen, Z. Lin, N. Gimelshein, L. Antiga, A. Desmaison, A. Köpf, E. Yang, Z. DeVito, M. Raison, A. Tejani, S. Chilamkurthy, B. Steiner, L. Fang, J. Bai, S. Chintala, PyTorch: an imperative style, high-performance deep learning library, arXiv, arXiv:1912.01703 [abs], 2019.

[66] M. Hossny, S. Nahavandi, D.C. Creighton, Comments on 'Information measure for performance of image fusion', Electronics Letters 44 (2008) 1066–1067.

[67] M. Haghighat, M.A. Razian, Fast-FMI: non-reference image fusion metric, in: 2014 IEEE 8th International Conference on Application of Information and Communication Technologies (AICT), Astana, Kazakhstan, 2014, pp. 1–3.

[68] C. Xydeas, V. Petroviä, Objective image fusion performance measure, Electronics Letters 36 (4) (2000) 308–309.

[69] G. Piella, H.J. Heijmans, A new quality metric for image fusion, in: Proceedings 2003 International Conference on Image Processing (Cat. No.03CH37429), 3, III-173, 2003.

[70] A. Mittal, A.K. Moorthy, A.C. Bovik, No-reference image quality assessment in the spatial domain, IEEE Transactions on Image Processing 21 (2012) 4695–4708.

[71] Om Prakash, Ashish Khare, CT and MR Images Fusion Based on Stationary Wavelet Transform by Modulus Maxima, 2015.

[72] X. Li, X. Guo, P. Han, X. Wang, H. Li, T. Luo, Laplacian redecomposition for multimodal medical image fusion, IEEE Transactions on Instrumentation and Measurement 69 (2020) 6880–6890.

[73] Z. Zhu, M. Zheng, G. Qi, D. Wang, Y. Xiang, A phase congruency and local Laplacian energy based multi-modality medical image fusion method in NSCT domain, IEEE Access 7 (2019) 20811–20824.

[74] Hui Li, Xiaojun Wu, Josef Kittler, Infrared and visible image fusion using a deep learning framework, in: 2018 24th International Conference on Pattern Recognition (ICPR), 2018, pp. 2705–2710.

[illegible]

[illegible]

[illegible]

CHAPTER

Data fusion in Internet of Medical Things: towards trust management, security, and privacy

12

Dipanwita Sadhukhan[a], Sangram Ray[b], and Mou Dasgupta[c]

[a]*Department of Computer Science and Engineering, National Institute of Technology Sikkim Ravangla, Sikkim, India*

[b]*Department of Computer Science and Engineering, National Institute of Technology Sikkim Ravangla, Sikkim, India*

[c]*Department of Computer Application, National Institute of Technology Raipur, Raipur, India*

12.1 Introduction

The Internet of Things (IoT) paradigm endeavors to configure a virtual world that facilitates the integration and fusion of physical "things" into cyber space [1]. IoT has led to the flourishing of different aspects of humankind and facilitates each and every corner of human life. With the assistance of IoT, problem identification has become easier and solutions to those problems have also become automated and optimized. The Internet of Medical Things (IoMT) platform is a type of IoT in the healthcare domain that provides a global infrastructure composed of a collection of medical appliances and wearable devices that are coordinated through information and communication technologies (ICT) [2,3]. IoMT incorporates data from visual sensors, humidity sensors, temperature sensors, accelerometer sensors, carbon dioxide sensors, ECG/EEG/EMG sensors, blood pressure sensors, gyroscope sensors, blood oxygen saturation sensors, respiration sensors, and blood pressure sensors for close observation and monitoring of the health condition of the patient in a continuous manner [4]. This process generates huge volumes of data, which is the first step of the healthcare services provided by IoMT. During the global pandemic, IoMT explores a new path for healthcare systems and eases many constraints, such as the exhaustive/limited medical resources, asymmetrical allocation of medical resources, remote patient monitoring, etc. IoMT provides a comprehensive solution for all the aforesaid issues. It is also convenient for doctors, facilitating decision-making processes according to the particular conditions of patients through an artificial intelligence visualization platform. Diagnosis, convalescence, long-term nursing, and precaution are the key features of IoMT.

Data Fusion Techniques and Applications for Smart Healthcare. https://doi.org/10.1016/B978-0-44-313233-9.00018-7

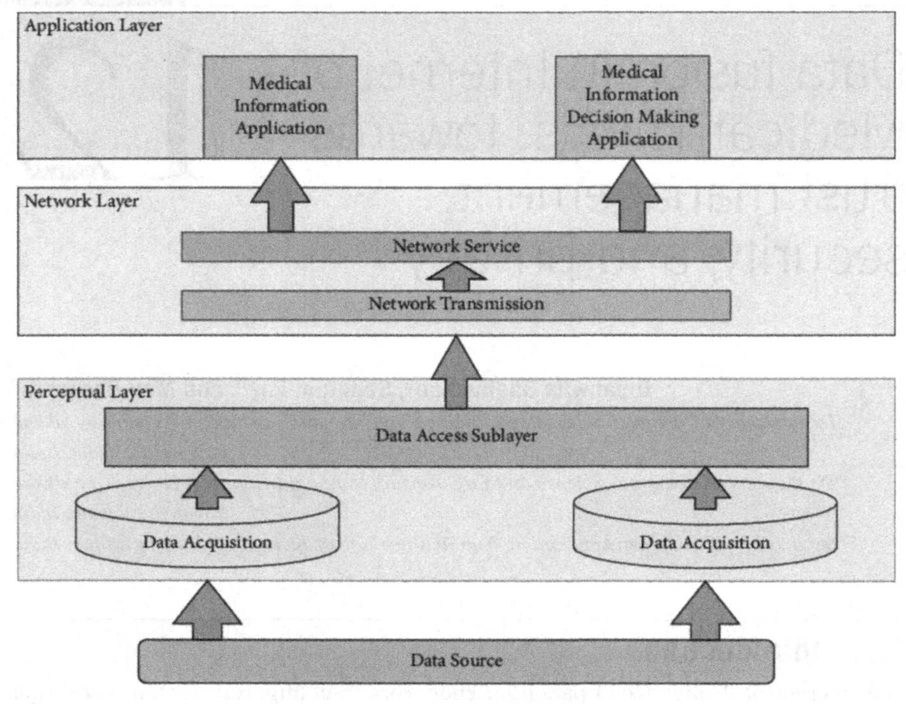

FIGURE 12.1

Basic architecture of IoMT [5].

The major benefits of IoMT are:

- It provides medical parameters in real-time for monitoring the actual condition of the patient to immediately react during any medical emergencies.
- It ensures adherence to the doctor's orders to enable better diagnosis and treatment plans.
- It improves drug management by keeping track of the chemical and physical reactions of the prescribed medicines in the patient's body.
- It reduces room for unwanted errors with the help of improved diagnosis accuracy, evidence-based treatment, and personalized medication.
- It has drastically stimulated chronic care management since IoMT provides insight into data derived from regular health checkups and condition assessments to determine trends and look for any red flags.
- It redefines tele-medicare systems by providing IoT-based mobility solutions for including video consultations when patients seek help in case of any medical emergency.
- It reduces costs incurred due to doctor visits, hospitalization, readmission, etc.

In IoMT, multiple sensors placed on different parts of the patient's body cover a broad surface to gather diverse data about the physical and mental condition of the patient [5–7]. This concept of multiple sensor deployment in a small bounded target region provides a way to enhance the quality of the information extracted from the raw data since single sensing modalities are insufficient to remotely monitor and diagnose patients [8,9]. However, multiple sources of data produce huge amounts of heterogeneous data, and hence it would be a complete waste of network bandwidth and energy/power resources of the limited resourced IoT nodes to transfer all the sensed data [10]. In this condition, data fusion happens to be the most significant solution to ascertain crucial data are extracted from the whole set of gathered/collected data to enhance data quality, which in turn facilitates the decision-making process during disease diagnosis in order to treat patients [11–13].

Bostrom et al. [14] write "Information fusion is the study of efficient methods for automatically or semi-automatically transforming information from different sources and different points in time into a representation that provides effective support for human or automated decision making." It is considered as the theory, methods, and tools which are required to combine sensor data into meaningful and significant information in a common representational format [15]. Data fusion extensively reduces the volume and dimensions of data and optimizes the quantity of data by eliminating imperfect and unnecessary sensory data. Many research works have been carried out in this domain. Although several research papers regarding multisensory data fusion have been published, as discussed in the previous paragraph, they are unsuccessful in providing a complete synopsis on IoMT data fusion [1–15] and abandon the most significant concerns of data fusion such as security and privacy issues in the fusion of data in IoMT.

Data fusion in IoMT is still in the infant stage, and several issues such as security, privacy protection, trust management, etc., need to be overcome. Data fusion in heterogeneous IoT environments like IoMT is challenging. Additionally, it is difficult to analyze relationships among different data collected by different sensors, even when the datasets are semantically correlated to each other [16]. Scalability is one of the major challenges in IoMT due to the frequent changes in the shape and size of the networks. Decision fusion methods provide a data formalism for combination using probability theory where conditional probability describes uncertainty.

In this book chapter, we describe the distinctiveness and features of IoT data, illustrate the fundamental knowledge and purposes of data fusion techniques. The chapter also discusses data fusion requirements and challanges for enabling the different fusion methods [17–25].

Chapter contributions

The key contributions of this chapter are:

i. A brief description of fundamental concepts in IoMT data fusion is given for clear understanding of the basics of IoMT.
ii. Basic knowledge and advanced applications of data fusion are discussed.
iii. Different fusion matrices and fusion methods of IoMT are discussed.

iv. The performance of the privacy-aware framework is evaluated and compared with recent intrusion detection models for IoMT.

12.2 Preliminaries of IoMT data fusion

This section illustrates the basic characteristics and properties of data, data fusion requirements, and fusion metrics for easy understanding of the data fusion process.

12.2.1 Characteristics of data in the context of IoMT

IoMT-enabled smart healthcare aims to improve intelligent diagnosis, provide error-free treatment, and control chronic medical situations, such as remote patient monitoring, identification of improper medical attributes, and drug management. Patient-related data are highly confidential, context-sensitive, and pivotal in making any crucial decision or executing context inference [26]. For example, data about glucose levels in a patient can indicate whether a remote patient requires automated insulin delivery or not. For another example, medical parameter analysis necessitates real-time monitoring of data for an actual representation of individual health information, which is extremely private and confidential.

Multiple sensors are placed in the target region such as the patient's body. The sensors accumulates high-dimensional complex data for continuous monitoring. However, the collected data are not always important but are nevertheless transferred over the channel, which in turn wastes network bandwidth. In the context of IoMT, data analysis is important to preserve the integrity and reliability of data.

Table 12.1 Data characteristics of IoMT.

Data characteristics	IoMT-enabled smart healthcare
Privacy sensitivity	High
Sensor distribution	Small area
Security goals	Data integrity, data confidentiality, timeliness, reliability

Low-quality, erroneous data fusion might lead to inaccurate monitoring of the patient's health, which may cause fatal risks. Thus, IoMT in the domain of smart healthcare particularly requires data confidentiality, timeliness, and fusion accuracy. Generally, in IoMT the number of sensors on the patient's body is fixed and seldom modified. Hence, scalability is usually not an issue in IoMT. Table 12.1 shows the key characteristics of IoMT data.

12.2.2 IoMT data properties

IoMT sensors are deployed in different places of the body to collect useful data from multiple sources that can be used as the input for data fusion and data analytics. Before going into the details of data fusion, it needs to be clearly conceptualized what are the different characteristics of IoMT data. The following characteristics of IoMT data hold in general cases.

- *Multimodality*: Various involved sensors (such as wearable sensors, microchips, microcontroller cameras, etc.) accumulate data of a different nature from the patient's body, amplifying the total volume and modality of data. The diversity of data in terms of category, scale, variety, representation, range, and density makes it infeasible to fuse data directly.
- *Big data*: In a similar way, since a large number of sensors are deployed in the target region for observation, large quantities of data ("big data") are generated. Consequently, big data fusion has become important in IoMT.
- *Data sensitivity*: After sensing, the transferred sensory data are comprised of various confidential parameters (medical attributes) of the patient. Sometimes the sensed data hold vital information, and if these data are compromised, this may cause serious harm to the patient's life. For example, if drug dosages, allergic reactions, or consumption habits of a patient are compromised, this increases the risk of privacy incursion and eventually might become life-threatening. However, it is extremely difficult to preserve user privacy and confidentiality while fusing multisensory data in an error-free manner.
- *Indefinite*: Data accumulated by various sensors may be vague, imperfect, and tentative because of data loss or data source unpredictability. This incurs additional challenges in case of data fusion due to data inaccuracy, data ambiguity, data inconsistency, and unpredictability.
- *Dynamic*: The collected data always depend on time, i.e., the data are time-variant, which means that the context of the data changes over time. Hence, the freshness of the data must be preserved. Hence, this conservation of timeliness of data is a critical issue during data fusion. It is a key factor for data fusion since it enhances the data quality, accuracy, and acceptance.

The maintain all the aforesaid characteristics of the data of IoMT, we are faced with a number of challenges due to the limited storage and energy of the sensor devices, which are: (a) to distinguish spurious data, (b) to prevent fusion inaccuracy caused by data dependency and correlation, (c) to deal with an extensive distribution of data, (d) to prevent high communication overhead, and (e) to reduce the delay sensitivity in centralized data fusion models.

12.2.3 IoMT data fusion requirements and challenges

This section illustrates the vital requirements and challenges of data fusion in IoMT.

12.2.3.1 IoMT data fusion requirements

The essential requirements mentioned in the previous section must be incorporated during the process to achieve secure, trustworthy, and reliable data fusion in IoMT as described below.

- *Reliable*: The final outcome of data fusion often directly leads to a specified decision, such as diagnosis or a medical emergency. Defective, error-prone data fusion outcomes may cause an unacceptable life-threatening danger to the patient. Hence, reliability is one of the most fundamental as well as significant requirements of data fusion for practical implementations.
- *Context awareness*: Context indicates the background or the situation of any operation to be characterized. Context awareness refers to the ability to anticipate the context or the immediate situational need of the system in order to adapt the interface and increase the precision of information retrieval. It is usually time-dependent and may not always intrinsically relate to the elements. For example, diabetes monitoring sensors must provide instantaneous results to the healthcare team for immediate insulin delivery. Data fusion is required to be context-aware to support adaptive and flexible services with high intelligence.
- *Privacy preservation*: Data fusion needs to be privacy preserving and must be treated as one of the most essential requirements, since it affects the data acceptance by the concerned authority for patient diagnosis. Trustworthiness of data fusion wholly depends on the privacy preservation of the fused data. Especially, medical data are extremely privacy-sensitive as any adversarial infringement may cause fatal threats to the patient.
- *Efficiency*: The efficiency of data fusion indicates how conveniently the computation overhead of the fusion center or sensors can be optimized to further provide a quick response after data analysis. Due to continual monitoring of the target environment by the sensor, maintaining high efficiency becomes the bottleneck. For performance assessment of data fusion methods, diverse detailed aspects need to be evaluated, including communication overhead for data collection, training and testing time, objective supervision, etc. Unlike others, we focus on the training and testing time in the following analysis.
- *Robustness*: Robustness in data fusion must be preserved to resist various security attacks (e.g., false data injection, data alteration) in IoMT due to data transfer through public channels which may be controlled by adversaries. Otherwise, it could result in imperfect collection of data and in turn generate some inappropriate results.
- *Verifiability*: The fusion outcome must be validated by the fusion authority or controller for quality checking. This requirement enables consumers to estimate the correctness of data fusion results. Particularly, data fusion with verifiability facilitates judging whether the gathered data contribute to taking the final decision in reality.

12.2.3.2 IoMT data fusion challanges

The major challenges that are faced by IoMT during data fusion are given below.

- Data deficiency: Data accumulated by sensors from remote places may be imperfect, inaccurate, and vague. These data deficiencies must be considered and dealt with by efficient data fusion methods.

- Inconsistencies of data: Data gathered by sensor nodes are ambiguous and inconstant as nodes may be installed in an unfavorable environment. Outlier detection, data substitution, and data attribution are essential in IoMT.
- Data conflict: The contradictory nature of the collected data rises the possibility of implausibility and counterintuitive fusion results. It requires critical care during data fusion.
  - Data correlation: This issue may occur due to sensor node registration. The data are passed through the sensor node to the sink node of the network and transformed from the local frame to a common frame prior to data fusion.
  - Iterative procedure: Data fusion is a dynamic process based on an iterative procedure, where regular modification is needed to process the data.

Studies in the field of data fusion have gradually moved forward, and the performance of the algorithms has improved. However, there is still a need to develop an ideal data fusion algorithm.

12.2.4 Fusion metrics

Evaluation of reliability and viability of data fusion requires different metrics. Four types of elements need to be included for measurement of the metrics [22]:

i. true positive (TP), which specifies the total number of positive samples considered to be positive;
ii. false positive (FP), which specifies the total number of negative samples considered to be positive;
iii. false negative (FN), which specifies the total number of positive samples considered to be negative;
iv. true negative (TN), which indicates the total number of negative samples considered to be negative.

The metrics for evaluation of the performance of data fusion methods are calculated as follows:

- accuracy (ACR): $ACR = (TP + TN)/(TP + FP + TN + FN)$;
- precision rate (PR): $PR = TP/(TP + FP)$;
- sensitivity (SN): $SN = TP/(TP + FN)$;
- false measure (FM): $FM = 2 * PR * SN/(PR + SN)$;
- false positive rate (FPR): $FPR = FP/(TN + FP)$;
- false negative rate (FNR) $FNR = FN/(FN + TP)$;
- true negative rate (TNR): $TNR = TN/(TN + FP)$.

12.3 Traditional data fusion methods

There exist several traditional methods based on many mathematical theories for typical data fusion [26]. From a detailed discussion of the previous state-of-the-art literature [27], we summarize all of the well-accepted data fusion techniques here in brief. In accordance with the previous discussion, they can be classified as probability-based methods, evidence reasoning methods, and knowledge-based methods.

a. *Probabilistic data fusion*: The probability distribution is fundamentally based on the density function or probability distribution [28–30]. This method was established to deal with data imperfection or uncertainty. It can be expressed as the interdependence among the random variables and different datasets. These probability-based data fusion methods mostly include the procedures illustrated in Table 12.2.
   However, these methods mostly suffer from: (i) the inability to attain a density function; (ii) inadequate performance in dealing with multivariate and complex data; and (iii) the inability to handle uncertainty.
b. *Knowledge-based methods*: These methods facilitate data fusion from imprecise/vague big data, from which density/distribution functions cannot be acquired [30–32]. They basically involve intelligent aggregation fuzzy logic, machine learning, etc. Fundamentally, machine learning is primarily classified into three different groups:
   - *Supervised learning*: These include classification algorithms such as K-nearest neighbor (KNN), support vector machines (SVM), and artificial neural network (ANN) to simulate any biological learning structure [33].
   - *Unsupervised learning*: These include clustering methods such as Density-Based Spatial Clustering of Applications with Noise (DBSCAN), hierarchical clustering, K-means partitioning, etc.
   - *Semi-supervised learning*: Fuzzy logic [34] adapts the theory of fuzzy sets to cope with uncertainty and imprecision to integrate numerical control yet does not support incorporation of prior knowledge.

12.4 Data fusion in IoMT

Observation of the status of the elderly or chronically sick patients using IoMT is a necessary requirement for delivering more effective pervasive healthcare, which can assist healthcare professionals as well as family members to thoroughly observe the condition of the patient independently, safely, and uninterruptedly. Various components of IoMT, such as wearable accelerometers, pulse oximeters, thermometers, and blood pressure monitors, point of care devices for analyzing specimens like blood, saliva, and skin cells, in-hospital devices and monitors, smart pills, etc., are deployed in smart environments around the patient to accumulate data, which are in turn forwarded to the gateway for fusion as presented in Fig. 12.1. The differences among the various traditional data fusion methods in terms of their characteristics, limitations and advantages are discussed in Table 12.3.

All the data collected by the sensors are significant factors during diagnosis of the physical and mental condition of the patient during healthcare monitoring. Be-

Table 12.2 Traditional data fusion methods.

Method	Description	Mathematical computations	Advantages	Limitations
Bayesian inference	A state-space representation of determining the posterior probability density/distribution of a hypothetical state.	Density $= p(x_n \mid y^n) = \frac{p(y_n \mid x_n)p(x_n \mid y^{n-1})}{p(y^k \mid y^{k-1})}$, where x_n is the hypothetical state at time n, $p(y_n \mid x_n)$ is the likelihood function, $p(x_n \mid y^{n-1})$ is the prior distribution, and $y^n = \{y_1 \ldots y_n\}$ (up to n times)	Easy, proficient, and superior performance for evaluating prior knowledge	Prior probability is required; complexity is high for multivariate data; huge volume of data; not capable of dealing with uncertainty
Kalman filtering	An exceptional case of the Bayes filter that provides an accurate analytical solution by enforcing simplified (rather impractical) constraints on the system dynamics to be linear Gaussian.			Susceptible to data corrupted with outliers. Unsuitable for applications whose error characteristics are not properly parameterized
Markov models	Required to solve problems regarding high-dimensional density estimation. This state space leads only on the random samples of the present state. It covers a unique fixed interest density.		Exact prediction	Not fit to continual prediction
Evidential belief	Based on the concept of assigning beliefs and plausibility to approximate measurement hypotheses combining with the rule to fuse them. Evidence from sensors is typically fused by means of Dempster's rule of combination.	The joint belief mass function $m_{1,2}$ is calculated as follows: $m_{1,2}(E) = (m_1 \oplus m_2) = \frac{1}{1-K} \sum_{B \cap C = E \neq \phi} m_1(B) \times m_2(C)$, where $K = \sum_{B \cap C = \phi} m_1(B) m_2(C)$; $m(E)$ stands for the part of accessible evidence that supports the actual system state x belongs to element E	Deals with uncertainty and vagueness with a theoretically evidential reasoning framework	-

Table 12.3 Differences among the various traditional data fusion methods in terms of their characteristics, limitations, and advantages.

Framework	Characteristics	Rules implemented	Capabilities	Limitations
Probabilistic	Sensory data are fused employing a probability distribution	Bayesian inference	Well-accepted and distinct concept to deal with data uncertainty	Incapable to address other data inconsistencies.
Evidential	Data are classified in terms of plausibilities and belief and fused using mass probability.	Dempsters' amalgamation rule	Well-accepted concept for enabling fusion of ambiguity and uncertainty of data.	Unable to address other data conflicts, data indistinctness
Fuzzy reasoning	Supports indistin-guishable/fuzzy data depiction	Fuzzy rules	Well-established approach to deal with human-generated data	Unable to address other data conflicts, data indistinctness
Possibilistic	Similar to probabilistic data fusion	Fuzzy rules	Supports fusion of incomplete data gathered from poorly informed environments	Usually not used in well-structured data fusion
Hybrid	More comprehensive and complete approach towards inconsistent data fusion	-	Facilitates complementary fusion	Huge computational burden
Random	Allows many features of imperfect data	Random subsets of measurement/ state space	potentially provides a coalesce structure for fusion of imperfect data	Comparatively novel approach and not well acknowledged

sides, the visual representation of data provides real-time information for detecting the human activity and achieves a detailed inference.

According to Fig. 12.2, typical IoMT data fusion consists of three steps – (i) data collection, (ii) data fusion at the gateway, and (iii) access control of the outcome of fusion. However, privacy and confidentiality of data are not sincerely preserved. In this research we propose a new step to be incorporated into the data fusion process in IoMT as the final step, trust management, to preserve the confidentiality of the fused data.

- ***Data collection*:** Accumulating adequate functional data for fusion is one of the most challenging tasks and eventually turns into a key concern since healthcare data are crucially privacy-sensitive and the safety of patients must be preserved. On the other hand, sensory data fusion amplifies the computation overhead due to extraction of features followed by reduction of the magnitude through principal component analysis. Moreover, sensory data provide sufficient information on activity detection and also contain private credentials, such as position, identity, appearance, etc. Some clinical sensors execute detection of health data without impairing confidentiality of the patient, but are unable to obtain rich information of the patients. Most commonly, the primary and most crucial task is to deploy appropriate sensors on the patient's body and in the in-house environment that can provide a proper equilibrium between fusion competence and privacy protection in IoMT.

- ***Data fusion*:** Feature extraction can be employed for lessening the huge communication overhead of data transmission caused by diverse sensors, particularly visual sensors, while preserving data reliability. Yet, balancing communication and computation overhead has become a new issue. Second, there exist several datasets which are accessible for training (such as the UCI repository [17], HOMECAD [18], UTD-MHAD [19], and the Aruba dataset [20]), while data gathered from tangible scenarios with limited volunteering actors makes the fusion accuracy unpersuasive. Additionally, the quality of the dataset has a great influence on the exactness of data fusion, but collecting enough data is not always possible. Hence, it becomes difficult to identify the behavior datasets for various objectives in IoMT, which can aid to conquer the limitations of experimental data such as imperfection of sensed data, undersized learning periods, and insufficient volunteers for training [21].
- ***Access control*:** Previous studies failed to enhance the exactness and correctness of integrated data and also did not consider the aspect of access control in data fusion. This aspect is also not taken into account during offline data fusion, which requires a lower level of result verifiability than real-time data integration. Basically, the fact is that fused data hold more significant, relevant, and consequential data than the previously accumulated data. This makes it more important to qualify the access control property. This access control becomes more essential in strongly distributed IoT networks [23–25].

12.5 Privacy-enhanced data fusion

This section illustrates the trust evaluated data fusion model, the trust evaluated data fusion procedure, and finally the algorithm for sensor node selection to fuse the data.

12.5.1 Diverse duties of various layers of IoMT in data fusion

The complete system of data fusion is presented in Fig. 12.2. Three individual participants are shown: user, resource, and service.

- *User* (U): The user desires to acquire any data in a relational table including n attributes, and it should satisfy any required condition such as data confidentiality, default ratio, and data accuracy.

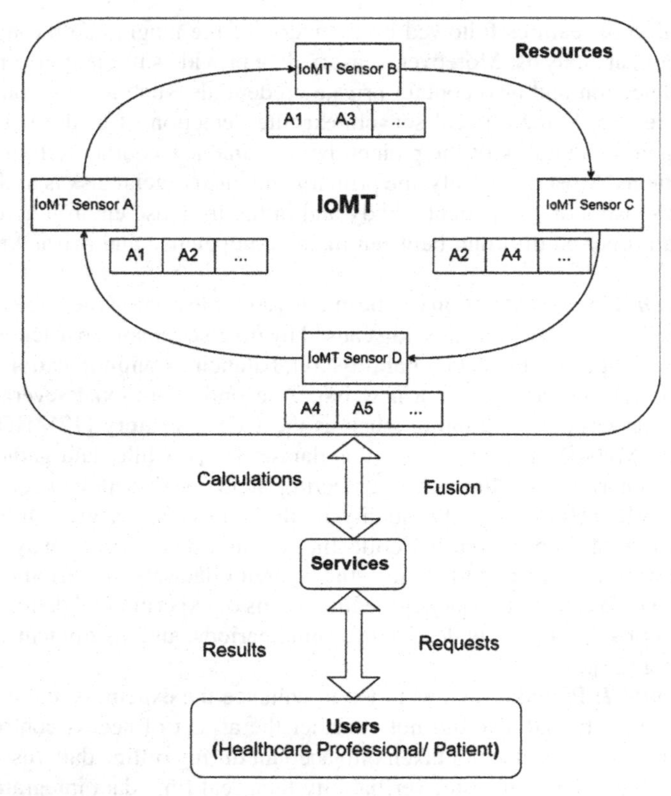

FIGURE 12.2

Duties of different layers of IoMT.

- *Service* (*S*): The intermediate participant service acts as the interface between U and R.
- *Resource* (*R*): The resource contains Sensor A, Sensor B, Sensor C, etc., which can be considered as decentralized data sources. Each source includes a series of attributes, i.e., party (Attribute 1, Attribute 2, etc.). Diverse sensors may contain similar attributes and may increase redundancy. Each party wants to participate in the data fusion process.

12.5.2 Trust evaluation model for data fusion

Fig. 12.3 represents the proposed trust evaluation model of data fusion in IoMT. This model can mainly be employed in the data perpetual layer of IoMT. The layer is shown in Fig. 12.2. In this layer, the sensor nodes acquire data. The proposed model is incorporated in this layer to evaluate the data quality before data fusion. After successful source selection of data and weight calculation of the data, it reaches the stage

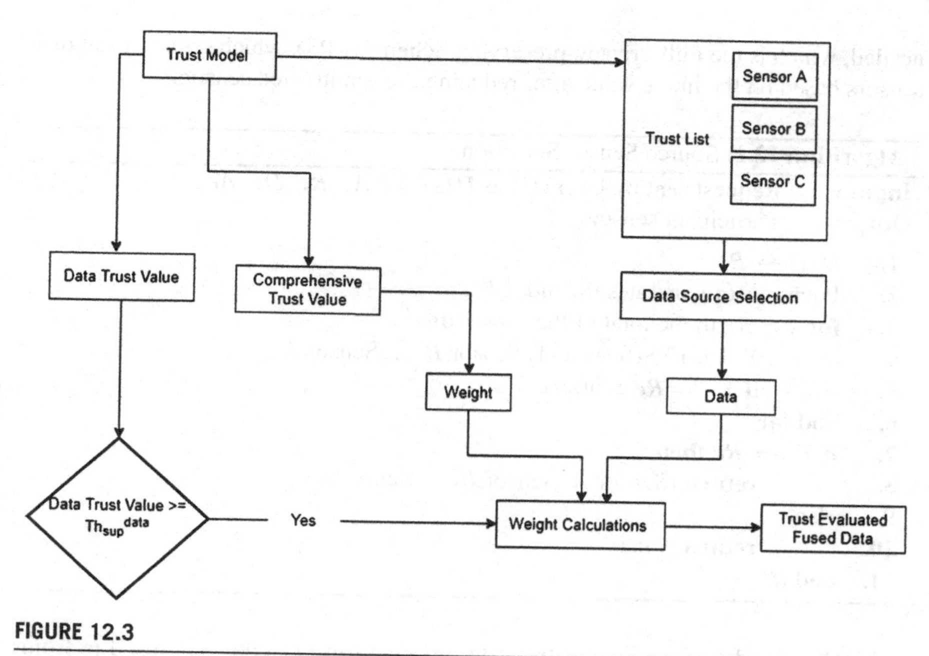

FIGURE 12.3

Trust evaluation model for data fusion.

of fusing the data. In the next subsection, we propose an algorithm to select the source sensor to collect the meaningful data. The proposed model guarantees data privacy and trust in data fusion by inducing a new parameter called the threshold value of trust. The trust threshold is calculated in the cluster heads of the IoMT network before data transmission. The trust value is mainly evaluated using three different types of trust: behavioral trust, historical, trust and data trust. The comprehensive value of trust is calculated in terms of each of the trust types. After evaluating whether the threshold value of trust is less than the calculated value of trust, if the condition is satisfied, the result is fed to the weight calculation step and finally the trust evaluated fused data are acquired. The data threshold value is always generated with the help of historical data trust of the source. With the proposed trust model and the specified threshold, abnormal data can be excluded from the fused data.

12.5.3 Source sensor selection

Another most important aspect of the proposed model is to find the reliable data source or sensors. We propose an algorithm to select the sensors that collect reliable and acceptable data (Algorithm 12.1). Proper sensor nodes are selected for extracting data to fuse. The inputs for the sensor selection are attribute (A), redundancy (Re), accuracy (Ac), and default (De) of data. For generating the nodes, another process is needed, which is the Full Privacy-preserving Scheme (FPS), which selects x out of y sensors based on the index value after reducing the semitrusted sensors.

Algorithm 12.1: Source Sensor Selection

Input : Request sent by User (U) as $U(s) = [(A), Re, De, Ac]$
Output : Participant sensors

1. $U(s) \xrightarrow{s} R$
2. Each sensor generates the index i
3. **for** $k = Re$ to the total of the sensor **do**
4. $Y =$ k-FPS(Sensor A, Sensor B, ... Sensor K)
5. **if** $Y >= Re$ continue
6. **end for**
7. **if** $Y >= Re$ **then**
8. **return** (Sensor A, Sensor B, ... Sensor K)
9. **else**
10. **return** None
11. **end if**

Finally, the data are acquired from the selected sensor nodes and fused to from trust evaluated privacy preserving data for further use in IoMT.

12.6 Conclusion

In this chapter, we discussed the basic properties of IoMT. Data are acquired and then the dimensions of data fusion are illustrated. We also considered essential requirements for trust computation, trust composition, trust propagation, trust aggregation, trust update, and trust formation for data fusion. All traditional methods of data fusion were also discussed with their advantages and limitations. Diverse duties of various layers of IoMT in data fusion were also presented. In particular, we proposed a trust evaluation model for data fusion and a sensor node selection algorithm for data fusion to extract meaningful information from the fused data. We also presented numerous challenges and limitations of the fusion model that need attention to improve its performance.

In the future we will assess the impact of automated and declarative proposed data fusion models on the eminence of medical services and to look into the limits and issues to implement the technique in IoMT data fusion.

Acknowledgment

The research work is supported by the Ministry of Education, Govt of India.

References

[1] Z. Yan, P. Zhang, A.V. Vasilakos, A survey on trust management for Internet of Things, Journal of Network and Computer Applications 42 (2014) 120–134.
[2] G. Manogaran, R. Varatharajan, D. Lopez, P.M. Kumar, R. Sundarasekar, C. Thota, A new architecture of Internet of Things and big data ecosystem for secured smart healthcare monitoring and alerting system, Future Generations Computer Systems 82 (2018) 375–387.
[3] G. Manogaran, D. Lopez, N. Chilamkurti, In-Mapper combiner based MapReduce algorithm for processing of big climate data, Future Generations Computer Systems 86 (2018) 433–445.
[4] D. Sadhukhan, S. Ray, G.P. Biswas, M.K. Khan, M. Dasgupta, A lightweight remote user authentication scheme for IoT communication using elliptic curve cryptography, Journal of Supercomputing 77 (2) (2021) 1114–1151.
[5] J. Srivastava, S. Routray, S. Ahmad, M.M. Waris, Internet of medical things (IoMT)-based smart healthcare system: trends and progress, Computational Intelligence and Neuroscience (2022) 2022.
[6] M. Li, X. Zhang, Information fusion in a multi-source incomplete information system based on information entropy, Entropy 19 (11) (2017) 570.
[7] D. Rangwani, D. Sadhukhan, S. Ray, M.K. Khan, M. Dasgupta, A robust provable-secure privacy-preserving authentication protocol for Industrial Internet of Things, Peer-to-Peer Networking and Applications 14 (3) (2021) 1548–1571.
[8] S. Stillman, I. Essa, Towards reliable multimodal sensing in aware environments, in: Proceedings of the 2001 Workshop on Perceptive User Interfaces, 2001, November, pp. 1–6.
[9] T.M. Ghazal, M.K. Hasan, S.N.H. Abdullah, K.A. Abubakkar, M.A. Afifi, IoMT-enabled fusion-based model to predict posture for smart healthcare systems, Computers, Materials & Continua 71 (2) (2022) 2579–2597.
[10] R. Chen, F. Bao, J. Guo, Trust-based service management for social Internet of things systems, IEEE Transactions on Dependable and Secure Computing 13 (6) (2015) 684–696.
[11] R. Chen, J. Guo, F. Bao, Trust management for SOA-based IoT and its application to service composition, IEEE Transactions on Services Computing 9 (3) (2014) 482–495.
[12] G. Pallapa, N. Roy, S. Das, Precision: privacy enhanced context-aware information fusion in ubiquitous healthcare, in: First International Workshop on Software Engineering for Pervasive Computing Applications, Systems, and Environments (SEPCASE'07), IEEE, 2007, May, p. 10.
[13] J. Pansiot, D. Stoyanov, D. McIlwraith, B.P. Lo, G.Z. Yang, Ambient and wearable sensor fusion for activity recognition in healthcare monitoring systems, in: 4th International Workshop on Wearable and Implantable Body Sensor Networks (BSN 2007), Springer, Berlin, Heidelberg, 2007, pp. 208–212.
[14] M. Al-Hawawreh, M.S. Hossain, A privacy-aware framework for detecting cyber attacks on Internet of medical things systems using data fusion and quantum deep learning, Information Fusion (2023) 101889.
[15] H. Medjahed, D. Istrate, J. Boudy, J.L. Baldinger, B. Dorizzi, A pervasive multi-sensor data fusion for smart home healthcare monitoring, in: 2011 IEEE International Conference on Fuzzy Systems (FUZZ-IEEE 2011), IEEE, 2011, June, pp. 1466–1473.
[16] F. Yang, Q. Wu, X. Hu, J. Ye, Y. Yang, H. Rao, R. Ma, B. Hu, Internet-of-things-enabled data fusion method for sleep healthcare applications, IEEE Internet of Things Journal 8 (21) (2021) 15892–15905.

[17] X. Hong, C. Nugent, M. Mulvenna, S. McClean, B. Scotney, S. Devlin, Evidential fusion of sensor data for activity recognition in smart homes, Pervasive and Mobile Computing 5 (3) (2009) 236–252.

[18] I. Belhajem, Y.B. Maissa, A. Tamtaoui, A robust low cost approach for real time car positioning in a smart city using Extended Kalman Filter and evolutionary machine learning, in: 2016 4th IEEE International Colloquium on Information Science and Technology (CiSt), IEEE, 2016, October, pp. 806–811.

[19] C. Yang, L. Feng, H. Zhang, S. He, Z. Shi, A novel data fusion algorithm to combat false data injection attacks in networked radar systems, IEEE Transactions on Signal and Information Processing over Networks 4 (1) (2018) 125–136.

[20] A. Fleury, M. Vacher, H. Glasson, J.F. Serignat, N. Noury, Data fusion in health smart home: preliminary individual evaluation of two families of sensors, in: ISG'08, 2008, June, p. 135.

[21] M.D. Alshehri, F.K. Hussain, A fuzzy security protocol for trust management in the Internet of things (Fuzzy-IoT), Computing 101 (7) (2019) 791–818.

[22] Z. Chen, L. Tian, C. Lin, Trust model of wireless sensor networks and its application in data fusion, Sensors 17 (4) (2017) 703.

[23] X. Su, K. Fan, W. Shi, Privacy-preserving distributed data fusion based on attribute protection, IEEE Transactions on Industrial Informatics 15 (10) (2019) 5765–5777.

[24] H. Lin, S. Garg, J. Hu, X. Wang, M.J. Piran, M.S. Hossain, Privacy-enhanced data fusion for COVID-19 applications in intelligent Internet of medical Things, IEEE Internet of Things Journal 8 (21) (2020) 15683–15693.

[25] N. Madaan, M. Ahad, S.M Sastry, Data integration in IoT ecosystem: information linkage as a privacy threat, Computer Law & Security Review 34 (1) (2018) 125–133.

[26] O. AlShorman, B. AlShorman, M. Alkhassaweneh, F. Alkahtani, A review of Internet of medical things (IoMT)-based remote health monitoring through wearable sensors: a case study for diabetic patients, Indonesian Journal of Electrical Engineering and Computer Science 20 (1) (2020) 414–422.

[27] J. Guo, R. Chen, J.J. Tsai, A survey of trust computation models for service management in Internet of things systems, Computer Communications 97 (2017) 1–14.

[28] A. Sharma, E.S. Pilli, A.P. Mazumdar, P. Gera, Towards trustworthy Internet of Things: a survey on Trust Management applications and schemes, Computer Communications 160 (2020) 475–493.

[29] W. Fang, W. Zhang, W. Yang, Z. Li, W. Gao, Y. Yang, Trust management-based and energy efficient hierarchical routing protocol in wireless sensor networks, Digital Communications and Networks 7 (4) (2021) 470–478.

[30] M. Frustaci, P. Pace, G. Aloi, G. Fortino, Evaluating critical security issues of the IoT world: present and future challenges, IEEE Internet of Things Journal 4 (5) (2017) 2483–2495.

[31] W. Fang, W. Zhang, W. Chen, Y. Liu, C. Tang, TMSRS: trust management-based secure routing scheme in industrial wireless sensor network with fog computing, Wireless Networks 7 (26) (2020) 3169–3182.

[32] N. Mohammed, D. Alhadidi, B.C. Fung, M. Debbabi, Secure two-party differentially private data release for vertically partitioned data, IEEE Transactions on Dependable and Secure Computing 11 (1) (2013) 59–71.

[33] J. Liu, J. Han, L. Wu, R. Sun, X. Du, VDAS: verifiable data aggregation scheme for Internet of Things, in: 2017 IEEE International Conference on Communications (ICC), IEEE, 2017, May, pp. 1–6.
[34] U. Javaid, M.N. Aman, B. Sikdar, A scalable protocol for driving trust management in Internet of vehicles with blockchain, IEEE Internet of Things Journal 7 (12) (2020) 11815–11829.

CHAPTER

Feature fusion for medical data

13

Nazanin Zahra Joodaki[a], **Mohammad Bagher Dowlatshahi**[a], **and Arefeh Amiri**[b]

[a]*Department of Computer Engineering, Faculty of Engineering, Lorestan University, Khorramabad, Iran*

[b]*Shahid Madani Hospital, Lorestan University of Medical Sciences, Khorramabad, Iran*

13.1 Introduction

The urgent need for data fusion is not excluded from the field of medicine. Due to the rapid progress in image processing methods and the variety of clinical data, in this field, these is a strong need for accurate and efficient data fusion techniques. With the recent advances in information storage and processing technologies, we face a lot of multimedia information, including text and images. Over the past 20 years, medical media content has been able to help the health field a lot in areas such as patient education and patient rehabilitation programs and in the area of diagnosing diseases and injuries to human organs and cells. Due to the progress in imaging models and their application in medical science, information and images in medicine are produced and used at high speed in large volumes.

Different types of imaging models have been introduced with the advancement of technology, which are able to record information on all the organs of the human body. Although each of these models alone can determine the condition of the body's organs to a large extent, the models together can provide more useful information to medical specialists. Together, the models strengthen each other's strengths and reduce each other's weaknesses. If different medical data, including images, text, and sound extracted from different models, are merged together, it will improve the ability to diagnose and treat diseases. Correct diagnosis is essential in the process of treating patients. In case of an incorrect diagnosis, irreparable damage may be caused. Each of these imaging methods can be used alone, but if they are combined with other types of images, they will support the diagnosis process better.

The large volume of multimedia and textual medical information, such as images, text, electrocardiograms (ECGs), etc., must be managed to facilitate retrieval, processing, and storage. Multimedia information retrieval frameworks, such as the General Multimedia Analysis Framework (GMAF), are adopted in medical applications [1]. Like other information, we are faced with a high volume of features in medical images and textual data. To reduce the cost and time of processing medical images and texts, we should use only the valuable features of images and texts instead of including irrelevant and useless features. Processing fewer and more rele-

Data Fusion Techniques and Applications for Smart Healthcare. https://doi.org/10.1016/B978-0-44-313233-9.00019-9

vant features leads to faster and more accurate diagnosis of disease and organ damage in medical processing.

The data fusion process is performed at three levels: the pixel level, the feature level, and the decision level. We investigated related works at the feature level. Feature fusion is a controversial and challenging topic in image and multimedia processing for medical applications. In the past few years, with the advancement of image and information processing technologies in the field of medicine, the amount of information has increased in such a way that the integration of information and images from several different sources has become a challenging and vital issue. The purpose of our work is to identify and categorize the methods and techniques used in medical databases to integrate images and other medical data. Here, we delve into the classification and examination of related works, taking into account the diverse landscape of machine learning methods, including notable progress in neural networks. We also explore the potential of combining these methods to create new approaches. In the present work, we will aim to present a more comprehensive and inclusive classification by examining the latest methods. Also, we have tried to compare the methods and their strengths and weaknesses. This chapter will overview the most common feature fusion methods for medical data. We will use these methods to investigate the combination of medical data features for medical applications.

We will review several common methods for merging the features of medical images. For each method, we have tried to provide an overview of the latest related articles. The remainder of the chapter is organized as follows: In Sections 13.2–13.7, we will review morphological methods, component substitution-based methods, multiscale decomposition-based methods, deep learning methods, fuzzy logic, sparse representation methods, and some related works. In Section 13.8, some works related to feature fusion on nonimage data are reviewed. In Section 13.9, methods are compared, and in Sections 13.10 and 13.11, future directions and conclusions are given.

13.2 Morphological methods

Morphological operators are widely used in image processing to extract image components and have been very useful for medical images [2]. In this method, morphological techniques navigate the pixels of images with the help of structural elements. This method can improve routes and remove extra points and noise. These techniques are based on a set of operations related to the features of shape and morphology in an image. Morphology algorithms are used as mathematical tools to extract components of images. The structural operator specifies the opening and closing operators, and in fact, the morphological operators are influenced by the structural operator. Morphological operators have been used in medical image models such as MRI and CT to diagnose brain-related issues. Operators such as averaging, morphology towers [3], K-L transforms, and morphological morphology wavelets [4] are used in image fusion.

In [5], a method is proposed that tries to keep useful information in the images during image integration. In this method, top-hat transfer is used to extract image areas, and then the contrast operator extracts the image detail. The multiscale contrast change operator and top-hat transformation operator are used for multiscale images and extracting their details and regions. In the final stage, the regions, edges, and details extracted from the previous stages are combined. It is an effective image fusion algorithm that has achieved much better results by creating images with more valuable information than other common methods. In [6], the nonsampled shearlet transform (NSST) method extracts the high- and low-frequency information of the images. The low-frequency information is integrated by saliency maps. A multiscaled morphology gradient method fuses high frequencies. This method preserves the details and edges of the image. Then the inverse NSST reconstructs the images. This method preserves contrast and more details of images than conventional methods.

13.3 Component substitution-based methods

Data-driven techniques, including statistical-based methods, provide a solid ability to shelter saliency [7]. Principal component analysis (PCA) uses mathematical formulas to reduce the dimensions of images and identify image patterns. PCA extracts image patterns by identifying differences and similarities. By using these patterns, images can be compressed without losing the original information [8]. PCA is performed with eigenvector and eigenvalue matrices. The term "rate of compression" refers to the ability to recover the original image. Using PCA, medical images can be compressed with approximately one fourth of the original size while preserving the original characteristics of the image. Using PCA, the images can preserve their texture and complexity. When recovering the original image from the compressed image, the number of main components at the time of compression affects the result of the work. In PCA, the data are stored in a computational structure that is depicted in a subspace by an orthogonal axes system, and its dimensions are reduced. PCA reduces data dimensionality by using a data covariance matrix, linear algebra, basic statistics, and a multimedia data representation.

13.3.1 How does PCA work?

In PCA, our goal is to find different patterns in the desired data; more precisely, we try to identify correlations between the data. If there is a strong correlation between our data in two or more specific dimensions, those dimensions can be converted into one dimension. By reducing the number of dimensions, the complexity and volume of data are greatly reduced. The ultimate goal of PCA is to find the direction (dimension) with the most variance of the data and reduce the dimension, so that the least amount of important data is lost [9].

The effect of PCA on the compression of a medical image is shown in Figs. 13.1 and 13.2. There are various combinations of Principal Component Analysis (PCA) with other image fusion methods, such as the combination of pyramid techniques and PCA, also known as hierarchical PCA, which obtains more satisfactory results when combining multimodal images [10]. Several PCA-based extended methods have been proposed. Due to the use of one dimension (1D) vectors instead of two dimension (2D) vectors and because the correlation is lost between columns and rows, 2D-PCA is proposed to avoid this drawback [11].

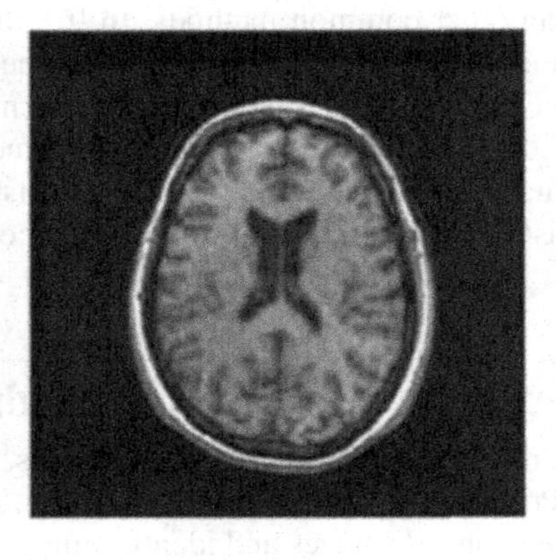

FIGURE 13.1

Original image [9].

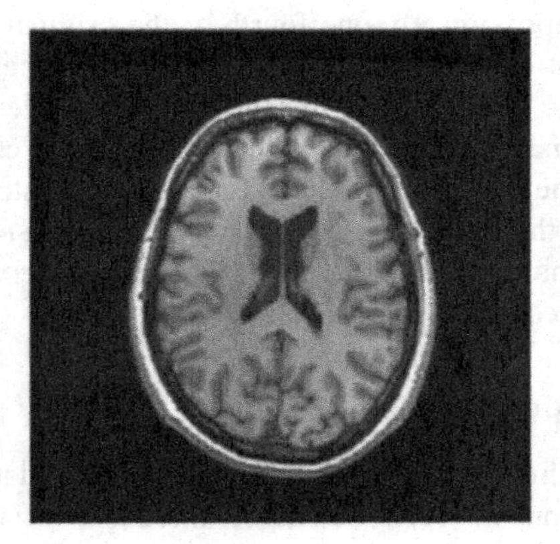

FIGURE 13.2

Recovered image with the PCA method [9].

The authors of [12] have been able to identify the points and positions of brain tumors through a method called robust PCA. In the next step, the task of the convolutional neural network (CNN) method is to identify tumors.

13.4 Multiscale decomposition-based methods

In this method, each image is decomposed into a set of coefficients. An example of the decomposition step for an image is shown in Fig. 13.3. In general, high frequencies show detailed information. Each image is characterized by a group of coefficients, each of which has the ability to provide specific features. By using the capabilities of coefficients, noise can be reflected in a coefficient, and other stable coefficients can be used for further processing. In fact, problematic coefficients can be removed, and useful coefficients can be injected into another image. The injection method can be used with math methods, ranging from simple addition operations and aggregator functions to more complex mathematical models. By injecting applicable coefficients of both images, the quality of the final image increases with the integration of appropriate features, and more decomposition steps are required to obtain a higher resolution.

13.4.1 Feature fusion with wavelet transform

Mathematical transformations such as the Fourier series have many applications in the processing and classification of various data, such as transmission signals and stationary time series. This simple technique will provide excellent performance and high accuracy for many problems. But a general rule about the Fourier transform is that it will perform excellently as long as the frequency spectrum of a signal is statistically stationary. In the case of signals and dynamic time series, wavelet transforms are much more powerful.

Wavelet transforms in medical science are applied in some fields, such as ECG signal processing, EEG signal processing, medical image compression, medical image reinforcing, edge detection, and medical image registration. The discrete wavelet transform (DWT) combines multiple modalities such as positron emission tomography (PET)/MRI [12], PET/CT [13], and MRI/CT [14]. The dual-tree complex wavelet transform (DTCWT) extracts valuable features from various modalities. The nonsubsampled dual-tree complex wavelet transform (NS DT-CxWT) extracts the directional features from coregistered CT and MRI slices [15]. Promising results have been obtained using the redundant discrete wavelet transform (RDWT) in terms of edge strength, mutual information fusion metrics, and shift-invariance [16]. Several applications of the developed wavelet transform have been reported, such as feature fusion using the lifting wavelet transform [17], fusion of CT/PET images using the multiwavelet transform [18], the stationary wavelet transform [19], and feature fusion applying the Daubechies complex wavelet transform [20]. One of the multiscale decomposition tools is the quaternion wavelet transform, which gives better results than the Discrete Wavelet Transform (DWT) and the wavelet transform [21].

13.4.2 Wavelet-based combination feature fusion

In these methods, a feature fusion method is integrated with the features extracted from the wavelet, making the feature fusion more efficient and effective. Combinations of wavelet and other methods, such as contourlet [19], curvelet [23], shearlet

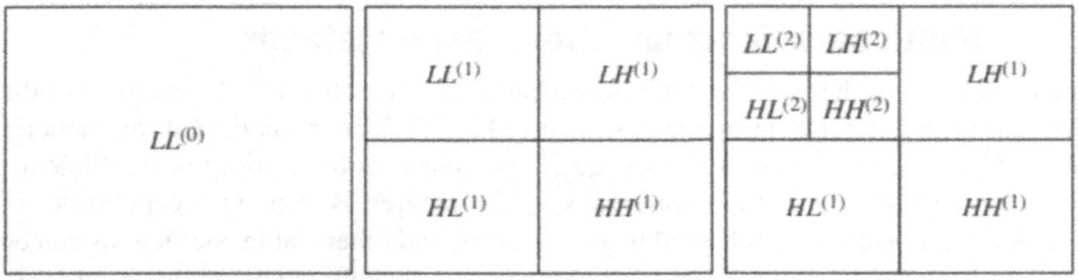

$LL^{(0)}$ | $LL^{(1)}$ $LH^{(1)}$ $HL^{(1)}$ $HH^{(1)}$ | $LL^{(2)}$ $LH^{(2)}$ $HL^{(2)}$ $HH^{(2)}$ $LH^{(1)}$ $HL^{(1)}$ $HH^{(1)}$

FIGURE 13.3

Wavelet decomposition [22].

[24], and ripplet [25], have been proposed to crush the restrictions of wavelets in choosing more salient directional features.

In previous studies, a method was proposed that combines the features of the wavelet methods and the features extracted from other methods. One of the methods that can be suitably combined with the wavelet method is the neural network. In processing medical images, various methods are combined with the wavelet method for feature fusion. For example, vector machine, magnetic resonance angiograms, entropy, independent component analysis, genetic methods, and fuzzy logic are combined with wavelets. Methods such as neural networks effectively reduce the number of features. Fusion neural networks and the wavelet transform are more effective than wavelet-based methods such as the multifeature fusion method for medical image retrieval using wavelet and bag-of-features [22].

The proposed method works as follows. First the wavelet method divides the original image into four coefficients and resolutions based on the resolution of the first level of the image, then two gray-based bag-of-feature methods extract related features, and in the third step, based on texture features from the second level the image resolution is selected. LBP features are obtained from the third level of resolution. Now, the features extracted from the previous steps are fused and delivered to the retrieval system to retrieve the desired image from the database. In the final step, based on similarity, the requested images are retrieved from the database, and the output of the retrieval system is checked. Fig. 13.4 is an overview of the proposed method. The aim of the work [26] was to watermark medical images for the authentication of medical data in healthcare applications. In this paper, RDWT is used to obtain the component with the highest entropy of multiple images merged by fast filtering. In the next step, a single-level discrete wavelet is applied to the decomposed components and embedded watermark. Finally, the final image is encrypted.

13.4.3 Feature fusion with contourlet methods

Mathematical transformations such as contourlet transform are methods to increase the efficiency of feature fusion. In fact, this method can capture the geometric structure in medical images. In image processing, the way to present an image is significant if the surrounding boundary can be extracted. Wavelets and Fourier transform

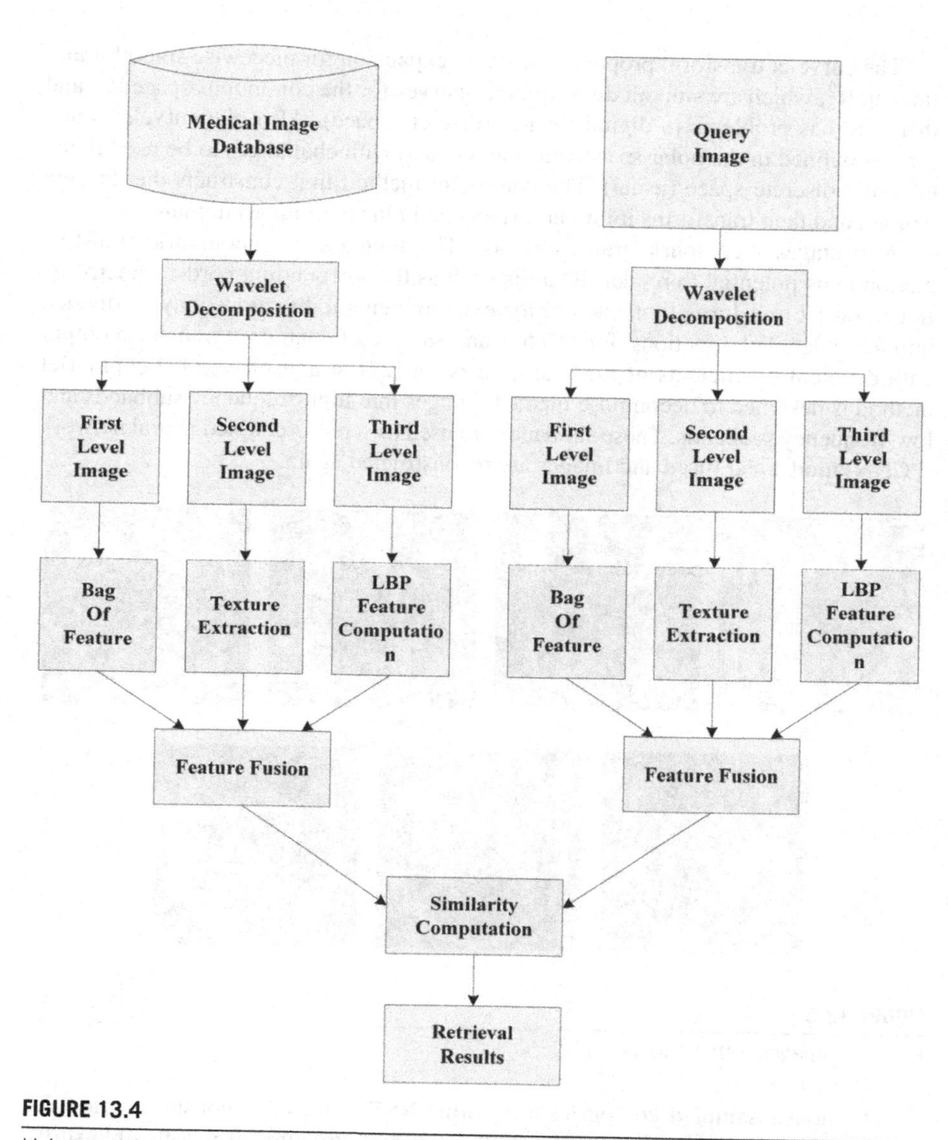

FIGURE 13.4

Using wavelet, bag-of-features, and multifeature fusion methods [22].

are common and distinct extensions of 1D transforms that are used to obtain edge geometry. Wavelets only deal with 1D units. For 1D piecewise smooth signals, a wavelet is a useful tool. For the 2D case, generally, discrete wavelets obtained by outer multiplication of 1D wavelets are used to render some areas, especially if they are not suitable along the perimeter. The contourlet transform solves this problem and is extended to the continuous domain.

The curvelet transform proposes a discrete expansion for piecewise smooth functions in R^2, which are smooth discontinuous curves for the continuous space R^2, and therefore has problems in digital images (discrete space). Also, the curvelet transform is defined in the polar space and is associated with challenges to be used in the nonpolar discrete space (usual). The contourlet method first constructs the discrete domain and then transforms it into an expansion in the continuous domain.

Advantages of contourlet transformation: The main feature of contourlet transformation is its potential to render 2D units such as the surrounding border effectively. In contourlet transformation, each of the existing general directions may be divided into more detailed directions. Fig. 13.5 is an example of contourlet transform output with different coefficients of medical images. In [27], a nonsubsampled contourlet method is designed to decompose medical images into high-frequency subbands and low-frequency subbands. These subbands are used by a pulse-coupled neural network (PCNN) model and fused and images are reconstructed again.

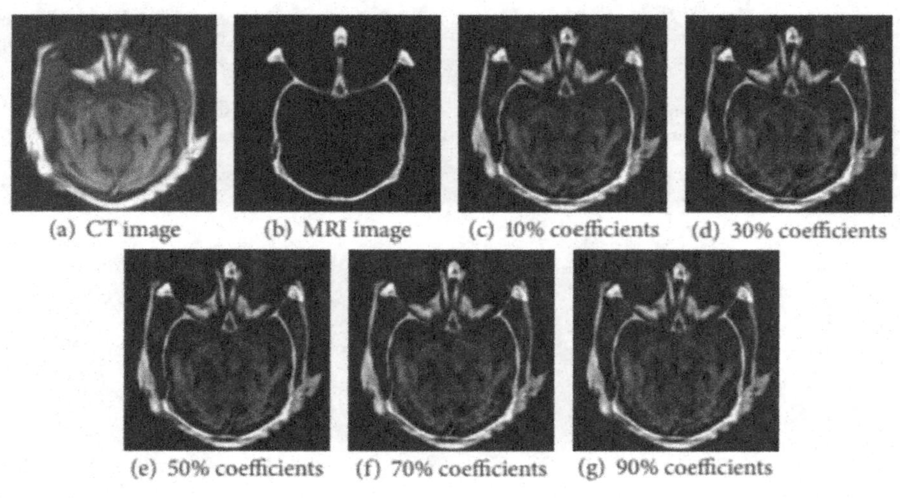

FIGURE 13.5

Fusion of images with contourlet [28].

The nonsubsampled contourlet transform (NSCT) includes nonsubsampled directional filter banks and a nonsubsampled pyramid structure. It is a flexible shift-invariant, multiscale, multidirectional image decomposition method that is implemented through an à trous algorithm [29]. Below we list several works that have used the contourlet transform.

NSCT fuses PET scans and MRIs using maximal energy and variance rules [30]. A dual-tree complex contourlet transform (DT-CCT) combines MRIs and CT scans to obtain an image including complementary information for diagnosis using fusion rules of PCA [31]. In [32], NSCT and perceptual high-frequency CNN (PHF-CNN) are used to integrate medical images. The task of NSCT is to decompose images into high- and low-frequency subbands, and the task of PHF-CNN, which is trained with the frequency domain, is to fuse the frequency subbands. Finally, an inverse NSCT reconstructs the images.

13.4.4 Ridgelet transform

Using some rules based on the wavelet functions, multiscale geometric decomposition (MGD) crushes the restriction of directionality and suggests exact treatment within wavelets. Directional or geometrical transforms are divided into two groups: nonadaptive transforms (curvelet, contourlet) and adaptive transforms (wedgelet, bandelet) [33]. Wavelets have a weak performance when decomposing signals in a two-dimensional context to distinguish 1D singularities against ridgelets. The ridgelet transform performs better than the wavelet transform in illustrating 1D singularities [34]. The ridgelet transform is aimed to discover a sparse indication of objects with straight discontinuities [35]. Another approach combinates the ridgelet transform and wavelet transform using PCA to eliminate their irrelevant features for CT scans and MRIs taken from "The Whole Brain Atlas" [36].

13.4.5 Curvelet transform

The curvelet transform was devised as an extension of the Ridgelet transform, aiming to overcome limitations associated with the Fourier transform by providing enhanced capabilities in capturing directional and multiscale information in signal processing [37]. The curvelet transform can denote curves as a superposition of bases with various lengths and widths. The curvelet transform proposes a discrete expansion for piecewise smooth functions in R^2, which are smooth discontinuous curves for the continuous space R^2. At first, curvelets were applied to maintain edges because PCA and DWT had weak performance in preserving edges [38]. Various CT, single-photon emission computed tomography (SPECT), PET, and MRI images were fused using the curvelet transform to extract important information from the original images [39]. In [40], a curvelet method named weighted fast discrete curvelet transform (W-FDCuT) is introduced to analyze high- and low-frequency subbands. The type-2 fuzzy entropy method integrates high-frequency subbands, and low-frequency subbands are fused by an averaging method. The final fused image is provided through the inverse of W-FDCuT.

13.5 Deep learning methods for feature fusion

Deep learning is the ability of multiple processing layers to learn different levels of abstraction. Neural networks are trained through an end-to-end architecture and can integrate data with high accuracy. Neural networks can adapt to different models through training. The network architecture is designed to be able to reduce input dimensions. Deep learning tries to imitate brain learning through artificial neural networks. The medical data category is vital for computer-aided diagnosis (CAD) methods. Classic fusion techniques rely mainly on form, color, and surface features. Current deep learning techniques provide a valuable method for designing an end-to-end standard that can estimate last category tags. The most common deep learning models are CNN, generative adversarial networks (GANs), and recurrent neural networks.

An example of a hybrid method [48] is the Coding Network with Multilayer Perceptron (CNMP), which combines features extracted from a trained deep convolutional network with features extracted by conventional methods from medical images. This efficient model is based on neural networks that are proposed to connect groups of different features together.

Process of the proposed model

Since 1990, CNN has been used to identify images and videos with acceptable results.

CNNs consist of three parts; the fully connected layers are used for feature extraction, and the softmax layer is in charge of classifying objects. The working steps of a deep model are as follows:

1- image preprocessing,
2- choosing the appropriate activator function,
3- appropriate initial weighting,
4- adding to the number of data, such as extracting random patches from the original images,
5- avoiding overfitting,
6- appropriate choice of learning rate in each epoch.

When we have extracted the high-level features from the neural network, we will merge these feature groups with traditional features in the next step. Opting for the merger of high-level and traditional features presents a pragmatic strategy, leveraging the contextual depth of high-level features alongside the reliability inherent in traditional ones, thereby enhancing the overall effectiveness of the method that can set the appropriate ratio between these two feature groups. This method uses a multilayer perceptron neural network that is trained to fuse traditional and high-level features in nonlinear space [41]. Fig. 13.6 depicts the work steps.

In [42], a CAD system is used to detect normal and abnormal tissues and pathological stages in breast cancer, but due to mammography images encounter challenges such as diminished contrast, variations in lesion locations, and diverse tumor shapes, the classification of these images is a challenging issue. The results of this method have been evaluated on two publicly available datasets, MIAS and BCDR, and this method achieved good results in terms of sensitivity, specificity, and accuracy. The general flow diagram of the CAD system for this method is shown in Fig. 13.7.

In [43], a shallow machine learning algorithm is defined that uses a PCNN and a steerable pyramid. First, the steerable pyramid decomposes the original images into subbands in different directions and levels. Then PCNN fuses the high weights, and the low-frequency subbands are also fused by weighting. Finally, the inverse steerable

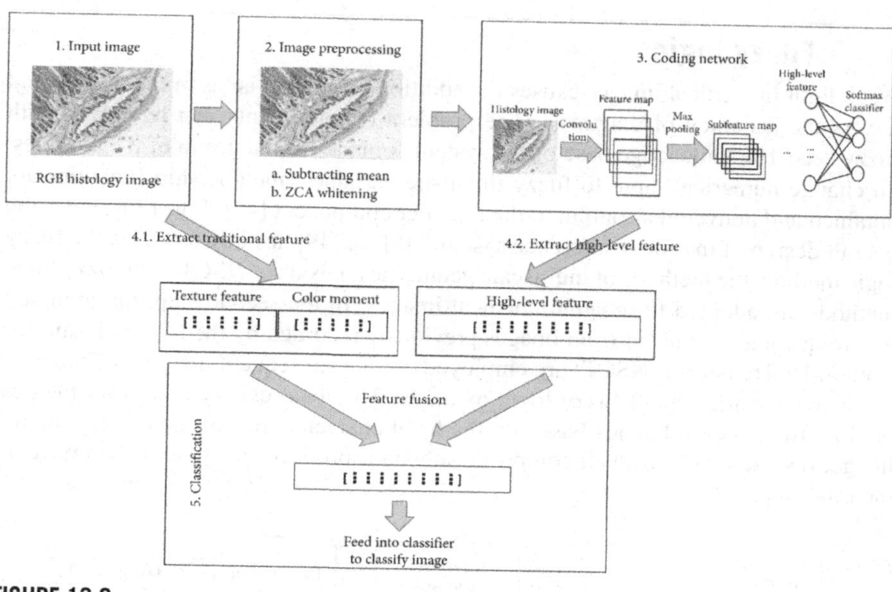

FIGURE 13.6

Feature fusion by a multilayer perceptron [41].

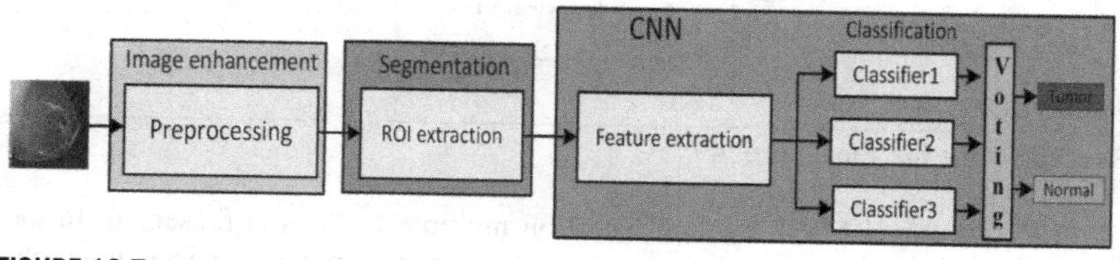

FIGURE 13.7

Classification of mammography images with a CAD system [42].

pyramid transform creates the fused image. In [44], a method based on CNN has shown better performance than existing 2D and 3D image integration methods and has provided a loss function that preserves more features. In [45], a new method for the fusion of medical images is introduced by combining the old shearlet method and deep learning methods. The advantage of this method compared to other methods is higher efficiency and correlation and a lower error rate. In [46], a self-supervised model based on a GAN model named U-patch GAN is presented, which has the ability to accurately preserve functional and structural information for multimodal brain images.

13.6 Fuzzy logic

Weak light in medical images causes the addition of unclear areas in the images and lowers the accuracy of diagnosis. The problem of uncertainty can be solved with fuzzy sets. The knowledge base of the system is made from a group of if-then rules. To change numerical input to fuzzy linguistic variables, membership functions are obtained and delivered as output to the fuzzifier component [47]. Fig. 13.8 shows the overall design of the fuzzy inference system [48]. In [49], a study based on the fuzzy logic method, the methods of multiscale geometric analysis of NSCT with fuzzy logic methods are adopted to integrate medical images. In the second step, the high and low frequencies extracted from images previously recorded by the Non-Subsampled Contourlet Transform (NSCT) are employed, with a subsequent merging of the high frequencies using type-2 fuzzy logic methods. The local energy algorithm merges the low-frequency subbands based on the local characteristics of the corresponding image. Inverse NSCT with all composite subbands produces the final fused image in the final step.

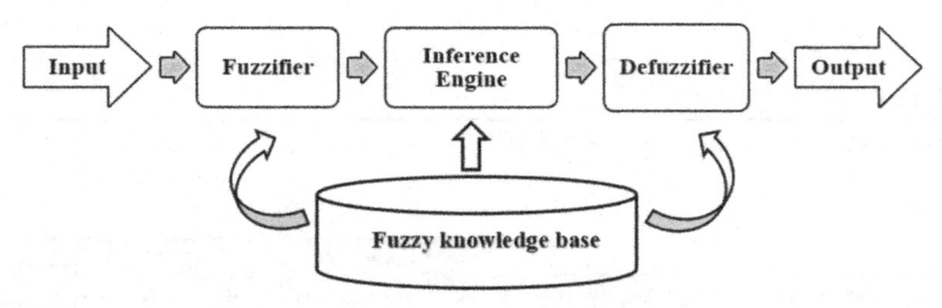

FIGURE 13.8

Structure of the fuzzy inference [49].

In [23], a novel MVIF method based on multiple features is presented. In the first step, 3DST decomposes the preregistered volumetric medical images into high- and low-frequency coefficients. Then local energy fuses the low-frequency subbands. CHMM statistical characteristics, type-2 fuzzy entropy, the gradient, and energy of the high-frequency coefficient together create a new fusion method that combines high-frequency subbands. Finally, inverse 3DST with all composite subbands creates the final fusion image. In [50], a new fuzzy metric is used to identify pixels. In fact, the input images are processed by fuzzy rules, and the results of this layer are given to the CNN. Then the hierarchical attention model extracts more effective features.

13.7 Sparse representation methods

It is possible to access the natural scattering of signals similar to the human visual system in sparse display methods. The objective of these methods is to find the solution for a system of linear equations. In [11], a combination of sparse representation methods along with NSCT is proposed to generate a powerful image fusion method. This method has achieved good results. The low-pass bands are merged using sparse representation-based fusion. The absolute values of the coefficients of the high-pass bands are fused as a measure of the level of activity. The inverse NSCT operates on the integrated coefficients, and the final fused image is created. In terms of visual quality and objective evaluations, this method performs better than NSCT and sparse representation methods. In [51], a new method with sparse representation and neighbor energy activity is designed for merging medical images. The merged output images have less error and higher quality than several methods in terms of color, brightness, and details. Each image is divided into detail layers and base layers. The base layers are fused together by sparse representation, and the task of the neighboring energy activity operator is to record the features of the image that have changed in the detail layer. Finally, by combining the selective layers, you get the final output. The working method of [52] is that, unlike standard sparse representations, it obtains several convolutional sparse representations and the texture of images according to prelearned dictionaries. Images are reconstructed by integrating sparse coefficients based on a relevant dictionary.

13.8 Feature fusion on medical data

In order to increase the accuracy of diagnostic results and improve treatment results, it is better to use other relevant accurate nonimaging data from patients' medical and laboratory histories in addition to pixel data. This helps to improve the interpretation of medical imaging data and increases the accuracy of clinical diagnoses. To obtain such a goal, text and nonpixel data from electronic health records (EHRs) should be processed [53]. Imaging pixel data, structured laboratory data, unstructured narrative data, and in some cases, audio or observational data are the data that together help in accurate diagnosis in modern medicine. If clinical and laboratory data are not used during image interpretation, the accuracy of the diagnosis process will decrease [53]. Clinical diagnosis accuracy based on medical images along clinical contexts, such as the patient's history, chief complaints, laboratory values, and previous diagnoses, goes up.

Although many studies have been conducted in the field of image and text processing in computer vision and natural language research, limited studies have been done to integrate images and EHR data. For concatenation of features in both models, early and intermediate fusion is performed. The ability of machine learning and deep learning methods to solve many problems in different fields, including the medical field, is a reason for combining different EHR data and medical images. CNN is one of the deep learning methods that perform the analysis and medical image classification. In [54], a method is proposed that utilizes Electronic Health Record (EHR) features from different subspaces through a multimodal attention mechanism module in the process of selecting important regions during the image feature extraction process performed by traditional CNN. Two modules, named multimodal attention and multihead GMU, are introduced in this work to combine EHR information and medical images. EHR data help GMU in dimension-reduced visual information for the final classification and multimodal attention in selecting important visual areas. The results show that the proposed method has better results than other methods on two datasets.

Accurate identification of the underlying cause is essential for appropriate treatment for patients with acute respiratory failure. With machine learning methods, it is possible to combine EHR data and chest X-ray images and achieve higher accuracy in diagnosis [55]. The results of this method showed that the combination of EHR data and radiographic images has a better performance than each of the methods alone.

In [56], with the goal of diagnosing coronary heart disease, a less dangerous and more accurate method than the old methods has been presented with the help of CNNs. The proposed method fuses ultrasound images and electronic medical records through CNNs.

13.9 Comparison of data fusion methods

In order to achieve better results in diagnosis, it is necessary to merge the medical data of different models. Examining the advantages and disadvantages of various methods will help researchers to choose the most suitable method for diagnosis. In this section, we compare the pros and cons of each data fusion method (Tables 13.1 and 13.2).

13.10 Future works

One of the important challenges of feature fusion in medical images is information loss, which can be solved with the help of deep learning and its power in feature extraction. Another issue is that suitable criteria and sufficient datasets in the field of medical image fusion are scarce. For example, although CNNs are widely used in the field of image processing, there are usually no suitable datasets for their training. This problem has made it impossible to use neural networks effectively alone. Also, the metrics of a particular method may not be usable for other methods, and this shows the importance of objective evaluation. Therefore, one of the future trends can be collecting large datasets and designing suitable criteria to check the efficiency of new methods. The popularity of deep learning in different fields will increase the motivation to use more of these methods in the field of feature fusion in medical images in future works.

13.11 Conclusion

In the current work, we discussed different up-to-date methods of medical data integration. According to our studies, scale analysis methods have a high ability to

Table 13.1 The pros and cons of each data fusion method.

Method	Advantages	Weaknesses
Multiscale decomposition	• High ability to preserve image details • Reduce geometric distortion • High ability to record geometric singularities, edges, and texts	• The inability of wavelets and pyramids to distinguish anisotropic features • The effect of fusing rules and decomposition level on the final efficiency of the method • Computational complexity
Fuzzy logic	• Acceptable performance on blurred images • Low impact of ambiguous information	• Need to define the optimal membership function
Sparse representation	• Providing more meaningful and stable information • Mis-registration control • It requires learning a dictionary • The need for time and high memory for the sparse coding process	• Unsuitable for real-time applications • Unable to retain details
Subspace and component substitution	• Dimension reduction of image features as preprocessing • Preservation of spatial details • Suitable for real-time applications due to its speed	• Presence of spatial distortion in some PCA applications
Morphological	• Simple and fast implementation	• The effect of pixel density on the efficiency of the method • Extreme sensitivity to noise
Deep learning	• High ability to fuse features	• The need for comprehensive and large datasets for network training • The need for special network design in new issues • A large number of hyperparameters and high computations are needed to tune them

preserve edges due to direction sensitivity and geometric compatibility. Complementary information from medical images cannot be extracted in the spatial domain, but it can be obtained from the transformation domain. Computational intelligence methods such as neural networks and fuzzy logic, along with multiscale analysis logic methods, achieve good results in the combination of features. Fuzzy logic methods are adept at addressing uncertainty issues, and these approaches demonstrate high efficiency not only in operator decision-making but also in tasks involving feature selection. Problems that require much time or contain high noise can be solved with statistical and component replacement methods. In the field of vision tasks, deep learning has been able to achieve acceptable results and improvements due to its

Table 13.2 Characteristics of each fusion method [57].

Method	Information loss	Information content	Difficulty	Need for preprocessing	Noise sensitivity
Morphological	Low	High	Low	Low	High
Subspace and component substitution	High	low	Medium	Low	High
Multiscale decomposition	Low	High	Low	Low	Medium
Deep learning	Low	High	High	High	Low
Sparse representation	High	low	High	High	Medium
Fuzzy logic	Low	High	High	Medium	Low

different concepts and architecture. In image fusion, CNN has the most suitable architecture despite the high computational requirements compared to other supervised and unsupervised models.

References

[1] S. Wagenpfeil, et al., Explainable multimedia feature fusion for medical applications, Journal of Imaging 8 (4) (2022) 104.

[2] P. Soille, Morphological Image Analysis: Principles and Applications, vol. 2, Springer, 1999.

[3] S. Mukhopadhyay, B. Chanda, Fusion of 2D grayscale images using multiscale morphology, Pattern Recognition 34 (10) (2001) 1939–1949.

[4] B. Yang, Z. Jing, Medical image fusion with a shift-invariant morphological wavelet, in: 2008 IEEE Conference on Cybernetics and Intelligent Systems, IEEE, 2008.

[5] X. Bai, Morphological image fusion using the extracted image regions and details based on multi-scale top-hat transform and toggle contrast operator, Digital Signal Processing 23 (2) (2013) 542–554.

[6] Y. Zhang, M. Jin, G. Huang, Medical image fusion based on improved multi-scale morphology gradient-weighted local energy and visual saliency map, Biomedical Signal Processing and Control 74 (2022) 103535.

[7] J. Du, et al., An overview of multi-modal medical image fusion, Neurocomputing 215 (2016) 3–20.

[8] K.C. Bhataria, B.K. Shah, A review of image fusion techniques, in: 2018 Second International Conference on Computing Methodologies and Communication (ICCMC), IEEE, 2018.

[9] R.d.E. Santo, Principal component analysis applied to digital image compression, Einstein (São Paulo) 10 (2012) 135–139.

[10] U. Patil, U. Mudengudi, Image fusion using hierarchical PCA, in: 2011 International Conference on Image Information Processing, IEEE, 2011.

[11] Q. Nawaz, et al., Multi-modal medical image fusion using 2DPCA, in: 2017 2nd International Conference on Image, Vision and Computing (ICIVC), IEEE, 2017.

[12] V. Bhavana, H. Krishnappa, Multi-modality medical image fusion using discrete wavelet transform, Procedia Computer Science 70 (2015) 625–631.

[13] S. Cheng, J. He, Z. Lv, Medical image of PET/CT weighted fusion based on wavelet transform, in: 2008 2nd International Conference on Bioinformatics and Biomedical Engineering, IEEE, 2008.

[14] Y. Zheng, et al., A new metric based on extended spatial frequency and its application to DWT based fusion algorithms, Information Fusion 8 (2) (2007) 177–192.

[15] S.N. Talbar, S.S. Chavan, A. Pawar, Non-subsampled complex wavelet transform based medical image fusion, in: Proceedings of the Future Technologies Conference, Springer, 2018.

[16] R. Singh, A. Khare, Redundant discrete wavelet transform based medical image fusion, in: Advances in Signal Processing and Intelligent Recognition Systems, Springer, 2014, pp. 505–515.

[17] O. Prakash, et al., Multiscale fusion of multimodal medical images using lifting scheme based biorthogonal wavelet transform, Optik 182 (2019) 995–1014.

[18] Y. Liu, J. Yang, J. Sun, PET/CT medical image fusion algorithm based on multiwavelet transform, in: 2010 2nd International Conference on Advanced Computer Control, IEEE, 2010.

[19] F. Shabanzade, H. Ghassemian, Combination of wavelet and contourlet transforms for PET and MRI image fusion, in: 2017 Artificial Intelligence and Signal Processing Conference (AISP), IEEE, 2017.

[20] R. Singh, A. Khare, Fusion of multimodal medical images using Daubechies complex wavelet transform–a multiresolution approach, Information Fusion 19 (2014) 49–60.

[21] P. Chai, X. Luo, Z. Zhang, Image fusion using quaternion wavelet transform and multiple features, IEEE Access 5 (2017) 6724–6734.

[22] L. Shuang, et al., Multi-feature fusion method for medical image retrieval using wavelet and bag-of-features, Computer Assisted Surgery 24 (sup1) (2019) 72–80.

[23] C. Kavitha, C. Chellamuthu, Fusion of SPECT and MRI images using integer wavelet transform in combination with curvelet transform, The Imaging Science Journal 63 (1) (2015) 17–23.

[24] C. Zhang, M. Fang, Brain MRI tumor image fusion combined with shearlet and wavelet, in: LIDAR Imaging Detection and Target Recognition 2017, SPIE, 2017.

[25] C. Kavitha, C. Chellamuthu, R. Rajesh, Medical image fusion using combined discrete wavelet and ripplet transforms, Procedia Engineering 38 (2012) 813–820.

[26] M. Sajeer, A. Mishra, A robust and secured fusion based hybrid medical image watermarking approach using RDWT-DWT-MSVD with hyperchaotic system-Fibonacci Q matrix encryption, Multimedia Tools and Applications (2023) 1–23.

[27] S.I. Ibrahim, M. Makhlouf, G.S. El-Tawel, Multimodal medical image fusion algorithm based on pulse coupled neural networks and nonsubsampled contourlet transform, Medical & Biological Engineering & Computing 61 (1) (2023) 155–177.

[28] H. Huang, X.a. Feng, J. Jiang, Medical image fusion algorithm based on nonlinear approximation of contourlet transform and regional features, Journal of Electrical and Computer Engineering 2017 (2017).

[29] A.L. Da Cunha, J. Zhou, M.N. Do, The nonsubsampled contourlet transform: theory, design, and applications, IEEE Transactions on Image Processing 15 (10) (2006) 3089–3101.

[30] N. Amini, E. Fatemizadeh, H. Behnam, MRI-PET image fusion based on NSCT transform using local energy and local variance fusion rules, Journal of Medical Engineering & Technology 38 (4) (2014) 211–219.

[31] N. Al-Azzawi, W.A.K. Wan Abdullah, Improved CT-MR image fusion scheme using dual tree complex contourlet transform based on PCA, International Journal of Information Acquisition 7 (02) (2010) 99–107.
[32] Z. Wang, et al., Medical image fusion based on convolutional neural networks and non-subsampled contourlet transform, Expert Systems with Applications 171 (2021) 114574.
[33] M. Zaouali, S. Bouzidi, E. Zagrouba, Review of multiscale geometric decompositions in a remote sensing context, Journal of Electronic Imaging 25 (6) (2016) 061617.
[34] A. Krishn, V. Bhateja, A. Sahu, PCA based medical image fusion in ridgelet domain, in: Proceedings of the 3rd International Conference on Frontiers of Intelligent Computing: Theory and Applications (FICTA) 2014, Springer, 2015.
[35] E.J. Candes, D. Donoho, A surprisingly effective nonadaptive representation for objects with edges, in: Curves and Surfaces, 1999.
[36] V. Bhateja, et al., Medical image fusion in wavelet and ridgelet domains: a comparative evaluation, International Journal of Rough Sets and Data Analysis 2 (2) (2015) 78–91.
[37] E. Candes, et al., Fast discrete curvelet transforms, Multiscale Modeling & Simulation 5 (3) (2006) 861–899.
[38] V. Bhateja, A. Krishn, A. Sahu, Medical image fusion in curvelet domain employing PCA and maximum selection rule, in: Proceedings of the Second International Conference on Computer and Communication Technologies, Springer, 2016.
[39] P. Mathiyalagan, Multi-modal medical image fusion using curvelet algorithm, in: 2018 International Conference on Advances in Computing, Communications and Informatics (ICACCI), IEEE, 2018.
[40] N. Nagaraja Kumar, T. Jayachandra Prasad, K.S. Prasad, An intelligent multimodal medical image fusion model based on improved fast discrete curvelet transform and Type-2 fuzzy entropy, International Journal of Fuzzy Systems 25 (1) (2023) 96–117.
[41] Z. Lai, H. Deng, Medical image classification based on deep features extracted by deep model and statistic feature fusion with multilayer perceptron, Computational Intelligence and Neuroscience 2018 (2018).
[42] I.U. Haq, et al., Feature fusion and ensemble learning-based CNN model for mammographic image classification, Journal of King Saud University: Computer and Information Sciences (2022).
[43] H. Deng, Y. Ma, Image fusion based on steerable pyramid and PCNN, in: 2009 Second International Conference on the Applications of Digital Information and Web Technologies, IEEE, 2009.
[44] Y. Liu, et al., Multimodal mri volumetric data fusion with convolutional neural networks, IEEE Transactions on Instrumentation and Measurement 71 (2022) 1–15.
[45] A. Mergin, M.G. Premi, Shearlet transform-based novel method for multimodality medical image fusion using deep learning, International Journal on Computational Intelligence and Applications (2023) 2341006.
[46] C. Fan, H. Lin, Y. Qiu, U-Patch GAN: a medical image fusion method based on GAN, Journal of Digital Imaging 36 (1) (2023) 339–355.
[47] B. Biswas, B.K. Sen, Medical image fusion technique based on type-2 near fuzzy set, in: 2015 IEEE International Conference on Research in Computational Intelligence and Communication Networks (ICRCICN), IEEE, 2015.
[48] M. Joodaki, M.B. Dowlatshahi, N.Z. Joodaki, An ensemble feature selection algorithm based on PageRank centrality and fuzzy logic, Knowledge-Based Systems 233 (2021) 107538.
[49] Y. Yang, et al., Multimodal sensor medical image fusion based on type-2 fuzzy logic in NSCT domain, IEEE Sensors Journal 16 (10) (2016) 3735–3745.

[50] A. Albahri, et al., A systematic review of trustworthy and explainable artificial intelligence in healthcare: assessment of quality, bias risk, and data fusion, Information Fusion (2023).
[51] X. Li, et al., Medical image fusion based on sparse representation and neighbor energy activity, Biomedical Signal Processing and Control 80 (2023) 104353.
[52] Y.-D. Zhang, et al., Advances in multimodal data fusion in neuroimaging: overview, challenges, and novel orientation, Information Fusion 64 (2020) 149–187.
[53] S.-C. Huang, et al., Fusion of medical imaging and electronic health records using deep learning: a systematic review and implementation guidelines, npj Digital Medicine 3 (1) (2020) 1–9.
[54] C. Jiang, et al., Fusion of medical imaging and electronic health records with attention and multi-head mechanisms, arXiv preprint, arXiv:2112.11710, 2021.
[55] S. Jabbour, et al., Combining chest X-rays and electronic health record (EHR) data using machine learning to diagnose acute respiratory failure, Journal of the American Medical Informatics Association 29 (6) (2022) 1060–1068.
[56] B. Yang, et al., M-US-EMRs: a multi-modal data fusion method of ultrasonic images and electronic medical records used for screening of coronary heart disease, in: Bioinformatics Research and Applications: 18th International Symposium, ISBRA 2022, Haifa, Israel, November 14–17, 2022, Proceedings, Springer, 2023.
[57] H. Hermessi, O. Mourali, E. Zagrouba, Multimodal medical image fusion review: theoretical background and recent advances, Signal Processing 183 (2021) 108036.

CHAPTER

Review on hybrid feature selection and classification of microarray gene expression data

14

L. Meenachi and S. Ramakrishnan
Department of Information Technology, Dr. Mahalingam College of Engineering and Technology, Pollachi, Tamilnadu, India

14.1 Introduction and motivation

The first methods used for cancer classification had limited diagnostic ability. Microarray gene expression data are applied for addressing such challenges. One of humanity's most serious and deadly diseases is cancer, in which cells grow abnormally and spread to different parts of the human body [1]. In the human body, normal cells grow during their lifetime and then die. Abnormal cells grow, multiply, and spread in the human body, which results in cancer [2]. Such abnormal cells, when they lead to death, are called a malignant tumor; if the disease is not fatal, it is called a benign tumor. Malignant tumors spread to different parts of the human body, and this process is accelerated when they enter the lymph system or the blood circulation. Benign tumors also consist of abnormal cells but they never migrate to other parts of the body.

Microarray gene expression data are presented in table format, wherein the row represents the sample and the column represents the gene [3]. The 2D structure of the microarray dataset is shown in Fig. 14.1 [4].

Selecting relevant features in order to pick the model through classification is the process of feature selection [5][34]. Before the classification phase, feature selection is a crucial step to select the relevant features and eliminate the irrelevant and redundant features from the microarray data [6]. Feature selection is performed prior to data classification because it makes the model easier to use for analysis, it takes less time and money to process the data, and it solves the issue of high-dimensionality [7]. For selecting pertinent features, a variety of machine learning techniques are available. These algorithms are more scalable, trustworthy, and exact. To select the subset, various evaluation metrics and search strategies are used.

Data Fusion Techniques and Applications for Smart Healthcare. https://doi.org/10.1016/B978-0-44-313233-9.00020-5

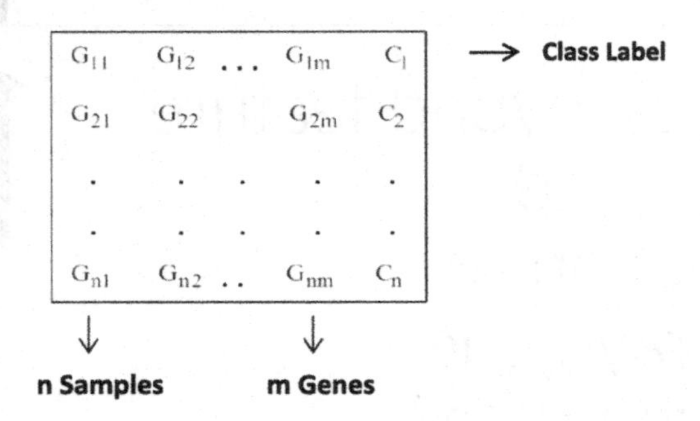

FIGURE 14.1

The structure of nXm microarray dataset.

A global optimal feature selection algorithm has been developed to alleviate the problems existing in feature selection methods. It selects the relevant features with improved accuracy and less computation time. Since there is a pitfall in selecting the neighborhood's features, the optimal feature selection methods are developed by combining the global and local optimal feature selection. These methods are evaluated based on their classification performance; the contributions are listed below.

- Feature selection involves a trade-off between the number of features and the cost and time of feature selection.
- Apparently there is a dearth in selection of feature subsets and a deficit in identification of suitable objective functions for solving the search optimization.
- Global optimal search algorithms lack in searching the neighborhood's features and local optimal search algorithms fall short in searching the whole possible solution space.
- Hybridization is the benchmark of this chapter to derive the results. Compatibility issues are identified in hybridizing neighborhood feature selection with global optimal feature selection.

Nature-inspired metaheuristics algorithms for feature subset search reduce the time required for selecting features from a large number of features and resolve the complexity in searching the features. This chapter describes various metaheuristic search techniques to search for the subsets of features and hybridization of search algorithms with the subset evaluation techniques. The selected feature subset's performance after classification is compared with other existing techniques based on different performance metrics.

14.2 Background, definitions, and notations

14.2.1 Dataset description

In this chapter, we compare the performance of the proposed feature selection method with current algorithms using benchmark datasets [9]. Four benchmark microarray datasets were employed, and Table 14.1 lists the features and instances for each class [18].

Small Round Blue-Cell Tumors (SRBCT) is one of the datasets for cancerous tissues. Such tissues are visible through the microscope.Breast cancer (Breast) can be seen as the growth of malignant cells on the lobules or ducts of the breasts. Leukemia is abnormal growth of bone marrow cells, which in turn increases the white blood cells in the body uncontrollably. Diffuse large B cell lymphoma (DLBCL) is an aggressive cancerous tumor which can grow in the entire body [8].The level of illness of a patient is associated with the rapidly growing mass of abnormal cells

Table 14.1 Dataset descriptions.

No.	Datasets
1	Small Round Blue-Cell Tumors (SRBCT) – It is one of the datasets for cancerous tissues. Such tissues are visible through the microscope. It contains 2309 features and 63 instances.
2	Breast cancer (Breast) – Breast cancer is the growth of malignant cells on the lobules or ducts of the breasts. This dataset contains 9217 features and 54 instances.
3	Leukemia – Leukemia is caused by abnormal growth of bone marrow cells, which in turn increases the white blood cells in the body uncontrollably. This dataset contains 12,583 features and 57 instances.
4	Diffuse large B cell lymphoma (DLBCL) – This is an aggressive cancerous tumor, which can grow in the entire body. The level of illness of a patient is associated with the rapidly growing mass of abnormal cells. This dataset contains 4027 features and 58 instances.

14.2.2 Feature selection and classification

Feature selection is the identification of significant features from the microarray gene expression data [25]. The performance of the classification increases when the contributing features are applied for classification. The feature selection starts with searching of feature subsets and among the subsets the best one is evaluated to select the contributing features. The metaheuristic search of features searches the subset of features either globally or locally and then the subsets are evaluated by the learning algorithm to assess their effectiveness [26]. Classification algorithms are used to evaluate the feature subsets so that the best subset could be chosen and also to evaluate how well the feature selection algorithms function [30]. As the microarray gene expression data consist of few instances and thousands of features, a few challenges occur in feature selection and classification [20][36], as shown in Fig. 14.2. The proposed hybrid feature selection techniques overcome the challenges that occur using the microarray gene expression data.

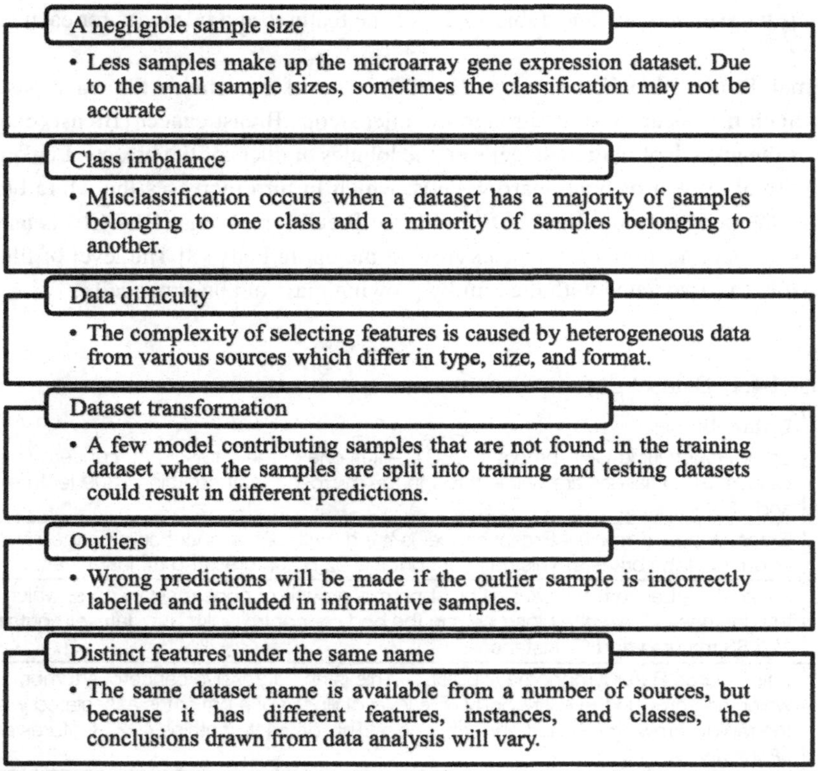

A negligible sample size

- Less samples make up the microarray gene expression dataset. Due to the small sample sizes, sometimes the classification may not be accurate

Class imbalance

- Misclassification occurs when a dataset has a majority of samples belonging to one class and a minority of samples belonging to another.

Data difficulty

- The complexity of selecting features is caused by heterogeneous data from various sources which differ in type, size, and format.

Dataset transformation

- A few model contributing samples that are not found in the training dataset when the samples are split into training and testing datasets could result in different predictions.

Outliers

- Wrong predictions will be made if the outlier sample is incorrectly labelled and included in informative samples.

Distinct features under the same name

- The same dataset name is available from a number of sources, but because it has different features, instances, and classes, the conclusions drawn from data analysis will vary.

FIGURE 14.2

Challenges in feature selection and classification.

14.3 System definition

Search techniques are broadly classified as continuous and combinatorial methods. Continuous methods are further classified as linear, nonlinear, and quadratic. Combinatorial methods are classified as approximate and exact. In feature selection, approximate methods are used to search the subsets of features with search algorithms to provide fruitful results [32]. The approximate search approach is further divided into metaheuristics, which have been successfully used on huge and real-world complicated issues, and heuristic methods, which seek to identify high-quality solutions within realistic computation durations. Metaheuristic algorithms are inclined-based methods, which are used to solve continuous, discrete, and mixed search spaces. Nature-inspired metaheuristic algorithms for subset search of features that resolve the complexity in searching the features are shown in Fig. 14.3. The feature subsets produced by the subset search techniques are assessed by the evaluation function or the learning algorithm.

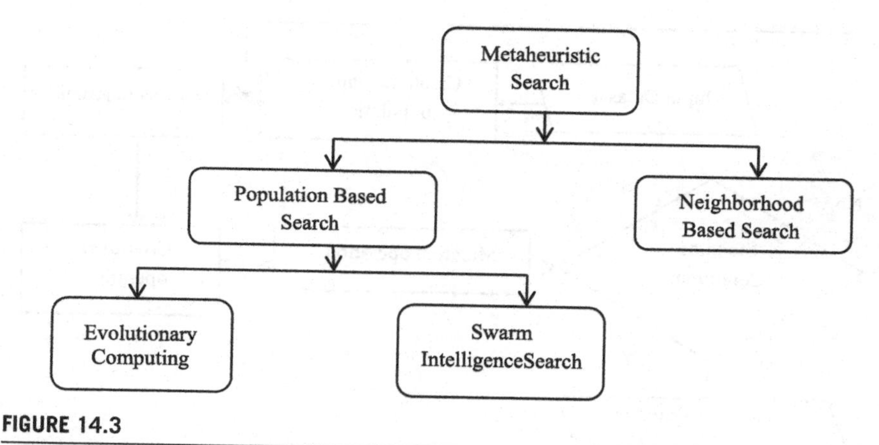

FIGURE 14.3

Classification of metaheuristic search techniques.

This section discusses various metaheuristic search algorithms and evaluation techniques which are commonly used in hybrid feature selection and classification algorithms.

14.3.1 Genetic algorithm (GA)

GA is a global optimal search method based on natural genetics and evolutionary principles which can search huge and complex spaces [21]. It searches the features simultaneously using probabilistic rules. However, a significant disadvantage of employing GA for feature selection is its high computational cost [37]. Four main steps are performed in GA to choose the features [10]. It searches the global optimal features using the genetic operators of crossover, mutation, and selection [11]. A flowchart of GA is shown in Fig. 14.4.

Initially the features are selected randomly from the population and a fitness function is applied in order to evaluate the subsets of features. The features are known as genes and subsets of features are generated using genetic operators as chromosomes. The subsets are generated and evaluated until the best subset is found. The parameters are initially set and then the features are selected using the algorithm. The procedure is shown in Algorithm 14.1.

14.3.2 Differential evolution (DE)

DE is a random search algorithm, which includes evolutionary operators to search the subsets of features systematically towards the global optimum [12]. It handles nonlinear multimodal functions. Typically, in evolutionary algorithms, the first generation of children is generated using the crossover operator, and the evaluation of those offspring is performed using the mutation operator. DE has encountered difficulties with rotation-invariance [29]. It includes crossover and mutation operations

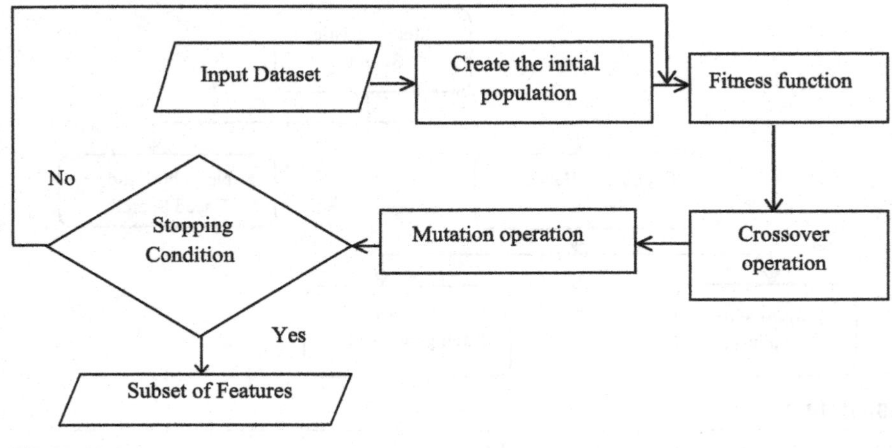

FIGURE 14.4

The flow of genetic algorithm.

Algorithm 14.1: Procedure for genetic algorithm

Input: Number of feature datasets
Output: Feature subsets

1. Set the population of the generation at random.
2. Evaluate and select the population using a fitness function
3. Define the selection probability
4. Apply the crossover probability to the population for the next generation
5. Apply the mutation probability for variation in the next generation
6. Repeat the process until the termination condition is satisfied

based on covariance matrices. Matrix inversion suffers from a lack of scalability in DE [38]. DE differs from conventional evolutionary algorithms in that the trial vector is first produced using the mutation operator, and then the offspring is generated by recombining the trial vector and parent vector using the crossover operator. The initial population is generated by

$$x_{ij} = x_j^{\min} + rand(x_j^{\max} - x_j^{\min}), \tag{14.1}$$

where $I = 1, 2, ..., P$, P is the population size, $j = 1, 2, ..., d$, d is the search space's dimensions, rand is a random variable in the range from 0 to 1, and min and max are the parameter j's specified values. The solution and the mutation that expands it are produced by

$$X = X_{v1} + WF(X_{v2} - X_{v3}), \tag{14.2}$$

where WF is the weighted factor between 0 and 1 and $v1$, $v2$, and $v3$ are random vectors where $v1 \neq v2 \neq v3$. The crossover rate is then used on the mutant vector and the initial solution to make the trial vector. Next, a subset of features from the trial vector and the current solution are chosen [13]. Algorithm 14.2 shows the procedure for DE which solves scalability problems.

Algorithm 14.2: Procedure for differential evolution

Input: Number of feature datasets

Output: Feature subsets

1. Set the population between the upper and lower bounds of the variable, named as target vector
2. Calculate the mutant vector based on the base vector and the weighted vector based on three random vectors
3. To produce the trial vector, use the crossover probability between the target vector and the mutant vector
4. Compare the target vector with the trial vector using the fitness function for the vector selection in the next generation
5. Repeat the same procedure until the termination condition is satisfied

14.3.3 Particle swarm optimization (PSO)

PSO is a population-based metaheuristic algorithm, where the initial phase is randomly generated and updated until the global optimum subsets of features are obtained from the dataset. But a premature convergence issue may occur, which easily traps it in restricted optimal regions [39]. Through cooperation, it creates an enormous number of homogeneous agents. Additionally, the particles' velocities, which represent the characteristics for the subset, are input features (swarm). It starts with random subsets and updates until it finds the best answer [14].

Each solution is a subset of the dataset, which is called swarm [15]. The working principle of PSO is as follows: the input features (particles) are initialized as candidate solution, particle position, velocity, personal position (pbest), and global position (gbest). After calculating the fitness value of each particle, evaluation takes place using the fitness function [31]. The velocity is calculated and the position is updated for each particle as follows:

$$Vel_{i+1} = Vel_i + L * rand() * (pbest_i - P_{val_i}) + L * rand() * (gbest_i - P_{val_i}) \tag{14.3}$$

$$P_{val_i+1} = P_{val_i} + Vel_i + 1, \tag{14.4}$$

where Vel_i is the particle velocity, P_{val_i} is the position of the particle, the *rand*() function produces a random number between 0 and 1, L is the learning factor, usually set as 2, and the individual and global solutions are pbest and gbest, respectively [33]. The best solution is obtained until it reaches the termination condition. The procedure for PSO is shown in Algorithm 14.3 [16].

Algorithm 14.3: Procedure for particle swarm optimization

Input: Number of feature datasets
Output: Feature subsets

1. Set swarm particles with position and velocity, let p_{best} be the personal position, and let g_{best} be the global swarm position
2. Determine each particle's fitness value
3. Check when the fitness value exceeds p_{best} and then assign the current value as p_{best}
4. Select the best values from p_{best} as g_{best}
5. For each particle,
   Determine the particle's velocity using Eq. (14.3)
   Update the particle's position using Eq. (14.4)
   Repeat the steps until the stop condition is satisfied

14.3.4 Ant colony optimization (ACO)

ACO is a population-based model that can be used to find solutions to hard optimization problems. It is inspired by the way a colony of ants works together to find food sources within a short distance [19][27]. By maintaining the current local and global ant pheromone trails, it creates the best subset of features [17]. Graphs are used to display the dataset. Each feature is represented as a node or vertex, and the edges or arcs connecting the nodes or vertices represent the next feature that can be selected from the dataset. After the graph is constructed, the subset that will be updated for local pheromone is created by counting the features that have not been observed yet during graph traversal. The global pheromone is created by updating the local pheromone collected from the various paths [14]. The subset construction by an ant is illustrated in Fig. 14.5, where F1 to F5 are features and the ant in node F1 traverses in the graph and generates a subset of {F1, F4, F5}. The dotted lines indicate the next path for ant traversal to find the next subset of features.

Step by step, the features are searched and updated to form the local subsets of features. Each local subset is evaluated globally and thereby the best subset is generated [18]. For each set of features, the pheromones are modified as follows:

$$T_i(t+1) = (1-\rho)T_i(t) + \sum_{j=1}^{m}(\Delta T_i^j(t)), \tag{14.5}$$

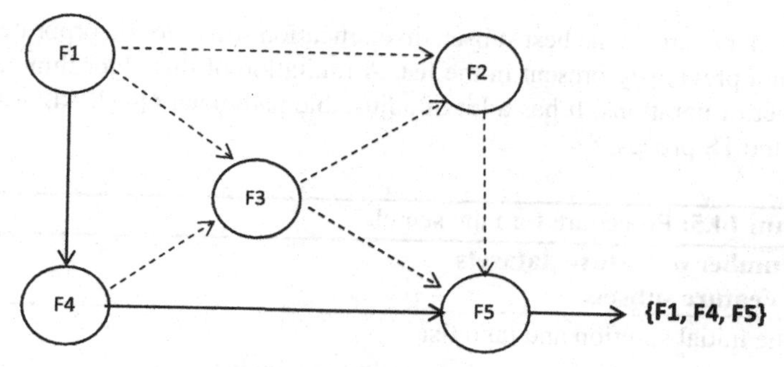

FIGURE 14.5

Subset construction by an ant.

where ρ is the decay coefficient, $\Delta T_i^j(t)$ is the pheromone deposited between edge i and edge j, and m is the number of ants. The algorithm's primary drawbacks include its tendency to stagnate, as well as the rate of exploration, exploitation, and convergence. Early convergence is an issue, which makes it possible to get trapped in local optimal regions. The procedure for ACO is shown in Algorithm 14.4 [40].

Algorithm 14.4: Procedure for ant colony optimization

Input: Number of feature datasets

Output: Feature subsets

1. Set the pheromone and parameters
2. Create the graph with m ants as nodes
3. Search the subset of features for each ant
4. Evaluate the feature subset
5. Update the global and local best subset of ants
6. Check for the culmination
7. If the condition is met, return the best subset
8. Update the pheromones using Eq. (14.5)
9. Generate the new ants for the next generation
10. Repeat the procedure until it reaches the end

14.3.5 Tabu search (TS)

TS is an algorithm for local best feature search. TS searches the search space and returns the recently visited neighbors [22]. To stop the search process from cycling, it incorporates changing memory and is reactive to explorations [23]. A balance between intensification and diversification is created by the search. Strategies for intensification are built around features that have historically been deemed to be beneficial. In order to create the best subset, diversification strategies incorporate features that were not previously present in the list. A limitation of this algorithm can be the high number of iterations. It has a lot of adjustable parameters [41]. Algorithm 14.5 illustrates the TS process.

Algorithm 14.5: Procedure for tabu search

Input: Number of feature datasets
Output: Feature subsets

1. Set the initial solution and tabu list
2. Generate the candidate set of features in the tabu list
3. Evaluate the candidate set of features
4. Choose the best subset of features
5. Update the tabu list with candidate item set until it reaches the end
6. When the final condition is met, the best subset is found.

14.3.6 Subset evaluation

The quality of feature selection depends on two components. The first one is the search component, where the feature subsets are searched. The second one is the evaluation component, where the subsets are evaluated by the classifier/learning algorithm. To choose the best subset of features, all feasible subsets are analyzed. The subsequent computing time will increase since microarray gene expression data subsets have more features [42]. The subsets are searched using search techniques; a few of them were discussed in the previous sections. The evaluation of subsets is done with classifiers. Popular classifiers include the decision tree classifier, nearest neighbor classifier, naive Bayesian classifier, random forest, etc. The best subset is the one that produces the most accurate findings. Different classifiers are compared in order to determine how well the feature selection algorithm performs.

14.3.7 Hybrid feature selection

Hybrid feature selection is done for two reasons. Firstly, hybridization of search techniques with classifiers leads to selection of the best subset of features for classification [24]. Secondly, the population-based global optimal feature selection methods select the relevant features in less time and with reduced costs. But there is a trap in searching the neighborhood's features. It searches the features simultaneously with limitations in converging to the neighborhood's features. Certain relevant features in the neighborhood may not be selected. Due to this, there may be a slight dip in performance because the contributing features are not selected. Hence, the neighborhood-based local search strategy was hybridized with the global feature selection algorithm to select the optimal subset of features. Even at this point the hybrid feature selection method selects features more slowly [43].

14.3.8 Support vector machine

The supervised learning method called support vector machine (SVM) is widely implemented in machine learning for classification. The primary objective of SVM is to create boundary lines or hyperplanes that classify the classes in the dataset so that it can accurately predict future data [35]. It selects a vector to create a hyperplane to classify classes, which is called support vector. It can linearly or nonlinearly separate the data and predict the classes well when the margins are clearly specified, but it requires more time for training [44]. It is employed for high-dimensional data prediction and is particularly well suited for microarray gene expression data prediction, where the number of features exceeds the number of samples. Fig. 14.6 displays an SVM with a separating hyperplane.

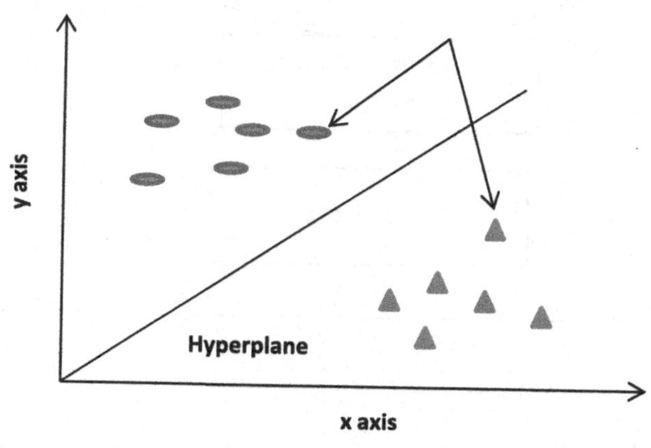

FIGURE 14.6

Support vector machine with separating hyperplane.

14.3.9 Random forest

Random forest is a machine learning algorithm that is supervised by a group and is used to group classes together. It is a group of decision tree algorithms put together. It can make better predictions because it can use both continuous and categorical data. Random forest considers *m* instances and *n* features. A decision tree is constructed for each instance and generates the output. The prediction is based on the average or the majority of voting from the individual decision tree output. The ensemble of decision tree algorithms in the random forest algorithm is shown in Fig. 14.7. The algorithm might become slow and ineffective for real-time predictions whenever there are a lot of trees [45].

14.4 Proposed solution

The flow of the feature selection and classification from microarray gene expression data is shown in Fig. 14.8, where the microarray dataset is used as input. Feature

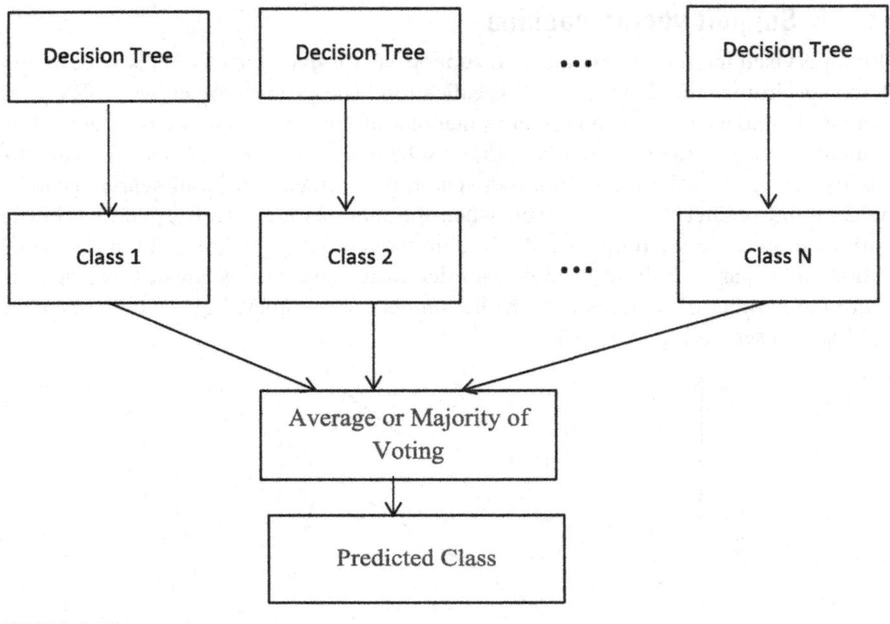

FIGURE 14.7

Random forest.

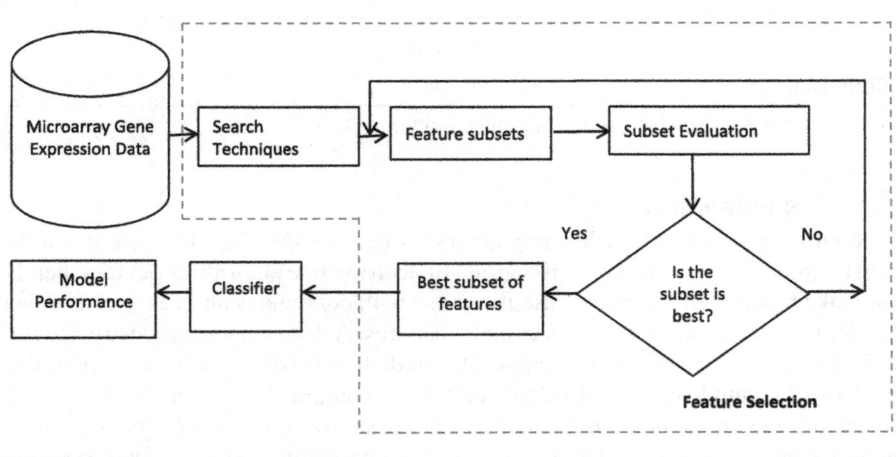

FIGURE 14.8

Feature selection and classification.

selection is performed before the classification to avoid overfitting and selects the relevant data only for classification. The process of choosing features is divided into two stages. The first stage involves searching the feature space for subsets of features, and the second involves evaluating those subsets to choose the best subset of features. The features from the feature space are searched using metaheuristic search strategies. Inclined-based approaches called metaheuristic algorithms are used to solve continuous, discrete, and mixed search spaces. In order to find the minimal or maximal result, in feature selection the subset search changes the input features. There should be a total of 2^N subsets, where N is the number of features in a dataset. Then the learning algorithm or evaluation function is used to evaluate the feature subsets. The appropriate optimal subset of characteristics is chosen using the feature selection algorithm. The unnecessary features are eliminated during feature selection. To classify and predict cancer and evaluate the effectiveness of the proposed algorithms, the reduced feature dataset is applied to the classification algorithm. The cancer data classification is categorized as class discovery and class prediction, which are used for diagnosing the cancer and decision support. Microarray gene expression data are high-dimensional with unrelated features, thereby reducing the classification performance. When selecting features, the unnecessary features are eliminated. Based on the input, the prediction accuracy is determined. The proposed method uses a hybrid feature selection algorithm which combines evaluation metrics and learning algorithms with metaheuristic search strategies to improve feature selection. When it comes to classification performance, hybrid algorithms outperform nonhybrid algorithms. Algorithm 14.6 illustrates the proposed approach.

Algorithm 14.6: Proposed feature selection and classification

Input: Microarray gene expression data
Output: Feature subsets and classification

1. Establish the starting conditions
2. The hybrid feature selection is performed using the dataset that has been provided
3. In order to find the optimal subset of features,
   (i) use a metaheuristic search method to find 2^N subsets of features from N features
   (ii) evaluate the subsets of features using evaluation metrics or learning algorithms
4. The classification method is applied for the best subset of features
5. The performance analysis for the proposed approach is carried out

14.5 Experimental analysis

SVM classification algorithms were used to evaluate the feature subsets so that the best subset could be chosen. The best subset was chosen based on how well each feature subset was able to be classified. The four microarray gene expression datasets are evaluated using the global and local optimal feature selection algorithm. The performance indicators for classifiers, which come from the confusion matrix, are used to measure how well the system works. The measurements used to compare performance are listed below.

- Classification accuracy is calculated as the number of well-predicted instances divided by the overall number of occurrences:

$$Classification\ accuracy = \frac{TP + TN}{TP + FN + TN + FP}. \tag{14.6}$$

- Specificity is calculated as the ratio between the number of well-predicted negative values and the number of actual negative values:

$$Specificity = \frac{TN}{TN + FP}. \tag{14.7}$$

- Recall, also known as sensitivity, is calculated as the ratio between the number of well-predicted positive values and the number of actual positive values:

$$Recall = \frac{TP}{TP + FN}. \tag{14.8}$$

- Precision is calculated as the ratio between the number of well-predicted positive values and the total number of predicted positive values:

$$Precision = \frac{TP}{TP + FP}. \tag{14.9}$$

- The F1-measure is the mean between precision and recall:

$$F1\ measure = 2\frac{Precision \times Recall}{Precision + Recall}. \tag{14.10}$$

- The receiver operating characteristic (ROC) curve is used to figure out how well the classifier can predict. The true positive rate is shown on the y-axis, and the false positive rate is shown on the x-axis. If the curve is in the upper left corner of the ROC area, the prediction was right [8].

DE, PSO, ACO, and GA are combined with SVM to make the DE-SVM, PSO-SVM, ACO-SVM, and GA-SVM global optimal feature selection algorithms. This is done so that the selected features from the SRBCT, Breast, Leukemia, and DLBCL datasets can be tested to see how well they work. These algorithms search for features at the same time, but they have limits on how close they can get to the local best features. Some important parts of the neighborhood might not be able to be chosen. Due to this, performance may drop a little if the contributing features are not chosen. So, to find the best subset of features, the local search strategy must be hybridized with the global feature selection algorithm. The local optimal search algorithm TS is combined with the four global methods for selecting the best features to produce DE-TS-SVM, PSO-TS-SVM, ACO-TS-SVM, and GA-TS-SVM.

14.5.1 Number of features selected

The number of features chosen by each of the four global optimal feature selection algorithms is shown in Table 14.2. GA-SVM chooses fewer features than the other three global optimal feature selection algorithms. During global optimal feature selection, the neighborhood feature problem came up so that the searching could be solved. The SVM algorithm is combined with the global and local search algorithms to find the best features. See Table 14.3.

Table 14.2 Number of features selected using the global optimal feature selection algorithm.

Dataset name	Total features	DE-SVM	PSO-SVM	ACO-SVM	GA-SVM
SRBCT	2309	185	126	87	56
Breast	9217	1507	1432	1345	1332
Leukemia	12583	489	423	328	203
DLBCL	4027	534	480	429	324

Table 14.3 Number of features chosen using the global and local best feature selection method.

Dataset name	Total features	DE-TS-SVM	PSO-TS-SVM	ACO-TS-SVM	GA-TS-SVM
SRBCT	2309	197	134	112	62
Breast	9217	1515	1444	1359	1343
Leukemia	12583	495	437	349	211
DLBCL	4027	543	487	438	335

The differences in the average number of features chosen using the global optimal feature selection algorithms and the average number of features chosen using the methods combining local and global feature selection are 13% for SRBCT, 11% for Breast, 12% for Leukemia, and 9% for DLBCL. This is because the global optimal feature selection method does not pick the neighborhood's features. Instead, it picks the global optimal features. The best features are chosen by global and local algorithms that choose the best features globally and locally. The number of features chosen is greater than the number of features chosen by the global method. The classifier uses its performance measures to look at the relevant features chosen by the global optimum feature selection method, as well as the hybrid global and local optimal feature selection algorithm [28].

14.5.2 Comparison of the performance of feature selection algorithms

The random forest classifier is used to compare how well global optimum algorithmic feature selection and hybrids of global and local optimal feature selection methods work. The datasets are split into ten parts, and a tenfold cross-validation is used to test the results of the categorization. Each one is used both for training and for testing. Performance metrics include the accuracy of the classifier, the time it takes to compute, its specificity, its recall, its precision, and its F1-measure. In Table 14.4, the reader can find measures of performance and the outcomes of both global and local feature selection algorithms.

In terms of accuracy, in comparison to global optimum feature selection techniques, the hybrid global and local optimal feature selection algorithms performed better. For all datasets (SRBCT, Breast, Leukemia, and DLBCL), the global optimum feature selection algorithms performed 11%, 10%, 10%, and 6% less well than the hybrid global and local optimal feature selection methods, respectively.

The hybrid global and local optimal feature selection algorithms have the ability to predict outcomes more accurately and quickly. The global optimum feature selection algorithms produce results with greater precision than the hybrid global and local optimal feature selection algorithms. This demonstrates that the majority of the positive outcomes are probably true, and just a small percentage of them are probably erroneous. Similar to this, hybrid global and local optimum feature selection methods have high values for sensitivity, specificity, and F1-measure. This implies that the predictions made by these algorithms are more precise. It is evident from the findings and discussions that the hybrid global and local optimum feature selection algorithms function effectively. GA-TS-SVM outperforms the other three suggested hybrid global and local optimal feature selection methods.

14.5.3 ROC analysis

The true positive rate is shown on the y-axis of the ROC curve and the false positive rate is shown on the x-axis. The hybrid global and local optimal feature selection methods are closer to the top left corner, according to the ROC results, indicating the prediction is accurate. The results shown in Fig. 14.9 imply that the feature selection methods' curves are located in the ROC space.

14.6 Conclusions

In this chapter, we talked about different ways to search for a subset of features and algorithms for evaluating a subset. We explained how to use microarray gene expression data to predict cancer by choosing which features to use. We described how important hybridization is in feature selection techniques. Four benchmark cancer datasets (SRBCT, Breast, Leukemia, and DLBC) were used to measure how well the feature selection algorithms work. The processing performance metrics are accuracy, computation time, recall, precision, specificity, F1-measure, and ROC. The global optimal feature selection algorithms use a combination of global search algorithms and SVM to find the best features from the microarray gene expression data. They do this by avoiding common problems while looking at the full dataset features. The global optimal algorithms choose the important features, but they do not look at the

Table 14.4 Analysis of the performance differences between hybrid global and local optimal feature selection algorithms and global optimal feature selection methods.

Dataset name	Performance measures	DE-SVM	DE-TS-SVM	PSO-SVM	PSO-TS-SVM	ACO-SVM	ACO-TS-SVM	GA-SVM	GA-TS-SVM
SRBCT	**Accuracy (%)**	68	82	73	83	78	86	79	89
	Computation time (ms)	91	100	51	69	30	70	13	20
	Specificity	84	93	86	93	90	97	90	96
	Recall	69	83	72	84	77	81	80	90
	Precision	70	81	73	83	78	92	81	89
	F1-measure	70	81	72	83	76	87	80	90
Breast	**Accuracy(%)**	64	75	68	77	69	79	73	82
	Computation time (ms)	134	143	126	118	64	92	41	49
	Specificity	88	92	89	92	90	93	94	94
	Recall	64	75	67	77	69	78	73	81
	Precision	69	77	73	78	74	81	82	84
	F1-measure	64	74	68	76	70	78	77	81
Leukemia	**Accuracy(%)**	74	88	80	90	83	92	87	94
	Computation time (ms)	112	120	99	135	79	78	52	64
	Specificity	84	94	87	95	92	96	86	96
	Recall	73	89	78	90	84	92	71	94
	Precision	72	88	79	90	81	92	73	94
	F1-measure	71	88	79	90	81	92	71	94
DLBCL	**Accuracy(%)**	82	92	87	94	90	96	95	96
	Computation time (ms)	85	65	84	94	34	49	21	28
	Specificity	92	98	95	98	94	98	96	99
	Recall	82	93	88	94	90	96	91	97
	Precision	83	94	89	95	91	97	88	97
	F1-measure	82	94	88	95	91	96	91	97

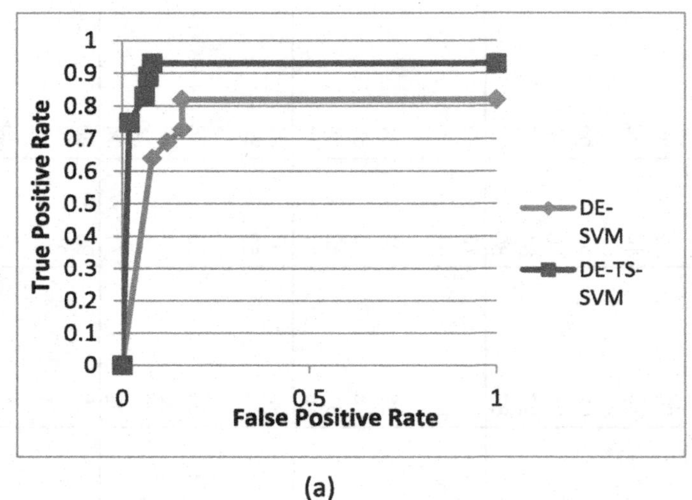

(a)

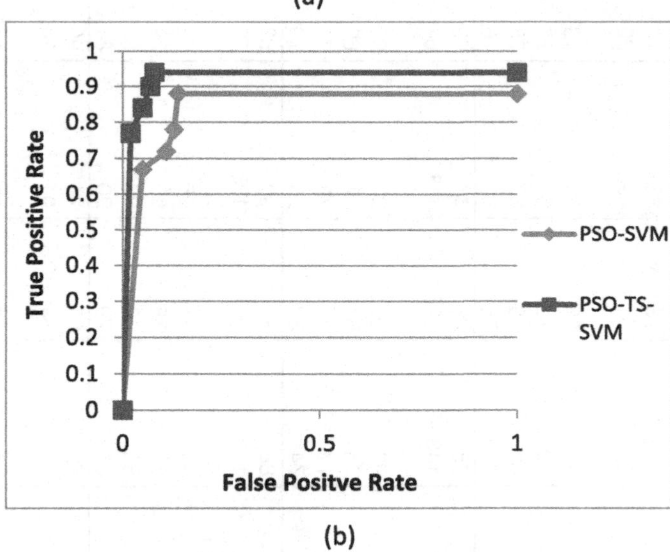

(b)

FIGURE 14.9

(a–d) Receiver operating characteristic curves using global optimal feature selection techniques and hybrid global and local optimal feature selection methods.

nearby features. In order to get past the drawbacks of the global optimum feature selection approaches and select the best subset of features from both the global and local space, the global optimal search algorithms were combined with a local optimal feature selection method. The four hybrid global and local optimal feature selection algorithms outperformed the global optimal feature selection algorithms in terms of

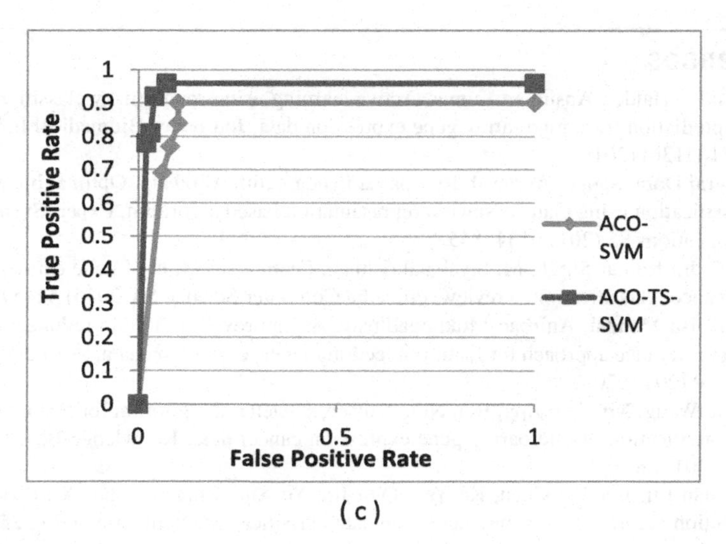

(c)

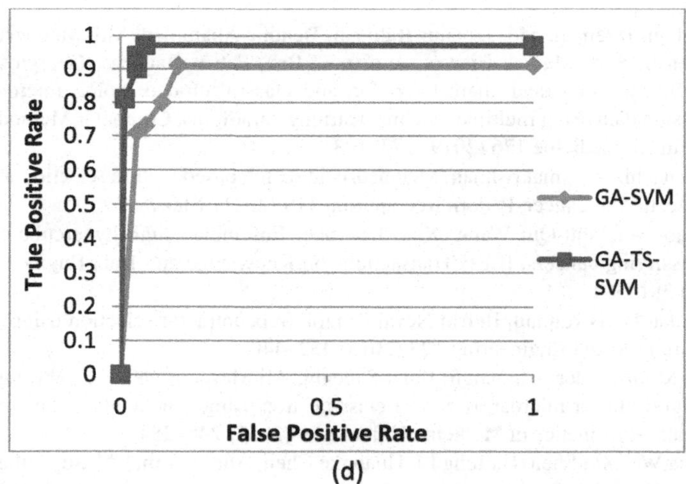

(d)

FIGURE 14.9

(continued)

performance. In summary, the accuracy, specificity, recall, precision, F1-measure, computation time, and ROC were all improved by the hybrid global and local optimum feature selection algorithms. Future research can focus on the prediction of other diseases, including heart disease, iris diseases, diabetes, etc., in a manner similar to how cancer can be predicted. Methods can also be tested with datasets from other fields.

References

[1] Anindya Halder, Ansuman Kumar, Active learning using rough fuzzy classifier for cancer prediction from microarray gene expression data, Journal of Biomedical Informatics 92 (103136) (2019).

[2] Lingraj Dora, Sanjay Agrawal, Rutuparna Panda, Ajith Abraham, Optimal breast cancer classification using Gauss–Newton representation based algorithm, Expert Systems with Applications 85 (2017) 134–145.

[3] Rabindra Kumar Singh, M. Sivabalakrishnan, Feature selection of gene expression data for cancer classification: a review, Procedia Computer Science 50 (2015) 52–57.

[4] Monalisa Mandal, Anirban Mukhopadhyay, An improved minimum redundancy maximum relevance approach for feature selection in gene expression data, Procedia Technology 10 (2013) 20–27.

[5] Hong Wang, Xingjian Jing, Ben Niu, A discrete bacterial algorithm for feature selection in classification of microarray gene expression cancer data, Knowledge-Based Systems 126 (2017) 8–19.

[6] Huijuan Lu, Junying Chen, Ke Yan, Qun Jin, Yu Xue, Zhigang Gao, A hybrid feature selection algorithm for gene expression data classification, Neurocomputing 256 (2017) 56–62.

[7] Md. Maniruzzaman, Md. Jahanur Rahman, Benojir Ahammed, Md. Menhazul Abedin, Harman S. Suri, Mainak Biswas, Ayman El-Baz, Petros Bangeas, Georgios Tsoulfas, Jasjit S. Sur, Statistical characterization and classification of colon microarray gene expression data using multiple machine learning paradigms, Computer Methods and Programs in Biomedicine 176 (2019) 173–193.

[8] L. Meenachi, S. Ramakrishnan, Metaheuristic search based feature selection methods for classification of cancer, Pattern Recognition 119 (2021) 108079.

[9] Jin-Mao Wei, Shu-Qin Wang, Xiao-Jie Yuan, Ensemble rough hypercuboid approach for classifying cancers, IEEE Transactions on Knowledge and Data Engineering 22 (3) (2010) 381–391.

[10] Gul Polat, Baris Kaplan, Befrin Neval Bingol, Subcontractor selection using genetic algorithm, Procedia Engineering 123 (2015) 432–440.

[11] Habib Motieghader, Ali Najafi, Balal Sadeghi, Ali Masoudi-Neja, A hybrid gene selection algorithm for microarray cancer classification using genetic algorithm and learning automata, Informatics in Medicine Unlocked 9 (2017) 246–254.

[12] Guohua Wu, Xin Shen, Haifeng Li, Huangke Chen, Anping Lin, P.N. Suganthan, Ensemble of differential evolution variants, Information Sciences 423 (2018) 172–186.

[13] Emrah Hancer, Bing Xue, Mengjie Zhang, Differential evolution for filter feature selection based on information theory and feature ranking, Knowledge-Based Systems 140 (2018) 103–119.

[14] Bing Xue, Mengjie Zhang, Will N. Browne, Xin Yao, A survey on evolutionary computation approaches to feature selection, IEEE Transactions on Evolutionary Computation 20 (4) (2016) 606–626.

[15] Laith Mohammad Abualigah, Ahamad Tajudin Khader, Essam Said Hanandeh, A new feature selection method to improve the document clustering using particle swarm optimization algorithm, Journal of Computational Science 25 (2018) 456–466.

[16] D. Ramyachitra, M. Sofia, P. Manikandan, Interval-value based particle swarm optimization algorithm for cancer-type specific gene selection and sample classification, Genomics Data 5 (2015) 46–50.

[17] Sina Tabakhi, Ali Najafi, Reza Ranjbar, Parham Moradi, Gene selection for microarray data classification using a novel ant colony optimization, Neurocomputing 168 (2015) 1024–1036.
[18] L. Meenachi, S. Ramakrishnan, Differential evolution and ACO based global optimal feature selection with fuzzy rough set for cancer data classification, Soft Computing 24 (2020) 18463–18475.
[19] Behrouz Zamani Dadaneh, Hossein Yeganeh Markid, Ali Zakerolhosseini, Unsupervised probabilistic feature selection using ant colony optimization, Expert Systems with Applications 53 (2016) 27–42.
[20] Yuanyu He, Junhai Zhou, Yaping Lin, Tuanfei Zhu, A class imbalance-aware relief algorithm for the classification of tumors using microarray gene expression data, Computational Biology and Chemistry 80 (2019) 121–127.
[21] Zhihua Chen, An Huang, Xiaoli Qiang, Improved neural networks based on genetic algorithm for pulse recognition, Computational Biology and Chemistry 88 (2020) 107315.
[22] Xiangjing Lai, Dong Yue, Jin-Kao Hao, Fred Glover, Solution-based tabu search for the maximum min-sum dispersion problem, Information Sciences 44 (2018) 79–94.
[23] L. Meenachi, S. Ramakrishnan, Random global and local optimal search algorithm based subset generation for diagnosis of cancer, Current Medical Imaging 16 (2020) 249–261.
[24] Sayantan Mitra, Sriparna Saha, Sudipta Acharya, Fusion of stability and multi-objective optimization for solving cancer tissue classification problem, Expert Systems with Applications 113 (2018) 377–396.
[25] Sara Haddou Bouazza, Khalid Auhmani, Abdelouhab Zeroual, Nezha Hamdi, Selecting significant marker genes from microarray data by filter approach for cancer diagnosis, Procedia Computer Science 127 (2018) 300–309.
[26] Sadia Sharmin, Mohammad Shoyaib, Amin Ahsan Ali, Muhammad Asif Hossain Khan, Oksam Chae, Simultaneous feature selection and discretization based on mutual information, Pattern Recognition 9 (2019) 162–174.
[27] M. Paniri, M.B. Dowlatshahi, H. Nezamabadi-pour, MLACO: a multi-label feature selection algorithm based on ant colony optimization, Knowledge-Based Systems 192 (2020) 105285.
[28] Loganathan Meenachi, Srinivasan Ramakrishnan, Evolutionary sequential genetic search technique-based cancer classification using fuzzy rough nearest neighbour classifier, Healthcare Technology Letters 5 (2018) 130–135.
[29] Mengnan Tian, Xingbao Gao, Differential evolution with neighborhood-based adaptive evolution mechanism for numerical optimization, Information Sciences 478 (2019) 422–448.
[30] Maisa Daouda, Michael Mayob, A survey of neural network- based cancer prediction models from microarray data, Artificial Intelligence in Medicine 97 (2019) 204–214.
[31] Ke Chena, Feng-Yu Zhoua, Xian-Feng Yuanb, Hybrid particle swarm optimization with spiral-shaped mechanism for feature selection, Expert Systems with Applications 128 (2019) 140–156.
[32] Kangfeng Zheng, Xiujuan Wang, Feature selection method with joint maximal information entropy between features and class, Pattern Recognition 77 (2018) 20–29.
[33] Indu Jain, Vinod Kumar Jain, Renu Jain, Correlation feature selection based improved-binary particle swarm optimization for gene selection and cancer classification, Applied Soft Computing 62 (2018) 203–215.
[34] Hong Wang, Lijing Tan, Ben Niu, Feature selection for classification of microarray gene expression cancers using bacterial colony optimization with multi-dimensional population, Swarm and Evolutionary Computation 48 (2019) 172–181.

[35] Chuan Liu, Wenyong Wang, Meng Wang, Fengmao Lv Martin Konan, An efficient instance selection algorithm to reconstruct training set for support vector machine, Knowledge-Based Systems 116 (2017) 58–73.

[36] Moshood A. Hambali, Tinuke O. Oladele, Kayode S. Adewole, Microarray cancer feature selection: review, challenges and research directions, International Journal of Cognitive Computing in Engineering 1 (2020) 78–97.

[37] Mohammed Ghaith Altarabichi, Sławomir Nowaczyk, Sepideh Pashami, Peyman Sheikholharam Mashhadi, Fast genetic algorithm for feature selection—a qualitative approximation approach, Expert Systems with Applications 211 (2023) 118528.

[38] Mohamad Faiz Ahmad, Nor Ashidi Mat Isa, Wei Hong Lim, Koon Meng Ang, Differential evolution: a recent review based on state-of-the-art works, Alexandria Engineering Journal 61 (2022) 3831–3872.

[39] T.M. Shami, A.A. El-Saleh, M. Alswaitti, Q. Al-Tashi, M.A. Summakieh, S. Mirjalili, Particle swarm optimization: a comprehensive survey, IEEE Access 10 (2022) 10031–10061.

[40] Patricia González, Roberto Prado-Rodriguez, Attila Gábor, Julio Saez-Rodriguez, Julio R. Banga, Ramón Doallo, Parallel ant colony optimization for the training of cell signaling networks, Expert Systems with Applications 208 (2022) 118199.

[41] S. Shanthi, V.S. Akshaya, J.A. Smitha, M. Bommy, Hybrid TABU search with SDS based feature selection for lung cancer prediction, International Journal of Intelligent Networks 3 (2022) 143–149.

[42] Tengyu Yin, Hongmei Chen, Zhong Yuan, Tianrui Li, Keyu Liu, Noise-resistant multi-label fuzzy neighborhood rough sets for feature subset selection, Information Sciences 621 (2023) 200–226.

[43] Mohammad Ahmadi Ganjei, Reza Boostani, A hybrid feature selection scheme for high-dimensional data, Engineering Applications of Artificial Intelligence 113 (2022) 104894.

[44] Qiuhao Huang, Chao Wang, Ye Ye, Lu Wang, Nenggang Xie, Recognition of EEG based on improved black widow algorithm optimized SVM, Biomedical Signal Processing and Control 81 (2023) 104454.

[45] Nour El Islem Karabadji, Abdelaziz Amara Korba, Ali Assi, Hassina Seridi, Sabeur Aridhi, Wajdi Dhifli, Accuracy and diversity-aware multi-objective approach for random forest construction, Expert Systems with Applications 225 (2023) 120138.

CHAPTER

MFFWmark: multifocus fusion-based image watermarking for telemedicine applications with BRISK feature authentication

15

Anurag Tiwari, Divyanshu Awasthi, and Vinay Kumar Srivastava
Electronics and Communication Engineering Department, Motilal Nehru National Institute of Technology, Prayagraj, Uttar Pradesh, India

15.1 Introduction

Patients' medical records contain sensitive and private information. These recordings can easily be made available via a variety of platforms for specific diagnostic purposes. Many medical datasets are being kept on virtual servers. However, if the medical data are sent to these well-known platforms, there can be security issues [1]. In [1], discrete wavelet transform (DWT)-based embedding, which conceals the system MAC address in the logo image, achieves dual watermarking and a high level of authentication. In [3], a fusion-based dual-tree complex wavelet transform (DTCWT)-based singular value decomposition (SVD)-based medical image watermarking system is created. Health information has generally become increasingly prone to alteration and illegal distribution, especially when it comes to images. In order to verify this type of fusion, CT scans and MRIs are fused utilizing a non-subsampled contourlet transform (NSCT)-based fusion technique and generating a single image in [4]. This type of fusion aims to improve the therapeutic experience by combining single modality medical images to produce a distinctive multimodality image. In order to simultaneously improve imperceptibility and robustness, performance comparisons of DWT- and lifting wavelet transform (LWT)-based methods are presented in [5], along with two additional optimization techniques. In [6], it is suggested to employ feature authentication along with Schur-based Digital Imaging and Communications in Medicine (DICOM) image watermarking as long as the input and watermarked images have the same essential characteristics. The suggested solution is less resistant to histogram equalization attacks, but it can still be utilized for feature authentication. Machine learning-based optimization techniques can also

Data Fusion Techniques and Applications for Smart Healthcare. https://doi.org/10.1016/B978-0-44-313233-9.00021-7

be used to increase robustness [10]. Due to the utilization of Zernike moments, the image watermarking method proposed by Dwivedi et al. [11] is extremely resistant to geometric type attacks. The robustness of the presented method is further improved by using features from the accelerated segment test, yet the proposed scheme exhibits less noise when subjected to different noise attacks. In [12], a watermarking method based on histogram shifting to increase the watermarking capacity and imperceptibility is proposed, but the recommended scheme failed to demonstrate improved robustness against higher-order affine attacks. The approach proposed by Hemdan et al. [13] can be used to provide multilevel security and is recommended for a variety of applications, including telemedicine systems, military systems, and copyright protection. The basic premise of the proposed method is to augment the embedded information payload by fusing two watermarks into a single fused watermark. The chaotic and Arnold encryption algorithms are used to encrypt the combined watermark. The three-level DWT and SVD techniques are used to apply the fused scrambled watermark on the cover image. The choice of the Arnold and chaotic watermark encryption techniques is said to have proved resilience, which withstands multiple multimedia attacks and raises the security level. According to Kang et al. [14], properties of discrete cosine transform (DCT) and SVD are utilized in the transform domain using a logistic chaotic map, and least-square curve fitting serves as the basis of their secure and blind image watermarking approach. This technique makes use of optimization to achieve a compromise between resilience and imperceptibility. The described method has less resistance to cropping attacks. A successful spread transform technique is proposed in [15] for robust multiple watermarking in the DWT domain. The trials' findings demonstrate that the suggested strategy enhances the watermark's durability and invisibility and ensures a high rate of watermark detection without altering the watermark's look in the watermarked image. The watermark is immune to large-scale signal distortions since it is distributed globally at low resolutions inside the host image using the spread transform in the discrete wavelet domain. The image watermarking method proposed by Kamble et al. [16,17] is resistant to all attacks other than cropping. A strong watermarking technique is put forth by Khare et al. [19,20] that makes use of the homomorphic transform's reflectance component's feature for increased resistance to various attacks. The approach proposed by Mahto et al. [21] employs NSCT to produce a fused watermark image. To further increase the security of the proposed task, hashing is applied. For some of the scaling values, the approach that was given is not very reliable. Mohammed et al. [22] suggest an image watermarking technique that provides more resilience and imperceptibility for copyright protection of digital images. The watermark data are first translated into the frequency domain for one pass before being embedded. The SVD method is then utilized to alter the values of the LL band. The diagonal matrix's cover pictures (S values) are ready. With the primary goal of compressing the watermarked fingerprint picture data in order to transport it over the network or store it in a lower amount of space, in [24] 2D-DCT is used for watermarking. The embedded watermark, fingerprint image compression, and network transmission could all have an impact on the fingerprint features, resulting in different sets of features or watermark data. An Arnold scrambling-based blind image watermarking strategy using redundant discrete wavelet transform (RDWT) is presented in [26] as a balance between the robustness and imperceptibility aspects of robust image watermarking systems. Each block of the grayscale cover image is initially subjected to RDWT after the image has been separated into fixed-size, nonoverlapping blocks. The binary watermark logo is encrypted and reorganized into a sequence using the Arnold chaotic map to strengthen the security of the logo. The performance of the reported work is improved in [27] using the particle swarm optimization approach. WatMIF, a multimodal image fusion-based safe watermarking system, was introduced by Singh et al. [28]. The host image must be encrypted, the fused mark must be injected and recovered, and multimodal medical images must be combined. These three components make up the suggested algorithm. In the proposed work, the host image is encrypted utilizing a key-based encryption method. The final mark image is created by fusing the images from the MRI and CT scans using an NSCT-based fusion technique. The study that was published is less nuanced in relation to several of the scale parameters. The fragile watermarking technique is another important technique proposed by [29]. The Speed up Robust Features (SURF) algorithm estimates the geometric distortion factor [31]. In [33], two different watermarking schemes are proposed; the first, based on the LWT-Schur decomposition algorithm, embeds the diagonal coefficients of the processed watermark into the processed cover image; the second, based on selected coefficients, proposes an additive watermarking method. Although both proposed schemes are more resilient to a variety of attacks, their resilience to the histogram equalization attack is very weak. An IWT-SD-SVD-based technique is proposed in [34] for maintaining the integrity of DICOM medical images, and it shows greater robustness against a variety of attacks. An IWT-SD-SVD and chaotic encryption-based watermarking approach with greater robustness against various assaults is proposed in [35]. In [7–9], image watermarking schemes based on wavelet domain are proposed, and different optimization techniques are utilized to balance the imperceptibility and robustness.

The above-presented watermarking approaches have many benefits, but there are also some problems that need to be resolved. The majority of techniques are unable to effectively optimize the resilience and visual quality. The existing methods have less embedding capacity even when the watermark is efficiently extracted. Additionally, the security of watermarked images, which must be secure for medical applications, is similarly crucial.

The discussion of the overview of previously used techniques leads to the discovery of a research gap that inspires us to suggest an effective watermarking method. The following are the significant contributions of the proposed technique:

(1) Integer wavelet transform (IWT) is preferred in the suggested scheme because, unlike ordinary wavelets, it offers an integer-to-integer sort of mapping, preventing information loss during forward and inverse transforms. Better resistance to different (affine, filtering, and compression) attacks is offered by IWT.

(2) Schur decomposition [32] is used as it requires a lower number of computations in comparison with SVD. Schur decomposes the image into two different matrices, and the upper triangular matrix is decomposed using SVD.

(3) Multiple decomposition (Schur-SVD) is used in the presented work to provide better resilience against higher-order affine attacks.

(4) MRI and CT scans are combined using a multifocus fusion technique, which produces the fused mark image. Rich information in this fused mark image makes it more suitable for diagnostic and assessment purposes than individual images.

(5) Binary robust invariant scalable key point (BRISK) feature [18] authentication is used as these features are not affected by checkmark attacks.

(6) In the suggested scheme, two watermark images – (i) a brain MRI (ii) a brain CT scan – are fused using the multifocus fusion technique, with the first technique using multifocus fusion without consistency verification (CV) and the second technique using multifocus fusion with CV in the DCT domain.

15.2 MFFWmark: proposed technique

Lifting techniques enable the correct construction of IWT [30]. One can use a lifting scheme to implement the reconstruction of an integer onto an IWT. IWT is utilized in the mapping process due to its benefits in data decomposition, including its reversibility trait and lack of rounding error. Moreover, IWT is quicker and more effective than the classical wavelet transform [25]. IWT consists of three fundamental processes, including split, prediction, and update, similar to a lifting method [30].

The 2D-DCT [5] for an $I \times I$ image is shown in Eq. (15.1):

$$D(x, y) = T(x) T(y) \sum_{p}^{I-1} \sum_{q}^{I-1} d(p, q) \times \left[cos\left(\frac{(p+1/2)\pi}{I} x \right) cos\left(\frac{(q+1/2)\pi}{I} y \right) \right]. \tag{15.1}$$

The inverse 2D-DCT [4] can be obtained as shown in Eq. (15.2):

$$d(p, q) = \sum_{p}^{I-1} \sum_{q}^{I-1} T(x) T(y) D(x, y) \times \left[cos\left(\frac{(p+1/2)\pi}{I} x \right) cos\left(\frac{(q+1/2)\pi}{I} y \right) \right], \tag{15.2}$$

where $T(x) = \begin{cases} \sqrt{\frac{1}{I}}, & x = 0, \\ \sqrt{\frac{2}{I}}, & x \neq 0, \end{cases} \quad T(y) = \begin{cases} \sqrt{\frac{1}{I}}, & y = 0, \\ \sqrt{\frac{2}{I}}, & y \neq 0. \end{cases}$

Schur decomposition is an important tool in numerical linear algebra [9]. For a real matrix R, the Schur decomposition can be defined as follows:

$$[S, D] = Schur\,(R), \tag{15.3}$$

where D is the upper triangular matrix and S is the unitary matrix. The matrix R can be obtained [6] as indicated in Eq. (15.4), where S' is the transpose matrix:

$$R = S \times D \times S'. \tag{15.4}$$

The dominant values of a matrix stipulate the properties of the data distribution and are often stable. The visual eminence of the image is unaffected by a slight change in the single value [9]. The SVD for an image matrix can be defined as follows:

$$[U, \varphi, V] = SVD\,(Image), \tag{15.5}$$

where U and V are the orthogonal matrices and φ is the dominant matrix.

15.2.1 MFFWmark: preprocessing watermark image of the proposed scheme

In the proposed work, two medical images – a 256 × 256 MRI of the brain and a 256 × 256 CT scan of the brain – are acquired and combined using the multifocus fusion approach before being embedded in the cover image. The watermark image is preprocessed using the following steps [23] (see Fig. 15.1):

i. In the multifocus fusion technique, firstly, an artificial blur image of both source images is created. Next, the input image is converted into 8 × 8 blocks, and the DCT of each individual block is computed.
ii. In the fusion process, the correlation coefficients between the input source blocks and their corresponding artificial blurred image source blocks are computed.
iii. Next, the blocks that have low correlation values are chosen from both images. To increase the output of an image's quality and decrease errors brought on by improper block selection, the CV of blocks, as given in [23], is applied as postprocessing, and an image with CV is also generated.
iv. So, two images are generated, one with multifocus fusion with DCT and one with multifocus fusion DCT with CV, which are used as watermark images in the presented work.
v. Both fused watermark images are further decomposed using IWT and SVD is applied into the LL subband, which embeds singular values (SVs) into the cover image.

15.2.2 MFFWmark: embedding procedure of the proposed scheme

The proposed scheme utilizes the properties of IWT and two decomposition techniques, Schur and SVD, to obtain SVs for embedding the watermark information. The following steps are used to embed the watermark (see Fig. 15.2):

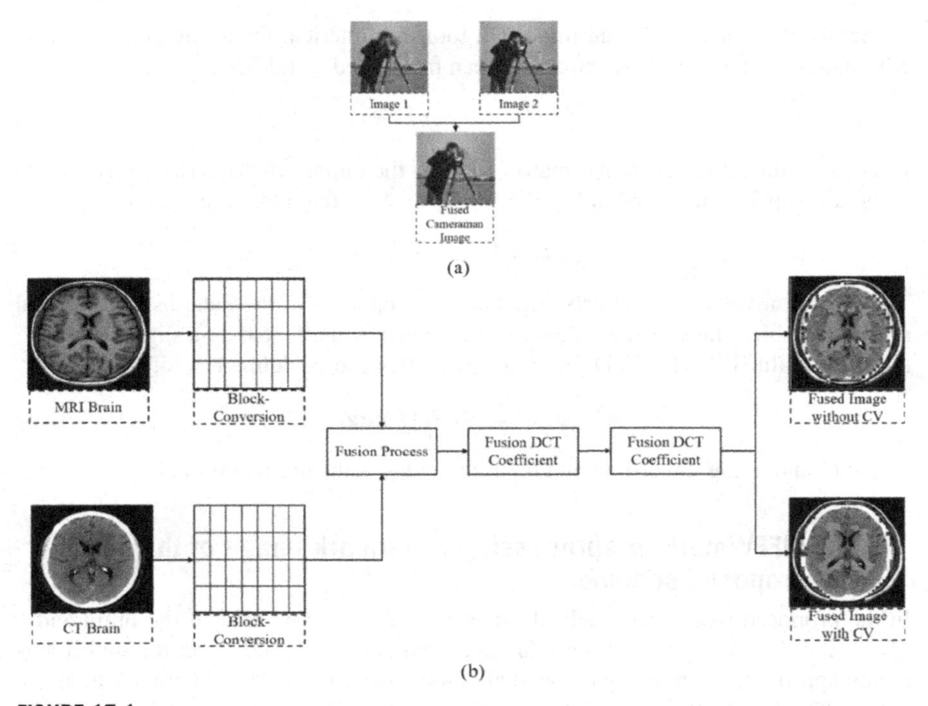

FIGURE 15.1

(a) Significance of the multifocus fusion technique. (b) Block diagram of preprocessing of watermark images using the multifocus fusion technique in the DCT domain.

i. The ultrasound image of the liver (I_c) is decomposed into different frequency subbands using level-3 IWT, as shown in Eqs. (15.6), (15.7), and (15.8):

$$\{LLI, HLI, LHI, HHI\} = IWT(I_c), \tag{15.6}$$

$$\{LLI1, HLI1, LHI1, HHI1\} = IWT(LLI), \tag{15.7}$$

$$\{LLI2, HLI2, LHI2, HHI2\} = IWT(LLI1). \tag{15.8}$$

ii. The corresponding $HHI2$ subband is decomposed into two matrices U_C and T_C using Schur decomposition as shown in Eq. (15.9), where U_C is a unitary matrix and the latter is the upper triangular matrix:

$$[U_{nC}, T_C] = schur(HHI2). \tag{15.9}$$

iii. T_C is further decomposed using SVD as follows:

$$[U_C, S_C, V_C] = svd(T_C). \tag{15.10}$$

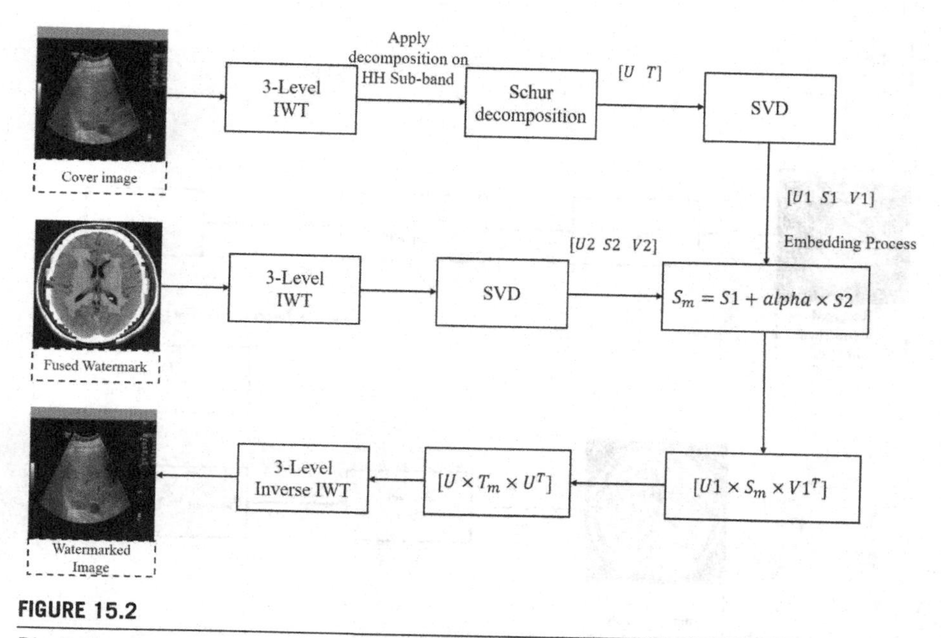

FIGURE 15.2

Block diagram representation of the watermark embedding process.

iv. The coefficient's SV matrix S_C is modified to embed the watermark by choosing a suitable scaling factor *alpha* using the following expression:

$$S_{mod} = S_C + alpha \times S_w. \tag{15.11}$$

v. The modified T_C matrix is obtained with the modified SV matrix using the expression $U_C \times S_{mod} \times V_c^T$, followed by $U_{nC} \times T_C \times U_{nc}^T$ to obtain the HH subband.

vi. Next, a three-level inverse IWT is applied to get the watermarked image.

15.2.3 MFFWmark: extraction procedure of proposed scheme

The steps for extracting the watermark image are as follows (see Fig. 15.3):

i. The watermarked image is decomposed into various subbands using IWT, as shown in Eqs. (15.12), (15.13), and (15.14):

$$\{LLI\_marked, HLI\_marked, LHI\_marked, HHI\_marked\} = IWT\left(I_{marked}\right), \tag{15.12}$$

$$\{LLI1_{marked}, HLI1_{marked}, LHI1_{marked}, HHI1_{marked}\} = IWT\left(LLI_{marked}\right), \tag{15.13}$$

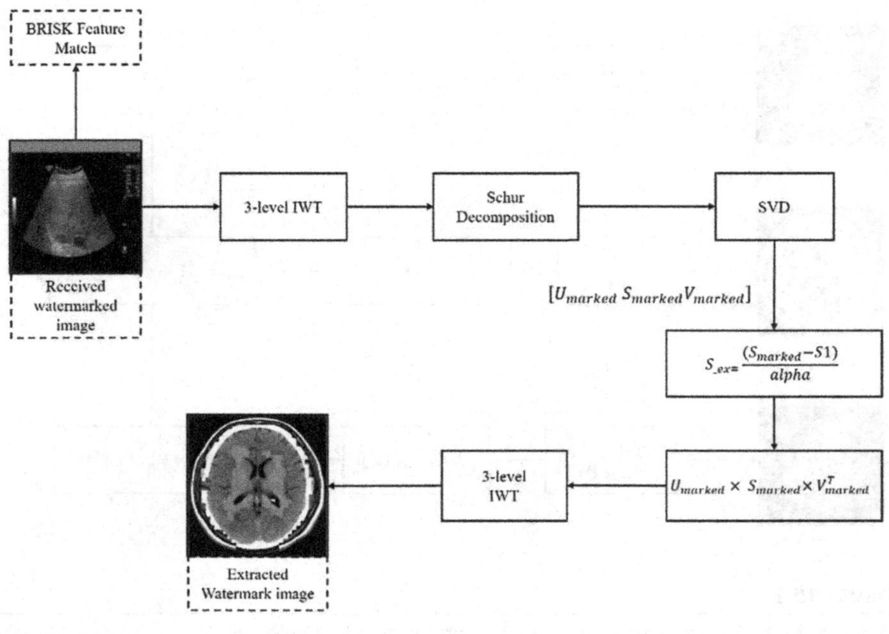

FIGURE 15.3

Block diagram representation of the watermark extraction process.

$$\{LLI2\_marked, HLI2\_marked, LHI2\_marked, HHI2\_marked\} = IWT\,(LLI1marked)\,. \tag{15.14}$$

ii. $HHI2\_marked$ is decomposed into two matrices using Schur decomposition as follows:

$$[U_{marked}, T_{marked}] = schur\,(HHI2\_marked)\,. \tag{15.15}$$

iii. SVD is applied to T_{marked} to obtain SVs of the watermarked image as shown in Eq. (15.16):

$$[U_{marked}, S_{marked}, V_{marked}] = \text{svd}\,(T_{marked})\,. \tag{15.16}$$

iv. The SVs of the watermark image are further extracted using Eq. (15.17):

$$S_{ext} = (S_{marked} - S_C)/alpha. \tag{15.17}$$

v. The low-low subband of the watermark image is generated using the expression $U_w \times S_{ext} \times V_w^T$, and further inverse IWT is applied to obtain the watermark image.

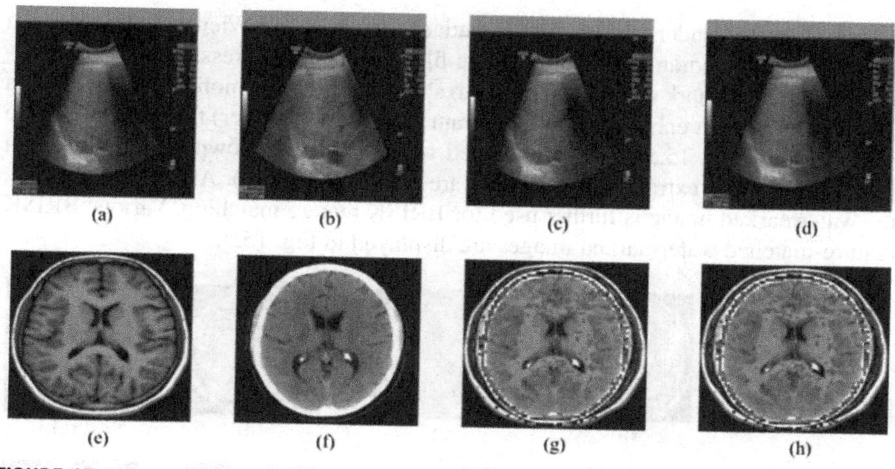

FIGURE 15.4

Test images. (a) Patient-1: P1 (512 × 512). (b) Patient-2: P2 (512 × 512). (c) Patient-3: P3 (512 × 512). (d) Patient-4: P4 (512 × 512). (e) Watermark CT scan (256 × 256). (f) Watermark MRI scan (256 × 256). (g) MFFWmark without CV. (h) MFFWmark with CV.

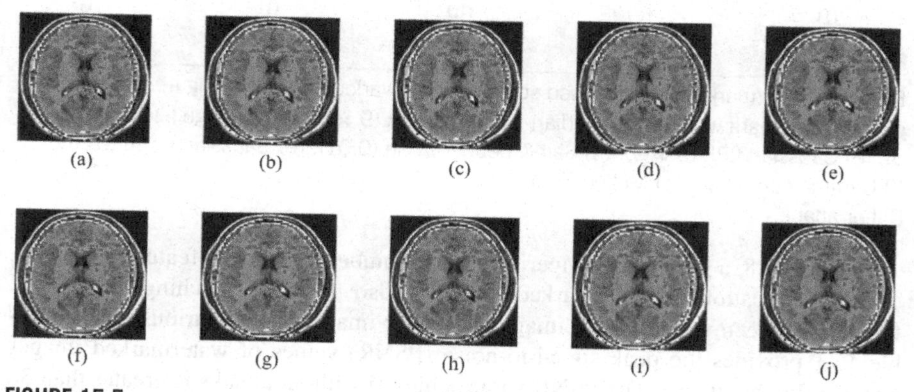

FIGURE 15.5

Extracted watermark of the proposed scheme using the preprocessed watermark without CV. (a) Median filtering attack (9 × 9). (b) Average filtering (9 × 9). (c) Gaussian LPF (9 × 9). (d) Salt & pepper noise (0.01). (e) Gaussian noise (0.01). (f) Motion & blur. (g) JPEG (Q = 10). (h) Speckle noise (0.01). (i) Sharpening attack. (j) Rotation (2°).

15.3 Simulation results and discussion

In this section, the performance parameters of the proposed method are evaluated under different attacks. In the proposed scheme, four different liver ultrasound images of size 512 × 512 are taken as host images and two different watermark images of size 256 × 256 are generated using a multifocus fusion process, as shown in Fig. 15.4. The scheme is tested under various filtering attacks (such as the Wiener filter, Gaussian low-pass filter, median filter, and average filter), image compression (JPEG, JPEG-2000), and checkmark attack (rotation, thresholding, flipping, motion & blur, gamma correction, contrast enhancement, histogram equalization, dither) for its performance evaluation. In Fig. 15.5, various extracted watermarks are shown (without CV). In Fig. 15.6, various extracted watermarks are shown (with CV). At the receiver end, the watermarked image is further used for BRISK feature matching. Various BRISK feature-matched watermarked images are displayed in Fig. 15.7.

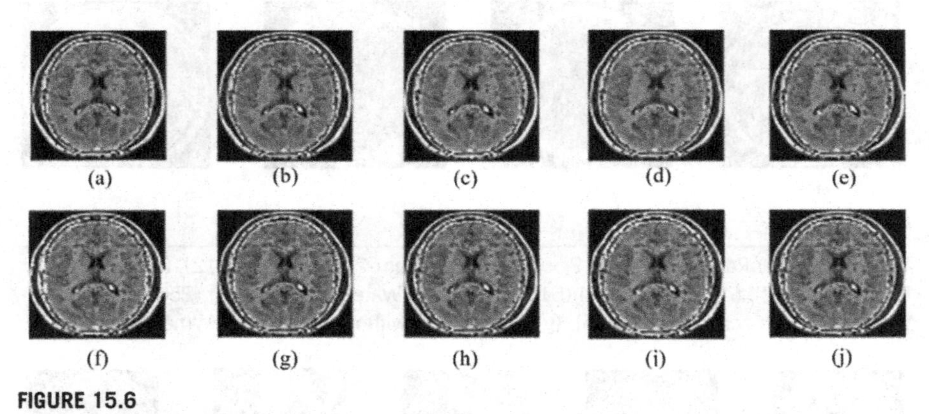

FIGURE 15.6

Extracted watermark of the proposed scheme under various attacks using the preprocessed watermark with CV. (a) Median filtering attack (9 × 9). (b) Average filtering (9 × 9). (c) Gaussian LPF (9 × 9). (d) Salt & pepper noise (0.01). (e) Gaussian noise (0.01). (f) Gamma correction. (g) Motion & blur. (h) Speckle noise (0.01). (i) JPEG (Q = 10). (j) Dither attack.

In Fig. 15.8, a graphical comparison of the number of matched features is shown. The authentication of watermarked image is also done by matching the BRISK features between watermarked image and cover image against various attacks. Table 15.1 provides the peak signal-to-noise (PSNR) values of watermarked images under different attacks. The PSNR value achieved without attacks is greater than 36 dB for each of the test images.

The Patient-2 test image provides higher PSNR values for all test images. The PSNR value obtained against histogram equalization attacks is in the range of 14 dB, which is the lowest. The PSNR value obtained against rotation attack (2°) is more than 25 dB for all test images. The imperceptibility of the proposed scheme against various attacks (motion blur, gamma correction, hard thresholding, JPEG, and JPEG-2000) is better as the achieved PSNR value is more than 27 dB. Additionally, the PSNR value against different filtering assaults is higher than 27 dB, which indicates improved imperceptibility against filtering attacks.

The NCC values obtained against various attacks of the multifocus fused watermark in the DCT domain without CV are given in Table 15.2. The NCC value achieved without attack is equal to 1. The NCC value against the filtering attack for

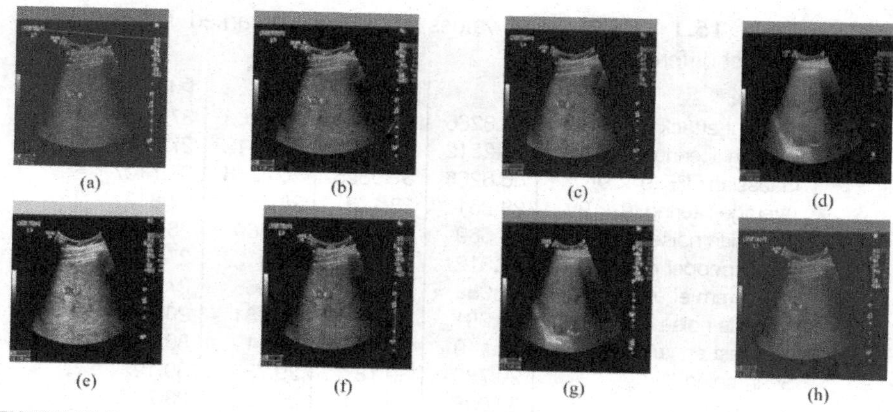

FIGURE 15.7

BRISK feature matching using the preprocessed watermark without CV. (a) Without attack. (b) Salt & pepper (0.01). (c) Gaussian noise (0.01). (d) Median filter (9 × 9). (e) Histogram equalization. (f) Contrast enhancement. (g) Speckle noise. (h) Sharpening attack.

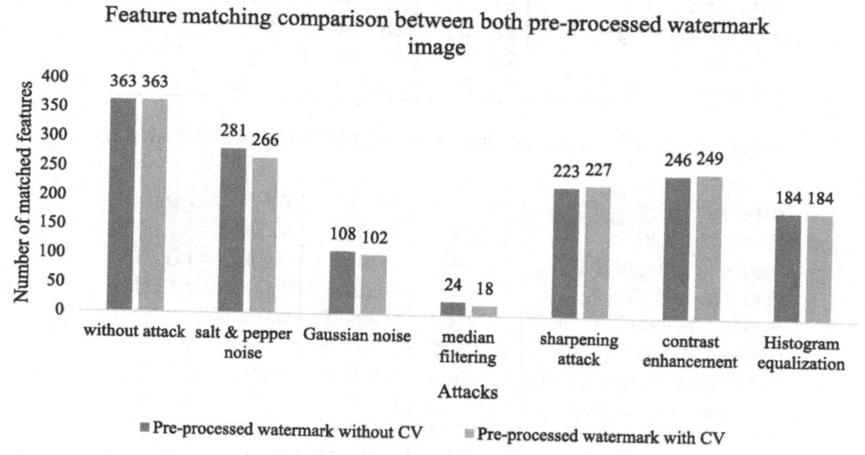

FIGURE 15.8

Comparison of BRISK feature matched points between both preprocessed watermark images.

9 × 9 is more than 0.99, which indicates that the proposed scheme has high robustness against filtering attacks. This scheme offers NCC values greater than 0.87 against all attacks. The achieved NCC value of 0.87127 indicates that the proposed method is least robust against the histogram equalization attack of all the attacks. The NCC value against rotation attack is more than 0.98 for 2° rotation. This method also shows better robustness under different checkmark attacks (thresholding, gamma cor-

Table 15.1 PSNR (dB) values of the watermarked image against different attacks.

Attack	P1	P2	P3	P4
Without attack	36.8266	37.9325	36.8266	37.7107
Median filtering (9 × 9)	28.2512	28.1778	28.2512	27.8122
Gaussian LPF (9 × 9)	36.8266	37.9325	36.8266	37.7107
Average filtering (9 × 9)	28.251	28.178	28.251	27.812
Gaussian noise (0.01)	26.569	26.49	26.569	26.515
Salt & pepper noise (0.01)	27.312	27.474	27.312	27.328
Histogram equalization	14.068	15.052	14.068	27.755
Speckle noise	29.981	29.914	29.981	29.105
Contrast enhancement	29.119	28.895	29.119	30.122
Sharpening	29.785	30.18	29.785	30.122
Motion blur	29.038	29.074	29.038	28.784
Rotation (2°)	25.977	25.967	25.977	25.701
JPEG compression	30.835	30.796	30.835	30.4861
JPEG-2000	30.835	30.796	30.835	30.4861
Gamma correction	27.132	26.811	27.132	26.93
Horizontal flipping	23.10	21.413	23.10	21.635
Vertical flipping	21.453	23.794	21.453	23.938
Hard thresholding	23.10	23.794	23.10	23.938
Dither attack	23.10	23.794	23.10	23.938
Wiener filtering (9 × 9)	28.632	28.905	28.632	29.176

Table 15.2 NCC values of the proposed scheme against various attacks.

Attack	P1	P2	P3	P4
Without attack	1.0000	1.0000	1.0000	1.0000
Median filtering (9 × 9)	0.9980	0.9984	0.9980	0.9979
Gaussian LPF (9 × 9)	1.0000	1.0000	1.0000	1.0000
Average filtering (9 × 9)	0.9976	0.99979	0.99976	0.99972
Gaussian noise (0.01)	0.99825	0.99865	0.99825	0.9887
Salt & pepper noise (0.01)	0.99809	0.99592	0.99809	0.99619
Histogram equalization	0.87348	0.88127	0.87348	0.88422
Speckle noise	0.99274	0.99671	0.99274	0.99194
Contrast enhancement	0.98201	0.98172	0.98201	0.98436
Sharpening	0.98387	0.98444	0.98387	0.98436
Motion blur	0.98368	0.98406	0.98368	0.98397
Rotation (2°)	0.98342	0.98421	0.98342	0.98427
JPEG compression	0.99549	0.99563	0.99549	0.99553
JPEG-2000	0.99549	0.99563	0.99549	0.99553
Gamma correction	0.95367	0.93966	0.95367	0.94522
Horizontal flipping	0.9818	0.98428	0.9818	0.98226
Vertical flipping	0.9818	0.98248	0.9818	0.98226
Hard thresholding	0.9818	0.98248	0.9818	0.98226
Dither attack	0.9818	0.98248	0.9818	0.98226
Wiener filtering (9 × 9)	0.98309	0.98335	0.98309	0.9833

rection, flipping, dither, motion blur). Additionally, the noise strength value of 0.01 indicates that the suggested scheme exhibits high robustness against noise (salt & pepper, speckle, and Gaussian noise).

In Table 15.3, SSIM values against different attacks are given. The proposed scheme provides a sufficient amount of imperceptibility against various filtering attacks (median, Gaussian, average, Wiener) as the obtained SSIM values are more than 0.88. This scheme also gives better SSIM values against different affine attacks and checkmark attacks. The SSIM value against Gaussian noise attack is the lowest among all the attacks. In comparison to attacks, the SSIM values after gamma correction, motion & blur, JPEG, JPEG-2000, and sharpening attacks is more than 0.90. The proposed scheme also achieves good SSIM values after salt & pepper and speckle noise. The PSNR values obtained against various attacks with preprocessed watermark using the multifocus fusion technique in the DCT domain with CV are given in Table 15.4.

Table 15.3 SSIM values of the proposed scheme against various attacks.

Attack	P1	P2	P3	P4
Without attack	0.9727	0.9747	0.9727	0.9733
Median filtering (9 × 9)	0.8436	0.8158	0.8436	0.8257
Gaussian LPF (9 × 9)	0.9727	0.9747	0.9727	0.9733
Average filtering (9 × 9)	0.84356	0.81577	0.84356	0.82572
Gaussian noise (0.01)	0.1464	0.16241	0.1464	0.15854
Salt & pepper noise (0.01)	0.73088	0.75004	0.73088	0.73888
Histogram equalization	0.20803	0.26285	0.20803	0.28516
Speckle noise	0.89355	0.88238	0.89355	0.89809
Contrast enhancement	0.85829	0.89815	0.85829	0.97299
Sharpening	0.9724	0.97437	0.9724	0.97299
Motion blur	0.9158	0.8994	0.9158	0.91998
Rotation (2°)	0.70895	0.69785	0.70895	0.69765
JPEG compression	0.95793	0.95424	0.95793	0.96135
JPEG-2000	0.95793	0.95424	0.95793	0.96135
Gamma correction	0.90096	0.92398	0.90096	0.9164
Horizontal flipping	0.56026	0.33939	0.56026	0.38552
Vertical flipping	0.56026	0.33939	0.56026	0.38552
Hard thresholding	0.37548	0.52294	0.37548	0.56911
Dither attack	0.56026	0.52294	0.56026	0.56911
Wiener filtering (9 × 9)	0.87019	0.84269	0.87019	0.86688

The proposed scheme provides better robustness against various attacks for both preprocessed watermark images. The proposed scheme gives PSNR values close to 30 dB against various attacks like image compression, speckle noise, sharpening, and motion blur. The proposed scheme also provides better robustness against various attacks for the preprocessed watermark using the multifocus fusion technique in the DCT domain with CV, as shown in Table 15.5. The robustness against various attacks for the preprocessed watermark with CV is better than that of the preprocessed

Table 15.4 PSNR values against various attacks for the multifocus fusion technique with CV.

Attack	P1	P2	P3	P4
Without attack	36.8225	37.9269	37.9269	37.7053
Median filtering (9 × 9)	25.8561	28.1779	28.1779	27.8123
Gaussian LPF (9 × 9)	36.8225	37.9269	37.9269	37.7053
Average filtering (9 × 9)	28.251	28.178	28.178	27.812
Gaussian noise (0.01)	26.569	26.494	26.494	26.505
Salt & pepper noise (0.01)	27.452	27.453	27.453	27.374
Histogram equalization	14.068	15.052	15.052	14.316
Speckle noise	29.984	29.915	29.915	29.756
Contrast enhancement	29.118	28.895	28.895	29.105
Sharpening	29.784	30.179	30.179	30.121
Motion blur	29.038	29.074	29.074	28.784
Rotation (2°)	25.977	25.967	25.967	25.701
JPEG compression	30.834	30.794	30.794	30.486
JPEG-2000	30.834	30.794	30.794	30.486
Gamma correction	27.132	26.81	26.81	26.93
Horizontal flipping	21.453	21.413	21.413	21.634
Vertical flipping	23.10	23.794	23.794	28.938
Hard thresholding	23.10	23.794	23.794	23.938
Dither attack	23.10	23.794	23.794	23.938
Wiener filtering (9 × 9)	28.632	28.905	28.632	28.632

Table 15.5 NCC values against various attacks for the multifocus fusion technique with CV.

Attack	P1	P2	P3	P4
Without attack	1.000	1.0000	1.0000	1.0000
Median filtering (9 × 9)	0.9975	0.9979	0.9979	0.9974
Gaussian LPF (9 × 9)	1.000	1.0000	1.0000	1.000
Average filtering (9 × 9)	0.99926	0.99929	0.99929	0.99923
Gaussian noise (0.01)	0.99282	0.99553	0.99553	0.99566
Salt & pepper noise (0.01)	0.99059	0.9957	0.9957	0.99579
Histogram equalization	0.87015	0.87766	0.87766	0.87312
Speckle noise	0.99684	0.99912	0.99912	0.9918
Contrast enhancement	0.98148	0.98121	0.98121	0.98145
Sharpening	0.98336	0.98396	0.98396	0.9839
Motion blur	0.98319	0.98358	0.98358	0.9835
Rotation (2°)	0.98288	0.9837	0.9837	0.98379
JPEG compression	0.99503	0.99518	0.99518	0.99505
JPEG-2000	0.99503	0.99518	0.99518	0.99505
Gamma correction	0.95751	0.94239	0.94239	0.98831
Horizontal flipping	0.98127	0.98198	0.98198	0.98178
Vertical flipping	0.98127	0.98198	0.98198	0.98178
Hard thresholding	0.98127	0.98198	0.98198	0.98178
Dither attack	0.98127	0.98198	0.98198	0.98178
Wiener filtering (9 × 9)	0.98559	0.98286	0.98259	0.98283

watermark without CV. The NCC values achieved against filtering attacks are more than 0.99. This scheme also provides better robustness against various noise attacks. SSIM values of the proposed scheme against various attacks for the preprocessed watermark using the multifocus fusion technique in the DCT domain with CV are given in Table 15.6.

Table 15.6 SSIM values against various attacks for the multifocus fusion technique with CV.

Attack	P1	P2	P3	P4
Without attack	0.9725	0.9725	0.9725	0.9731
Median filtering (9 × 9)	0.8436	0.8158	0.8158	0.8257
Gaussian LPF (9 × 9)	0.9725	0.97446	0.97446	0.9731
Average filtering (9 × 9)	0.84356	0.81577	0.81577	0.82571
Gaussian noise (0.01)	0.1466	0.16271	0.16271	0.15847
Salt & pepper noise (0.01)	0.74407	0.7475	0.7475	0.74407
Histogram equalization	0.20801	0.26278	0.26278	0.24354
Speckle noise	0.89381	0.88183	0.88183	0.88527
Contrast enhancement	0.85801	0.89785	0.89785	0.89778
Sharpening	0.97216	0.97413	0.97413	0.97271
Motion blur	0.91575	0.89936	0.89936	0.91992
Rotation (2°)	0.7088	0.67836	0.67836	0.69746
JPEG compression	0.95775	0.95398	0.95398	0.96126
JPEG-2000	0.95775	0.95398	0.95398	0.96126
Gamma correction	0.90064	0.92382	0.92382	0.91602
Horizontal flipping	0.97547	0.33938	0.33938	0.9855
Vertical flipping	0.56011	0.52277	0.52277	0.56893
Hard thresholding	0.56011	0.52277	0.52277	0.56893
Dither attack	0.56011	0.52277	0.52277	0.56893
Wiener filtering (9 × 9)	0.87016	0.84267	0.87016	0.86683

The PSNR values of the watermarked image under different attacks are shown in Fig. 15.9. The proposed scheme offers better imperceptibility (the watermark is undetectable) against various attacks as SSIM values against various attacks are more than 0.7 except for a few attacks like Gaussian noise, histogram equalization, dither, and hard thresholding. Against various attacks, the NCC values of extracted watermarks fused with CV and without CV are shown in Fig. 15.10.

Fig. 15.11 provides the NCC values of extracted fused watermarks with CV and without CV against different attacks. The suggested approach provides SSIM values over 0.9 for various attacks, including the Gaussian low-pass filter, sharpening, motion blur, JPEG compression, and flipping using the fused watermark with CV. This approach with CV yields lower SSIM values (between 0.6 and 0.50) under dither, vertical flipping, and severe thresholding attacks.

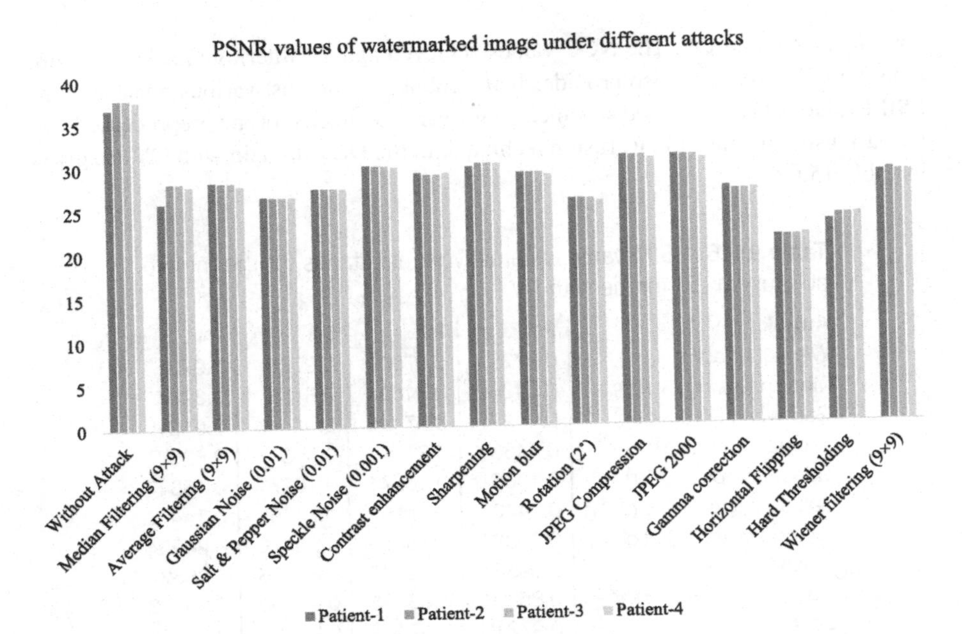

FIGURE 15.9

PSNR values of the watermarked image under different attacks.

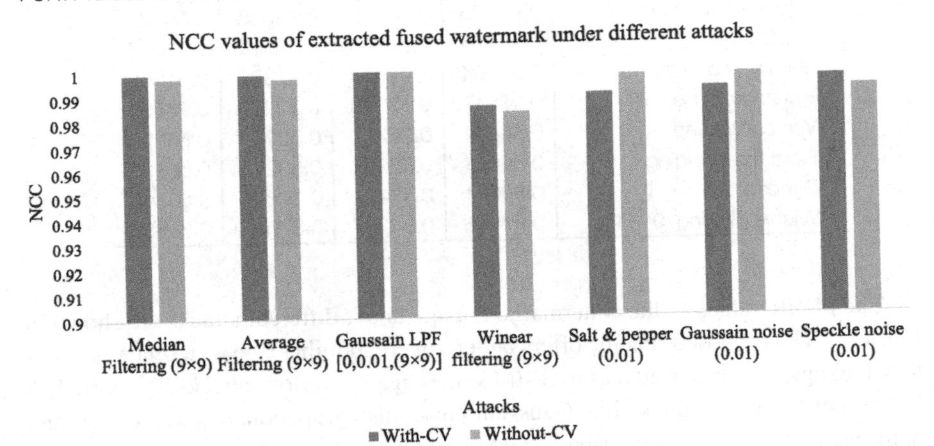

FIGURE 15.10

NCC values with and without CV.

15.4 Comparison of results

In Table 15.7, the performance of the existing scheme and the proposed scheme is compared. The performance of the presented scheme against various attacks is pre-

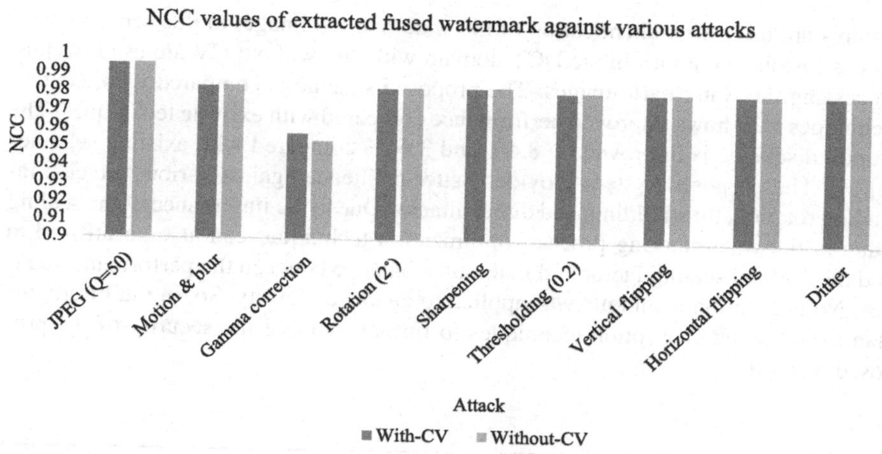

FIGURE 15.11

NCC values with and without CV.

Table 15.7 Comparison of NCC values with existing schemes.

Attack	[2]	[3]	[20]	Proposed scheme without CV
Speckle noise	0.9927	0.9936	-	0.99274
Salt & pepper noise (0.01)	0.8934	0.9908	-	0.99809
Gaussian [0, 0.001]	-	-	0.9983	0.99924
Histogram equalization	0.8491	0.9125	0.9984	0.87348
Rotation (45°)	0.9209	0.8799	-	0.95446
Gaussian LPF (variance, 0.04)	-	0.9902	-	1.00000

sented, and the proposed work shows better results than some of the schemes. The proposed scheme shows better performance than the existing schemes against rotation (45°), Gaussian LPF (variance, 0.04), and salt & pepper noise (0.01). The suggested scheme provides an improvement in NCC value of more than 8.4% and 3.64% compared to the existing scheme [2] [3] against rotation attack (45°). The proposed scheme enhances the NCC value against salt & pepper noise and Gaussian noise compared with existing techniques. Based on the increased NCC value, the suggested technique outperforms existing schemes against various attacks.

15.5 Conclusion

In the proposed scheme, two watermark images are fused using two different fusion techniques, and the generated fused images are separately used as watermarks. IWT-Schur-SVD is used to embed the watermark into the high frequency, and BRISK features are used for authentication of the watermarked image. In this scheme, multifocus fusion techniques in the DCT domain with and without CV are used for preprocessing the watermark images. The proposed scheme is compared with existing techniques and shows improved performance compared with existing techniques. The proposed scheme is improved by 8.4% and 3.64% compared with existing schemes [2] [3]. The proposed system provides better resilience against various attacks, including rotation, thresholding, and dither attacks. Due to the importance of the scaling value in the watermarking process, optimization techniques can also be utilized to find the optimal scaling factor and maintain a balance between the performance metrics. No encryption technique was applied to enhance security. So, in the future, we plan to use some encryption techniques to further increase the security of the proposed method.

References

[1] A. Anand, A.K. Singh, Cloud based secure watermarking using IWT-Schur-RSVD with fuzzy inference system for smart healthcare applications, Sustainable Cities and Society 75 (2021) 103398.

[2] A. Anand, Amit Kumar Singh, Dual watermarking for security of COVID-19 patient record, IEEE Transactions on Dependable and Secure Computing (2022).

[3] A. Anand, A.K. Singh, Health record security through multiple watermarking on fused medical images, IEEE Transactions on Computational Social Systems (2021).

[4] A. Anand, A.K. Singh, SDH: secure data hiding in fused medical image for smart healthcare, IEEE Transactions on Computational Social Systems (2021).

[5] D. Awasthi, V.K. Srivastava, LWT-DCT-SVD and DWT-DCT-SVD based watermarking schemes with their performance enhancement using Jaya and Particle swarm optimization and comparison of results under various attacks, Multimedia Tools and Applications (2022) 1–25.

[6] D. Awasthi, V.K. Srivastava, Robust, imperceptible and optimized watermarking of DICOM image using Schur decomposition, LWT-DCT-SVD and its authentication using SURF, Multimedia Tools and Applications (2022) 1–35.

[7] D. Awasthi, V.K. Srivastava, Performance enhancement of SVD based dual image watermarking in wavelet domain using PSO and JAYA optimization and their comparison under hybrid attacks, Multimedia Tools and Applications (2023) 1–33.

[8] D. Awasthi, V.K. Srivastava, Hessenberg decomposition-based medical image watermarking with its performance comparison by particle swarm and JAYA optimization algorithms for different wavelets and its authentication using AES, Circuits, Systems, and Signal Processing (2023) 1–32.

[9] D. Awasthi, P. Khare, V.K. Srivastava, BacterialWmark: telemedicine watermarking technique using bacterial foraging for smart healthcare system, Journal of Electronic Imaging 32 (4) (2023) 042107.

[10] R. Dwivedi, V.K. Srivastava, Fundamental optimization methods for machine learning, 2023, pp. 227–247.

[11] R. Dwivedi, V.K. Srivastava, Geometrically robust digital image watermarking based on Zernike moments and FAST technique, in: Advances in VLSI, Communication, and Signal Processing, Springer, Singapore, 2022, pp. 671–680.

[12] R. Dwivedi, V.K. Srivastava, Reversible digital image watermarking scheme using histogram shifting method, in: 2021 IEEE 8th Uttar Pradesh Section International Conference on Electrical, Electronics and Computer Engineering (UPCON), IEEE, 2021, November, pp. 1–5.

[13] E.E.D. Hemdan, An efficient and robust watermarking approach based on single value decompression, multi-level DWT, and wavelet fusion with scrambled medical images, Multimedia Tools and Applications 80 (2) (2021) 1749–1777.

[14] X.B. Kang, F. Zhao, G.F. Lin, Y.J. Chen, A novel hybrid of DCT and SVD in DWT domain for robust and invisible blind image watermarking with optimal embedding strength, Multimedia Tools and Applications 77 (11) (2018) 13197–13224.

[15] S. Kamble, V. Maheshkar, S. Agarwal, V.K. Srivastava, DWT-based multiple watermarking for privacy and security of digital images in e-commerce, in: 2011 International Conference on Multimedia, Signal Processing and Communication Technologies, IEEE, 2011, December, pp. 224–227.

[16] S. Kamble, V. Maheshkar, S. Agarwal, V.K. Srivastava, DWT-SVD based robust image watermarking using Arnold map, International Journal of Information Technology and Knowledge Management 5 (1) (2012) 101–105.

[17] S. Kamble, V. Maheshkar, S. Agarwal, V.K. Srivastava, DWT-SVD based secured image watermarking for copyright protection using visual cryptography, in: Computer Science & Information Technology (CS & IT), 2012.

[18] M. Kashif, T.M. Deserno, D. Haak, S. Jonas, Feature description with SIFT, SURF, BRIEF, BRISK, or FREAK? A general question answered for bone age assessment, Computers in Biology and Medicine 68 (2016) 67–75.

[19] P. Khare, V.K. Srivastava, A reliable and secure image watermarking algorithm using homomorphic transform in DWT domain, Multidimensional Systems and Signal Processing 32 (1) (2021) 131–160.

[20] P. Khare, V.K. Srivastava, A secured and robust medical image watermarking approach for protecting integrity of medical images, Transactions on Emerging Telecommunications Technologies 32 (2) (2021) e3918.

[21] D.K. Mahto, O.P. Singh, A.K. Singh, FuSIW: fusion-based secure RGB image watermarking using hashing, Multimedia Tools and Applications (2022) 1–17.

[22] A.A. Mohammed, D.A. Salih, A.M. Saeed, M.Q. Kheder, An imperceptible semi-blind image watermarking scheme in DWT-SVD domain using a zigzag embedding technique, Multimedia Tools and Applications 79 (43) (2020) 32095–32118.

[23] Mostafa Amin Naji, Ali Aghagolzadeh, Multi-focus image fusion in DCT domain based on correlation coefficient, in: 2015 2nd International Conference on Knowledge-Based Engineering and Innovation (KBEI), IEEE, 2015.

[24] M. Lebcir, S. Awang, A. Benziane, Robust blind watermarking approach against the compression for fingerprint image using 2D-DCT, Multimedia Tools and Applications (2022) 1–23.

[25] J. Pan, J. Bisht, R. Kapoor, A. Bhattacharyya, Digital image watermarking in integer wavelet domain using hybrid technique, in: International Conference on Advances in Computer Engineering, 2010, 2010, pp. 163–167.

[26] S. Roy, A.K. Pal, A robust blind hybrid image watermarking scheme in RDWT-DCT domain using Arnold scrambling, Multimedia Tools and Applications 76 (3) (2017) 3577–3616.

[27] V.S. Rao, R.S. Shekhawat, V.K. Srivastava, A DWT-DCT-SVD based digital image watermarking scheme using particle swarm optimization, in: 2012 IEEE Students' Conference on Electrical, Electronics and Computer Science, IEEE, 2012, March, pp. 1–4.

[28] K.N. Singh, O.P. Singh, A.K. Singh, A.K. Agrawal, Watmif: multimodal medical image fusion-based watermarking for telehealth applications, Cognitive Computation (2022) 1–17.
[29] G.D. Su, C.C. Chang, C.C. Lin, Effective self-recovery and tampering localization fragile watermarking for medical images, IEEE Access 8 (2020) 160840–160857.
[30] W. Sweldens, The lifting scheme: a construction of second generation wavelets, SIAM Journal on Mathematical Analysis 29 (2) (1998) 511–546.
[31] C. Tian, R.H. Wen, W.P. Zou, L.H. Gong, Robust and blind watermarking algorithm based on DCT and SVD in the contourlet domain, Multimedia Tools and Applications 79 (11) (2020) 7515–7541.
[32] A. Tiwari, V.K. Srivastava, Imperceptible digital image watermarking based on discrete wavelet transform and Schur decomposition, in: Sustainable Technology and Advanced Computing in Electrical Engineering, Springer, Singapore, 2022, pp. 119–128.
[33] A. Tiwari, V.K. Srivastava, Novel schemes for the improvement of lifting wavelet transform-based image watermarking using Schur decomposition, Journal of Supercomputing (2023) 1–38.
[34] A. Tiwari, V.K. Srivastava, A chaotic encrypted reliable image watermarking scheme based on integer wavelet transform-Schur transform and singular value decomposition, in: 2022 International Conference on Computing, Communication, and Intelligent Systems (ICCCIS), IEEE, 2022, November, pp. 581–586.
[35] A. Tiwari, V.K. Srivastava, Integer wavelet transform and dual decomposition based image watermarking scheme for reliability of DICOM medical image, in: 2022 IEEE 9th Uttar Pradesh Section International Conference on Electrical, Electronics and Computer Engineering (UPCON), IEEE, 2022, December, pp. 1–6.

CHAPTER

16 Distributed information fusion for secure healthcare

Jaya Pathak and Amitesh Singh Rajput

Department of Computer Science & Information Systems, Birla Institute of Technology & Science, Pilani, Rajasthan, India

16.1 Introduction

Artificial intelligence (AI) has become ubiquitous in various fields for the betterment of society, including humanitarian aid [12,13], agriculture [1], and clinical diagnosis [4,15]. Healthcare is one of the primary beneficiaries of AI development. To train machine learning (ML) models, data are essential, and due to the rapid progress in technology and digitization, vast amounts of data are readily available for processing. These data can be transformed into crucial information to support healthcare informatics. However, one of the significant obstacles in healthcare is the secure management and maintenance of extensive medical data.

The application of artificial neural networks (ANNs) in the medical field has proven to be pivotal for enhancing diagnosis, treatment, and disease prevention. Deep neural networks are particularly effective at extracting valuable insights from vast datasets, which has resulted in the increased use of deep learning in the medical industry. For instance, deep learning-based ANN models have facilitated the early diagnosis of stomach cancer using sensitivity and specificity as standard metrics for trained models [23]. Although AI has amazing potential in the healthcare sector, its deployment in clinical procedures raises data privacy problems, ethical concerns, and medical errors.

Medical data present a problem due to their massive size, and extracting knowledge from raw data is hectic yet important. Data fusion plays a significant role in enhancing the collection and processing of medical data. It is a technique of integrating data and knowledge from several sources into a compact representation in an aggregated form. The data obtained from several sources are heterogeneous, i.e., these data may vary with respect to size, distribution, owners, collection and sampling techniques, etc. Still, the main issue arises while keeping and storing the data safely. This distributed nature of healthcare data makes it difficult to provide security and privacy guarantees required to safeguard patients' personal data. Despite the substantial advantages of leading-edge medical technologies, patients must be protected against inaccurate diagnoses. Federated learning (FL) can be of great help in these

Data Fusion Techniques and Applications for Smart Healthcare. https://doi.org/10.1016/B978-0-44-313233-9.00022-9

scenarios as FL works on privacy-preserving ML and works with heterogeneous data as well.

Data fusion and FL share a common purpose in enhancing ML models by aggregating data from diverse sources. Data fusion merges data from various sources, while FL is an ML technique that trains models on distributed data without centralizing them, thereby preserving data security and privacy. By adopting FL as a data fusion strategy, data from numerous sources can be pooled to boost the dependability and precision of ML models while preserving data privacy and security.

16.1.1 Healthcare data sources

Traditionally, medical data were gathered and collected via reliable, systematic, and compiled sources; however, the overall process took months or years before their release. Some of the traditional sources include:

- **Censuses** are taken every 10 years. This is defined as the process of gathering, collating, and disseminating socioeconomic, demographic, and political statistics relevant to all people in a defined region in a certain period. However, major drawbacks of this data source include the extensive preparation period and the amount of time it takes to assess the data once the census has been conducted. Among the primary issues with using census as a data source is how slowly the results come in.
- **Hospital records** are a reliable data source but do not offer data on a representative sample of the population. Instead, they only provide information on individuals seeking medical assistance. The admission policy may vary from hospital to hospital, and hence data may not provide accurate information with respect to the demographics and extent of the disease.
- **Population surveys** that focus on any aspect of health, such as morbidity, mortality, nutritional status, etc., are referred to as health surveys. This data source needs planning for developing a population's health services. Data collection methods include health examinations, health interviews, and questionnaires.

Since the advancement of digitization and technology, healthcare data are readily available for access. The underlying authenticity and reliability of the data are two important factors that influence credibility [20]. This implies that sources with data that adhere to established professional standards are more likely to be trustworthy than those that do not. Some of the modernized data sources include the following:

- **Wearable technologies:** Wearable electronics or gadgets with multiple sensors for tracking various body functions and diseases are becoming more and more common. The development of the Internet of Things (IoT) has made data collection tasks more convenient [34]. Wearables are ideal data-collecting tools for the healthcare industry since they can even communicate data to a general practitioner. This method will help doctors to diagnose illnesses earlier and recommend the best treatment option. However, these devices have limited capabilities and may have some technical and security concerns.

- **Telemedicine:** Despite telemedicine's existence for many years, it has only recently realized its full potential as a result of the development of smartphones, wireless portable devices, and video conferencing. It promotes convenience and wellness by enabling video or phone appointments between patients and their healthcare providers. More medical professionals now offer to "see" patients through video calls.
- **Electronic medical records:** The conventional medical and clinical information obtained from patients is kept in an electronic medical record. These elements offer the potential to lower healthcare expenditures while also increasing the quality and efficiency of services. These kinds of data sources are hugely dependent on information technology. Big data from the healthcare industry hold the potential to enhance health outcomes and reduce expenses.

Regardless of where the data come from, we need to process them before we use them for building ML models.

16.1.2 Data fusion for secured healthcare

Data fusion, also known as data combination, is a technique that combines data and information from several sources into a condensed representation that represents all sources together. According to Hall et al. [2], data fusion is defined as "A study of efficient methods for automatically or semi-automatically transforming information from several sources and different points in time into a representation that effectively supports human or automated decision-making."

Data fusion and information fusion are often used interchangeably. However, in some contexts, information fusion is linked with processed data while data fusion is related to raw data. The data received from various sources will be heterogeneous, i.e., data may vary with respect to size, distribution, owner, collection and sampling techniques, etc. Data gathered from many heterogeneous sources should always be combined to create better and more effective healthcare solutions. The most important elements of data fusion include the following:

- Data sources: Data fusion involves single or multiple data sources from various locations and times throughout distinct periods.
- Operations on data: Information is refined after combining the data from different sources.
- Fusion's purpose: The purpose of data fusion can include estimation, decision making, prediction, etc.

Recent research is focused on the application of data fusion techniques in the medical field. To interpret the complex multidimensional information provided by wearable sensors, data fusion techniques are employed to provide a meaningful representation of sensor outputs. Data fusion techniques and algorithms are used to interpret wearable sensor data in the context of health monitoring applications [3]. However, we still need a mechanism that utilizes information fusion in a private and secure environment to work with healthcare data.

16.1.3 Federated learning for healthcare informatics

Since healthcare data are maintained in a distributed manner, it is challenging to provide the security and privacy assurances that are necessary to safeguard patients' personal data. In a healthcare system, patient data remain stored on hospitals' servers, allowing the hospital agency to retain ownership of the data while enabling the training of algorithms on the data. It is very likely that this practice poses a significant threat to the patients' personal and sensitive information. An attack on the hospital's primary server (where data of every patient are stored) may lead to detrimental effects. In order to address the risk involved in maintaining the cost and privacy of storing these increasing healthcare data, the need for a new field in ML is raised, known as private and secure AI. This new field of research deals with privacy-preserving ML models [24] which aim to perform training without the need for centralized storage of the data while upholding privacy concerns, termed as FL [9,14].

The FL setup consists of a global server whose task is to store a globally untrained model which is shared with all the participating users. The user stores their own data and signals if they are ready for processing. The FL process is as follows:

1. The initial untrained model is stored on the server, which is broadcast to all the participating users.
2. After receiving the global model, each user trains the model using its own private data. After the training, users only send updates of the model back to the global server.
3. The server, after receiving updates from all users, combines all the updates to improve the existing model.
4. This process is repeated until the model is converged and achieves optimal performance by keeping the data intact with users.

To improve model quality while providing security and privacy advantages for users, the FL paradigm sends the ML model to users instead of collecting their data. However, FL is not without its challenges when performed across a network [19,25]. While it is helpful for addressing data privacy concerns, the challenges associated with FL will be briefly discussed in the following subsections.

16.1.3.1 Client selection

The selection of clients is a crucial issue in the FL paradigm, as it is necessary to choose clients that can provide efficient and cost-effective training. Several studies have demonstrated that appropriate client selection in each iteration is crucial in achieving desired accuracy and reducing overall network latency. In FL, each client is expected to participate in only a few rounds to avoid data imbalance problems. However, client selection is a practical challenge due to the heterogeneous and diverse data content within clients, where each client may have different system configurations such as processing speed, memory size, and power level.

16.1.3.2 Non-Independent and Identically distributed data

Federated systems function with nonindependent and identically distributed (non-iid) data, which implies that local data within one client do not accurately represent the entire distribution of the system. Clients within FL setup have a non-iid data distribution, that is, the data may vary from one client to another. This also implies that local data do not perfectly represent the distribution. One of the major issues with non-iid data is to have a convergence guarantee in a federated setup.

16.1.3.3 Heterogeneity

The federated network is comprised of numerous devices that have varied data distribution and volumes, as well as varying computational power. The FL paradigm encounters a unique challenge of heterogeneity in two forms:

- Statistical or data heterogeneity, which refers to the diversity in the distribution of data across participating clients,
- System or device heterogeneity, which refers to the diversity in hardware utility, such as computational power, battery life, CPU, memory size, etc., of the participating clients' devices.

Due to the heterogeneity in the FL setting, some clients with low system utilities fail to send their local updates to the server on time and are excluded from the learning process. This results in biases in the overall system, as only clients with sufficient system utilities can participate in learning. Experiments in [36] have shown that heterogeneity in an FL environment can accelerate performance degradation by reducing model accuracy, increasing training time, and affecting model fairness.

16.1.3.4 Fairness

Fairness refers to the impartial and equitable treatment of all parties involved. In federated environments, data distribution and size are heterogeneous across devices, making it possible for some devices to perform worse than the average performance of all other participating devices. According to Vasileva [28], heterogeneity and bias in training data are the two most significant factors that affect fairness. Our research focuses on fairness in FL, and Section 16.2 will dive deeper into these issues with regard to imbalanced data.

16.1.3.5 Privacy

To achieve distributed and secure learning, FL employs certain mechanisms. First, the server sends the global parameters to each client, and each client uses its local data to run the model. After computation, the client sends the updates back to the server, which aggregates them by computing a weighted average and adding them to the model at the previous time step. This process continues until convergence. Importantly, the server cannot reconstruct the data from updates since the updates are encrypted with a key that an attacker cannot easily read. Secure aggregation is used on the server side to combine the encrypted results. Additionally, FL uses differential privacy to evaluate and regulate the amount of model memorization that may occur if a client has unique data.

16.1.3.6 Types of federated systems

FL was originally developed based on data partitioning, with two main mechanisms: model-centric and data-centric. The model-centric approach aims to improve the model, while the data-centric approach is focused on hosting data rather than the model.

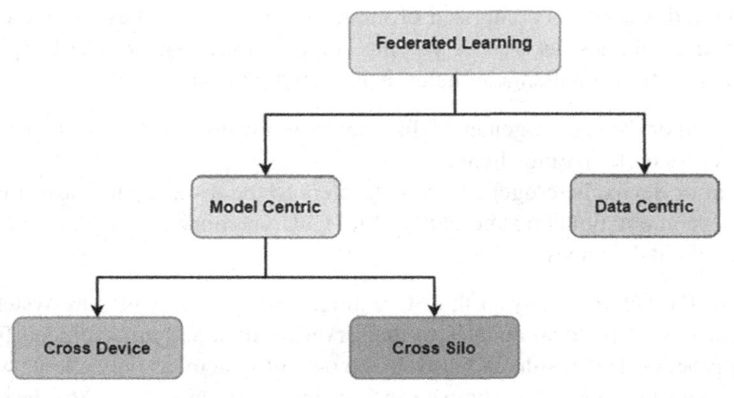

FIGURE 16.1

Categorization in federated learning with respect to data split [32].

Data-centric FL utilizes distributed data from end-users to enhance the centralized model. Model-centric methods can be further divided into cross-device and cross-silo, as illustrated in Fig. 16.1.

The first model-centric type is cross-device FL, also known as horizontal or sample-based FL. This method is used when data share the same feature space but a different sample space. Data are partitioned so that rows consist of multiple data entries and columns contain features. This approach is useful when data reside on multiple devices, as depicted in Fig. 16.2(a). The data on various devices can have different features for each user, such as name, age, and gender.

The second model-centric type is cross-silo FL, also known as vertical or feature-based FL. This approach is used when the organization shares the same sample space, but different divisions within the organization have a different feature space and are distributed within the same organization. Cross-silo partitioning is useful when data are distributed among branches of the same institution, as shown in Fig. 16.2(b). For example, within the same organization, data for the same set of users can be stored, but each division stores different features for each user. If both the feature space and sample space do not overlap, it is called federated transfer learning.

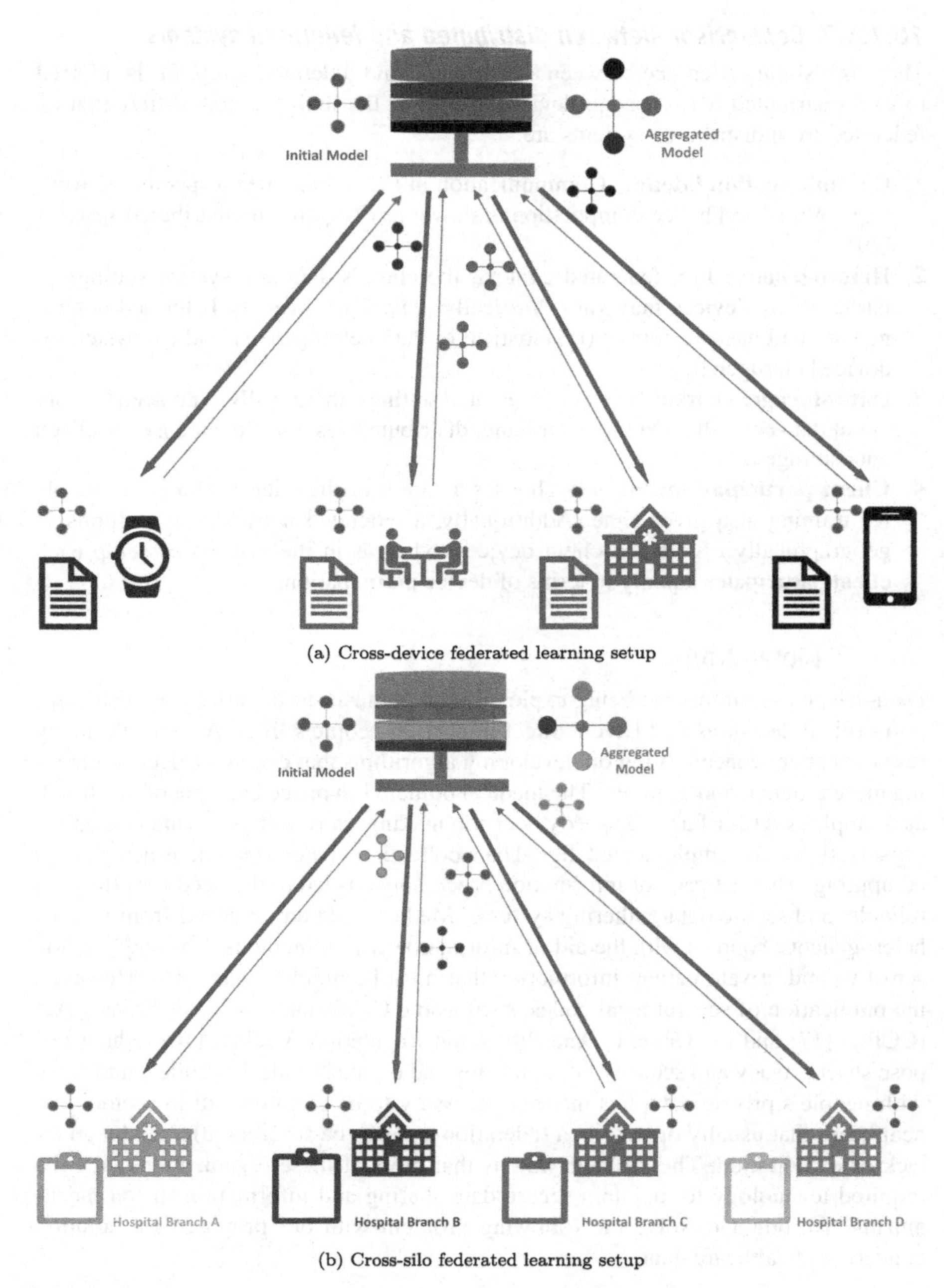

(a) Cross-device federated learning setup

(b) Cross-silo federated learning setup

FIGURE 16.2

Cross-device vs. crosss-silo federated learning setup.

16.1.3.7 Comparison between distributed and federated systems

There is a slight difference between a distributed and federated setup. FL is referred to as a distributed ML on edge devices. Some of the major factors differentiating federated from distributed systems are:

1. **Communication latency:** Communication in FL is done over a specific network (e.g., WiFi), and hence computation is slower as compared to distributed systems [29].
2. **Heterogeneity:** In a federated context, the client's data and system settings of participating devices may vary drastically [21]. This diversity is termed heterogeneity and has two forms: (i) statistical or data heterogeneity and (ii) system or device heterogeneity.
3. **Data storage:** To train models in federated settings, there really is no need to store client data centrally. On the other hand, distributed systems do need a centralized data storage unit.
4. **Client participation:** FL only chooses a subset of the clients who are available for training at a given time. Additionally, a federated network has millions of geographically distributed client devices, whereas in the distributed setup each client participates equally in terms of device participation.

16.1.4 Motivation

Data-driven algorithms are being explored with enthusiasm in various fields to support critical decisions that have a direct impact on people's lives. As a result, many researchers are concentrating on developing algorithms that can make decision making more efficient and accurate. The medical domain is a prime example of this trend, as it employs AI for fair and speedy decision making. In recent years, many medical organizations have implemented digital data collection in their operations using cloud computing. The fast pace of information processing has led to the need for effective, reliable, and secure data-gathering systems. Medical data are obtained from various heterogeneous sources with the aid of information fusion methods, and they include sensitive and private patient information that must be highly secure [16]. However, the publication of several legal codes, such as the California Consumer Privacy Act (CCPA) [17] and the General Data Protection Regulation (GDPR) [10], which impose strict privacy and security laws on firms and organizations that collect and work with people's private data, has made it necessary to be extra careful in sectors like healthcare that usually operate in a federation and can be severely affected by an attack or exploitation. The problem now is that most of these organizations lack the required technology to maintain secure data sharing and information fusion mechanisms. We ought to solve the following problem with our proposed FL solution concerning healthcare data.

Problem: In the healthcare industry, vast amounts of sensitive data are collected from multiple sources, leading to a potential issue of data imbalance that can cause bias in ML models. Furthermore, it is essential to ensure that the use of ML does not violate any data protection regulations, such as GDPR and CCPA. Privacy concerns can be addressed using FL in these scenarios. However, deploying FL-based solutions cannot simply resolve all problems. The fairness problem still remains crucial for ML and the utilization of heterogeneous data from various sources exacerbates this problem in an FL setup. This may cause inadequate model behavior, which forms one of the major problems.

Proposed solution: We use FL to process distributed and heterogeneous data. FL can act as a secure data fusion technique that combines knowledge through global aggregation without the need to maintain patients' private information centrally. Our focus in this problem is to address demographic group fairness issues in FL, where the heterogeneity and biased nature of data within clients exacerbate fairness issues. To tackle these data biases, we aim to reduce data imbalance and promote fairness in FL. Our goal is to provide a fair solution for minorities in demographic groups with respect to factors such as gender and race in the healthcare sector, where people have different medical care requirements. For example, a cardiology study revealed that females have a higher chance of being misdiagnosed with heart attacks compared to males. This highlights the importance of providing healthcare tailored to the specific needs of each demographic group.

Our approach involves leveraging sensitive information such as the class ratio for attributes like gender or race and using it to address the data imbalance problem within each client's dataset. We aim to increase the ratio of positive minorities within each client's data distribution to achieve fairness. Rather than applying this to the entire dataset, we apply it to each participating client in the FL setup to improve the overall fairness of the system. This approach is novel as it is a data-driven approach to combat the data bias problem in an FL setup, which improves the fairness criterion.

The remainder of the chapter is organized as follows. Section 16.2 provides the necessary background related to FL and its underlying fairness definitions. Section 16.3 presents the motivation and proposed method for the fairness scheme. Section 16.4 describes the dataset we worked on and the results achieved. Finally, Section 16.5 provides the conclusion of this chapter.

16.2 Background study

The increasing demand for algorithmic fairness has led to a new trend in which researchers are emphasizing the importance of data transparency and standardized documentation procedures to describe essential features of datasets.

Before working on making a decision-making process fair, we ought to know the deep-rooted reason for its unfair behavior. Biased and imbalanced data are the primary factors that affect model fairness. Imbalanced datasets are those that have an unequal data distribution across a target class, resulting in one class having more data than the other. Bias, on the other hand, is described as a disposition for or against one individual or group, especially when expressed unfairly. In ML models, biases are frequently introduced by the underlying data. For instance, bank loan applications tend to favor men over women because the data used to evaluate applicants' income are often biased against women [12]. Data are a human-influenced phenomenon shaped by a set of independent assumptions about measurement, sampling, and classification, which affects how and by whom data will be collected and annotated. Although ML contributes to knowledge creation, learning algorithms cannot differentiate between real information and societal prejudices, resulting in information that contains societal and moral judgments.

In the healthcare sector, individuals of various ethnical backgrounds, sexual orientations, and ages require different medical care. For example, a study in cardiology showed that females are more likely to receive an incorrect diagnosis of heart attacks than males. Another study [35] discovered that the oximeter device used for measuring the amount of oxygen in the blood measures it incorrectly for people with darker skin tones. This leads to cases of undetected hypoxemia in people with dark skin tones. Personalized reviews are of high importance when medical records are concerned. Several healthcare studies report racial and gender disparities with imbalanced and biased data. According to a medical trial [27], from 2001 to 2018, black and female patients were consistently underreported. Males are more likely than females to be referred to a cardiologist for care even though the proportion of men and women who report chest discomfort is the same. Compared to white patients, black patients in the emergency room are 40% less likely to be given pain medicine. These types of medical data are used to train AI systems and aid in preserving disparities.

Bias damages model quality and is a major contributor to social inequality [39]. A number of preprocessing methods are employed to deal with imbalanced data, such as resampling (either oversampling or undersampling techniques). One method is to increase or decrease the threshold values. An appropriate evaluation metric is also beneficial for controlling data imbalance. However, metrics like accuracy, precision, and recall only operate effectively in the case of a balanced dataset. Researchers have introduced certain definitions that provide standard fairness schemes and aid in evaluating model fairness.

16.2.1 State-of-the-art techniques

The integration of ML into healthcare systems to extract knowledge that can improve healthcare decision making is gaining popularity. FL has emerged as a viable approach to enhance ML systems in this context by enabling them to comply with regulatory standards, increase credibility, and ensure data sovereignty. In [37], the authors present a systematic literature review of current research on FL in the context of electronic health records for healthcare applications. They highlight the importance of FL for healthcare applications, as medical data are often sensitive by nature and cannot be easily shared. The article outlines the main research topics, proposed solutions, case studies, and respective ML methods. Additionally, it discusses a general architecture for FL applied to healthcare data.

In the healthcare domain, the system being fair and personalized are critical aspects. However, data heterogeneity has an impact on both these factors simultaneously. In FL, clients with low system utilities may fail to send their local updates to the server on time and are subsequently excluded from the learning process. This can lead to biases in the overall system, as only clients with sufficient system utilities are able to participate in learning, resulting in unfair models. To address this, researchers are working on developing innovative aggregate solutions that can promote both local and global fairness. The FedAvg algorithm [14] is the default aggregation function, which randomly selects $\lceil K * C \rceil$ clients (where K is the number of clients and C is the fraction of clients being selected in one iteration) and aggregates their updates every iteration.

There are several standards to define a fair FL approach. Uniformness in accuracy levels among clients is utilized in [22,33] to ensure the model performs fairly for every client. Their method utilizes a reweighted hyperparameter to regulate fairness interpolation between local and global model fairness. However, this technique may result in slower convergence. Other fairness standards prioritize treating FL clients based on their contribution during model training, for example, incentive mechanisms [26,31]. Collaborative fairness aims to reward high-contributing clients with a better performing model. In [26], this is achieved by utilizing a reputation mechanism that assesses the contributions of participants in the learning process and updating their reputations iteratively. A reputation list for all participating clients is stored and managed on the server. However, this incentive mechanism still results in bias. Some fairness standards rely on model performance concerning demographic groups within clients [30]. The FairFed algorithm is a fairness-aware aggregation method designed to promote fair model performance for demographic groups. This algorithm assigns higher weights to clients that have similar local fairness to the global fairness metric. All the above fairness mechanisms are model-based solutions.

However, a data-driven solution may help in reducing the bias produced mainly from imbalanced data distributions. Unlike the previous state-of-the-art approaches that are model-based solutions, our approach is based on data to address the issue of group fairness.

16.2.2 Fairness definitions

Most fairness measures rely on standard metrics and are defined using a confusion matrix as shown in Table 16.1. For binary classification, both predicted and actual classes have positives and negatives, depending on the classification problem. For example, individuals are divided into *Dominant (D) and Suppressive (S)*. Cells of a confusion matrix are defined as:

- True positives (TP): Individuals that are predicted as and actually are of the positive class.
- False positives (FP): Individuals that are predicted as positives but are of the negative class.
- False negatives (FN): Individuals that are predicted as negatives but are actually positives.
- True negatives (TN): Individuals that are predicted as and actually are of the negative class.

Table 16.1 A confusion matrix [18].

	Actual positives	Actual negatives
Predicted positives	**TP** PPV TPR	**FP** FDR FPR
Predicted negatives	**FN** FOR FNR	**TN** NPV TNR

- Positive predicted value (PPV): The ratio of true positives to all the predicted positives.
- False discovery rate (FDR): The ratio of false positives to all the predicted positives.
- False omission rate (FOR): The ratio of false negatives to all the predicted negatives.
- Negative predicted value (NPV): The ratio of true negatives to all the predicted negatives.
- True positive rate (TPR): The ratio of true positives to all the actual positives.
- False positive rate (FPR): The ratio of false positives to all the actual negatives.
- False negative rate (FNR): The ratio of false negatives to all actual positives.
- True negative rate (TNR): The ratio of false negatives to all the actual negatives.

Some of the most prominent fairness definitions that are used and studied over several domains are provided below.

1. **Statistical parity:** Also known as demographic parity, this definition is usually applied in a group fairness scheme. If both the Dominant (D) and Suppressive (S) groups have equal probabilities of being in the positive predicted class, then according to the definition of statistical parity, the model works fairly.
2. **Predictive parity:** If for both S and D groups, the positive predicted values (PPVs) are equal, then according to the definition of predictive parity, the model is fair.
3. **Predictive equality:** If the S and D groups have equal false predicted rates (FPRs), the model is fair according to the definition of predictive equality.
4. **Equal opportunity:** If groups S and D have equal false negative rates (FNRs), the model is fair according to the definition of equal opportunity.
5. **Equalized odds:** The model is fair when in the S and D groups, both the true positive rate (TPR) and the false positive rate (FPR) are equal.

   These concepts perform according to their intended purposes as well as depending on the dataset they have been applied to in the majority of earlier literature on fairness. Hence, if the dataset has imbalance issues, these definitions do not guarantee fairness. We have adopted a new data-driven approach that can be useful in reducing data imbalance, which is discussed in Section 16.3.

16.3 Methodology

There are concerns about the fairness of algorithms in every field where automated decision-making systems can potentially affect human well-being. The criminal justice system, education, search engines, online marketplaces, emergency responses, social media, medicine, and employment are some of the domains where automatic decision-making systems are being researched. That is why algorithmic fairness is a developing field that looks at algorithms from the perspectives of justice, equality, prejudice, power, and damages.

Algorithmic approaches like data mining claim that certain methods remove human bias from decision-making processes [8]. On the contrary, the quality of an algorithm is directly related to the data it handles. The demographic disparities in our society are encoded using the available training data. We work to improve the model by modifying and upgrading it. Still, if the data we feed the model have disparities, there is a good chance that the ML model will also be negatively impacted. Disparities in the data on which the model works are among the leading sources of injustice or unfairness.

Fairness can be achieved through (i) modification of the classifier or (i) modification of the data used by the classifier. According to computer scientist and researcher Andrew Ng, 80% of the ML models depend on the quality and build-up of data. However, less than 1% of the researchers focus on making the data better to feed them to the model because data are the food for AI. Hence, our work in this chapter concentrates on achieving fairness by modifying the data. The amount of fair decision making for a model is computed using disparate impact (DI). Previous methods use DI as a notion for discrimination [11] or as a fairness scheme [6]. DI is defined as follows.

Definition 16.1. Disparate impact (DI). Given a dataset $D = (X, Y, C)$ with X being the protected or sensitive attributes (e.g., age, gender, race), Y being all the other attributes in the dataset, and C being the binary classification (e.g., YES or NO), DI is defined as the ratio of the positive classification ($C = YES$) for the unprotected class ($X = 0$) over the positive classification for the protected ($X = 1$) class:

$$\frac{P(C = YES|X = 0)}{P(C = YES|X = 1)}. \tag{16.1}$$

The value of this ratio is directly correlated to the degree of model fairness.

Sensitive attributes are defined as attributes or features that can cause discrimination against an individual. For our convenience, we assume the protected or sensitive attribute is a binary class and can only take two values, say positive ($X = 1$) or negative ($X = 0$), termed as "majority class" and "minority class," respectively. In the case of the sensitive attribute being *gender*, it has values of "male," mostly considered as protected class ($X = 1$), and "female," the unprotected class ($X = 0$). The numerator $P(C = YES|X = 0)$ in Eq. (16.1) calculates the probability of the minority class in the positive class, whereas the denominator $P(C = YES|X = 1)$ represents the probability of the majority class in the positive class. The ratio of these values is the DI score. In the case of *race* being a sensitive attribute, usually we consider the dominant or protected class to be "White" or "Caucasian" and the submissive or unprotected class to be "Black" or "non-Caucasian." In the event of a multivalued attribute, such as *race*, our standard practice is to presume the majority class is the one with the highest number of individuals. The other class is a composite class for all of the classes that are not the majority class as the unprotected group or minority class.

Our method includes repopulating the data within each client by replicating the positive class minorities, such that the ratio of positive classification for the minority class over the majority class increases (the value for DI). We try to reduce the bias in the data by replicating the positive minorities to reduce the imbalance within each class. The outline of our work is described using a workflow diagram presented in Fig. 16.3.

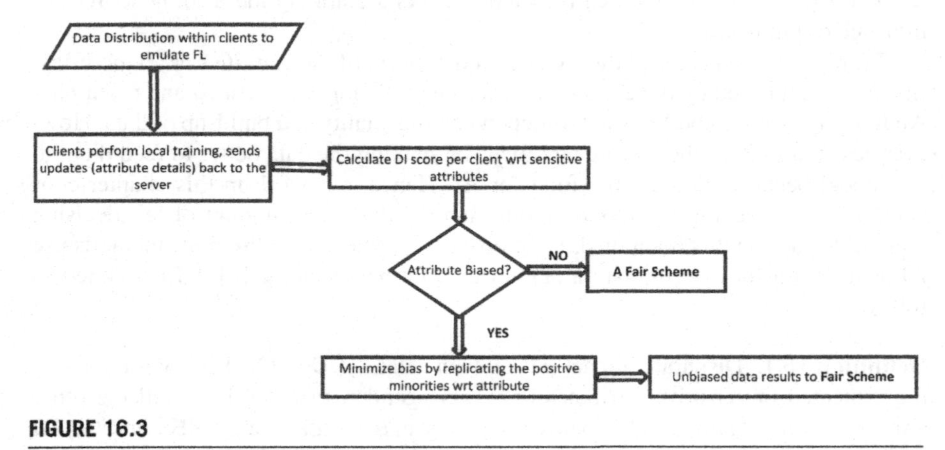

FIGURE 16.3

Proposed method workflow.

16.3.1 Implementation

We use a multiparty TensorFlow [7] setup to emulate an FL environment. To emulate a realistic imbalanced data distribution for the FL setup, we use non-iid data selection. We achieve this using random allotment of data to each client. We train a convolutional neural network (CNN) model on both medical datasets. The model consists of a CNN layer with a kernel size of 3 and a max-pooling layer. We use Adam as the optimizer and ReLU as the activation function. The model is trained for 10 learning epochs over 10 iterations. The train–test split is 30% for training the model. The training data are randomly divided among individual clients to introduce a non-iid distribution.

16.4 Evaluation

16.4.1 Datasets

We work with two healthcare datasets: (i) the Diabetes 130 US hospitals 1999-2008 dataset [5] and (ii) the Personal Key Indicators of Heart Disease dataset [40].

Diabetes 130 US hospitals Dataset: The Diabetes dataset describes the clinical care at 130 US hospitals and integrated delivery networks from 1999 to 2008. The classification task is to predict whether a particular patient will be readmitted within 30 days.

The original Diabetes dataset is a three-class classification problem, with more than 50 attributes. However, we have used the clean version of the dataset, which is a binary classification problem with labels "<30" and ">30." For our understanding, we consider "<30" days as the positive class label and ">30" days as the negative class label. The dataset contains a total of 45,715 entries, with 11,066 entries as the positive class (<30 days) and 34,649 entries as the negative class (>30 days). The dataset is imbalanced with respect to the positive and negative class labels, as shown in Fig. 16.4(a).

The Diabetes dataset has gender, race, and age as sensitive attributes. The attribute *gender* has two categories: *male* and *female*; the class distribution with respect to *gender* is shown in Fig. 16.4(c). The category *male* has a total of 20,653 entries, out of which 5064 are classified as positive. The category *female* has 25,062 entries, out of which 6002 are classified as positive. The total number of positives is 11,066. These numbers suggest that this dataset does not have a gender imbalance problem. The attribute *race* has five categories: *Caucasian, African-American, Others, Hispanic*, and *Asian*. The category *Caucasian* has the highest number of entries (35,386), among which 8512 are classified as positive. The class distribution is shown in Fig. 16.4(b). We observe that the attribute race is imbalanced by examining the class distribution graphs. Hence, the dataset is biased towards the race *Caucasian*. We consider race as a binary attribute for our experimentation with positive and negative labels as *Caucasian* and *non-Caucasian*, respectively. All individuals other than *Caucasian* are considered to be *non-Caucasian*, with a total of 2554 out of the 11,066 total positives, which means that only 2554 *non-Caucasians* will be readmitted, as shown in Fig. 16.5(a) and (b). This is incorrect since diabetes affects people of all colors, not just Caucasians.

Heart Disease dataset: The 2020 Heart Disease dataset contains annual CDC survey data of 400,000 adults related to their heart health status. This dataset has approximately 320,000 instances, with the binary classification task of determining whether an individual suffers from a heart disease or not [38]. We consider the label "Heart Disease = YES" as the positive class and "Heart Disease = NO" as the negative class. The data distribution with respect to the above classes is 292,422 and 27,373, respectively. The dataset contains 18 attributes that are associated with heart health like body mass index (BMI), smoking, alcohol consumption, physical and mental health, etc.

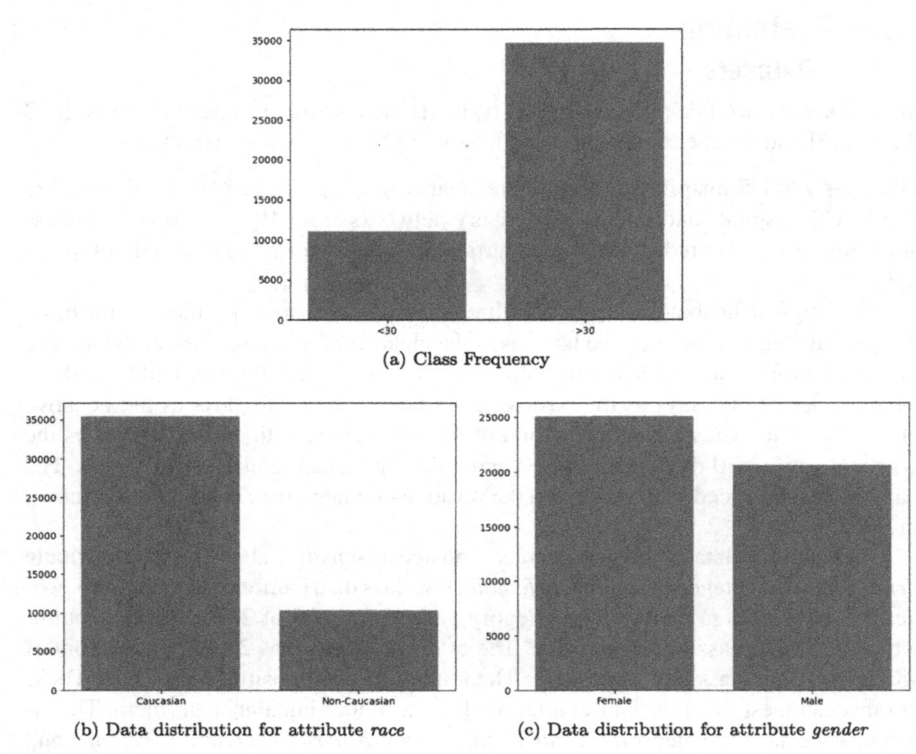

(a) Class Frequency

(b) Data distribution for attribute *race*

(c) Data distribution for attribute *gender*

FIGURE 16.4

Data distribution with respect to sensitive attributes for the Diabetes dataset.

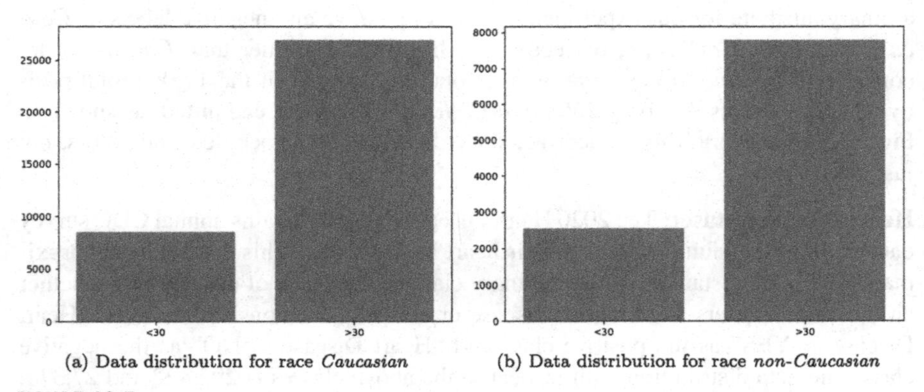

(a) Data distribution for race *Caucasian*

(b) Data distribution for race *non-Caucasian*

FIGURE 16.5

Data distribution for attribute *race* for the Diabetes dataset.

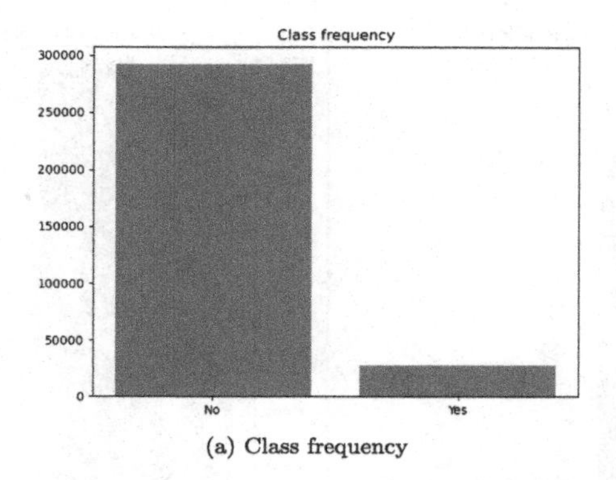

(a) Class frequency

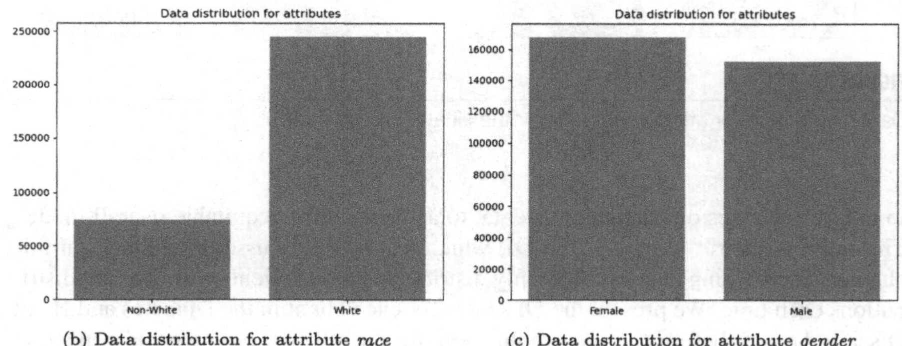

(b) Data distribution for attribute *race* (c) Data distribution for attribute *gender*

FIGURE 16.6

Data distribution with respect to sensitive attributes for the Heart Disease dataset.

This dataset has the same three sensitive attributes, that is, gender, race, and age. The class frequencies and data distribution along some of the sensitive attributes are presented in Fig. 16.6(a)–(c). The *gender* attribute has 151,990 male and 167,805 female entries, with 16,139 entries for males in the positive class and 11,234 entries for females in the positive class. This dataset is more biased with respect to the attribute *gender*, as shown in Fig. 16.7.

16.4.2 Experimental results

16.4.2.1 Working with non-iid and sensitive data

Our approach aims to address bias and unfairness towards underrepresented groups by increasing the number of positive classifications for minority groups. The dataset under consideration is often biased towards majority groups such as *Caucasians* and *males*, based on sensitive attributes like race and gender. However, in an FL setup, data are distributed across clients and are non-iid. Thus, it is not practical to increase overall fairness across the dataset. Instead, we propose balancing the ratio of minority

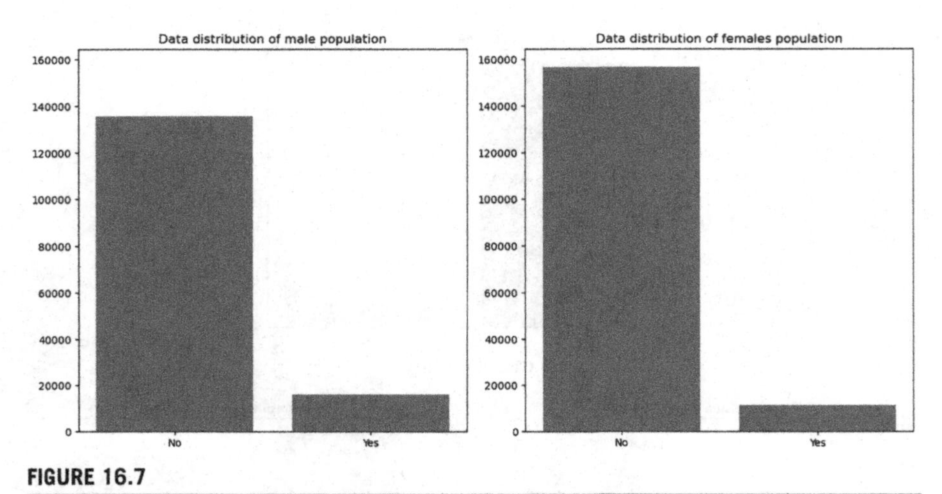

FIGURE 16.7

Data distribution for attribute *gender* for the Heart Disease dataset.

to majority classes on each client's data, to achieve a more equitable overall model. The fairness metric we use is the DI, which has been discussed previously in this chapter. The training data are randomly distributed among clients with varying distributions each time. We present the DI scores for each client in the Diabetes and Heart Disease datasets before and after replication, for the attributes of race and gender, in Figs. 16.8 and 16.9, respectively. Table 16.2 provides an overview of the total data distributed within each client, along with the numbers in the majority and minority classes before and after replication.

16.4.3 Discussion

In this chapter, we aimed to minimize the disparities which are introduced because of data imbalance. We used two medical domain datasets that are imbalanced with respect to sensitive attributes.

Table 16.2 demonstrates the replication of positive minorities, where the majorities (*Caucasian* in the Diabetes dataset and *male* in the Heart Disease dataset) and minorities (*non-Caucasian* in the Diabetes dataset and *female* in the Heart Disease dataset) are gradually equalized after replication. A graphical comparison between the DI for original data and the DI for replicated data for the positive minorities for the Diabetes and Heart Disease datasets is depicted in Figs. 16.8 and 16.9, respectively. However, according to our experimental results, there should be a limit on the amount of replication to prevent further bias that can be produced after replication of minorities. It is worth mentioning that a client with a high DI score implies a significant data imbalance reduction with respect to a certain group. For instance, with reference to Table 16.2, Client 1 has 4890 data, with a total of 954 positives in the majority class and 273 positives in the minority class. Its DI after replication is

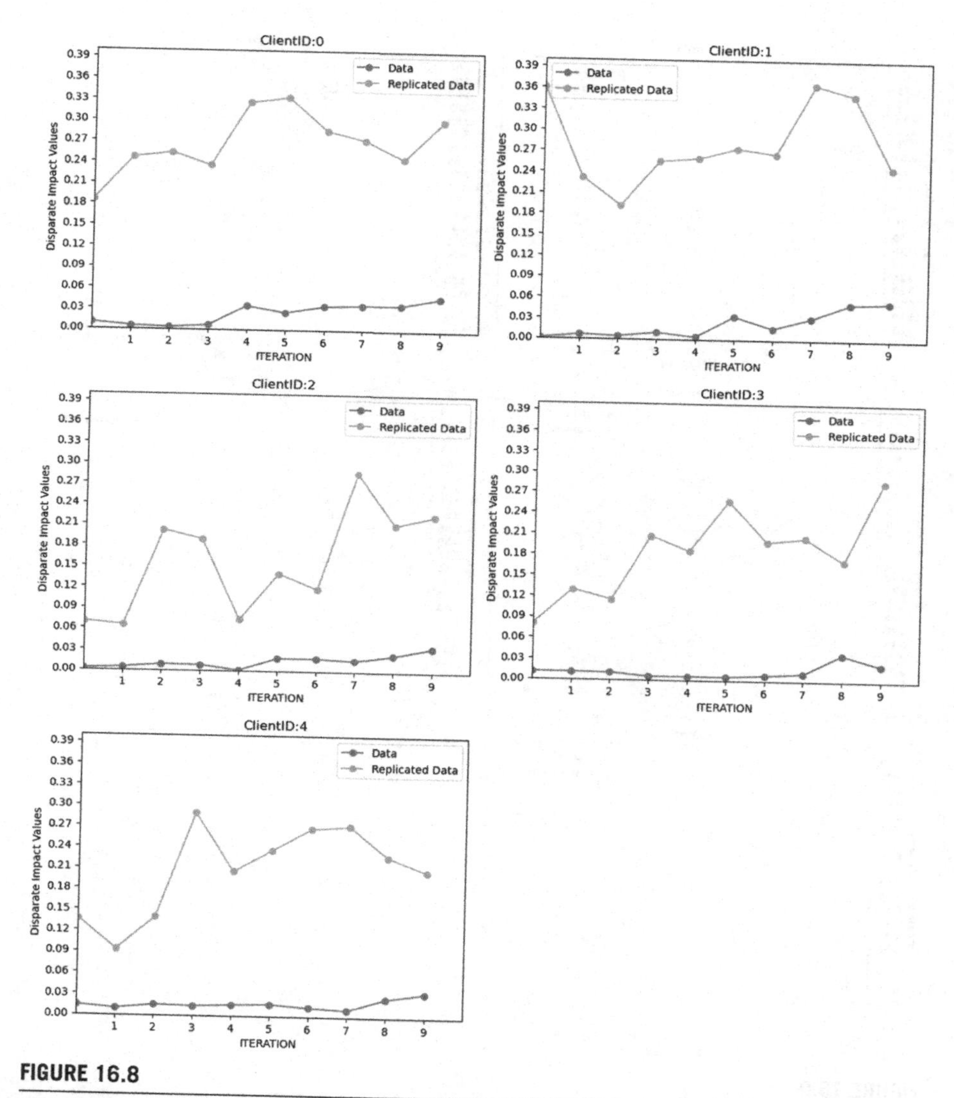

FIGURE 16.8

Disparate impact per client for the original vs. replicated data for *Caucasian* vs. *non-Caucasian* for the Diabetes dataset.

highest at 0.38, which is the maximum among the clients. This demonstrates how, in a federated environment, we can detect disparities in client data by employing DI without looking at user data.

Our main objective is to achieve model fairness through a data-driven technique. Thus, we concentrate on data rather than the model itself because this would ul-

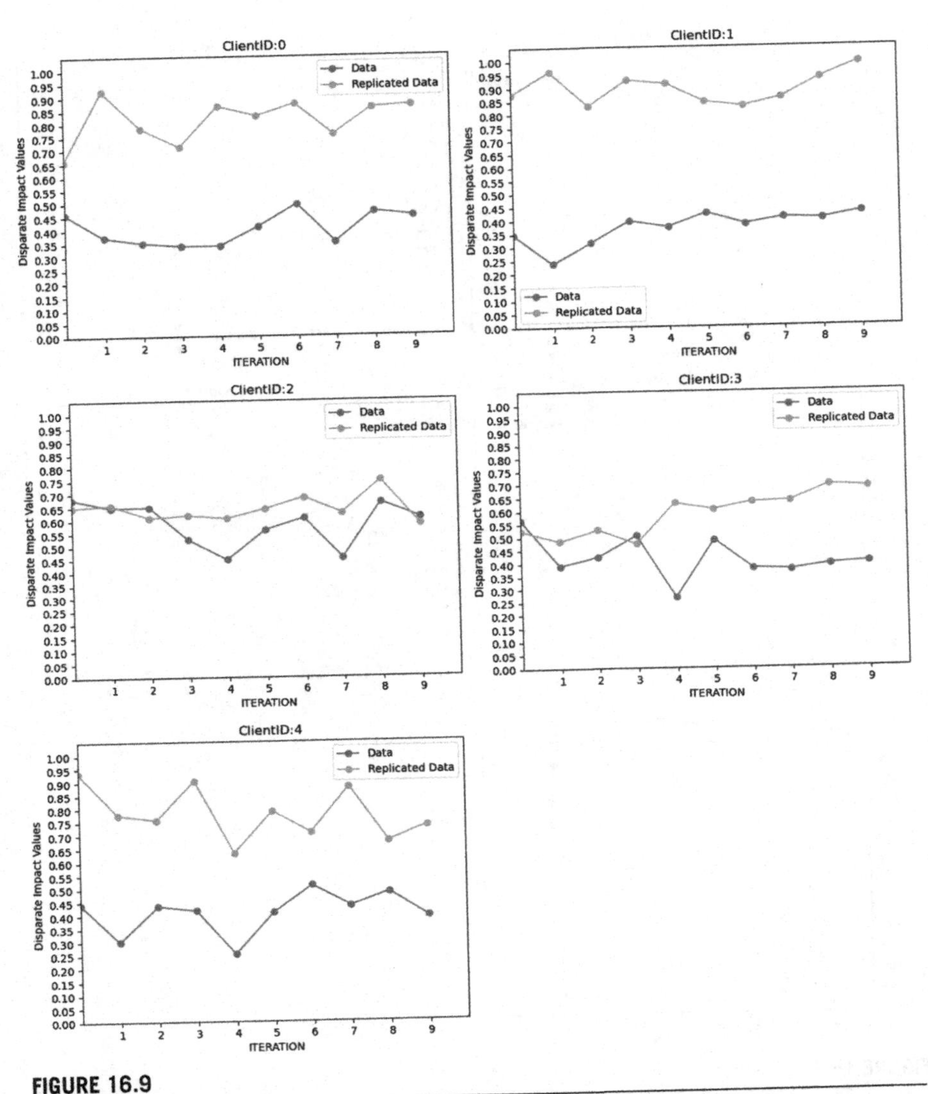

FIGURE 16.9

Disparate impact per client for the original vs. replicated data for *male* vs. *female* for the Heart Disease dataset.

timately make the model fair. However, the proposed methodology assumes zero client dropouts during model training, which may not hold in a real FL scenario. This limitation can be further addressed by including client dropouts and using the methodology at an extended scale.

Table 16.2 Data replication for positive minorities.

Diabetes dataset	Client 0	Client 1	Client 2	Client 3	Client 4
Original total data	6471	4890	4934	5396	7109
#Majority (Caucasians) in positive class	1173	954	953	1005	1324
#Minority (non-Caucasians) in positive class	372	273	234	288	412
After replication					
#Majority (Caucasians) in positive class	1173	954	953	1005	1324
#Minority (non-Caucasians) in positive class	**744**	**546**	**468**	**576**	**824**
Heart Disease dataset	**Client 0**	**Client 1**	**Client 2**	**Client 3**	**Client 4**
Original total data	39,651	32,355	48,506	36,534	44,424
#Majority (males) in positive class	1960	1675	2379	1858	2233
#Minority (females) in positive class	1364	1098	1730	1274	1514
After replication					
#Majority (males) in positive class	1960	1675	2379	1858	2233
#Minority (females) in positive class	**2728**	**2196**	**3460**	**2548**	**3028**

16.5 Conclusion

The importance of data-driven algorithms in influencing people's welfare has led research communities to focus on creating fair and effective decision-making algorithms. In the medical field, future work is aiming to develop AI regulations that promote transparency and data sharing while protecting patient privacy. One of the prominent factors influencing fairness is data bias, and this chapter addresses this issue in the US Diabetes and Heart Disease datasets using FL to preserve patient privacy while maintaining a distributed and heterogeneous setting. We have employed the DI metric to measure fairness and modified the data to improve fairness levels rather than altering the existing model, making our solution a data-driven technique that handles the bias within data well. For future work, the problem can be further expanded to incorporate multivariate attributes so that the solutions lead to understanding of the data in more and better ways.

References

[1] K.A. Gomez, A.A. Gomez, Statistical Procedures for Agricultural Research, John Wiley & Sons, 1984.

[2] D.L. Hall, J. Llinas, An introduction to multisensor data fusion, Proceedings of the IEEE 85 (1) (1997) 6–23.
[3] F. Castanedo, A review of data fusion techniques, The Scientific World Journal 2013 (2013) 2–19.
[4] B. Strack, J.P. DeShazo, C. Gennings, J.L. Olmo, S. Ventura, K.J. Cios, J.N. Clore, Impact of hba1c measurement on hospital readmission rates: analysis of 70,000 clinical database patient records, BioMed Research International 2014 (2014).
[5] B. Strack, J.P. DeShazo, C. Gennings, J.L. Olmo, S. Ventura, K.J. Cios, J.N. Clore, Impact of hba1c measurement on hospital readmission rates: Analysis of 70,000 clinical database patient records, vol. 2014, Hindawi, 2014, pp. 1–12.
[6] M. Feldman, S.A. Friedler, J. Moeller, C. Scheidegger, S. Venkatasubramanian, Certifying and removing disparate impact, in: Proceedings of the 21th ACM SIGKDD International Conference on Knowledge Discovery and Data Mining, 2015, pp. 259–268.
[7] M. Abadi, P. Barham, J. Chen, Z. Chen, A. Davis, J. Dean, M. Devin, S. Ghemawat, G. Irving, M. Isard, et al., Tensorflow: a system for large-scale machine learning, in: 12th USENIX Symposium on Operating Systems Design and Implementation (OSDI 16), 2016, pp. 265–283.
[8] S. Barocas, A.D. Selbst, Big data's disparate impact, California Law Review (2016) 671–732.
[9] J. Konečný, H.B. McMahan, D. Ramage, P. Richtárik, Federated optimization: distributed machine learning for on-device intelligence, arXiv preprint, arXiv:1610.02527, 2016, https://doi.org/10.48550/arXiv.1610.02527.
[10] P. Regulation, Regulation (eu) 2016/679 of the European Parliament and of the council, Regulation (EU) 679 (2016).
[11] S. Barocas, M. Hardt, A. Narayanan, Fairness in Machine Learning, vol. 1, 2017.
[12] D. Dua, C. Graff, UCI machine learning repository, [Online], available: http://archive.ics.uci.edu/ml/datasets/Adult, 2017. (Accessed 8 May 2023).
[13] S.M. Julia Angwin Jeff Larson, P. Lauren Kirchner, Machine bias: there's software used across the country to predict future criminals. and it's biased against blacks, [Online], available: https://www.propublica.org/article/machine-bias-risk-assessments-incriminal-sentencing, 2017. (Accessed 8 May 2023).
[14] B. McMahan, E. Moore, D. Ramage, S. Hampson, B.A. y Arcas, Communication-efficient learning of deep networks from decentralized data, in: Artificial Intelligence and Statistics, PMLR, 2017, pp. 1273–1282.
[15] X. Wang, Y. Peng, L. Lu, Z. Lu, M. Bagheri, R.M. Summers, Chestx-ray8: hospitalscale chest x-ray database and benchmarks on weakly-supervised classification and localization of common thorax diseases, in: Proceedings of the IEEE Conference on Computer Vision and Pattern Recognition, 2017, pp. 2097–2106.
[16] C. Cobb, S. Sudar, N. Reiter, R. Anderson, F. Roesner, T. Kohno, Computer security for data collection technologies, Development Engineering 3 (2018) 1–11.
[17] L. de la Torre, A guide to the California consumer privacy act of 2018, available at SSRN 3275571, 2018, pp. 1–17.
[18] S. Verma, J. Rubin, Fairness definitions explained, in: 2018 Ieee/Acm International Workshop on Software Fairness (Fairware), IEEE, 2018, pp. 1–7.
[19] K. Bonawitz, H. Eichner, W. Grieskamp, D. Huba, A. Ingerman, V. Ivanov, C. Kiddon, J. Konečný, S. Mazzocchi, B. McMahan, et al., Towards federated learning at scale: system design, in: Proceedings of Machine Learning and Systems, vol. 1, 2019, pp. 374–388.
[20] S. Dash, S.K. Shakyawar, M. Sharma, S. Kaushik, Big data in healthcare: management, analysis and future prospects, Journal of Big Data 6 (1) (2019) 1–25.

[21] A. Ghosh, J. Hong, D. Yin, K. Ramchandran, Robust federated learning in a heterogeneous environment, arXiv preprint, arXiv:1906.06629, 2019, https://doi.org/10.48550/arXiv.1906.06629.
[22] T. Li, M. Sanjabi, A. Beirami, V. Smith, Fair resource allocation in federated learning, arXiv preprint, arXiv:1905.10497, 2019, https://doi.org/10.48550/arXiv.1905.10497.
[23] Y. Li, L. Deng, X. Yang, Z. Liu, X. Zhao, F. Huang, S. Zhu, X. Chen, Z. Chen, W. Zhang, Early diagnosis of gastric cancer based on deep learning combined with the spectral-spatial classification method, vol. 10(10), Optica Publishing Group, 2019, pp. 4999–5014.
[24] S. Truex, N. Baracaldo, A. Anwar, T. Steinke, H. Ludwig, R. Zhang, Y. Zhou, A hybrid approach to privacy-preserving federated learning, in: Proceedings of the 12th ACM Workshop on Artificial Intelligence and Security, 2019, pp. 1–11.
[25] T. Li, A.K. Sahu, A. Talwalkar, V. Smith, Federated learning: challenges, methods, and future directions, IEEE Signal Processing Magazine 37 (2020) 50–60.
[26] L. Lyu, X. Xu, Q. Wang, H. Yu, Collaborative fairness in federated learning, in: Federated Learning, Springer, 2020, pp. 189–204.
[27] E. Tat, D.L. Bhatt, M.G. Rabbat, Addressing bias: artificial intelligence in cardiovascular medicine, The Lancet Digital Health 2 (12) (2020) e635–e636.
[28] M.I. Vasileva, The dark side of machine learning algorithms: how and why they can leverage bias, and what can be done to pursue algorithmic fairness, in: Proceedings of the 26th ACM SIGKDD International Conference on Knowledge Discovery & Data Mining, 2020, pp. 3586–3587.
[29] J. Wang, Q. Liu, H. Liang, G. Joshi, H.V. Poor, Tackling the objective inconsistency problem in heterogeneous federated optimization, Advances in Neural Information Processing Systems 33 (2020) 7611–7623.
[30] Y.H. Ezzeldin, S. Yan, C. He, E. Ferrara, S. Avestimehr, Fairfed: enabling group fairness in federated learning, corr, CoRR, arXiv:2110.00857 [abs], 2021, https://doi.org/10.48550/arXiv.2110.00857.
[31] L. Gao, L. Li, Y. Chen, W. Zheng, C. Xu, M. Xu, Fifl: a fair incentive mechanism for federated learning, in: 50th International Conference on Parallel Processing, 2021, pp. 1–10.
[32] A. Gooday, Understanding the types of federated learning, [Online], available: https://blog.openmined.org/federated-learning-types/, 2021. (Accessed 8 May 2023).
[33] T. Li, S. Hu, A. Beirami, V. Smith, Ditto: fair and robust federated learning through personalization, in: International Conference on Machine Learning, PMLR, 2021, pp. 6357–6368.
[34] Z. Lv, R. Lou, J. Li, A.K. Singh, H. Song, Big data analytics for 6g-enabled massive Internet of things, IEEE Internet of Things Journal 8 (7) (2021) 5350–5359.
[35] N. Norori, Q. Hu, F.M. Aellen, F.D. Faraci, A. Tzovara, Addressing bias in big data and ai for health care: a call for open science, Patterns 2 (10) (2021) 1–9.
[36] C. Yang, Q. Wang, M. Xu, Z. Chen, K. Bian, Y. Liu, X. Liu, Characterizing impacts of heterogeneity in federated learning upon large-scale smartphone data, in: Proceedings of the Web Conference, 2021, 2021, pp. 935–946.
[37] R.S. Antunes, C. André da Costa, A. Küderle, I.A. Yari, B. Eskofier, Federated learning for healthcare: Systematic review and architecture proposal, 4, vol. 13, ACM, New York, NY, 2022, pp. 1–23.
[38] M. Mamun, M.M. Uddin, V.K. Tiwari, A.M. Islam, A.U. Ferdous, Mlheartdis: can machine learning techniques enable to predict heart diseases?, in: 2022 IEEE 13th Annual Ubiquitous Computing, Electronics & Mobile Communication Conference (UEMCON), IEEE, 2022, pp. 0561–0565.

[39] A. Momenzadeh, A. Shamsa, J.G. Meyer, Bias or biology? Importance of model interpretation in machine learning studies from electronic health records, JAMIA Open 5 (3) (2022) 2–13.
[40] K. Pytlak, Personal key indicators of heart disease, Version 1 (2022) 636–638.

CHAPTER

Deep learning for emotion recognition using physiological signals 17

Sakshi Indolia[a,b], **Swati Nigam**[a,b], **and Rajiv Singh**[a,b]
[a]*Department of Computer Science, Banasthali Vidyapith, Rajasthan, India*
[b]*Centre for Artificial Intelligence, Banasthali Vidyapith, Rajasthan, India*

17.1 Introduction

The human body generally exhibits some physical changes as a result of environmental events. Different emotional states are triggered in the human body as a response to these physical changes. Various physiological signals can be analyzed to monitor the emotional response to these physical changes. Although human beings can express their emotions through various visual factors, including body language and facial expressions, depending on the situation, they may intentionally hide these emotions. Therefore, evaluating physiological data (which cannot be intentionally modified) gathered from various sensors can help determine a person's feelings for a variety of applications [1–3]. There are four distinct parts of the cerebrum in the human brain; these parts control cognitive skills and conscious thoughts, handle vision-related tasks, integrate and process information from various sensory organs, and handle long- and short-term memory and process language.

In recent years, various machine learning and signal processing methods have been proposed to perform emotion recognition through electroencephalography (EEG) signals. However, machine learning algorithms require handcrafted features, which is time consuming. To address this issue, many deep learning techniques have been proposed that exhibit better classification performance as compared to machine learning techniques. Furthermore, studies show that EEG signals are highly contaminated by either environmental or subject-related artifacts. These artifacts generate electric potential, which contaminates the original EEG data. Thus, effective removal and preprocessing of raw EEG data play a vital role in accurate emotion recognition.

The current research on emotion recognition based on EEG signals focuses on feature fusion techniques, while the potential of deep models remains underexplored. Therefore, the performance accuracy and effectiveness of emotion recognition models are limited. In order to evaluate the effectiveness of deep learning models for EEG-based emotion recognition, comparative analysis of models utilizing benchmark datasets is also required.

Data Fusion Techniques and Applications for Smart Healthcare. https://doi.org/10.1016/B978-0-44-313233-9.00023-0

Thus, this study introduces a deep learning model designed specifically to overcome the mentioned challenges. The chapter provides an explanation of various physiological signals, particularly EEG, that can be explored for emotion recognition. It also presents new research findings and experimental results that demonstrate the effectiveness and potential applications of deep learning techniques for emotion recognition using EEG signals. It explores fast Fourier transform (FFT) integrated with a deep learning approach to enhance emotion recognition accuracy. These findings contribute to the field and provide practical insights.

The major contributions of this chapter are:

- EEG signals are contaminated by unconscious movements such as eye blinking. Thus, FFT has been applied in the proposed model as a preprocessing technique for enhancing performance of the model.
- Literature shows that EEG signals contain both spatial and temporal information, yet current deep learning models like convolutional neural networks (CNNs) can only detect spatial information. Thus, in this work, bidirectional long short-term memory (Bi-LSTM) has been incorporated in the proposed model to exploit spatial as well as temporal information carried by EEG signals.
- The performance of the proposed model has been evaluated based on two benchmark datasets, i.e., DEAP and SEED, and comparisons with existing methods show effectiveness of the proposed model.

17.2 Background and literature

An event that occurs in the environment may result in physical changes in various parts of the human body. These physical changes often lead to different emotional states. Physiological signals are specific signals unconsciously produced by the human body in response to these external events. Although emotions can be recognized through different visible features like facial expressions and body gestures, humans may intentionally hide these features. Therefore, analyzing physiological signals collected through different sensors (shown in Fig. 17.1) can assist in determining emotions of a person without intentional modification of emotions [2].

In addition to this, various types of physiological signals, such as heart rate variability (HRV), respiratory sinus arrhythmia (RSA), breaths per minute, electromyography (EMG), skin conductance, the galvanic skin response, EEG, and PET, can be used to identify human emotions, as shown in Table 17.1. We mainly focus on different approaches which are used to study brain activities through EEG as EEG signals are easy to record as compared to other physiological signals. Therefore, this study will discuss various approaches related to emotion elicitation, data preprocessing, feature extraction, and classification techniques utilized for the classification of human emotions based on EEG signals.

The cerebrum of a human brain is essentially divided into four parts (shown in Fig. 17.2). The frontal lobe controls cognitive skills and conscious thoughts; the

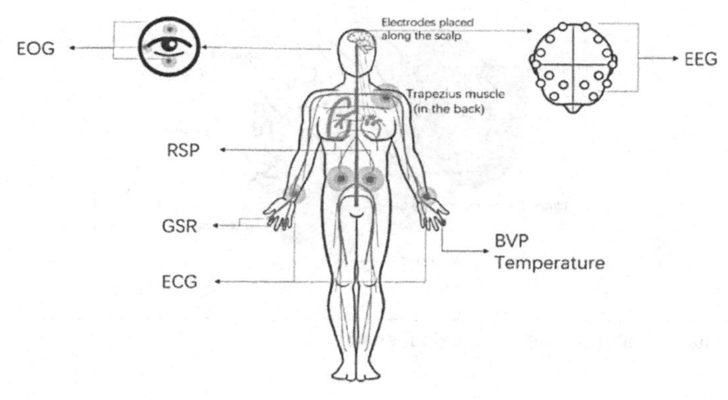

FIGURE 17.1

Biosensors used to collect physiological signals [3].

Table 17.1 Different physiological signals for emotion recognition.

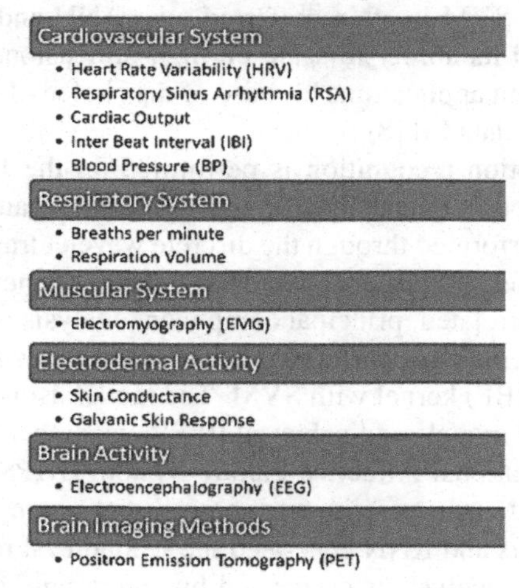

System	Signals
Cardiovascular System	• Heart Rate Variability (HRV) • Respiratory Sinus Arrhythmia (RSA) • Cardiac Output • Inter Beat Interval (IBI) • Blood Pressure (BP)
Respiratory System	• Breaths per minute • Respiration Volume
Muscular System	• Electromyography (EMG)
Electrodermal Activity	• Skin Conductance • Galvanic Skin Response
Brain Activity	• Electroencephalography (EEG)
Brain Imaging Methods	• Positron Emission Tomography (PET)

parietal lobe is responsible for integrating and processing information received from various sensory organs; the occipital lobe carries out vision-related tasks; and the temporal lobe is responsible for maintaining long-term and short-term memory and language processing [4]. However, studies on different cerebral lobes proved that both frontal and parietal lobes can provide relevant features associated with emotion processing [5].

In the past few years, EEG signals have been used to recognize human emotions by implementing various machine learning algorithms. Some widely adopted machine learning algorithms to perform human emotion recognition are support vector machine (SVM), multilayer perceptron (MLP), k-nearest neighbor (KNN), random

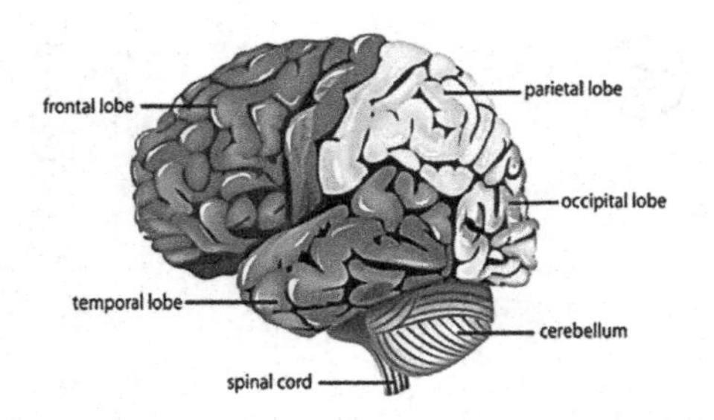

FIGURE 17.2

Parietal, frontal, temporal, and occipital lobes [4].

forest, artificial neural networks (ANNs), Bayesian networks, decision tree, hidden Markov model, deep belief network, etc.

SVM is a supervised learning algorithm capable of linear or nonlinear classification. In the case of linear classification, SVM exploits an n-dimensional hyperplane, whereas, in the case of nonlinear classification, a kernel function is used to determine the decision boundary. SVM is often preferred over ANN and KNN due to its low computational cost and its ability to scale on high-dimensional data [7]. However, SVM is limited in certain applications because of the choice of filters and inability to handle large amount of data [6] [8].

In [9], human emotion recognition is performed on the DEAP dataset by exploiting five decomposed frequency bands, i.e., alpha, beta, gamma, theta, and noise. Feature extraction is performed through the discrete wavelet transform (DWT) by extracting entropy and energy from each window of all frequency bands. To make the extracted features uncorrelated, principal component analysis is incorporated. Moreover, different experiments were performed by alternatively using KNN, ANN, and a radial basis function (RBF) kernel with SVM. The model using SVM with RBF kernel outperformed other models and achieved 91.3% accuracy. Similarly, the authors of [10] used the International Affective Picture System (IAPS) dataset for eliciting emotions in the subjects and experiments showed that among five methods, SVM produced the best results and KNN generated the second best results.

In [11], quantitative analysis is performed by comparing different features such as power spectral density, entropy, wavelet, statistical features, and fractal dimension. The authors experimented with different machine learning techniques such as KNN, decision tree, and SVM. However, studies showed that statistical features assist in quantifying the intraclass correlation and thus help in identifying whether the data belong to the same group or not [12]. These findings complement the experimental results performed in [11] and demonstrate that SVM exploiting statistical features performs better than any other experimental setup. Similarly, the authors of [13,14] performed automatic seizure detection through EEG signals by employing SVM.

KNN is also a supervised machine learning technique which essentially classifies the input data based on k neighbors. This classification is performed by calculating the Euclidean distance of the input sample to all the data points. Then, k neighbors are selected, and the input is classified based on the distance between the inputs and k neighbors. Here, the value of k is decided on the basis of dataset size. Emotion classification was performed in the 2D valence-arousal plane on the DEAP dataset in [15]. They performed feature extraction by incorporating DWT. The model specifically exploits mean, standard deviation, and power spectral density features. Several experiments were performed by altering the value of k, and the results demonstrated that k = 10 resulted in the best classification accuracy of 87.1%. In [16], classification of fear and relaxation emotions is performed by exploiting mean energy and the power spectrum of differential asymmetry features through short-time Fourier transform. The proposed model was tested with SVM and KNN. The experiments show that KNN outperformed SVM with a classification accuracy of 94%.

ANN is inspired by the highly dense and complex network architecture of the human brain, where the neurons (also called nodes) are connected to each other. In the past few years, ANN has been exploited by many researchers for human emotion recognition through EEG signals. In [17], experiments are performed on the DEAP dataset by comparing quantitative results generated by ANN and KNN classifiers. Their proposed model exploits DWT for feature extraction and based on performance, five channels, i.e., P3, FC2, AF3, O1, and FP1, were selected among 32 channels through a dynamic channel selection procedure. ANN outperformed KNN with an accuracy of 77.14%. Additionally, the authors observed that the performance of the model is dependent on the channel and feature selection. In [18], features obtained from variance, DWT, and FFT are tested for emotion recognition on the DEAP and SEED datasets. The authors employ a spiking neural network (SNN) for classification, and experiments showed that the model using variance outperformed the model exploiting both DWT and FFT.

Deep learning models, especially variants of CNN, are tested in [19]. Initially, filtered EEG signals are converted to images using a time-frequency representation. The authors employ a smoothed pseudo-Wigner–Ville distribution for converting EEG signals to images. Thereafter, the images are fed to variants of CNN. The classification accuracies of the variants were compared with the proposed configurable CNN. It was observed that the proposed configurable CNN model (93.01%) outperformed AlexNet (90.98%), ResNet50 (91.91%), and VGG16 (92.71%). A CNN-based model which uses power spectral density features extracted from EEG signals outperforms an SVM-based architecture in [20]. Additionally, the authors compared three architectures, i.e., CNN-2, CNN-5, and CNN-10, in terms of accuracy, and found that CNN-5 performed best, because CNN-2 could not extract a significant number of features whereas CNN-10 suffered from overfitting. In [21], 3D-CNN is used for extracting relevant features from EEG signals; an accuracy of 88% was achieved for valence and arousal.

A modified version of CNN by incorporating a convolution operation and a PST attention module is proposed in [27]. The proposed model captures the active brain regions through a positional attention module and the most informative EEG features are learned through a spectral module, whereas a temporal attention module is used to learn the time dimension mask. The proposed PST attention model was tested on the

SEED and SEED IV datasets. Experiments showed that different attention modules can complement each other and hence enhance the classification accuracy. However, it was seen that P-Attention and S-Attention help more than T-attention in mining relevant information related to emotion classification.

A model exploiting LSTM and a stacked autoencoder to perform human emotion recognition on the DEAP dataset is proposed by [23]. This model decomposes the collected EEG signals by using a stack autoencoder and LSTM is used to recognize emotions through correlations among EEG sequences decomposed by the stacked autoencoder. The model also reduces the chances of overfitting by incorporating dropout, regularization, and 10-fold cross-validation.

Experiments focused on channel selection and experimental analysis have shown that channel selection plays a vital role in improving the performance of a model [17,24]. In [25], Attention-based LSTM with Domain Discriminator (ATLDD-LSTM) is proposed, which exploits LSTM along with the attention mechanism. ATLDD-LSTM automatically identifies the suitable EEG channels required for emotion recognition. The model was evaluated on the DEAP, SEED, and CMEED databases and outperformed the state-of-the-art existing models. Similarly, spatial and temporal attention are applied to model the complex features of EEG signals through spatial-temporal transformer [22].

It can be observed that incorporation of deep learning techniques has shown continuous improvement as compared to machine learning techniques. Although many models have been proposed for effective emotion recognition through EEG signals, below we list some research gaps that need to be addressed:

(a) Lack of large standardized datasets: For the creation and assessment of deep learning models for emotion recognition, large standardized datasets are essential. Currently, there is a scarcity of publicly available large EEG datasets specifically designed for emotion recognition tasks.

(b) Limited generalization across individuals: Deep models are often trained on datasets which are collected by capturing brain waves of individuals. However, humans have different brain anatomy and scalp characteristics, which may lead to the generation of unsatisfactory datasets through different individuals with different brain structures.

(c) Feature extraction and selection: Deep learning models require training using labeled datasets, which is a challenging task as it requires laboratory-controlled environmental conditions. To address this issue, better feature extraction and selection methods need to be explored which ultimately reduce the reliance on large labeled datasets.

17.3 Preliminaries

In this section, the preliminaries, i.e., Bi-LSTM and datasets required for training the proposed model, are discussed in detail.

Bidirectional long short-term memory

Recurrent neural networks (RNNs) are neural networks that are designed specifically to handle temporal input. The major advantage of using RNNs is that they retain the order of the data and make predictions based on input data patterns. RNNs are able to understand long-term dependencies in data. The RNN module achieves this by efficiently capturing and analyzing sequential data since the module is made up of various layers that interact with one another. Through these interactions, RNNs are able to recall and use data from earlier time steps, which makes it possible to identify long-term dependencies in the data. The long short-term memory (LSTM) module, shown in Fig. 17.3, is a specific kind of RNN architecture. Traditional RNNs suffer from the vanishing gradient problem, which makes it difficult for them to detect long-term dependencies in sequential data. LSTM addresses this issue by including memory cells and gating mechanisms, which enhance information retention and provide selective updating of the hidden state.

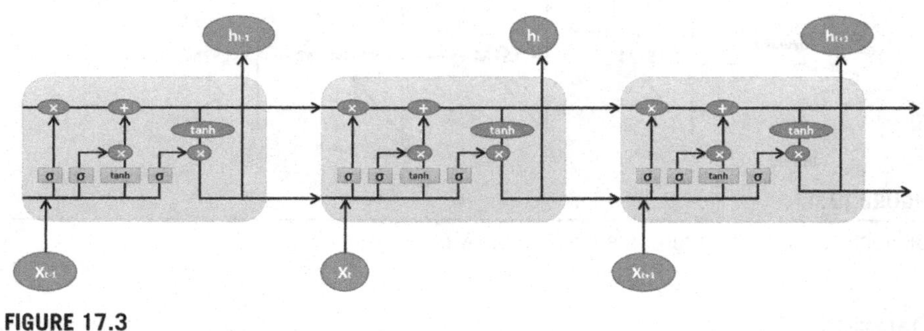

FIGURE 17.3

Long short-term memory (LSTM) nodes.

Each LSTM cell contains an input gate, a forget gate, and an output gate. The input gate chooses which input values and prior hidden state values should be saved as new data. It applies a sigmoid activation function to the current input and the previous hidden state to produce a value between 0 and 1. This value determines the amount of new information to be stored in the memory cell. The data to be discarded from the memory cell are decided by the forget gate. It creates a forget vector by applying a sigmoid activation function to the current input and the previous hidden state. The previous memory cell state is then multiplied elementwise with this vector to enable the LSTM cell to forget unimportant information. The LSTM cell's memory is represented by the cell state. It is updated by combining the information from the input gate and the forget gate. The new values are chosen by the input gate, and a tanh activation function scales them. Through elementwise multiplication of the forget gate output and the previous cell state and elementwise addition of the input gate output by the scaled candidate values, the new cell state is generated. The output gate regulates which aspects of the cell state are displayed as the LSTM cell's output. It takes the current input and the previous hidden state as inputs and applies a sigmoid activation function to generate an output vector. This output vector is multiplied elementwise with the updated cell state, which produces the hidden state of the LSTM cell. The hidden state represents the summarized information that can be passed to the next time step or used for downstream tasks.

A sequence processing model called Bi-LSTM (shown in Fig. 17.4) is made up of two LSTMs, one of which gets input forward and the other of which receives it backward. The network performs better because Bi-LSTM has access to more data.

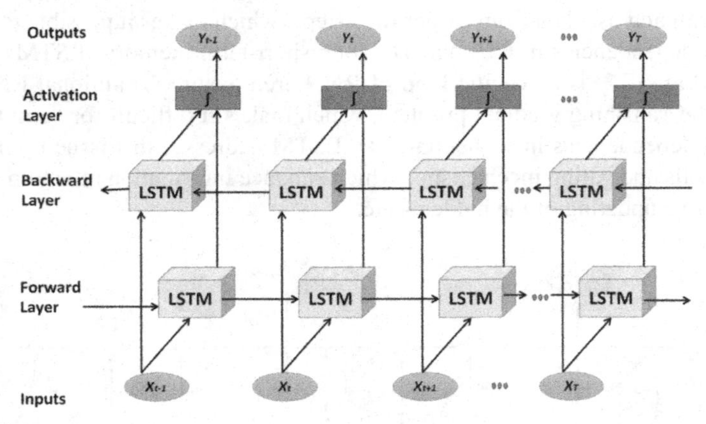

FIGURE 17.4

Bidirectional long short-term memory (Bi-LSTM).

Datasets

In this work, two publicly available databases, i.e., DEAP [30] and SEED [31], have been used for performance evaluation of the proposed model. The DEAP dataset was generated by showing 40 music videos to 32 volunteers. The dataset contains 48 channels recorded at 512 Hz, which consists of 32 EEG, 12 peripheral, 3 unused, and 1 status channel. The SEED dataset was recorded by showing approximately 4-minute movie clips to 15 participants. The dataset contains 62 channels recorded at 1000 Hz. A detailed description of these datasets is given in Table 17.2.

17.4 Proposed methodology

The proposed model includes three stages: (i) data acquisition, (ii) preprocessing and feature extraction, and (iii) classification. In this section, a detailed description of these stages is given. The framework is shown in Fig. 17.5.

Data acquisition and selection

Data acquisition can be performed through a variety of stimuli such as pictures, video clips, memories, music, and self-induction. It is observed that video clips are pre-

Table 17.2 Description of the DEAP and SEED datasets.

Property	DEAP	SEED
Emotion elicitation	40 music video clips	15 movie clips
Subjects	32	15
Channels	48	64
Emotion classes	4	3
Emotions	valence, arousal, dominance, liking	positive, neutral, negative
Data dimension description	video × channel × data	subjects × trials × channel × data
Data dimension	40 × 40 × 8064	15 × 3 × 62 × 45000

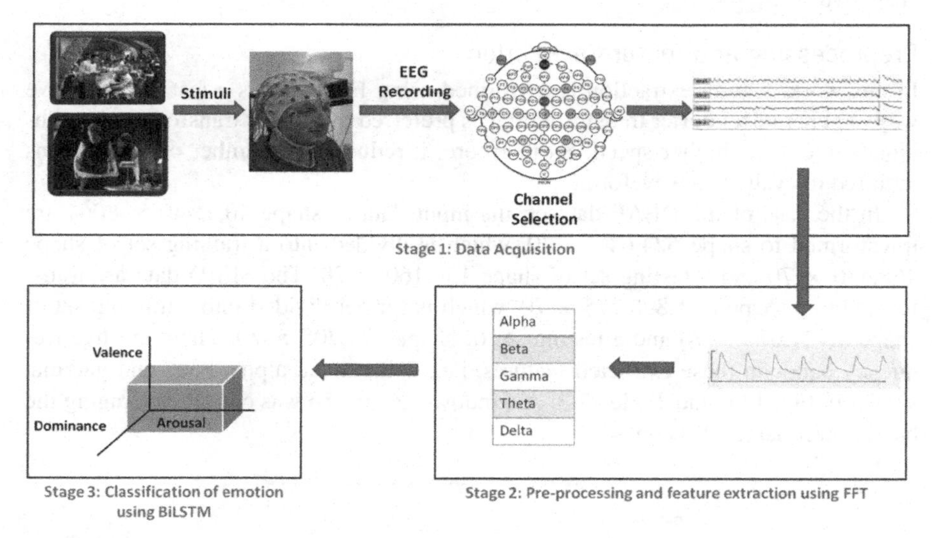

FIGURE 17.5

Human emotion recognition through the proposed model based on EEG signals.

ferred over other kinds of stimuli because they effectively stimulate human emotions. Furthermore, strong stimulation of emotions generates a better database which ultimately results in an accurate emotion recognition model. The performance of the proposed model has been evaluated on two benchmark datasets i.e., the DEAP and SEED datasets. For simplification of the proposed pipeline, preprocessed data made available by the owners have been used.

As mentioned in Table 17.2, the publicly available DEAP dataset contains 48 channels. However, among 48 channels, only 32 channels are specifically available as EEG signals. Thus, for experimentation purposes, we have exploited the first 32 channels. Furthermore, the DEAP dataset contains EEG signals downsampled to 128 Hz, which are already segmented and preprocessed in .dat (python) as well as .mat (MATLAB®) format. The proposed model utilizes .dat files as input data. Two arrays are included in each participant file: (i) data of shape 40 × 40 × 8064 (video × channel × data) and (ii) a label of shape 40 × 4 (video/trial × label {valence, arousal, dominance, liking}).

Similarly, the publicly available SEED dataset contains raw data as well as downsampled data to 200 Hz. The downsampled data are further processed by applying a bandpass frequency filter (0–75 Hz). Furthermore, three trials have been conducted for each subject, generating 15 (subjects) × 3 (trials) = 45 .mat files, where each subject file contains 16 arrays including 15 trial arrays (EEG1–EEG15, channel × data) and 1 emotional label array (1, "positive"; 0, "neutral"; −1, "negative"). These emotion labels were converted to one-hot encoding; thus, class 2 corresponds to label "negative."

Preprocessing and feature extraction

In this work, feature extraction is performed using FFT. FFT is a fast and effective way to generate a Fourier transform. FFT is preferred over other transformation techniques due to its higher speed. Furthermore, it reduces the number of calculations required to evaluate a waveform.

In the case of the DEAP dataset, the input data of shape 40 × 40 × 8064 are transformed to shape 624,640 × 70, which is divided into a training set of shape 468,480 × 70 and a testing set of shape 156,160 × 70. The SEED data are transformed to a shape of 1,898,775 × 70, which is further divided into a training set of shape 1,424,070 × 70 and a testing set of shape 474,705 × 70. There are five frequency bands in these extracted features, i.e., delta, theta, alpha, beta, and gamma, shown in Fig. 17.6 and Table 17.3. A window size of 256 was chosen, averaging the band power across 2 seconds.

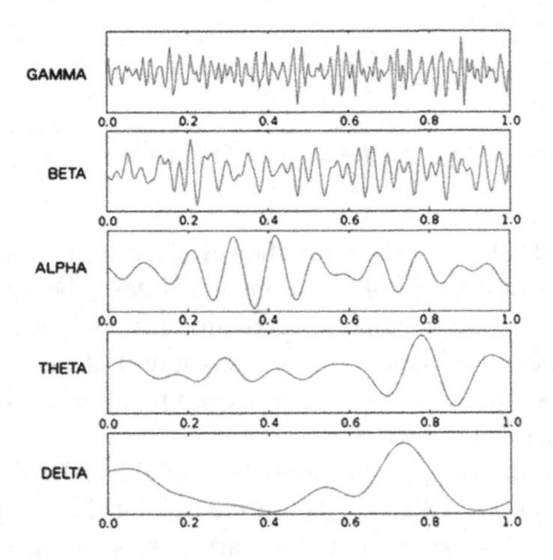

FIGURE 17.6

Five brain waves: delta, theta, alpha, beta, and gamma [6].

Table 17.3 Specific ranges of five brain waves: delta, theta, alpha, beta, and gamma.

Band	Specific range
Delta	1–4 Hz
Theta	4–7 Hz
Alpha	8–13 Hz
Beta	13–30 Hz
Gamma	>30 Hz

Classification

To perform classification of emotions, the input data are passed to a deep learning block consisting of a Bi-LSTM node with a dropout probability of 0.6, followed by four consecutive LSTM and dropout layers, as shown in Fig. 17.7. The number of nodes in the Bi-LSTM block is 128 and four LSTM layers contain 256 (dropout: 0.6), 64 (dropout: 0.6), 64 (dropout: 0.4), and 32 (dropout: 0.4) nodes. Thereafter, these nodes are flattened and provided as input to a dense layer with 16 nodes and ReLU activation function. This flattened dense layer (consisting 16 nodes) is connected to another dense layer (consisting node count matching number of classes i.e., 4 nodes for DEAP and 3 nodes for SEED dataset) with a softmax activation function for the probability distribution among all the classes. For optimization of the proposed model, the Adam optimizer is used. Batch size is set to 256 and 300 epochs are used to train the model for the DEAP dataset, whereas the SEED dataset requires only 5 training epochs.

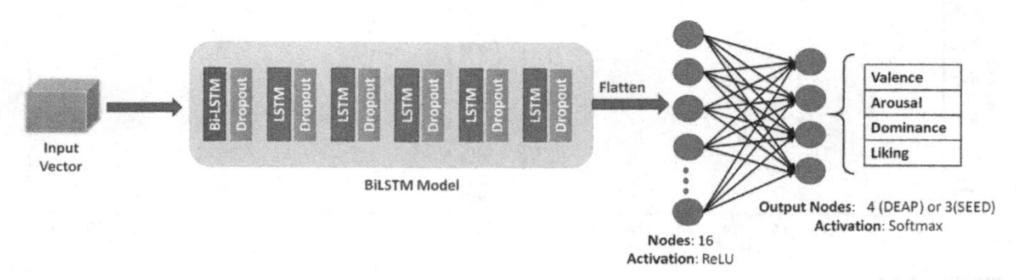

FIGURE 17.7

Proposed classification phase using Bi-LSTM.

17.5 Experiments and results

This section discusses the experimental findings drawn from the abovementioned proposed model. The classification accuracy obtained by the proposed Bi-LSTM model on the DEAP dataset is 90.47% and classification accuracy obtained on the SEED dataset is 98.68%. Fig. 17.8 shows the accuracy and loss curve obtained dur-

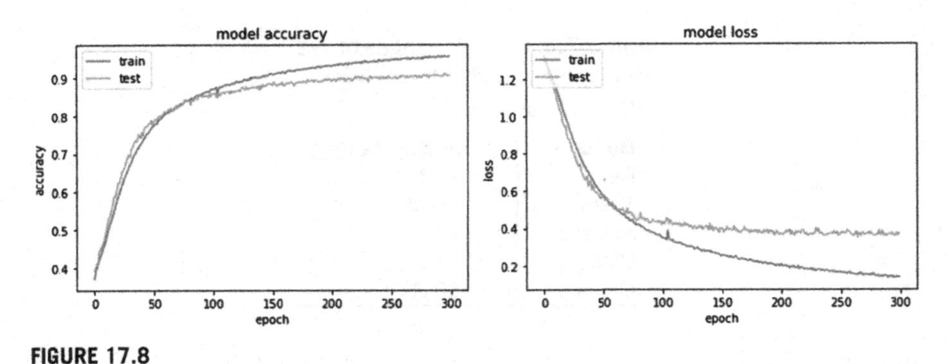

FIGURE 17.8

Accuracy and loss curves generated by the proposed model on the DEAP dataset.

ing training of the model on the DEAP dataset. Here, the training curve is shown in blue (dark gray in print version), whereas the testing curve is shown in orange (gray in print version). Fig. 17.9 depicts the accuracy and loss curve obtained during training of the model on the SEED dataset. These curves follow the pattern expected from the literature. The train and test accuracy curves increase with the increase in number of epochs, whereas the train and test loss curves decrease with the increase in epochs. Additionally, the training loss is marginally less than the test loss curve, while the train accuracy is marginally greater than the test accuracy curve.

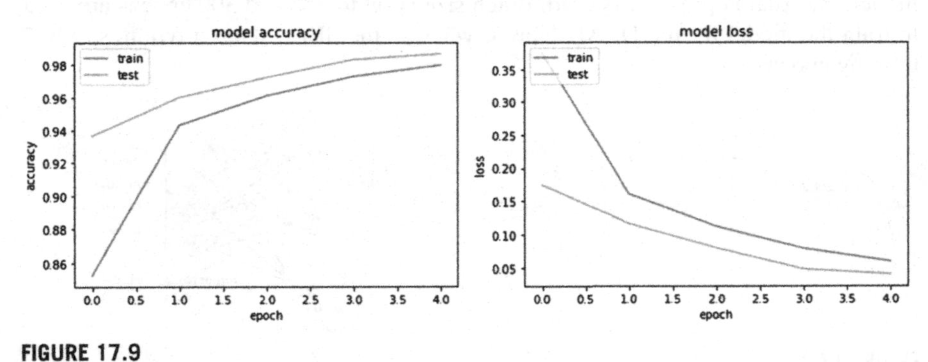

FIGURE 17.9

Accuracy and loss curves generated by the proposed model on the SEED dataset.

It can be observed that the classification accuracy obtained by the proposed model on the DEAP dataset is lower than that obtained on the SEED dataset; however, the number of epochs for training the model on the DEAP dataset is 300, compared to 5 epochs for the SEED dataset. Thus, it can be inferred that a higher number of training epochs does not always improve performance of the model.

Furthermore, an abrupt change in accuracy and loss curves for the SEED dataset can be observed in Fig. 17.9, indicating early saturation of the model in terms of accuracy. The particular reason behind this is the large number of training samples, which does not overfit the model and generates higher accuracy. Moreover, the model incorporates a dropout mechanism which helps the model to avoid overfitting.

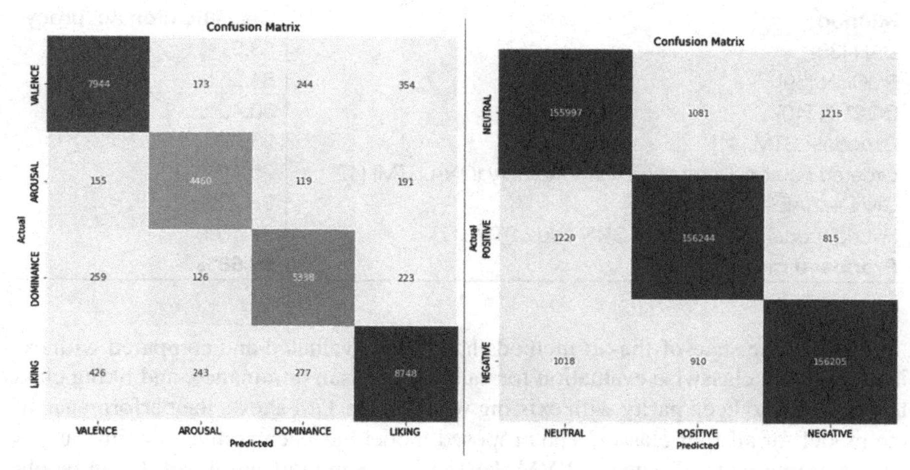

FIGURE 17.10

Confusion matrix generated by the proposed model on the DEAP and SEED datasets.

The confusion matrix is a useful tool for evaluating model performance in multi-class classification. Fig. 17.10 shows the confusion matrix generated by the proposed model on the DEAP and SEED datasets. In case of the DEAP dataset, the classification accuracy for valence, arousal, dominance, and liking is 91.53%, 90.55%, 89.77%, and 90.24%, respectively. It can be observed from Fig. 17.10 that more liking samples are misclassified as valence. Similarly, the proposed model achieves an classification accuracy of 98.54%, 98.71%, and 98.78% for the neutral, positive, and negative classes on the SEED dataset. Furthermore, to validate the performance of the proposed model, it is compared with existing state-of-the-art models on the DEAP and SEED datasets, as shown in Tables 17.4 and 17.5, respectively.

Table 17.4 Comparison between the proposed model and existing models for the DEAP dataset in terms of classification accuracy.

Method	Valence	Arousal	Dominance	Liking
SVM and KNN [32]	86.75%	84.05		
LSTM-RNN [33]	85.45%	85.65%	-	87.99%
GCNN + LSTM + Dense Layer [34]	90.45%	90.60%	-	-
Spiking Neural Net [35]	78.00%	74.00%	80.00%	86.27%
ScalingNet [36]	71.32%	71.65%	72.89%	-
DCNN + ConvLSTM + Attention [37]	87.84%	87.69%	-	-
SAE + LSTM [26]	81.10%	74.38%	-	-
Proposed model	**91.53%**	**90.55%**	**89.77%**	**90.24%**

Table 17.5 Comparison between the proposed model and existing models for the SEED dataset in terms of classification accuracy.

Method	Classification accuracy
DAN [38]	83.81%
PGCNN [39]	84.35%
DGCNN [40]	90.40%
Bimodal-LSTM [41]	93.97%
Ordered neuronal long short-term memory (ON-LSTM) [42]	95.49%
CNN + SAE + DNN [38]	96.77%
Hybrid model comprised of CNN and LSTM [17]	97.16%
Proposed model	**98.68%**

Most of the state-of-the-art methods have been evaluated and compared with existing work in classwise evaluation for valence, arousal, dominance, and liking class labels. Thus, to keep parity with existing work, Table 17.4 shows the performance of the model for all four classes. The proposed model has been compared with various deep learning methods such as SVM, LSTM, and Spiking Neural Net. It can be observed that the proposed model performs better than these algorithms. However, the results generated by the model comprised of GCNN in conjunction with LSTM were comparable to the results of our proposed model.

Similarly, the proposed model outperforms several existing methods, such as DAN, Bimodal-LSTM, CNN in conjunction with SAE, and DNN on the SEED database. However, the obtained results are comparable to those of a hybrid model comprised of CNN and LSTM. Thus, it can be inferred that emotion recognition through EEG signals can be performed with higher accuracy if CNN is used with LSTM. The particular reason for this observation might be that CNN allows the model to learn spatial information, whereas LSTM assists in learning temporal features from the data.

17.6 Conclusions and future works

Human emotion recognition has different applications, such as medical treatment, sociable robots, human–computer interaction, and recommendation systems. However, existing machine learning and deep learning algorithms for human emotion recognition are inefficient in terms of performance, due to which they are not widely adopted in widespread applications. Human emotion recognition can be performed through facial expressions; however, facial expressions can be faked and it is easy to hide true emotions. Thus, physiological signals can be used to detect human emotions. This work proposes a deep learning framework for emotion recognition through EEG signals. In this proposed work, the datasets are preprocessed using FFT and classification of emotions is performed using a Bi-LSTM architecture which exploits temporal information in EEG signals. The proposed model is evaluated on two benchmark datasets and compared with existing state-of-the-art deep learning models. The comparative analysis shows that the proposed model performs better than existing frameworks. However, future research can be performed in the following directions: (i) development of hybrid deep learning architectures such as [29] which can combine CNN and RNN to capture both spatial and temporal data in an effective manner, (ii) transfer learning through pretrained models can be performed by training a deep model with a large dataset and fine-tuning it on a given task-specific dataset, which will reduce the dependence of the model on large labeled datasets, and (iii) multimodal fusion techniques can be incorporated such as [28] to exploit different modalities such as speech and text for improved performance of the model.

References

[1] B. He, A. Sohrabpour, E. Brown, Z. Liu, Electrophysiological source imaging: a noninvasive window to brain dynamics, Annual Review of Biomedical Engineering 20 (2018) 171–196.

[2] S.M. Alarcao, M.J. Fonseca, Emotions recognition using EEG signals: a survey, IEEE Transactions on Affective Computing 10 (3) (2017) 374–393.

[3] L. Shu, J. Xie, M. Yang, Z. Li, Z. Li, D. Liao, et al., A review of emotion recognition using physiological signals, Sensors 18 (7) (2018) 2074.

[4] T. Alotaiby, F.E.A. El-Samie, S.A. Alshebeili, I. Ahmad, A review of channel selection algorithms for EEG signal processing, EURASIP Journal on Advances in Signal Processing 2015 (2015) 1–21.

[5] Y. Liu, O. Sourina, M.K. Nguyen, Real-time EEG-based human emotion recognition and visualization, in: 2010 International Conference on Cyberworlds, IEEE, 2010, October, pp. 262–269.

[6] S. Bhattacharyya, A. Khasnobish, A. Konar, D.N. Tibarewala, A.K. Nagar, Performance analysis of left/right hand movement classification from EEG signal by intelligent 37 algorithms, in: 2011 IEEE Symposium on Computational Intelligence, Cognitive Algorithms, Mind, and Brain (CCMB), IEEE, 2011, April, pp. 1–8.

[7] R. Palaniappan, K. Sundaraj, S. Sundaraj, A comparative study of the svm and k-nn machine learning algorithms for the diagnosis of respiratory pathologies using pulmonary acoustic signals, BMC Bioinformatics 15 (2014) 1–8.

[8] Q. Huang, C. Wang, Y. Ye, L. Wang, N. Xie, Recognition of EEG based on improved black widow algorithm optimized SVM, Biomedical Signal Processing and Control 81 (2023) 104454.

[9] O. Bazgir, Z. Mohammadi, S.A.H. Habibi, Emotion recognition with machine learning using EEG signals, in: 2018 25th National and 3rd International Iranian Conference on Biomedical Engineering (ICBME), IEEE, 2018, November, pp. 1–5.

[10] A.T. Sohaib, S. Qureshi, J. Hagelbäck, O. Hilborn, P. Jerčić, Evaluating classifiers for emotion recognition using EEG, in: Foundations of Augmented Cognition: 7th International Conference, AC 2013, Held as Part of HCI International 2013, Las Vegas, NV, USA, July 21-26, 2013. Proceedings 7, Springer Berlin Heidelberg, 2013, pp. 492–501.

[11] R. Nawaz, K.H. Cheah, H. Nisar, V.V. Yap, Comparison of different feature extraction methods for EEG-based emotion recognition, Biocybernetics and Biomedical Engineering 40 (3) (2020) 910–926.

[12] Z. Lan, O. Sourina, L. Wang, Y. Liu, Real-time EEG-based emotion monitoring using stable features, The Visual Computer 32 (2016) 347–358.
[13] E.H. Houssein, A. Hamad, A.E. Hassanien, A.A. Fahmy, Epileptic detection based on whale optimization enhanced support vector machine, Journal of Information & Optimization Sciences 40 (3) (2019) 699–723.
[14] A. Hamad, E.H. Houssein, A.E. Hassanien, A.A. Fahmy, S. Bhattacharyya, A hybrid gray wolf optimization and support vector machines for detection of epileptic seizure, in: Hybrid Metaheuristics, in: Series in Machine Perception and Artificial Intelligence, vol. 27, 2018, pp. 197–225.
[15] S. Shukla, R.K. Chaurasiya, Emotion analysis through eeg and peripheral physiological signals using knn classifier, in: Proceedings of the International Conference on ISMAC in Computational Vision and Bio-Engineering 2018 (ISMAC-CVB), Springer International Publishing, 2019, pp. 97–106.
[16] A. Jalilifard, E.B. Pizzolato, M.K. Islam, Emotion classification using single-channel scalp-EEG recording, in: 2016 38th Annual International Conference of the IEEE Engineering in Medicine and Biology Society (EMBC), IEEE, 2016, August, pp. 845–849.
[17] M.S. Özerdem, H. Polat, Emotion recognition based on EEG features in movie clips with channel selection, Brain Informatics 4 (4) (2017) 241–252.
[18] L. Jiang, J. He, H. Pan, D. Wu, T. Jiang, J. Liu, Seizure detection algorithm based on improved functional brain network structure feature extraction, Biomedical Signal Processing and Control 79 (2023) 104053.
[19] S.K. Khare, V. Bajaj, Time–frequency representation and convolutional neural network-based emotion recognition, IEEE Transactions on Neural Networks and Learning Systems 32 (7) (2020) 2901–2909.
[20] S.E. Moon, S. Jang, J.S. Lee, Convolutional neural network approach for EEG-based emotion recognition using brain connectivity and its spatial information, in: 2018 IEEE International Conference on Acoustics, Speech and Signal Processing (ICASSP), IEEE, 2018, April, pp. 2556–2560.
[21] E.S. Salama, R.A. El-Khoribi, M.E. Shoman, M.A.W. Shalaby, EEG-based emotion recognition using 3D convolutional neural networks, International Journal of Advanced Computer Science and Applications 9 (8) (2018).
[22] J. Liu, H. Wu, L. Zhang, Y. Zhao, Spatial-temporal transformers for EEG emotion recognition, in: 2022 the 6th International Conference on Advances in Artificial Intelligence, 2022, October, pp. 116–120.
[23] Y. Deng, S. Ding, W. Li, Q. Lai, L. Cao, EEG-based visual stimuli classification via reusable LSTM, Biomedical Signal Processing and Control 82 (2023) 104588.
[24] R. Qiao, C. Qing, T. Zhang, X. Xing, X. Xu, A novel deep-learning based framework for multi-subject emotion recognition, in: 2017 4th International Conference on Information, Cybernetics and Computational Social Systems (ICCSS), IEEE, 2017, July, pp. 181–185.
[25] X. Du, C. Ma, G. Zhang, J. Li, Y.K. Lai, G. Zhao, et al., An efficient LSTM network for emotion recognition from multichannel EEG signals, IEEE Transactions on Affective Computing 13 (3) (2020) 1528–1540.
[26] X. Xing, Z. Li, T. Xu, L. Shu, B. Hu, X. Xu, SAE+ LSTM: a new framework for emotion recognition from multi-channel EEG, Frontiers in Neurorobotics 13 (2019) 37.
[27] J. Liu, Y. Zhao, H. Wu, D. Jiang, Positional-spectral-temporal attention in 3D convolutional neural networks for EEG emotion recognition, in: 2021 Asia-Pacific Signal and Information Processing Association Annual Summit and Conference (APSIPA ASC), IEEE, 2021, December, pp. 305–312.

[28] J.M. Vala, U.K. Jaliya, Analytical review and study on emotion recognition strategies using multimodal signals, in: Advancements in Smart Computing and Information Security: First International Conference, ASCIS 2022, Rajkot, India, November 24–26, 2022, Revised Selected Papers, Part I, Springer Nature Switzerland, Cham, 2023, January, pp. 267–285.
[29] M.Y. Zhong, Q.Y. Yang, Y. Liu, B.Y. Zhen, B.B. Xie, EEG emotion recognition based on TQWT-features and hybrid convolutional recurrent neural network, Biomedical Signal Processing and Control 79 (2023) 104211.
[30] S. Koelstra, C. Muhl, M. Soleymani, J.S. Lee, A. Yazdani, T. Ebrahimi, I. Patras, Deap: a database for emotion analysis; using physiological signals, IEEE Transactions on Affective Computing 3 (1) (2011) 18–31.
[31] W.L. Zheng, B.L. Lu, Investigating critical frequency bands and channels for EEG-based emotion recognition with deep neural networks, IEEE Transactions on Autonomous Mental Development 7 (3) (2015) 162–175.
[32] Z. Mohammadi, J. Frounchi, M. Amiri, Wavelet-based emotion recognition system using EEG signal, Neural Computing & Applications 28 (2017) 1985–1990.
[33] S. Alhagry, A.A. Fahmy, R.A. El-Khoribi, Emotion recognition based on EEG using LSTM recurrent neural network, International Journal of Advanced Computer Science and Applications 8 (10) (2017).
[34] Y. Yin, X. Zheng, B. Hu, Y. Zhang, X. Cui, EEG emotion recognition using fusion model of graph convolutional neural networks and LSTM, Applied Soft Computing 100 (2021) 106954.
[35] Y. Luo, Q. Fu, J. Xie, Y. Qin, G. Wu, J. Liu, X. Ding, EEG-based emotion classification using spiking neural networks, IEEE Access 8 (2020) 46007–46016.
[36] J. Hu, C. Wang, Q. Jia, Q. Bu, R. Sutcliffe, J. Feng, ScalingNet: extracting features from raw EEG data for emotion recognition, Neurocomputing 463 (2021) 177–184.
[37] Y. An, N. Xu, Z. Qu, Leveraging spatial-temporal convolutional features for EEG-based emotion recognition, Biomedical Signal Processing and Control 69 (2021) 102743.
[38] H. Li, Y.M. Jin, W.L. Zheng, B.L. Lu, Cross-subject emotion recognition using deep adaptation networks, in: Neural Information Processing: 25th International Conference, ICONIP 2018, Siem Reap, Cambodia, December 13–16, 2018, Proceedings, Part V 25, Springer International Publishing, 2018, pp. 403–413.
[39] Z. Wang, Y. Tong, X. Heng, Phase-locking value based graph convolutional neural networks for emotion recognition, IEEE Access 7 (2019) 93711–93722.
[40] T. Song, W. Zheng, P. Song, Z. Cui, EEG emotion recognition using dynamical graph convolutional neural networks, IEEE Transactions on Affective Computing 11 (3) (2018) 532–541.
[41] H. Tang, W. Liu, W.L. Zheng, B.L. Lu, Multimodal emotion recognition using deep neural networks, in: Neural Information Processing: 24th International Conference, ICONIP 2017, Guangzhou, China, November 14–18, 2017, Proceedings, Part IV 24, Springer International Publishing, 2017, pp. 811–819.
[42] Q. Li, Y. Liu, Q. Liu, Q. Zhang, F. Yan, Y. Ma, X. Zhang, Multidimensional feature in emotion recognition based on multi-channel EEG signals, Entropy 24 (12) (2022) 1830.

Index

D

E

G

H

I

J

K

L

M

N

R

S

T

Printed in the United States
by Baker & Taylor Publisher Services